Physics 12

Nelson

College Preparation

Physics *12*

College Preparation

Author

Alan J. Hirsch
Formerly of Peel District School Board

Program Consultant

Maurice Di Giuseppe
Toronto Catholic District School Board

THOMSON

NELSON

Australia   Canada   Mexico   Singapore   Spain   United Kingdom   United States

Nelson Physics 12: College Preparation

Author
Alan J. Hirsch

Program Consultant
Maurice Di Giuseppe

Director of Publishing
David Steele

Publisher
Kevin Martindale

**Executive Managing Editor,
Development and Testing**
Cheryl Turner

Program Managers
Tony Rodrigues
John Yip-Chuck

Developmental Editor
Betty Robinson

Editorial Assistants
Matthew Roberts
Jennie Worden

**Executive Managing Editor,
Production**
Nicola Balfour

Senior Production Editor
Joanne Close

Copy Editor
Denyse O'Leary

Proofreader
Dawn Hunter

Indexer
Noeline Bridge

Senior Production Coordinator
Sharon Latta Paterson

Creative Director
Angela Cluer

Art Director
Ken Phipps

Art Management
Suzanne Peden

Illustrators
Andrew Breithaupt
Steven Corrigan
Deborah Crowle
Margo Davies LeClair
John Fraser
Irma Ikonen
Norman Lanting
Dave Mazierski
Dave McKay
Linda Neale
Frank Netter

Peter Papayanakis
Ken Phipps
Marie Price
Myra Rudakewich
Gabriel Sierra
Katherine Strain
Bart Vallecoccia
Jane Whitney

Composition
Nelson Gonzalez

Interior Design
Kyle Gell
Allan Moon

Cover Design
Peter Papayanakis

Cover Image
SPL/Photo Researchers

Photo Research and Permissions
Cindy Howard

Set-up Photos
Dave Starrett

Printer
Transcontinental Printing Inc.

**National Library of Canada
Cataloguing in Publication Data**

Main entry under title:
Physics 12 : college preparation / Alan J. Hirsch.

Includes index.
ISBN 0-17-626530-9

1. Physics—Textbooks. I. Title:
Physics twelve.

QC23.H585 2003   530
C2003-902523-3

Reviewers

▸ CONTENTS

▸ Unit 1
Mechanical Systems

▸ Unit 2
Energy Transformations

▸ **Unit 3
Hydraulic and Pneumatic Systems**

▶ Unit 4
Electricity and Electronics

▶ **Appendices**

Mechanical Systems

A mechanical system can be as simple as the parts of a nail clipper, or it can be more complex, like the parts of your arm needed to lift this book off the desk or the pulley systems used by rock climbers. In this unit, you will review the principles of motion and forces and see how they apply to mechanical systems. You will also design and perform investigations on forces, friction, and machines, leading to the Unit Performance Task, in which you will design and build a machine that performs a specific task.

The unit is divided into two chapters. The first chapter begins with a review of motion and then presents forces, with an emphasis on friction. The second chapter applies the principles from Chapter 1 to machines.

What you learn about the mechanical systems you examine in this unit will be useful as you consider a variety of careers, such as the one depicted. Other possible careers tied closely to specific parts of the unit will be highlighted in the unit.

▶ Overall Expectations

In this unit, you will be able to

- describe and apply concepts related to motion, forces, Newton's laws of motion, friction, simple machines, torques, and mechanical advantage
- design and carry out experiments to investigate motion, forces, friction, and simple machines
- identify and analyze applications of forces and simple machines in real-world machines and the human body
- identify and describe science- and technology-based careers related to concepts presented in this chapter

▸ **Prerequisites**

Concepts

- distance, speed, average speed
- vector quantities versus scalar quantities
- position, displacement, velocity, acceleration
- machines, the lever, mechanical advantage
- pulleys, wheels, axles, gears
- friction

Skills

- solve an equation with one unknown
- the metric system, metric conversions, rounding off
- sketch and analyze distance–time and position–time graphs
- draw simple vector diagrams
- work safely in a laboratory environment
- write lab reports for investigations

Knowledge and Understanding

1. Copy **Table 1** into your notebook and complete it. The first row has been completed for you.

Table 1

Unit	Full name	Quantity	Vector or scalar
m	metre	length	scalar
km	?	?	?
km/h	?	?	?
m/s [E]	?	?	?
(km/h)/s [W]	?	?	?
m/s² [N]	?	?	?
kg	?	?	?

2. **Figure 1** shows the motion of a car along a straight road. The images are taken every 1.0 s. Describe the motion of the car using your vocabulary of motion.

start   stop

Figure 1

3. A dog runs along the path shown in **Figure 2**, starting at A and following the direction of the arrows. The dog takes 16 s to complete the circuit.

Figure 2

N
↑
⊕→ E

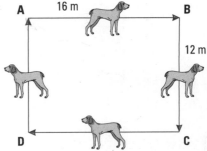

   (a) State the compass direction the dog is following in each part of the run.
   (b) Calculate the total distance travelled by the dog.
   (c) Calculate the dog's average speed.
   (d) What is the dog's net or total displacement over the entire path?

4. Calculate the average acceleration of a bus that accelerates from 15 km/h [E] to 75 km/h [E] in 24 s.

5. Can you open a heavy door more easily if you push at the handle or halfway between the handle and the hinges? Why?

6. Draw a sketch to show how a pulley system can be used to raise a flag to the top of a pole.

7. How does friction help you walk?

Inquiry and Communication

8. Describe your pen qualitatively and quantitatively.

9. (a) Describe the sources of error you would encounter when measuring the length of this page with a metre stick.
   (b) Use a centimetre ruler to measure the length of this page, and compare your value to the values found by other students.

Math Skills

10. Use your calculator to determine each answer, and then round off your answer to the correct number of significant digits:
    (a) $12.0 \div 1.70$
    (b) 1.6×1.7
    (c) $1.6 \times 10^6 \div 8.9 \times 10^2$
    (d) the ratio of circle circumference 682 cm to its diameter 217 cm
    (e) the ratio of 95 km to 1.1 h

11. The motion of three cars, L, M, and N, is illustrated in the graph in **Figure 3**. Compare the times of travel and the average speeds of the three cars.

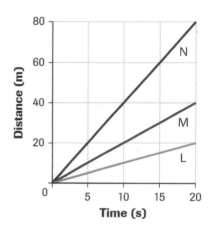

Figure 3

12. Starting with the equation $v = \dfrac{d}{t}$, solve for d and then solve for t.

Technical Skills and Safety

13. Draw a sketch of your classroom, and label all the safety equipment in the room. Beside each piece of equipment, state its function.

Making Connections

14. Suppose you are asked to test a new design for an artificial human arm. Describe three tasks you think the arm should be able to perform.

1

Motion and Forces

▶ **In this chapter, you will be able to**

- define and describe the concepts and units related to velocity, acceleration, force, and coefficients of friction
- state Newton's laws of motion and apply them to practical situations
- analyze the forces acting on an object and describe the resulting motion of the object
- verify Newton's second law of motion in an experiment
- determine, through experimentation, some factors that affect static and kinetic friction
- describe advantages and disadvantages of friction in everyday situations, as well as methods used to increase or reduce friction in these situations

Getting Started

The tires and the racing track surface in **Figure 1** are designed to provide the maximum possible friction for race conditions. Without the force of friction, racing cars would not be able to speed up, travel around curves, or come to a safe stop. Passenger cars and other moving objects also need friction.

Friction is one of several forces presented in this chapter. Other forces you will explore are gravity and tension. By performing force experiments and observing demonstrations of forces, you will learn how three laws of physics, called Newton's laws of motion, explain how forces relate to motion.

Forces are used in many technologies. For example, you will discover ways in which desired friction can be increased and undesired friction can be decreased.

The link between forces and motion cannot be well understood without a review of the physics of motion. Therefore, we begin with a review of the concepts of displacement, velocity, and acceleration.

💡 REFLECT on your learning ▼

1. Examine **Figure 1**. In a table using the titles listed below, list all the forces you can think of that relate to one of the racing cars.

Name of Force	Direction of Force	Object Exerting the Force	Object on which the Force Is Acting

2. Is it safe to pile small bags and cases inside the car on the rear window shelf and then go for a drive? Explain your answer, taking into consideration what happens when you start and stop the car.

3. Three boats of different masses have the same size of motor. You take each boat out and start it from rest. Do you think the difference in mass will affect the acceleration of each boat? Explain your answer.

4. When you drop a golf ball, it reaches the ground fairly quickly. Will a heavier ball, say a baseball, drop faster? Why or why not?

5. Name two situations in which friction can be useful and two situations in which friction is unwanted.

6. When you brake to stop your bike, what factors affect the stopping distance?

7. A skater standing next to the boards of a hockey rink pushes westward against the boards.
   (a) In what direction does the skater move?
   (b) What do you think exerted the force that made the skater move in that direction?
   (c) Give a similar example that illustrates the same type of behaviour.

Figure 1
The design of tires for racing cars is very different from the design of tires for passenger cars. However, both designs are based on the same type of research.

▶ **TRY THIS** activity **Balancing a Metre Stick**

Balance a metre stick horizontally on your index fingers so that your left index finger is at the 10-cm mark and your right index finger is at the 70-cm mark (**Figure 2**).

Figure 2
Balance the metre stick as shown.

(a) Predict where your fingers will meet on the metre stick when you gently and slowly move them toward each other. Try it, and then repeat the procedure two or three times, with your fingers starting at different points.

(b) Describe how friction affects the motion of your fingers beneath the metre stick.

(c) Based on your results from this activity, do you think there is a difference between starting friction and moving friction?

From the tiny particles that make up all matter, through Earth travelling around the Sun, to our solar system moving through space in the Milky Way Galaxy, everything in our universe is in a state of motion (**Figure 1**). All of this motion can be analyzed in physics.

Figure 1
The motion of these stunt skydivers as they move toward the ground is one of countless examples of motion that can be analyzed in physics.

kinematics the study of motion itself, not the forces that produce motion

dynamics the study of the forces that produce motion

scalar quantity a quantity that has magnitude (or size) only (including any units)

Scientists call the study of motion **kinematics**. This word comes from the Greek word for motion, *kinema*. A "cinema" is a place where people watch motion pictures. An understanding of kinematics is important as a background to **dynamics**, the study of the forces that produce motion. In other words, kinematics is about observing motion; dynamics is about explaining motion.

Scalar and Vector Quantities

Speeds we encounter in our daily lives are usually given in kilometres per hour (km/h) or metres per second (m/s). From this you can see that speed involves both distance and time. Speed, distance, and time are all examples of **scalar quantities**, quantities that have magnitude (size) only. The magnitude is made up of a number and often a unit of measure. For example, a distance of 2.5 m, a time interval of 15 s, a mass of 2.4 kg, and a temperature of 12 °C are all scalar quantities.

A **vector quantity** is one that has magnitude (including any units) and direction. In this text, a vector quantity is indicated by a symbol with an arrow above it; the direction is stated in square brackets after the unit. A common vector quantity is **position**, which is the distance and direction of an object from a reference point. For example, the position of a friend in your class could be a distance of 2.2 m in the west direction relative to your desk. Using symbols, we would describe your friend's position as follows: $\vec{d} = 2.2$ m [W] relative to the desk.

Displacement, another vector quantity, is the change in position of an object in a given direction. The symbol for displacement is $\Delta\vec{d}$, where Δ, the Greek letter delta, means "change." For example, **Figure 2** shows the displacement of a person as a result of moving from position $\vec{d}_i$ to position $\vec{d}_f$ relative to a reference point, where $\vec{d}_i$ is the initial position, and $\vec{d}_f$ is the final position.

vector quantity a quantity that has magnitude (including any units) and direction

position the distance of an object from a reference position, including the object's direction; it is a vector quantity, symbol $\vec{d}$

displacement the change in position of an object, including the object's direction; it is a vector quantity, symbol $\Delta\vec{d}$

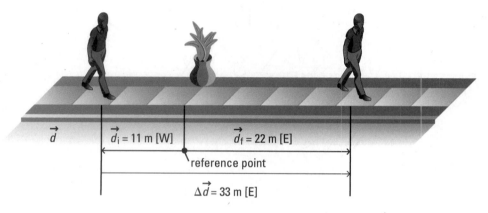

Figure 2
A person walks from position $\vec{d}_i = 11$ m [W] of the reference point to another position, $\vec{d}_f = 22$ m [E] of the reference point. The person's displacement for this motion is 33 m [E], as shown in Sample Problem 1.

LEARNING TIP

SI Units
The international scientific communities of many countries, including Canada, have agreed on a system of measurement called the Système international d'unités, or simply the SI. In this system, all physical quantities can be expressed as a combination of seven SI units, called *base units* (metre, kilogram, and second, for example, are three of the base units). Units written in terms of the base units are called *derived units* (e.g., the unit for speed, metres per second, is a derived unit). Refer to **Table 1** in Appendix C for a list of the SI base units and their symbols.

▶ *SAMPLE* problem *1*

Displacement

Calculate the displacement of the person walking in **Figure 2**.

Solution
$\vec{d}_i = 11$ m [W]
$\vec{d}_f = 22$ m [E]
$\Delta\vec{d} = ?$
$\Delta\vec{d} = \vec{d}_f - \vec{d}_i$
$\quad = 22$ m [E] $- 11$ m [W]
$\quad = 22$ m [E] $+ 11$ m [E]
$\Delta\vec{d} = 33$ m [E]

The person's displacement during the motion is 33 m [E]. Notice that negative west is changed to positive east before the mathematical calculation is made.

Understanding Concepts

1. Classify the following measurements as scalar or vector quantities:
   (a) 12 s
   (b) 440 Hz
   (c) 4.2 m [down]
   (d) 85 km/h [W]
   (e) 12 cm²
   (f) 2.5 L

2. (a) Name at least five scalar quantities presented so far.
   (b) Name at least two other scalar quantities you can think of.

3. A curling rock leaves a curler's hand at a point 2.1 m from the end of the ice and travels southward [S]. Calculate the displacement of the curling rock from its point of release to a point 9.7 m from the same end. (Draw a diagram showing the position and displacement vectors in this situation.)

4. A dog, initially at a position 2.8 m [W] of its owner, runs to retrieve a stick that is 12.6 m [E] of its owner.
   (a) Draw a diagram showing the position and displacement vectors in this situation.
   (b) Determine the displacement the dog needs to reach the stick.
   (c) Repeat (a) and (b) where the dog's initial position is 2.8 m [E] of its owner.

5. **Table 1** gives the position–time data of a ball that has left a bowler's hand and is rolling forward. Determine the displacement between the following times:
   (a) $t = 0$ s and $t = 1.0$ s
   (b) $t = 1.0$ s and $t = 2.0$ s
   (c) $t = 1.0$ s and $t = 3.0$ s

Table 1   For Question 5

Time (s)	Position (m [forward])
0.0	0.0
1.0	4.4
2.0	8.7
3.0	13.0

Answers

3. 7.6 m [S]

4. (b) 15.4 m [E]
   (c) 9.8 m [E]

5. (a) 4.4 m [forward]
   (b) 4.3 m [forward]
   (c) 8.6 m [forward]

Measuring Time Intervals

Time is an important quantity in the study of motion. In physics classrooms, various instruments are used to measure intervals of time:

- A digital timer is an electronic instrument that measures time intervals to a fraction of a second.

- A computer with appropriate sensor software can be used to measure time intervals.

- Spark timers and ticker-tape timers are often used for student experimentation. These devices produce dots on paper at a set number of dots per second, or frequency. A ticker-tape timer, shown in **Figure 3**, has a metal arm that vibrates at constant time intervals. A needle on the arm strikes a piece of carbon paper and records dots on a paper tape pulled through the timer. The faster the motion of the paper tape, the greater the space between the dots.

Figure 3
A ticker-tape timer

Some spark timers and ticker-tape timers make 60 dots each second. They are said to have a frequency of 60 vibrations per second, or 60 Hz. The symbol Hz stands for hertz and means "cycles per second." This unit is named after the German scientist Heinrich Hertz (1857–94). The period of vibration (T), or the time between successive dots, is related to the frequency (f) in the following way:

$$T = \frac{1}{f} \quad \text{and} \quad f = \frac{1}{T}$$

Figure 4 shows a tape with dots produced by a spark timer with a frequency of 60 Hz. The figure illustrates how to find T, as well as the total time for six intervals (seven dots).

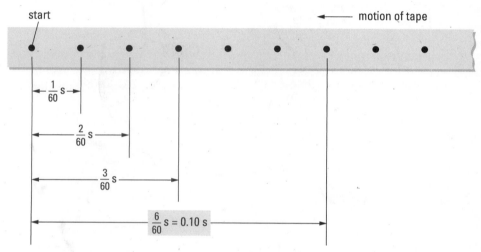

Figure 4

Measuring time intervals with a spark timer. In this case, the frequency of vibration is 60 Hz, and the period of vibration is the reciprocal of the frequency, that is, $\frac{1}{60\,\text{Hz}} = \frac{1}{60}\,\text{s} = 0.017\,\text{s}$, to two significant digits. There are six intervals being measured, so we have $6 \times \frac{1}{60}\,\text{s} = \frac{6}{60}\,\text{s}$, or 0.10 s.

To practise working with ticker-tape times, do Activity 1.2, page 17.

To practise working with ticker-tape times, do Activity 1.2, page 17.

> ▶ **Practice**

Understanding Concepts

6. Calculate the period of a spark timer set at (a) 60.0 Hz and (b) 30.0 Hz.
7. Determine the frequency of a spark timer set at a period of 0.010 s.

Answers

6. (a) 0.0167 s
   (b) 0.0333 s
7. 1.0×10^2 Hz

LEARNING TIP

Significant Digits and Calculators

Using a calculator is very convenient. However, it is important to understand how to treat the answers that the calculator displays. For example, if you perform the calculation shown in **Figure 4**, that is, 1 ÷ 60, your calculator will yield a value with many digits, for example, 0.0166667. You must round off the number to an appropriate number of significant digits, such as 0.02, 0.017, or 0.0167. To learn more about significant digits and rounding off answers, refer to Appendix A1.

Average Speed

Everyone participating in a race (**Figure 5**) must run the same distance, and the winner is the person who finishes first. The winner does not have to be the fastest at every point in the race. At some points during the race, other runners may reach a greater **instantaneous speed**, the speed at a particular instant. However, the winner has the greatest *average speed*. **Average speed** is the total distance travelled divided by the total time of travel. (The symbol for average speed, v_{av}, is taken from the word "velocity," defined shortly.) The equation for average speed is

$$v_{av} = \frac{d}{t}$$

where d is the total distance travelled in a total time t.

instantaneous speed the speed at a particular instant; it is a scalar quantity, symbol v

average speed the ratio of the total distance travelled to the total time of travel; it is a scalar quantity, symbol v_{av}

Figure 5
Every runner covers the same distance, so the person with the shortest time has the greatest average speed.

> ► **SAMPLE** problem **2**
>
> **Average Speed**
>
> In 47.8 s, a sprinter runs 400 m. (Assume three significant digits in this distance.) What is the sprinter's average speed for this motion?
>
> **Solution**
>
> $d = 400\ m = 4.00 \times 10^2\ m$
>
> $t = 47.8\ s$
>
> $v_{av} = ?$
>
> $$v_{av} = \frac{d}{t}$$
>
> $$= \frac{4.00 \times 10^2\ m}{47.8\ s}$$
>
> $$v_{av} = 8.37\ m/s$$
>
> The sprinter's average speed is 8.37 m/s.

Average Velocity

Many people think that speed and velocity are the same. However, in physics there is an important difference because velocity has a direction. **Instantaneous velocity** is the rate of change of position, so it is a vector quantity, $\vec{v}$. The **average velocity**, symbol $\vec{v}_{av}$, of a motion is the change of position divided by the time interval for that change. The equation for average

velocity is $\vec{v}_{av} = \dfrac{\Delta \vec{d}}{\Delta t}$

where $\Delta \vec{d}$ is the displacement (or change of position), and

Δt is the time interval.

instantaneous velocity the velocity at a particular instant; it is the rate of change of position; it is a vector quantity, symbol $\vec{v}$

average velocity the ratio of the displacement to the time interval for the displacement; it is a vector quantity, symbol $\vec{v}_{av}$

▶ **SAMPLE** problem **3**

Average Velocity

The world's fastest coconut tree climber takes only 4.9 s to climb 9.0 m [up] a coconut tree. Calculate the climber's average velocity for this motion, assuming that the climb was straight upward.

Solution

$\Delta \vec{d} = 9.0$ m [up]
$\Delta t = 4.9$ s
$\vec{v}_{av} = ?$

$$\vec{v}_{av} = \frac{\Delta \vec{d}}{\Delta t}$$

$$= \frac{9.0 \text{ m [up]}}{4.9 \text{ s}}$$

$$\vec{v}_{av} = 1.8 \text{ m/s [up]}$$

The climber's average velocity is 1.8 m/s [up].

LEARNING TIP

The Magnitude of Average Velocity
For constant velocity motion, the magnitude of the average velocity is the same as the speed. However, if the direction of the velocity changes, even if the speed remains constant, the magnitude of the average velocity differs from the average speed. For example, if an athlete runs around a 200-m circular track at an average speed of 7.5 m/s, upon arriving at the starting position the average velocity is zero (because the displacement is zero).

▶ **Practice**

Understanding Concepts

8. Calculate the average speed in each of the following:
   (a) A jogger takes 14 s to run 88 m.
   (b) A truck travels 425 km in 4.65 h.
   (c) A cyclist travels 120 m in the first 18 s of a trip and then 440 m in the last 55 s of the trip.

9. While running on his hands, an athlete sprinted 50.0 m [forward] in a record 16.9 s.
   (a) Determine the average velocity for this feat.
   (b) Convert your answer in (a) to kilometres per hour.

10. A jogger takes 3.5 min to run once around a square city block that is 220 m on each side.
    (a) Draw a sketch of the motion.
    (b) Determine the jogger's average speed in metres per minute and metres per second.
    (c) What is the jogger's average velocity upon returning to the starting position?

Answers

8. (a) 6.3 m/s
   (b) 91.4 km/h
   (c) 7.7 m/s

9. (a) 2.96 m/s [forward]
   (b) 10.7 km/h [forward]

10. (b) 2.5×10^2 m/min; 4.2 m/s

▶ **TRY THIS** activity   **Motion at Constant Velocity**

How difficult is it to create motion at a constant velocity? You can find out in this activity.

For the first two steps described below, sketch a position–time graph predicting what you will observe.

• Use a motion sensor connected to a graphics program to determine how close to constant velocity you can walk. Try more than one constant speed, and try moving toward and away from the sensor. Explain any irregularities on the graph.

• Repeat the procedure using a different moving object, such as a glider on an air track, a battery-powered toy vehicle, or a low-friction cart on a ramp slightly elevated at one end.

• Try creating motion at a constant velocity using tape pulled through a ticker-tape timer. How can you judge whether you are successful?

Graphing Motion with Constant Velocity

In experiments involving motion, the variables that can be measured directly are usually time and either position or displacement. The third variable, velocity, is often obtained by calculation.

When motion involves constant velocity, the displacement is the same during equal time intervals. For example, say that an ostrich, the world's fastest bird on land, runs 18 m straight west per second for 8.0 s. The ostrich's velocity is constant at 18 m/s [W]. We can record the data for this movement in a position–time table, starting at 0.0 s; see **Table 2**.

Figure 6 shows a graph of this motion, with position plotted on the vertical axis. Notice that for motion with constant velocity, a position–time graph is a straight line.

Table 2   Position–Time Data

Time (s)	Position (m [W])
0.0	0.0
2.0	36
4.0	72
6.0	108
8.0	144

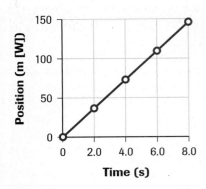

Figure 6
A position–time graph of constant velocity using the data from **Table 2**

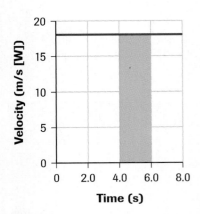

Figure 7
A velocity–time graph showing constant velocity. (The shaded region is for Sample Problem 5.)

▶ **SAMPLE** problem **4**

Slope of the Line on a $\vec{d}$-t Graph

Calculate the slope of the $\vec{d}$-t line on the graph in **Figure 6**, and state what that slope represents.

Solution
The data points used to find the slope are taken from the graph, not from the data table.

$\vec{d}_i$ (initial position) = 0 m
$\vec{d}_f$ (final position) = 144 m [W]
t_i (initial time) = 0.0 s
t_f (final time) = 8.0 s
slope = ?

$$\text{slope} = \frac{\Delta \vec{d}}{\Delta t}$$

$$= \frac{\vec{d}_f - \vec{d}_i}{t_f - t_i}$$

$$= \frac{144 \text{ m [W]} - 0 \text{ m [W]}}{8.0 \text{ s} - 0.0 \text{ s}}$$

$$\text{slope} = 18 \text{ m/s [W]}$$

The slope of the line is 18 m/s [W]. The value of the slope is the same as the ostrich's average velocity.

The slope calculation in Sample Problem 4 can be used to plot a velocity–time graph of the motion. Because the slope of the line is constant, the velocity is constant from $t = 0.0$ s to $t = 8.0$ s. The resulting velocity–time graph is shown in **Figure 7**.

A velocity–time graph can be used to find the displacement during various time intervals. This is done by finding the area under the line on the velocity–time graph.

▶ **SAMPLE** problem **5**

Area under the Line on a $\vec{v}$-t Graph

Find the area of the shaded region on the $\vec{v}$-t graph in **Figure 7**. What does that area represent?

Solution

For a rectangular shape,

$$A = lw$$
$$= \vec{v}(\Delta t)$$
$$= \left(18\,\frac{m}{s}[W]\right)(2.0\,s)$$
$$A = 36\text{ m [W]}$$

The area of the shaded region is 36 m [W]. This quantity represents the ostrich's displacement from $t = 4.0$ s to $t = 6.0$ s. In other words, $\Delta\vec{d} = \vec{v}_{av}\Delta t$. This value agrees with **Table 2**, which shows that the displacement during the time interval was 108 m [W] − 72 m [W] = 36 m [W].

▶ **Practice**

Understanding Concepts

11. An ice skater on a frozen canal moves at a steady 5.5 m/s [S]. At time zero, the skater passes a post, which is the initial position for this question.
    (a) Set up a table to record the skater's position relative to the starting position at the end of each second for 8.0 s.
    (b) Use the data from your table to construct a position–time graph.
    (c) Find the slope of the line on the position–time graph. What does that slope represent?
    (d) Plot a velocity–time graph of the skater's motion.
    (e) Calculate the total area under the line on the velocity–time graph. What does that area represent?

Answers

11. (c) 5.5 m/s [S]
    (e) 44 m [S]

SUMMARY ## Kinematics: The Study of Motion

- A scalar quantity has magnitude (including any units). Examples are distance, time, and speed.

- A vector quantity has magnitude (including any units) and direction. Examples are position, displacement, and velocity.

- A ticker-tape timer is just one of several instruments used to measure time intervals.

- Average speed is the ratio of the total distance travelled to the total time, or $v_{av} = \dfrac{d}{t}$.

- Average velocity is the ratio of the displacement to the time interval, or $\vec{v}_{av} = \dfrac{\Delta\vec{d}}{\Delta t}$.

- A straight line on a position–time graph means constant velocity, and the slope of the line represents the average velocity between any two times.

- The area under the line on a velocity–time graph represents the displacement between any two times.

▶ *Section 1.1* *Questions*

Understanding Concepts

1. In Australia's Ironman Triathlon, Canada's Lori Bowden set a course record for women by swimming 3.8 km, then biking 180 km, and then running 42.2 km, all in an astonishing 8.9 h. Calculate Lori's average speed in setting this record.

2. In one indoor event, an athlete sprints 50.0 m [forward] in 5.96 s. In a second indoor event, the same athlete sprints 60.0 m [forward] in 6.92 s.
   (a) Calculate the athlete's average velocity in each sprint.
   (b) Why do you think there is a difference in the values in (a)?

3. State what each of the following represents:
   (a) the slope of a line on a position–time graph
   (b) the area under the line on a velocity–time graph

4. A student, using a stopwatch, finds that a ticker-tape timer produces 139 dots (i.e., 138 intervals) in 2.50 s.
   (a) Determine the frequency of vibration according to these results.
   (b) Calculate the percent error of the student's value of the frequency, assuming that the accepted value is 60.0 Hz.

Applying Inquiry Skills

5. (a) Design and perform a simple experiment to determine the average speed that your index and middle fingers of one hand "walk" from one edge of your desk to another. Show your measurements and calculations.
   (b) Describe the main sources of error in the experiment you performed in (a).

6. **Table 3** shows the results of a motion experiment.
   (a) Plot a position–time graph of the motion and draw the line of best fit.
   (b) Determine the slope of the line on the graph, and then plot a corresponding velocity–time graph.
   (c) On the velocity–time graph, determine the area between 0.20 s and 0.60 s.

Making Connections

7. What quantities are measured by a car's odometer and speedometer? Are these quantities scalar or vector?

8. **Figure 8** shows four possible ways of indicating speed limits on roads. Which one communicates the information in the clearest way? Why?

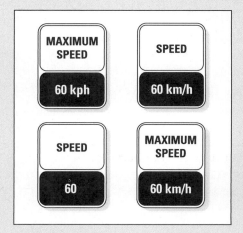

Figure 8
For question 8

Table 3   Data for Question 6

Time (s)	Position (cm [E])
0.00	0.00
0.20	21
0.40	45
0.60	65
0.80	88

Calibrating a Ticker-Tape Timer

This activity will introduce you to using a ticker-tape timer. You will practise pulling ticker tape through the timer at a steady rate, and you will calibrate the timer by determining its frequency and period. Before performing the activity, find out how to use the ticker-tape timer (**Figure 1**) you are given.

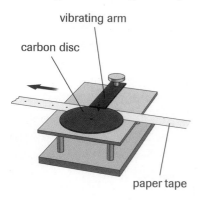

Figure 1
The main parts of a ticker-tape timer

Materials

For each group of three or four students:
stopwatch
ticker-tape timer
ticker tape
clamp
additional timer materials

Procedure

1. Use the clamp to secure the timer to the lab bench. Obtain a piece of ticker tape about 120 cm long, and position it in the timer. With the timer off and the instrument held firmly in place on the lab bench, practise pulling the tape through it so the motion takes about 3 s. Repeat until you can do it smoothly.

 CAUTION: Follow the safety guidelines for the use of electrical apparatus as described in Appendix B2.

2. Connect the timer to an electrical source, remembering safety guidelines. As you begin pulling the tape through the timer at a steady rate, have your partner start the stopwatch and turn on the timer at the same instant. Just before the tape leaves the timer, have your partner simultaneously turn off both the stopwatch and the timer. Record the time interval, and count the number of dots in that time interval.

Analysis

(a) Calibrate the timer by determining the frequency (the number of dots divided by the time interval in seconds) and the period (the time interval divided by the number of dots).

Evaluation

(b) Calculate the percent error of your measurement of the period of vibration of the timer. Your teacher will tell you what the "accepted" value is. (To review percent error, refer to Appendix A1.)

(c) Describe the major sources of error that may affect your measurements and calculations of the period. (To review sources of error in physics experiments, refer to Appendix A1.)

On the navy aircraft carrier shown in **Figure 1**, a steam-powered catapult can cause an aircraft to accelerate from zero to 265 km/h in 2.0 s! Stopping the aircraft so that it lands on the carrier also involves a high magnitude of acceleration. However, in this case the craft is slowing down. A hook extends from the tail section of the aircraft and catches a steel cable stretched across the deck of the carrier. That causes the aircraft, which is moving with a speed of about 240 km/h, to stop in about 100 m. Motion equations can be used to analyze these motions, which leads to an understanding of the causes of *acceleration*.

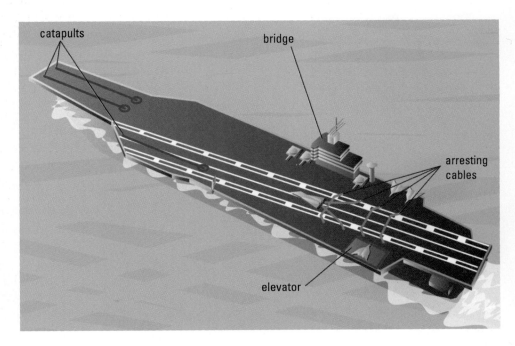

Figure 1
The catapults dramatically boost velocity in takeoff, and the arresting cables dramatically slow the velocity in landing.

Constant Acceleration

We saw in section 1.1 that since velocity is a vector quantity, constant velocity is motion at a constant speed in a straight line. We also saw that a velocity–time graph that shows this type of motion is a horizontal straight line.

However, most moving objects do not move with constant velocity. Any change in an object's speed, direction, or both means a change in velocity. This change in velocity is called **accelerated motion**. Any car ride where the car speeds up, slows down, and turns corners, is an example of accelerated motion. The rate of change of velocity is called **acceleration**.

When an object travelling in a straight line changes its speed by the same amount each second, the motion is **constant acceleration**. For example, the motion of a motorcycle that starts from rest and increases its speed by 6.0 m/s every second while travelling straight westward is constant acceleration.

accelerated motion motion with changing velocity

acceleration the rate of change of velocity; it is a vector quantity, symbol $\vec{a}$

constant acceleration straight-line motion in which the speed changes by the same amount each second

Sample data for this example are listed in **Table 1**. The velocity–time graph in **Figure 2** uses the data from **Table 1**.

Table 1	Constant Acceleration of a Motorcycle

Time (s)	Velocity (m/s [W])
0.0	0.0
1.0	6.0
2.0	12.0
3.0	18.0
4.0	24.0

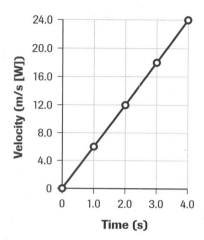

Figure 2
The velocity–time graph of the constant acceleration of a motorcycle that is speeding up

An object that travels in a straight line and slows down by the same amount each second is also said to move with constant acceleration. **Table 2** lists some sample data for a car slowing down from 24.0 m/s [E] to 0.0 m/s in 4.0 s; this car, like the motorcycle, is also experiencing constant acceleration. The graph in **Figure 3** uses the data from **Table 2** to depict the car's constant acceleration.

Table 2	Constant Acceleration of a Car

Time (s)	Velocity (m/s [E])
0.0	24.0
1.0	18.0
2.0	12.0
3.0	6.0
4.0	0.0

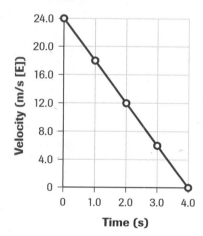

Figure 3
The velocity–time graph of the constant acceleration of a car that is slowing down

▶ *TRY THIS* activity *Analyzing Motion Graphs*

A cart is pushed so that it travels up a straight ramp, stops briefly, and then travels back down to the point where it was first pushed. Consider the motion from the point just after the force pushing the cart upward is removed until the cart is caught on the way down. For this activity, assume that "up the ramp" is the positive direction.

(a) Sketch the velocity–time graph for the motion of the cart for the entire motion.

(b) Set up a motion sensor or a "smart pulley" to generate the graph by computer or graphing

calculator as the cart undergoes the motion. Compare your graph from (a) with the graph generated by computer.

(c) If an interactive computer software program is available, use it to observe the same types of graphs. Describe what you observe.

 Be sure the moving cart does not hit the detector as it completes its downward motion.

(a)

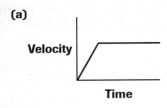

(b)

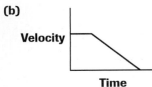

(c)

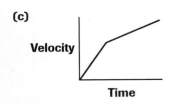

Figure 4
Velocity–time graphs for question 2

average acceleration the change of velocity divided by the time interval for the change; it is a vector quantity, symbol $\vec{a}_{av}$

> **Practice**

Understanding Concepts

1. **Table 3** shows five different sets of velocities at times of 0.0 s, 1.0 s, 2.0 s, and 3.0 s. Which of them involve constant acceleration with an increasing velocity for the entire time? Describe the motion of the other sets.

Table 3   Motions for Question 1

Time (s)	0.0	1.0	2.0	3.0
(a) Velocity (m/s [E])	0.0	8.0	16.0	24.0
(b) Velocity (cm/s [W])	0.0	4.0	8.0	8.0
(c) Velocity (km/h [N])	58	58	58	58
(d) Velocity (m/s [W])	15	16	17	18
(e) Velocity (km/h [S])	99	66	33	0

2. Using terms such as constant velocity, constant acceleration, and increasing or decreasing velocity, describe the motion illustrated in each velocity–time graph in **Figure 4**. In **(c)**, compare the magnitudes of the accelerations.

Calculating Acceleration

The **average acceleration** of an object is found using the equation

$$\text{average acceleration} = \frac{\text{change of velocity}}{\text{time interval}}$$

$$\vec{a}_{av} = \frac{\Delta \vec{v}}{\Delta t}$$

Since the change of velocity of a moving object is the final velocity minus the initial velocity ($\Delta \vec{v} = \vec{v}_f - \vec{v}_i$), the equation for average acceleration can also be written as follows:

$$\vec{a}_{av} = \frac{\vec{v}_f - \vec{v}_i}{\Delta t}$$

> **SAMPLE problem 1**
> **Average Acceleration**

A motorcycle starting from rest and undergoing constant acceleration reaches a velocity of 21.0 m/s [N] in 8.4 s. Calculate the average acceleration.

Solution

$\vec{v}_i = 0.0$ m/s

$\vec{v}_f = 21.0$ m/s [N]

$\Delta t = 8.4$ s

$\vec{a}_{av} = ?$

$$\vec{a}_{av} = \frac{\vec{v}_f - \vec{v}_i}{\Delta t}$$

$$= \frac{21.0 \text{ m/s [N]} - 0.0 \text{ m/s}}{8.4 \text{ s}}$$

$$\vec{a}_{av} = 2.5 \text{ (m/s)/s [N], or } 2.5 \text{ m/s}^2 \text{ [N]}$$

The motorcycle's average acceleration is 2.5 m/s² [N].

LEARNING *TIP*

Acceleration Units
Acceleration can be stated using a variety of units, for example, (km/h)/s. In SI, the units are (m/s)/s, which is equivalent to m/s²:

$$\left(\frac{m}{s}\right) \div s = \left(\frac{m}{s}\right) \times \left(\frac{1}{s}\right) = \frac{m}{s^2}$$

In Sample Problem 1, the constant acceleration of 2.5 m/s² [N] means that the velocity of the motorcycle increases by 2.5 m/s [N] every second. Thus, the motorcycle's velocity is 2.5 m/s [N] after 1.0 s, 5.0 m/s [N] after 2.0 s, 7.5 m/s [N] after 3.0 s, and so on.

If an object is slowing down, its acceleration is opposite in direction to the velocity. If the velocity is positive, the acceleration is negative, or if the velocity is negative, the acceleration is positive. This is illustrated in Sample Problem 2.

▶ **SAMPLE** problem **2**

Negative Acceleration

A cyclist, travelling initially at 14 m/s [S], brakes smoothly and stops in 4.0 s. What is the cyclist's average acceleration?

Solution

$\vec{v}_i = 14 \text{ m/s [S]}$

$\vec{v}_f = 0.0 \text{ m/s}$

$\Delta t = 4.0 \text{ s}$

$\vec{a}_{av} = ?$

$$\vec{a}_{av} = \frac{\vec{v}_f - \vec{v}_i}{\Delta t}$$

$$= \frac{0.0 \text{ m/s} - 14 \text{ m/s [S]}}{4.0 \text{ s}}$$

$$= -3.5 \text{ m/s}^2 \text{ [S]}$$

$$\vec{a}_{av} = 3.5 \text{ m/s}^2 \text{ [N]}$$

The cyclist's average acceleration is 3.5 m/s² [N]. Notice that the direction positive north is the same as the direction negative south.

The acceleration of an object moving with constant acceleration can also be found by calculating the slope of the line on a velocity–time graph of the motion. Consider the graph shown in **Figure 2**, page 19. The slope of the line is constant and is equal to

$$\text{slope} = \frac{\Delta \vec{v}}{\Delta t}$$

$$= \frac{24.0 \text{ m/s [W]} - 0.0 \text{ m/s}}{4.0 \text{ s} - 0.0 \text{ s}}$$

$$\text{slope} = 6.0 \text{ m/s}^2 \text{ [W]}$$

Understanding Concepts

Answers

3. (a) 12 m/s² [W]
   (b) 1.4 m/s² [E]
   (c) 5.8 m/s² [N]
   (d) 28 (km/h)/s [N]

4. (a) 13 m/s²
   (b) 47 (km/h)/s

5. 11 (km/h)/s [S]

6. 0.10 m/s² [W]

7. 1.3 × 10² (km/h)/s

8. 7.1 m/s² [E]

9. 75 m/s² [E]

3. Calculate the acceleration in each case:
   (a) $\Delta\vec{v}$ = 72 m/s [W]; Δt = 6.0 s
   (b) $\Delta\vec{v}$ = 8.4 m/s [E]; Δt = 6.0 s
   (c) $\Delta\vec{v}$ = −35 m/s [S]; Δt = 6.0 s
   (d) $\Delta\vec{v}$ = 42 km/h [N]; Δt = 1.5 s

4. A motorcycle took only 6.0 s to go from rest to 78 m/s (2.8 × 10² km/h), earning the world record for motorcycle acceleration. Calculate the magnitude of the record average acceleration in (a) m/s² and (b) (km/h)/s.

5. A car, travelling initially at 55 km/h [S], changes its velocity to 103 km/h [S] in 4.5 s. Calculate the average acceleration.

6. Calculate the average acceleration needed by a train travelling initially at 12 m/s [E] to stop in 120 s.

7. Determine the magnitude of the average acceleration of the aircraft that takes off from the aircraft carrier described in the first paragraph of this section.

8. A ball is rolling with a velocity of 2.4 m/s [W] when it hits a wall. After bouncing off the wall, the ball's velocity 0.62 s later is 2.0 m/s [E]. Calculate the ball's average acceleration during the time interval.

9. Plot the velocity–time graph that corresponds to the data given in **Table 4**, and determine the average acceleration.

Table 4 Data for Question 9

Time (s)	Velocity (m/s [E])
0.0	0.0
0.20	15
0.40	30
0.60	45
0.80	60

Applying Inquiry Skills

10. **Figure 5** illustrates two designs of an accelerometer, a device used to determine horizontal acceleration.
    (a) Based on the diagrams, what do you think would happen in each accelerometer if the object to which it is attached accelerates to the right? Explain why in each case.
    (b) If you have access to a horizontal accelerometer, discuss its safe use with your teacher. Then use it to test your answers in (a). Describe what you discover.

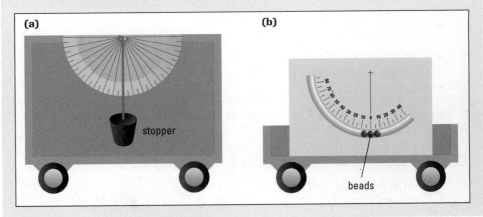

Figure 5
(a) A horizontal accelerometer using a stopper suspended from a protractor
(b) A horizontal accelerometer using beads in clear tubing

Acceleration Due to Gravity

Now that we have looked at horizontal acceleration, we can take a look at vertical acceleration: acceleration near Earth's surface, or acceleration due to gravity. Consider this: If two rubber stoppers of different masses, 15 g and 75 g, are dropped from the same height above the floor, they land at the same time. Therefore, the acceleration of falling objects near Earth's surface does not depend on mass. (Air resistance does affect the acceleration, but for solid objects falling short distances, air resistance is too small to notice.)

It was the famous scientist Galileo Galilei who first determined that, if we ignore the effect of air resistance, the acceleration of falling objects is constant. Measurements show that the acceleration due to gravity near Earth's surface is approximately 9.8 m/s² [down]. This is an average value because the measurements vary slightly; for example, the value is slightly less on the top of a mountain than at the bottom. We use the symbol $\vec{g}$ to represent this average value because the acceleration is due to gravity. (Do not confuse $\vec{g}$ with the symbol for gram, g.)

▶ **TRY THIS** activity

Predicting Acceleration Due to Gravity

For this activity, you will need a sheet of notepaper and a rubber stopper. Predict what you will observe before each step, and then try it.

- Drop the paper and the stopper from the same height above the floor. What happens?

- Crunch the paper into a small ball, then drop the ball of paper and the stopper again from the same height. Describe what you observe.

When solving problems involving the acceleration due to gravity, we use $\vec{a}_{av} = \vec{g} = 9.8$ m/s² [down] if air resistance is zero or nearly zero. When air resistance on a falling object is negligible, the object is said to be in **free fall**. Thus, the acceleration during free fall near Earth's surface is 9.8 m/s² [down].

Some very popular amusement park rides allow passengers to experience free fall (**Figure 6**). Riders accelerate toward the ground, but a braking system slows them down over a small distance. The concept is similar to the arresting cables on the aircraft carrier in **Figure 1**, page 18.

free fall the acceleration of an object near Earth's surface if air resistance is ignored; the average value is $\vec{g} = 9.8$ m/s² [down]

Figure 6
The Drop Zone at Paramount Canada's Wonderland, north of Toronto, allows the riders to accelerate toward the ground, experiencing free fall for approximately 3.0 s. Then the braking system dramatically, and safely, slows them down.

▶ **Practice**

Understanding Concepts

11. An astronaut on the Moon drops a tool from rest. After 1.3 s the tool has a velocity of 2.1 m/s [down]. Determine the acceleration due to gravity on the Moon.

Applying Inquiry Skills

12. A vertical accelerometer (**Figure 7**) can be used to measure acceleration in the vertical direction. The one shown here is calibrated so it reads 1*g* when it is held still.

(a) Predict the reading on the accelerometer when it is
   (i) moved vertically upward at a constant speed
   (ii) moved vertically downward at a constant speed
   (iii) accelerated suddenly upward
   (iv) accelerated suddenly downward
(b) If an accelerometer is available, use it to test your predictions in (a).

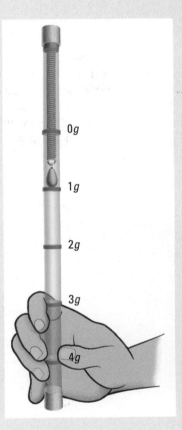

Figure 7
A typical vertical accelerometer for student use

SUMMARY   *Constant Acceleration and Acceleration Due to Gravity*

- An object travelling in a straight line experiences constant acceleration when it changes its speed by the same amount each second. The object can be speeding up or slowing down.

- The equation for average acceleration is $\vec{a}_{av} = \dfrac{\Delta \vec{v}}{\Delta t}$, or $\vec{a}_{av} = \dfrac{\vec{v}_f - \vec{v}_i}{\Delta t}$.

- On a velocity–time graph of constant acceleration, the slope of the line represents the average acceleration.

- The acceleration of an object undergoing free fall at Earth's surface varies slightly with location and has an average value of $\vec{g} = 9.8$ m/s² [down].

▸ *Section 1.3* *Questions*

Understanding Concepts

1. When can an object have
   (a) a northward velocity and a southward acceleration?
   (b) a southward velocity and a southward acceleration?
   (c) an upward velocity and a downward acceleration?

2. At the start of its descent on the first hill, one of the world's fastest roller coasters has a velocity of 12 km/h [forward]. It takes 4.4 s to reach the bottom of that hill, where its velocity is 96 km/h [forward]. Calculate the roller coaster's average acceleration (in kilometres per hour per second) for this motion.

3. Calculate the average acceleration in each of the following:
   (a) A sprinter, starting from rest, reaches a maximum velocity of 11 m/s [forward] in 4.8 s.
   (b) A truck moving initially at 11 m/s [E] changes its velocity to 12 m/s [W] in 96 s.

4. Explain how you can determine the average acceleration of an object using its velocity–time graph.

5. An apple drops from a tree and undergoes free fall toward the ground.
   (a) Sketch the shape of the velocity–time graph of the apple's motion, assuming that downward is positive.
   (b) What is the slope of the line on your graph, and what does that slope represent?

Applying Inquiry Skills

6. **Figure 8** shows two accelerometers attached to carts that are in motion. In each case, describe two possible motions that would create the condition shown.

7. Describe how you would design and build a vertical accelerometer, using everyday materials, that measures vertical acceleration directly.

Making Connections

8. During a head-on collision, a certain car's airbag activates and increases the stopping time of a passenger from 0.010 s to 0.30 s.
   (a) Calculate the magnitude of the acceleration of a person travelling initially at 28 m/s (approximately 100 km/h) for each stopping time. What do you conclude?
   (b) Express each acceleration in (a) as a multiple of g, the magnitude of the acceleration due to gravity.

9. Think about the greatest accelerations you have experienced. Where did they occur? Did they involve speeding up or slowing down? What effects did they have on you?

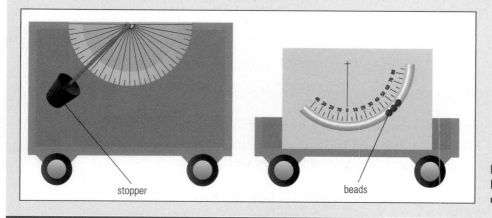

stopper   beads

Figure 8
Horizontal accelerometers for question 6

To make the dogsled in **Figure 1** accelerate forward, the dogs must exert a backward force against the ground. The ground pushes with a reaction force that helps them accelerate forward. Tension forces in the ropes cause the sled to accelerate along with the dogs. After an introduction to forces, you will learn how to use diagrams to help analyze the forces acting on objects, such as this dogsled.

Figure 1
The forces between the dogs' feet and the ground help accelerate the dogs, and thus the dogsled, forward. What other forces are evident in this photograph?

force a push or a pull; it is a vector quantity, symbol $\vec{F}$

force of gravity the attraction between all objects, only noticeable if at least one object is huge; symbol $\vec{F}_g$

In simple terms, a **force** is a push or a pull. Forces speed things up, slow them down, push them around corners or up hills, or just hold them still. Forces can also distort matter by compressing, stretching, or twisting it. Force is a vector quantity, and its direction can be stated in various ways, such as up, down, east, northeast, and so on.

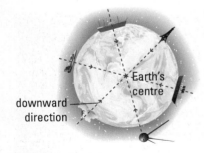

Figure 2
The force of gravity between Earth and objects at or near its surface is directed in a line to Earth's centre. This direction defines what we mean by vertical at any location on Earth.

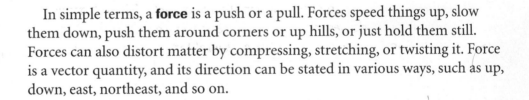

> ▶ **Practice**

Understanding Concepts

1. State an everyday example in which a force causes an object to (a) increase its speed, (b) become compressed, (c) become stretched.

2. You are facing east, in front of a gate that can swing. In what direction is your force if you (a) pull on the gate? (b) push on the gate?

Forces

We experience several types of forces daily. An obvious one is the **force of gravity**, which is the force of attraction between all objects. The force of gravity between Earth and objects at or near its surface is directed toward Earth's centre. This direction is referred to as vertically downward (**Figure 2**).

The force of gravity is an *action-at-a-distance force,* which means that contact between the objects is not needed. However, the force of gravity is extremely small unless at least one of the objects is very large. For example, the force of gravity between two pens is extremely small; the force of gravity between a pen and Earth is more significant; and the force of gravity between Earth and the Moon is huge.

If your pen is resting on your desk, the force of gravity is pulling down on it, but since it is at rest, it must be experiencing an opposite force to balance the force of gravity. The opposing force is the desk pushing upward. This force is called the **normal force**, which is a force that is perpendicular to the surfaces of the objects in contact. Normal means perpendicular in this context. (See **Figure 3**.)

An important force in our lives is **friction**, which is a force between objects in contact and parallel to the contact surfaces. If you shove your pen across your desk, it soon comes to rest because the force of friction between the desktop and the pen causes the pen to slow down. Friction on an object acts in a direction opposite to the object's motion or attempted motion. For example, if you shove your pen eastward across your desk, the force of friction is westward. **Static friction** is the force that prevents a stationary object from starting to move. **Kinetic friction** is the force that acts against an object's motion. **Air resistance** is friction that acts on an object moving through air; it becomes noticeable at high speeds; it is a special case of kinetic friction.

Another common force is **tension**, which is exerted by strings, ropes, fibres, and cables. A spider web (**Figure 4**) consists of numerous fine strands that pull on each other. These strands are under tension. In our bodies, muscle fibres experience tension when they contract.

Normal force, friction, and tension are all examples of *contact forces.* That is, they are forces that exist when objects are in direct contact with each other. There are several other names for various pushes, pulls, thrusts, and so on. The term **applied force**, $\vec{F}_A$, can be used as a general term for any contact force.

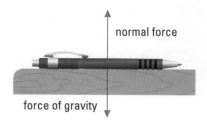

Figure 3
The normal force is perpendicular to both surfaces in contact.

normal force the force at right angles to objects in contact; symbol $\vec{F}_N$

friction the force between objects in contact; it is parallel to the contact surfaces and acts in a direction opposite to any motion or attempted motion; symbol $\vec{F}_f$

static friction the force that prevents a stationary object from starting to move; symbol $\vec{F}_S$

kinetic friction the force that acts against an object's motion; symbol $\vec{F}_K$

air resistance friction on an object moving through air; symbol $\vec{F}_{air}$

tension the force exerted by strings, ropes, fibres, and cables; symbol $\vec{F}_T$

applied force a general name for any contact force; symbol $\vec{F}_A$

> ▶ *Practice*

Understanding Concepts

3. Draw a sketch to show the force of gravity, the normal force, tension, and friction in each of the following:
   (a) A toboggan on a horizontal surface is pulled by a rope, and the rope is also horizontal.
   (b) A toboggan is pulled by a rope up a hill, and the rope is parallel to the hillside.

4. Describe, with examples, the difference between a contact force and an action-at-a-distance force.

5. Engineers say "You can't push a rope." What do they mean?

Figure 4
Tension keeps this spider web together.

newton the SI unit of force; it is derived from the base units of metres, kilograms, and seconds; symbol N

Measuring Force

The SI unit of force is the **newton**, symbol N, named after one of the greatest scientists in history, Sir Isaac Newton. The newton is a derived unit because it is a combination of the base units metres, kilograms, and seconds. When you study forces, it is important that your calculations express distance in metres, mass in kilograms, and time in seconds. In other words, all units must conform to the preferred SI units of metres, kilograms, and seconds.

▶ **TRY THIS** activity

Measuring and Estimating Forces

For this activity, you will need either a force sensor or a spring scale calibrated in newtons (N), both shown in **Figure 5**. You will also need graph paper, various objects that can be hung from the sensor or the scale, and the following masses: one 100-g mass, two 200-g masses, and one 500-g mass. Be sure that the objects you choose are not too large for the force scale available.

Carefully pull on the sensor or spring so that you can "feel" forces of 1 N, 2 N, and so on. Pick up various objects, one at a time, and estimate the magnitude (or size) of the force in newtons required to hold the object. After each trial, test your estimation in order to improve your estimating skills.

(a) Pick up the 100-g mass and estimate the magnitude of the force needed to hold it. Hang it from the sensor or scale and record the value.

(b) Measure and record the magnitude of the force needed to hold masses of 200 g, 300 g, 400 g, and so on, up to the maximum suggested by your teacher.

(c) Plot a graph of the magnitude of the force (on the vertical axis) as a function of mass. Determine the slope of the line of best fit on the graph. (This value is the force per unit mass on Earth's surface, a value that remains constant at a particular location.)

(d) Based on your results, apply your skills of interpolation and extrapolation to calculate the magnitude of the force required to hold a mass of

(i) 350 g   (ii) 1200 g   (iii) 2.0 kg

✋ **Protect the masses from dropping or recoiling violently.**

(a)

(b)

Figure 5
(a) Using a force sensor
(b) Using a spring scale

LEARNING TIP

Graphing Skills
To review drawing and analyzing line graphs, including interpolation and extrapolation, refer to Appendix A1.

> ▶ **Practice**

Understanding Concepts

6. State the magnitude of the force needed to hold each of the following masses steady:
   (a) 1.0 kg of sugar
   (b) a stapler of mass 0.20 kg
   (c) a student of mass 65 kg

7. The force exerted by gravity on a book resting on a desk is 8.5 N [down].
   (a) State the magnitude and direction of the normal force acting on the book.
   (b) If the same book is hanging from a string instead, what is the direction of the tension force acting on the book?

Answers

6. (a) 9.8 N
    (b) 2.0 N
    (c) 6.4×10^2 N

Drawing Force Diagrams

When analyzing forces and the effects they have on objects, it is helpful to use force diagrams. Two types of force diagrams, called *system diagrams* and *free-body diagrams,* are used.

A **system diagram** is a drawing of all the objects under analysis. A **free-body diagram** (FBD) is a drawing in which only the object being analyzed is drawn, with arrows showing all the forces acting on the object. **Figure 6** shows three examples of FBDs. The arrows in the FBDs are force vectors; they are drawn with their lengths proportional to the magnitudes of the forces. In each FBD, a positive direction is indicated, such as $+y$ for the vertical direction. (It doesn't matter which direction is chosen for $+y$ or $+x$ as long as the choice remains the same during the analysis.) The symbols for some of the most common

system diagram a drawing of all the objects in the situation under analysis

free-body diagram (FBD) a drawing of just the object being analyzed, not the entire situation, that shows all the forces acting on the object

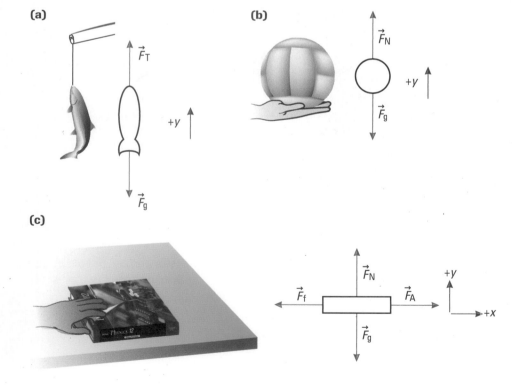

(a)

(b)

(c)

Figure 6
Notice that the arrows in the FBDs are drawn facing away from the body.
(a) A fish is held by a fishing line. (The only direction labelled is $+y$ because there are no horizontal forces.)
(b) A volleyball is held by a hand.
(c) A book is pushed across a desk by a horizontal applied force.

Table 1 Symbols for Common
Forces

Force	Symbol
gravity	$\vec{F}_g$
normal	$\vec{F}_N$
tension	$\vec{F}_T$
friction	$\vec{F}_f$
kinetic friction	$\vec{F}_K$
static friction	$\vec{F}_S$
applied force	$\vec{F}_A$
air resistance	$\vec{F}_{air}$

forces are listed in **Table 1**. Notice that each symbol has an arrow above it to indicate that the quantity is a vector.

▶ **TRY THIS** *activity*   **Forces on Spring**

Four identical spring scales, A, B, C, and D, are arranged in two different ways to hold a 1.0-kg mass (**Figure 7**). The readings are not shown on the scales.

(a) Assume that each spring scale has a negligible mass. Predict the readings on all four scales.

(b) Now assume that the spring scales have a mass that cannot be ignored. Predict the readings on all four scales.

(c) Set up a demonstration to check your predictions. Describe what you discover.

(d) Draw a system diagram and an FBD for each of the following:

(i) the 1.0-kg mass held by springs A and B

(ii) the 1.0-kg mass held by springs C and D

(iii) spring D

(iv) spring C

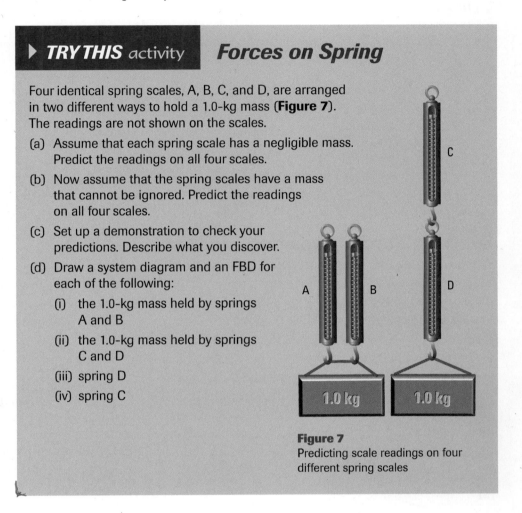

Figure 7
Predicting scale readings on four different spring scales

▶ **Practice**

Understanding Concepts

8. For each of the situations below, draw a system diagram and an FBD for each object in italics. (Be careful when deciding what forces are acting on each object. If you cannot decide what would cause a force, the force may not exist.)

(a) Your *notebook* is resting on your desk.

(b) A *tennis ball* is falling through the air from the server's hand. Neglect air resistance.

(c) A fully loaded *dogsled*, moving slowly along a flat, snowy trail, is being pulled horizontally by dogs attached to it by rope.

9. At a particular instant, the force of gravity on an elevator and the people in it is 2.80×10^4 N [down], and the tension in the cable holding the elevator is 3.20×10^4 N [up]. Draw an FBD of the elevator at the instant described.

SUMMARY   *Forces and Free-Body Diagrams*

- A force, which is a push or a pull, is a vector quantity; its SI unit is the newton (N).

- Some common forces we experience are gravity, friction, tension, and the normal force.

- Drawing free-body diagrams is an important skill that helps when analyzing problems involving forces.

▶ Section 1.4 Questions

Understanding Concepts

1. Explain why gravity is called an action-at-a-distance force.

2. In the following, state the direction of each force acting on the object in italics.
   - (a) A *puck* experiences friction on rough ice while sliding southward.
   - (b) The force of gravity exerts a force on *you*.
   - (c) The force of gravity acts on the *Moon,* keeping it in orbit around Earth.
   - (d) The force of the wind pushes against a *cyclist* who is cycling eastward.

3. Draw a system diagram and an FBD for each object in italics:
   - (a) A batter pops a baseball, and the *ball* is rising vertically in the air. (Neglect air resistance.)
   - (b) A *cup* is hanging from a hook.
   - (c) A *person* is standing in an elevator that is moving downward at a constant speed.
   - (d) A *person* is standing in an elevator that is moving upward at a constant speed.

   - (e) A *curling rock* is sliding in a straight line on a rink.
   - (f) A *crate* is being dragged along a floor by a rope that is horizontal.

4. A golfer is raising a pail of golf balls by exerting an applied force of 44 N [up]. The force of gravity on the pail is 32 N [down]. Draw an FBD of the pail.

5. A student pushes with a force of 55 N [E] on a desk, but the desk does not move. The force of gravity on the desk is 360 N.
   - (a) Draw a system diagram and an FBD of the desk.
   - (b) Explain why the desk does not move.

Applying Inquiry Skills

6. (a) Estimate the force in newtons required to hold up this physics book.
   - (b) Describe how you would test your estimation.

Making Connections

7. Explain how ocean tides provide evidence of action-at-a-distance forces.

Each year in Canada there are more than 150 000 traffic collisions, resulting in over 200 000 injuries and almost 3000 deaths. Of these deaths, the greatest number in any single category is the 15- to 25-year age group. Although the average number of deaths on Canada's highways has decreased in the past 20 years, there is still a great need for improved safety.

Sadly, many of these deaths and injuries could easily have been prevented, simply by wearing seatbelts and being in cars with airbags. Understanding *inertia* and, therefore, Newton's first law of motion, will help you appreciate how wearing a seatbelt can save your life.

Inertia

If you are riding a bicycle on a horizontal road and you stop pedalling, you will gradually slow to a stop. Some force must be causing you to slow down. In this case, it is the force of friction, mainly between the tires and the road, but also within the moving parts of the bike.

What would happen if there were no friction? To answer this question, let's imagine a frictionless ball on a frictionless ramp, as illustrated in **Figure 1**. In **(a)**, the ball speeds up as it rolls down the ramp, then it moves at a constant velocity along the horizontal surface, and finally rolls up the far ramp to the same level it started from. In **(b)**, the same thing happens, but this time the ball rolls farther because the far ramp is not as steep. In **(c)**, the ball continues to roll along the horizontal slope because it never reaches another ramp. In this experiment, we conclude that, in the absence of friction, the moving ball will continue moving at a constant velocity in an attempt to reach its original height.

Galileo Galilei

Galileo Galilei (1564–1642) was the first scientist to use controlled experiments to investigate objects in motion. The experiment featured in **Figure 1** is a version of a thought experiment he performed to help understand how forces affect motion. His ideas had an important influence on Sir Isaac Newton, whose laws of motion are presented in this chapter.

Figure 1
(a) The ascending ramp has a steep slope.
(b) The ascending ramp has a shallow slope.
(c) There is no ascending ramp.

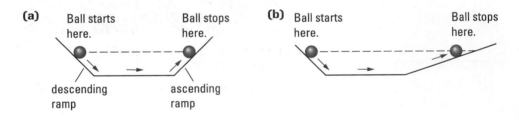

inertia a property of matter that causes an object to resist changes in its state of motion; it is directly proportional to the mass of the object

Every object possesses a property that Newton called *inertia*. **Inertia** is a property of matter that causes a body to resist changes in its state of motion. In the above example, the ball would keep on moving unless it was stopped by some force. If at rest, the ball would stay at rest unless some force caused it to move. In other words, inertia causes the ball to resist changes to its motion.

You can experience this force as a passenger on a bus. Suppose you are standing in the aisle of a stopped bus. As the bus starts to move, your body wants to stay at rest because of inertia. As a result, you may fall toward the back of the bus if you do not brace yourself. When the bus reaches a constant velocity, you have no trouble standing in the aisle because you are also moving with constant velocity. What happens if the bus driver suddenly applies the brakes? The bus will slow down. But your body has a tendency to keep moving because of inertia. Again, you will need to brace yourself or you will fall toward the front of the bus. The amount of inertia an object has depends directly on its mass: *the greater the mass, the greater the inertia an object possesses.*

▶ **Practice**

Understanding Concepts

1. (a) The following objects are at rest; rank their inertia in order from the least to the greatest: a school bus, a small child, a compact car, and you.
   (b) Repeat (a) assuming all the objects are moving at the same velocity.

Net Forces

As you learned in section 1.4, there can be several forces acting on an object at the same time. The vector sum of all the forces acting on an object is the **net force**. (The net force can also be called the *resultant force.*) In this text, the symbol used for net force is $\vec{F}_{net}$. We can use force diagrams to analyze the effects of forces, or the net force, acting on any object. It is important to remember that the net force is *not* an actual force or a separate force of nature; it is the *sum* of actual forces.

net force the vector sum of all the forces acting on an object; also called the resultant force; symbol $\vec{F}_{net}$

(a)

▶ **SAMPLE** problem *1*

Force Diagrams for Net Forces

A weightlifter holds a weight above the head by exerting a force of 1.6 kN [up]. The force of gravity acting on the weight is 1.6 kN [down].

(a) Draw a system diagram and an FBD of the weight, and state the net force at that instant.

(b) If the weightlifter's applied force changes to 1.8 kN [up] for an instant, what is the net force?

Solution

(a) The diagrams are shown in **Figure 2**. The upward force exerted by the weightlifter can be called the normal force or the applied force. The net force at the instant shown is zero.

(b)

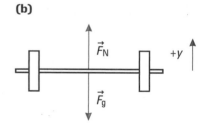

$$\vec{F}_{net} = \vec{F}_N + \vec{F}_g$$
$$= 1.6 \text{ kN } [\uparrow] + 1.6 \text{ kN } [\downarrow]$$
$$\vec{F}_{net} = 0$$

Figure 2
(a) The system diagram
(b) The FBD

(b) The net force is the vector sum of the forces.

$$\vec{F}_N = 1.8 \text{ kN [up]}$$
$$\vec{F}_g = 1.6 \text{ kN [down]}$$
$$\vec{F}_{net} = ?$$

$$\begin{aligned}
\vec{F}_{net} &= \vec{F}_N + \vec{F}_g \\
&= 1.8 \text{ kN [up]} + 1.6 \text{ kN [down]} \\
&= 1.8 \text{ kN [up]} - 1.6 \text{ kN [up]} \\
\vec{F}_{net} &= 0.2 \text{ kN [up]}
\end{aligned}$$

The net force is 0.2 kN [up].

▶ *TRY THIS* activity   *Observing Objects at Rest and in Motion*

Your teacher will set up demonstrations of objects initially at rest and others initially in motion. Set up a table in your notebook using the headings in **Table 1**. For each demonstration, predict what you think will happen and record your prediction. Then observe the result of each motion and summarize your observations in your table. In the last column, state whether the object remains at rest or moves at a constant velocity.

Figure 3
This ballistics cart shoots a ball vertically upward from a spring-loaded barrel as the cart moves with a constant forward velocity.

Table 1   Demonstrations of Objects (in *italics*) at Rest and in Motion

Object Observed (Instruction in Brackets)	Initial State	Predicted Result	Observed Result	Motion of Object
Coin on card on a beaker (Snap the card horizontally.)	at rest	?	?	?
Plastic flask on paper on desk (Jerk the paper horizontally.)	at rest	?	?	?
Teddy bear on cart (Jerk the cart forward quickly.)	at rest	?	?	?
Ball launched from ballistics cart, **Figure 3** (Push the cart straight forward so the ball is launched vertically as the cart moves at a constant velocity.)	in motion	?	?	?
Teddy bear on moving cart (Allow the cart to hit the wall, but not hard enough to damage the cart.)	in motion	?	?	?

> ▶ **Practice**

Understanding Concepts

2. Calculate the net force when each of the following sets of forces act on the same object:
   (a) 54 N [up], 65 N [down], and 92 N [up]
   (b) 8.6 N [S], 1.8 N [N], and 2.4 N [N]
   (c) 13.5 N [E], 21.2 N [W], 33.0 N [E], and 25.3 N [W]

3. A store clerk pushes a parcel on a counter with a force of 7.6 N [W]. The kinetic friction on the parcel is 6.5 N [E]. Both the force of gravity and the normal force have a magnitude of 9.9 N. Draw an FBD of the parcel and determine the net force acting on it.

Answers

2. (a) 81 N [up]
   (b) 4.4 N [S]
   (c) 0.0 N
3. 1.1 N [W]

Newton's First Law of Motion: The Law of Inertia

The observations made in the example presented in **Figure 1** on page 32 and in the *Try This* Activity on page 34 were summarized by Newton (**Figure 4**) in what is now called *Newton's first law of motion*:

Newton's First Law of Motion
If the net force acting on an object is zero, the object will maintain its state of rest or constant velocity.

A common way to express this law is to say that an object at rest or moving with a constant velocity maintains its state of rest or constant velocity unless acted upon by a net force. Note that the net force must be external in order to change an object's velocity. Internal forces have no effect on an object's motion. For example, pushing with your arms on the dashboard of a car (an internal force) does not change the car's velocity.

The law of inertia helps us to understand the principles behind using seatbelts and airbags. Once an object is moving, it tends to keep moving at a constant velocity because of its inertia. When a car suddenly slows down, the people in the car continue to move forward. If they are not wearing seatbelts, they may crash through the windshield and be fatally injured. A passenger who wears a seatbelt properly not only stays in place in the car, reducing the risk of injury, but is also protected from injuries that can result when an airbag deploys. The operation of one type of seatbelt is shown in **Figure 5**.

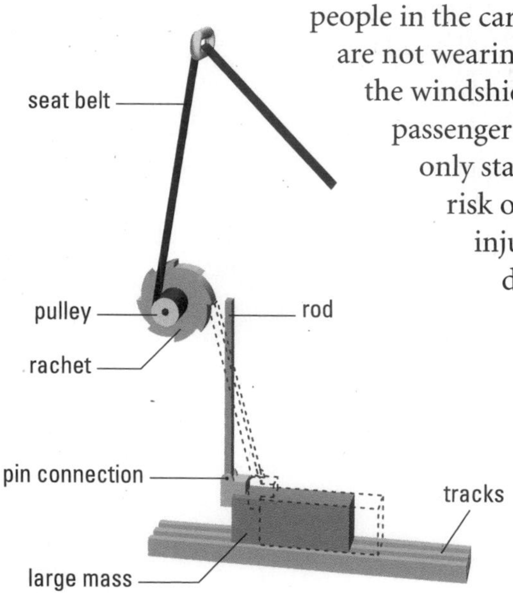

Figure 4
Sir Isaac Newton (1642–1727) had made great discoveries in mathematics, mechanics, and optics by the age of 26. His important book, *Principia Mathematica,* laid the foundations of physics that still apply today, including the physics presented in this text.

seat belt —

pulley — — rod

rachet —

pin connection —

— tracks

large mass —

Figure 5
The operation of the seatbelt shown here relies on inertia. Normally, the ratchet turns freely, allowing the seatbelt to wind or unwind whenever the passenger moves. If the car moves forward (to the right in this case), then quickly slows, the large mass on the tracks shown continues to move forward because of inertia. This causes the rod to turn around the pin connection, which in turn locks the ratchet wheel and keeps the belt firmly in place.

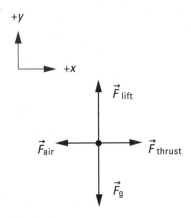

+y

+x

$\vec{F}_{lift}$

$\vec{F}_{air}$ $\vec{F}_{thrust}$

$\vec{F}_{g}$

Figure 6
The FBD of the aircraft in Sample Problem 2

SAMPLE problem 2

FBDs for Net Forces

A 12-passenger jet aircraft of mass 1.6×10^4 kg is travelling at a constant velocity of 850 km/h [E] while maintaining a constant altitude. Besides gravity and air resistance, the aircraft also experiences an upward force called "lift" and a forward force of the engines called "thrust." Draw an FBD of the aircraft, and state the net force acting on the aircraft.

Solution
Figure 6 shows the FBD of the aircraft. According to Newton's first law of motion, the net force on the aircraft must be zero because it is moving with a constant velocity.

We can summarize the first law of motion by stating four important results:

(a) Objects at rest remain at rest unless acted upon by a net force.

(b) Objects in motion remain in motion unless acted upon by a net force.

(c) If the velocity of an object is constant (or zero), the net external force acting on it must be zero.

(d) If the velocity of an object is changing either in magnitude, direction, or both, the change must be caused by a net external force on the object. (You will explore this result by performing Investigation 1.6.)

Practice

Understanding Concepts

4. Explain why Newton's first law of motion can also be called the law of inertia.

5. You exert a force of 46 N [up] on your backpack, causing it to move upward with a constant velocity. Draw an FBD of the backpack and determine the force of gravity on the pack.

6. Explain how to apply the first law of motion when trying to get a heap of snow off a shovel. Try to give more than one solution.

Applying Inquiry Skills

7. (a) Describe how you could use a piece of paper, a coin, and your desk to demonstrate the law of inertia for an object initially at rest.
   (b) Describe how you could safely demonstrate, using objects of your choosing, the law of inertia for an object initially in motion.

Making Connections

8. Explain the danger of stowing heavy or sharp objects in the rear window space of a car.

| SUMMARY | *Inertia and Newton's First Law of Motion* |

- Inertia is a property of matter that causes a body to resist changes in its state of motion; it is directly proportional to mass.
- The net force acting on an object, $\vec{F}_{net}$, is the vector sum of all the forces acting on the object.
- Newton's first law of motion, also called the law of inertia, states that if the net force on an object is zero, the object will maintain its state of rest or constant velocity.
- The first law of motion is observed and applied in many situations, including the use of restraint systems such as seatbelts and airbags in automobiles.

▶ *Section 1.5 Questions*

Understanding Concepts

1. A curling rock moving along a rink eventually comes to a stop. Does this disagree with Newton's first law of motion? Explain your answer.

2. Draw a system diagram and an FBD for the birdie in each case:
   (a) A badminton racket makes contact with a badminton birdie.
   (b) The birdie goes straight up, just above the racket.
   (c) The birdie stops for a brief instant at the top of its flight.
   (d) The birdie begins to fall straight down; air resistance exists but has not reached maximum value.

3. State the value of the net force acting on an object that is (a) at rest and (b) moving with constant velocity.

Applying Inquiry Skills

4. Describe how you would demonstrate Newton's first law of motion, relating to objects at rest and in motion, to a class of elementary school students using toys or recreational devices (e.g., inline skates, scooter, skateboard, ice skates). Include safety considerations.

Making Connections

5. You are helping a friend move furniture. The friend asks you to stand inside the back of a pickup truck to hold onto a piano because there is no rope available to tie it to the truck. Use physics principles from this chapter to explain why you should refuse this request.

6. Explain why the application of Newton's first law of motion is important in transportation safety.

Inquiry Skills

○ Questioning	● Conducting	● Evaluating
● Predicting	● Recording	● Communicating
○ Planning	● Analyzing	● Synthesizing

What Factors Affect Acceleration?

If you observe vehicles at a stoplight accelerating from rest, it is obvious that more than one factor affects acceleration. For instance, the acceleration of a car is greater than the acceleration of a large truck, and the acceleration of a car with a powerful engine is greater than the acceleration of a car with a small engine.

To discover the relationship between several variables, a controlled experiment must be carried out. A *controlled experiment* is an investigation in which only one variable or factor is changed at a time; the other variables are kept constant. In this investigation, you will determine how the dependent variable, the acceleration, depends on two independent variables: the net force applied to the object (with a constant mass) and the mass of the object (with a constant net force). (You can review controlled experimentation in Appendix A2.)

To determine how the acceleration of a cart depends on the net force applied to the cart, you will vary the applied force while keeping the mass constant, as shown in **Figure 1(a)**. To determine how the acceleration of the cart depends on the total mass of the cart, you will vary the mass while keeping the applied force constant, as shown in **Figure 1(b)**.

Although a spring scale is used in **Figure 1** to measure the magnitude of the applied force, a force sensor can be used.

The instructions for this investigation require qualitative observations. For quantitative observations, you must make sure the applied force is constant. This can be done by using a device such as a smart pulley. You will also need a ticker-tape timer or similar apparatus to determine the acceleration of the cart in each trial.

Question

How does the acceleration of an object depend on (i) the net force applied to the object and (ii) the mass of the object?

Prediction

(a) Communicate in various ways (such as words, graphs, and mathematical variation statements) your answer to the Question. (For information about writing mathematical variation statements, see Appendix A1.)

(a)

(b)

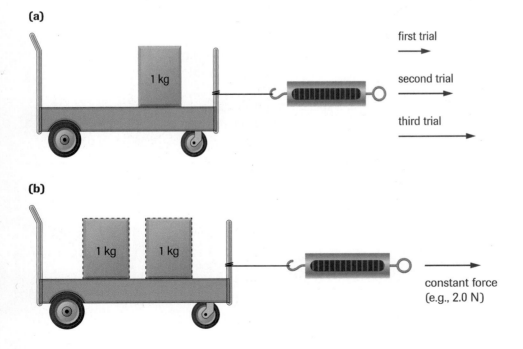

Figure 1
(a) Increasing the force applied to a constant mass. Use three different forces, for example, 1.0 N, 2.0 N, 3.0 N.
(b) Increasing the mass while keeping the applied force constant. Use three different masses, for example, cart plus 2.0 kg, cart plus 1.0 kg, cart only.

Materials

For each group of three or four students:
mass scale
cart
two 1.0-kg masses
masking tape
spring scale (to 10 N) or force sensor

Procedure

1. Determine the mass of the cart. Secure the 1.0-kg mass to the cart with masking tape. Record the total mass.

2. Clear an area on the lab bench so there will be no obstacles to the cart's motion. With the force scale or sensor attached to the cart as shown in **Figure 1(a)**, apply a force of magnitude 1.0 N and observe the cart's acceleration. Be sure to stop the cart before it reaches the end of the lab bench.

- **Wear safety goggles.**
- **Make sure the spring on the force scale does not recoil.**
- **Make sure all masses on the cart are secure.**

3. Repeat step 2 using applied forces of magnitude 2.0 N and then 3.0 N.

4. Secure the second 1.0-kg mass to the cart with masking tape. Apply a force of magnitude 2.0 N and observe the acceleration. Record the total mass and your observations.

5. Repeat step 4 using only one 1.0-kg mass on the cart, and then with no mass on the cart.

Analysis

(b) Which of the graphs and mathematical statements in **Figure 2** show most accurately what you discovered in this investigation? Record them in your lab report.

(c) Answer the Question.

Evaluation

(d) Describe any difficulties or sources of error in this investigation.

(a)

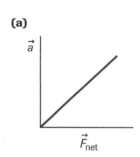

(b)

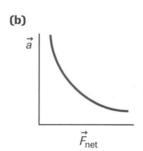

(c)

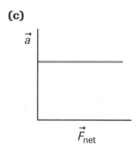

(d)

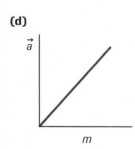

(e)

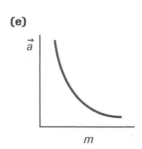

(f)

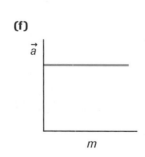

Figure 2
Graphs for Analysis (b). The symbol $\propto$ is read as "is proportional to."

(a) $\vec{a} \propto \vec{F}_{net}$ (with a constant mass)

(b) $\vec{a} \propto \dfrac{1}{\vec{F}_{net}}$ (with a constant mass)

(c) $\vec{a}$ does not depend on $\vec{F}_{net}$ (with a constant mass)

(d) $\vec{a} \propto m$ (with a constant $\vec{F}_{net}$)

(e) $\vec{a} \propto \dfrac{1}{m}$ (with a constant $\vec{F}_{net}$)

(f) $\vec{a}$ does not depend on m (with a constant $\vec{F}_{net}$)

Figure 1
The space shuttle *Endeavour* is launched, carrying astronauts to the *International Space Station*.

For the space shuttle (**Figure 1**) to accelerate upward, the upward thrust caused by the engines must be greater than the force of gravity pulling downward on the shuttle. In other words, the net force on the shuttle is upward, and the shuttle accelerates upward. In mathematical terms, we represent the effect of force on acceleration as $\vec{a} \propto \vec{F}_{net}$.

Net force isn't the only factor that affects an object's acceleration; the object's mass must also be considered. In a controlled investigation, if the net force is kept constant, the acceleration decreases as the mass increases. In mathematical terms, the effect on acceleration of a constant net force applied to an object is $\vec{a} \propto \dfrac{1}{m}$. These two variation statements, $\vec{a} \propto \vec{F}_{net}$ and $\vec{a} \propto \dfrac{1}{m}$, are combined in *Newton's second law of motion.*

> **Newton's Second Law of Motion**
> If the external net force on an object is not zero, the object accelerates in the direction of the net force. The magnitude of the acceleration is directly proportional to the net force and inversely proportional to the object's mass.

If the units of the net force, acceleration, and mass are all SI units, the second law of motion can be summarized in equation form:

$$\vec{a} = \frac{\vec{F}_{net}}{m}$$

This equation is often rearranged and written as

$$\vec{F}_{net} = m\vec{a},$$

where $\vec{F}_{net}$ is the net force measured in newtons (N),
  m is the mass measured in kilograms (kg), and
  $\vec{a}$ is the acceleration in metres per second squared (m/s²).

newton the magnitude of the net force needed to give a 1-kg object an acceleration of magnitude 1 m/s²

The equation $\vec{F}_{net} = m\vec{a}$ allows scientists to define the SI unit of force, the newton, in terms of base SI units. One **newton** (N) is the magnitude of the net force needed to give a 1-kg object an acceleration of magnitude 1 m/s². (Note that this definition is more detailed than the one given in section 1.4.) Substituting into the second law equation we have

$$1\ N = 1\ kg \left(1\ \frac{m}{s^2}\right),\ \text{or } 1\ N = 1\ \frac{kg \cdot m}{s^2}.$$

Does Newton's second law agree with his first law of motion? According to the second law, $\vec{a} = \dfrac{\vec{F}_{net}}{m}$ so the acceleration is zero if the net force is zero. This is in exact agreement with the first law.

▶ **SAMPLE** problem **1**

Calculating Acceleration

A net force of 58 N [W] is applied to a water polo ball of mass 0.45 kg. Calculate the ball's acceleration.

Solution

$\vec{F}_{net}$ = 58 N [W]

m = 0.45 kg

$\vec{a}$ = ?

$$\vec{a} = \frac{\vec{F}_{net}}{m}$$

$$= \frac{58\ N\ [W]}{0.45\ kg}$$

$$= \frac{58\ kg \cdot m/s^2\ [W]}{0.45\ kg}$$

$$\vec{a} = 1.3 \times 10^2\ m/s^2\ [W]$$

The ball's acceleration is 1.3 × 10² m/s² [W].

▶ **SAMPLE** problem **2**

Calculating Net Force

In an extreme test of its braking system under ideal road conditions, a sports car, travelling initially at 26.9 m/s [S], comes to a stop in 2.61 s. The mass of the car with the driver is 1.18×10^3 kg. Calculate (a) the car's acceleration and (b) the net force needed to cause that acceleration.

Solution

(a) $\vec{v}_i$ = 26.9 m/s [S]

$\vec{v}_f$ = 0.0 m/s

Δt = 2.61 s

$\vec{a}$ = ?

$$\vec{a} = \frac{\vec{v}_f - \vec{v}_i}{\Delta t}$$

$$= \frac{0.0\ m/s - 26.9\ m/s\ [S]}{2.61\ s}$$

$$= -10.3\ m/s^2\ [S]$$

$$\vec{a} = 10.3\ m/s^2\ [N]$$

The car's acceleration is 10.3 m/s² [N].

(b) m = 1.18×10^3 kg

$\vec{F}_{net}$ = ?

$$\vec{F}_{net} = ma$$

$$= (1.18 \times 10^3\ kg)(10.3\ m/s^2\ [N])$$

$$\vec{F}_{net} = 1.22 \times 10^4\ N\ [N]$$

The net force on the car is 1.22 × 10⁴ N [N].

LEARNING TIP

Two-Part Solutions
Sample Problem 2 is the first sample problem we have encountered in which the numerical answer to part (a) is used in part (b). Whenever this occurs, write the rounded-off answer for part (a), but leave the numerical, or unrounded, value in your calculator and use it in part (b). After calculating the value for part (b), round off the final answer to the correct number of significant digits.

> **Practice**

Understanding Concepts

1. Explain how the direction of an object's acceleration relates to the direction of the net force causing the acceleration.

2. Calculate the acceleration in each of the following:
   (a) A net force of 27 N [W] is applied to a cyclist and bicycle having a total mass of 63 kg.
   (b) A bowler exerts a net force of 18 N [forward] on a 7.5-kg bowling ball.

3. Calculate the net force in each of the following situations:
   (a) A cannon gives a 5.0-kg shell an acceleration of 5.0×10^3 m/s^2 [forward] before it leaves the muzzle.
   (b) A 28-g arrow is given an acceleration of 2.4×10^3 m/s^2 [E].

4. (a) Rewrite the second law equation to solve for mass.
   (b) Calculate the mass of a regulation shot in the women's shot-put event (**Figure 2**) if a net force of 7.2×10^2 N [forward] gives the shot an average acceleration of 1.8×10^2 m/s^2 [forward].

5. Assume that for each pulse, a human heart accelerates 21 g of blood from 18 cm/s to 28 cm/s during a time interval of 0.10 s. Calculate the magnitude of (a) the acceleration of the blood and (b) the force needed to cause that acceleration.

Answers

2. (a) 0.43 m/s^2 [W]
   (b) 2.4 m/s^2 [forward]
3. (a) 2.5×10^4 N [forward]
   (b) 67 N [E]
4. (b) 4.0 kg
5. (a) 1.0 m/s^2
   (b) 2.1×10^{-2} N

Figure 2
The shot is made of iron or brass.

weight the force of gravity on an object; it is a vector quantity measured in newtons, symbol $\vec{F}_g$

mass the quantity of matter in an object; it is a scalar quantity measured in kilograms (kg) in SI

Mass and Weight

Newton's second law equation, $\vec{F}_{net} = m\vec{a}$, can be applied to objects in free fall near Earth's surface. During free fall, the net force is $\vec{F}_g$ and the acceleration is the acceleration due to gravity, $\vec{g}$, so the equation is written $\vec{F}_g = m\vec{g}$, where $\vec{g} = 9.8$ m/s^2 [down].

This force of gravity on an object is called **weight**. Being a force, weight is measured in newtons, not in kilograms. In the laboratory, a force sensor or a spring scale can be used to measure weight. Because the force of gravity can vary, the weight of an object will vary, depending on its location. The magnitude of an object's weight is equal to the magnitude of the force needed to hold the object steady or to raise or lower the object without acceleration.

Notice that, even though weight and mass are used interchangeably in daily transactions, they are quite different quantities. **Mass** is the quantity of matter in an object. As long as the amount of matter in an object remains the same, its mass stays the same. Mass is measured using a balance, an instrument that compares an unknown mass to a standard (the kilogram, for example). The label on a kilogram of ground beef should read "mass: 1 kg."

In the *Try This* Activity on page 28, you learned that the force of gravity acting on a 1.0-kg mass is 9.8 N/kg [down]. To see how this relates to weight, we rearrange the equation for weight to express $\vec{g}$ by itself:

$$\vec{g} = \frac{\vec{F}_g}{m}$$

$$= \frac{9.8 \text{ N [down]}}{1.0 \text{ kg}}$$

$$\vec{g} = 9.8 \frac{\text{N}}{\text{kg}} \text{ [down]}$$

Substituting for the newton, N:

$$\vec{g} = 9.8 \frac{\text{N}}{\text{kg}} \text{ [down]}$$

$$= 9.8 \frac{\text{kg} \cdot \frac{\text{m}}{\text{s}^2}}{\text{kg}} \text{ [down]}$$

$$\vec{g} = 9.8 \frac{\text{m}}{\text{s}^2} \text{ [down]}$$

Thus, the value of $\vec{g}$ at Earth's surface can be written in either way:

$$\vec{g} = 9.8 \frac{\text{m}}{\text{s}^2} \text{ [down]} = 9.8 \frac{\text{N}}{\text{kg}} \text{ [down]}$$

This value is called the *gravitational constant* at Earth's surface. It agrees with the slope of the line on the graph you plotted in the *Try This* Activity on page 28.

▶ **SAMPLE** problem **3**

Determining Weight

The maximum train load pulled through the Chunnel, the train tunnel under the English Channel that links England and France, is 2434 t. Determine the weight of this load.

Solution

$m = 2434 \text{ t}$

$\vec{g} = 9.8 \text{ N/kg [down]}$

$\vec{F}_g = ?$

First, convert the mass in tonnes to kilograms:

$$2434 \text{ t} = 2434 \cancel{t} \times \frac{1000 \text{ kg}}{1 \cancel{t}} = 2.434 \times 10^6 \text{ kg}$$

Now,

$$\vec{F}_g = m\vec{g}$$

$$= (2.434 \times 10^6 \cancel{\text{kg}})(9.8 \frac{\text{N}}{\cancel{\text{kg}}} \text{ [down]})$$

$$\vec{F}_g = 2.4 \times 10^7 \text{ N [down]}$$

The weight of the load is 2.4×10^7 N [down].

Understanding Concepts

6. Summarize the differences between mass and weight by setting up and completing a table using these titles: Quantity; Type of Quantity; Definition; Symbol; SI Unit; Method of Measuring.

	Mass	Weight
Quantity		
Type of Quantity		
Definition		
Symbol		
SI Unit		
Method of Measuring		

Answers

7. (a) 1.9×10^2 N [down]
   (b) 1.9×10^2 N [up]

8. 4.8×10^2 N [down]

9. 18 kg

7. (a) Calculate the weight of a 19-kg curling stone.
   (b) Calculate the force required to raise the curling stone upward without acceleration.

8. Calculate the weight of a 54-kg robot on the surface of Venus where the gravitational constant is 8.9 N/kg [down].

9. Calculate the mass of a backpack whose weight is 180 N [down].

SUMMARY **Newton's Second Law of Motion and Weight**

- Newton's second law of motion states that the acceleration of an object is in the direction of the net force acting on the object, and the magnitude of the acceleration is directly proportional to the magnitude of the net force and inversely proportional to the mass. The second law is represented by

$$\vec{a} = \frac{\vec{F}_{net}}{m} \quad \text{or} \quad \vec{F}_{net} = m\vec{a}$$

- Mass is the quantity of matter in an object (measured in kilograms), and weight is the force of gravity on an object (measured in newtons and calculated using the equation $\vec{F}_g = m\vec{g}$).

▶ *Section 1.7* Questions

Understanding Concepts

1. A net force of 5.0 N [S] is applied to a toy electric train of mass 2.5 kg. Calculate the train's acceleration.

2. Calculate the net force needed to give a 250-kg boat an acceleration of 2.8 m/s² [W].

3. Calculate the net force needed to cause a 1310-kg sports car to accelerate from zero to 28.6 m/s [forward] in 5.60 s.

4. A net force of 29.4 N [down] causes an object to accelerate at 9.8 m/s² [down]. Calculate the object's mass.

5. Calculate the weight at Earth's surface of each of the following:
   (a) a person of mass 64 kg
   (b) a pop can of mass 450 g
   (c) a load of gravel of mass 2.9 t

6. Calculate the mass of each of the following objects:
   (a) a magazine of weight 0.98 N [down]
   (b) an infant of weight 42 N [down]
   (c) a car of weight 18 kN [down]

7. Using your mass in kilograms, calculate your weight.

Applying Inquiry Skills

8. **Table 1** contains the data from an experiment that measured the acceleration of an object.
   (a) Plot a graph of the acceleration (vertical axis) versus the net force for a mass of 1.0 kg.
   (b) Plot a graph of the acceleration (vertical axis) versus the mass for a net force of 4.0 N [E].
   (c) Do the graphs support Newton's second law of motion? Explain your answer.

Table 1 Data for Question 8

Question	Mass (kg)	Net Force (N [E])	Acceleration ($\frac{m}{s^2}$ [E])
8(a)	1.0	1.0	1.0
	1.0	2.0	2.0
	1.0	3.0	3.0
	1.0	4.0	4.0
8(b)	1.0	4.0	4.0
	2.0	4.0	2.0
	3.0	4.0	1.3
	4.0	4.0	1.0

Newton's third law of motion, often called the *action–reaction law,* is the last of Newton's laws of motion. The first and second laws analyzed one object at a time, but the third law analyzes forces that act in pairs on two objects. For example, if the *skater* (object one) in **Figure 1** pushes with a force of 75 N [W] against the *boards* (object two), she accelerates eastward, away from the boards. Newton's third law tells us that the force causing this acceleration is 75 N [E], which is the reaction to her initial force. The skater's force against the boards is called the *action force,* and the force of the boards on the skater is called the *reaction force.* Notice that the magnitude of the action force is the same as the magnitude of the reaction force, and that the directions are opposite.

75 N [W]

Figure 1
When the skater pushes on the boards in one direction, the boards push on the skater in the opposite direction.

> **Newton's Third Law of Motion**
> For every force, there is a reaction force equal in magnitude but opposite in direction.

We begin the analysis of this law by examining action–reaction pairs of forces on an object at rest. Consider an apple hanging in a tree, as shown in **Figure 2**. **Figure 2(a)** and **(b)** show two pairs of action–reaction forces:

- The apple pulls downward on the stem while the stem pulls upward on the apple.
- The pull of Earth's gravitational force pulls downward on the apple, while the apple's gravitational force pulls upward on Earth.

(a)

force of stem on apple

force of apple on stem

(b)

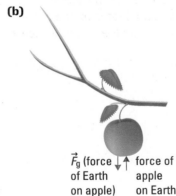

$\vec{F}_g$ (force of Earth on apple)   force of apple on Earth

(c)

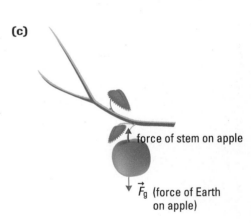

force of stem on apple

$\vec{F}_g$ (force of Earth on apple)

Figure 2
(a) The force of the stem on the apple and the force of the apple on the stem comprise an action–reaction pair of forces.
(b) The force of Earth's gravity on the apple and the force of the apple on Earth comprise a second pair of action–reaction forces.
(c) The downward force of Earth's gravity on the apple is balanced by the upward force exerted by the stem. This is not an action–reaction pair of forces because both forces are acting on the apple.

To understand why the apple remains at rest, we must consider only the forces acting *on the apple*. As shown in **Figure 2(c)**, the only two forces acting on the apple are the downward force of gravity and the upward force of the stem. Since the apple isn't moving, the net force acting on the apple is zero; therefore, the forces are equal in magnitude and opposite in direction.

Next, we analyze a situation in which the net force acting on the objects is not zero. Consider the ball and the cart in **Figure 3**. When the ball is pushed into the cart, the spring compresses. When the spring is released, the spring (and thus the cart) pushes forward on the ball. This is the action force. At the same instant, the ball pushes backward on the spring (and thus the cart). This is the reaction force of the ball on the cart. The action and reaction forces are equal in size but opposite in direction, and they act on different objects. Note that because both forces act at the same instant, it does not matter which is called the action force and which is called the reaction force. The names can be interchanged with no effect on the description.

Figure 3
When the spring is released, it exerts a forward force on the ball. At the same instant, the ball exerts a force on the spring (and the cart) in the opposite direction. The ball moves forward and the cart moves backward.

▶ **SAMPLE** problem **1**

The Effects of Action–Reaction Forces

Assume that the cart in **Figure 3** has a mass of 1.2 kg and the ball has a mass of 0.072 kg. During the short time interval that the spring is released, the horizontal force on the ball is 1.8 N [E].

(a) Identify three action–reaction pairs of forces involving the cart as the spring is released.

(b) Draw an FBD of the ball, and calculate its acceleration while the spring is pushing on it.

Solution

(a) The action–reaction pairs of forces involving the cart are the following:
  • The cart pushes downward with a normal force on the surface beneath it, while the surface pushes upward with a normal force on the cart.
  • Earth's gravity pulls downward on the cart while the cart pulls upward on Earth.
  • The cart pushes on the ball with a force of 1.8 N [E] while the ball pushes on the cart with a force of 1.8 N [W].

(b) The FBD is shown in **Figure 4**.

$\vec{F}_{net} = 1.8$ N [E]

$m = 0.072$ kg

$\vec{a} = ?$

$$\vec{a} = \frac{\vec{F}_{net}}{m}$$

$$= \frac{1.8 \text{ N [E]}}{0.072 \text{ kg}}$$

$$\vec{a} = 25 \text{ m/s}^2 \text{ [E]}$$

The acceleration of the ball is 25 m/s² [E].

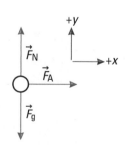

Figure 4
The FBD of the ball

Understanding Concepts

1. Describe the reaction force in each of the following:
   (a) The action force is a westward force of a balloon on compressed air being released from the nozzle.
   (b) The action force is the force of your foot backward on the floor.
   (c) A skater applies a force of 250 N [S] against the boards of a hockey rink.

2. (a) See **Figure 2**, page 46. Draw an FBD of the apple hanging from the stem.
   (b) Are there any action–reaction pairs of forces in the FBD?

3. (a) Describe three action–reaction pairs of forces involving the ball in **Figure 3**, page 47, while the spring is pushing on the ball.
   (b) Draw an FBD of the cart in **Figure 3** as the spring is being released, and calculate the acceleration of the cart.

4. Explain, with an example, why the following statement is false: "All the action–reaction pairs of forces on two objects are equal in magnitude but opposite in direction, so neither object can undergo acceleration."

Answers

3. (b) 1.5 m/s² [W]

▶ **TRY THIS** activity **Demonstrating Newton's Third Law**

Your teacher will set up demonstrations of the third law of motion. In each case, predict what you think will occur, then observe what happens. Summarize your observations in a table with titles as shown in **Table 1**. Examples of demonstrations are the following:

- Release an inflated balloon.
- Launch a water rocket (**Figure 5(a)**), outdoors and away from people.
- Turn on a rotating water sprinkler or a tap from which a hose hangs freely.

- Carefully step forward off a low chair or trolley.
- Operate a propeller-driven cart (**Figure 5(b)**) with no sail attached.
- Operate the propeller-driven cart with a sail attached.
- Punch a hole in the end of the carbon dioxide container inserted in the propulsion device in **Figure 5(c)**.

 Always release moving objects away from people.

(a) 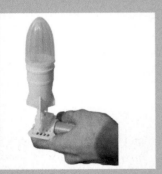 **(b)**  **(c)**

Figure 5
(a) After the rocket is partially filled with water, air is pumped into it, then the trigger is released.
(b) After observing the direction in which air is forced away from the propeller, you can predict which way the cart will move when it is released.
(c) In this propulsion device, each carbon dioxide cylinder can be used only once.

Table 1 Third-Law Demonstrations

Object Observed	Predicted Result	Observed Result	Description of the Action Force(s)	Description of the Reaction Force(s)	Diagram of the Action–Reaction Pairs

Applications of the Third Law of Motion

Newton's third law of motion has many interesting applications. As you read the following descriptions, remember there are always two objects to consider. One object exerts the action force, and at the same instant the other object exerts the reaction force. In some cases, one of the "objects" may be a gas such as air.

- When swimming, your arms and legs exert an action force backward against the water. The water exerts a reaction force forward against your arms and legs, pushing you forward.

- When a car accelerates forward, the tires exert a backward force on the road and the road exerts a forward reaction force on the tires (**Figure 6**). (These forces are friction forces between the tires and the road; we assume that the tires are not spinning.)

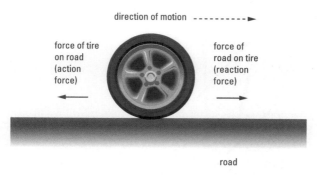

direction of motion --------►

force of tire on road (action force)

force of road on tire (reaction force)

road

Figure 6
The tire pushes on the road, and the road pushes back on the tire.

- Helicopter propeller blades are designed to force air in one direction as they spin rapidly. Thus, the blades exert an action force downward on the air. The air exerts a reaction force upward on the blades, sending the helicopter in a direction opposite to the motion of the air.

- A squid is a marine animal whose body size ranges from about 3 cm to 6 m. It propels itself by taking in water and then expelling it in sudden bursts. The squid exerts the action force backward on the discharged water. The discharged water exerts the reaction force forward on the squid, pushing it forward.

- A jet engine allows air to enter a large opening at the front of the engine. The engine compresses the air, heats it, and then expels it rapidly out the rear (**Figure 7**). The engine exerts the action force backward on the expelled air. The expelled air exerts the reaction force forward on the engine, pushing the engine and, along with it, the entire airplane forward. ⚓█

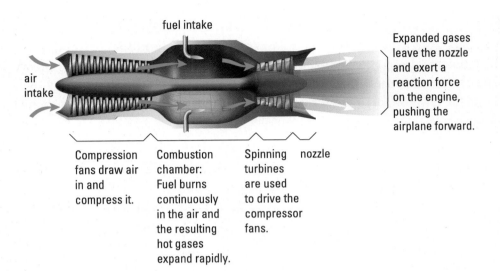

fuel intake

air intake

Expanded gases leave the nozzle and exert a reaction force on the engine, pushing the airplane forward.

Compression fans draw air in and compress it.

Combustion chamber: Fuel burns continuously in the air and the resulting hot gases expand rapidly.

Spinning turbines are used to drive the compressor fans.

nozzle

Figure 7
The design of a turbo jet engine

> **Practice**

Understanding Concepts

5. Explain each event described below in terms of Newton's third law. In each case, include a system diagram of the situation and label the action and reaction forces.
   (a) A paddle is used to propel a canoe.
   (b) A motorcycle accelerates forward.
   (c) A space shuttle, like the one shown in **Figure 1** of section 1.7, is launched. (Hint: This is not the same as a jet engine.)

6. You are a passenger standing in a small rowboat that is not moving. You are about to step onto a nearby dock. Explain why you may end up in the water instead.

7. The total applied horizontal force of a car's tires on the road is 2.1×10^3 N [W]. The car's mass is 1.4×10^3 kg.
   (a) What is the horizontal force of the road on the car?
   (b) Calculate the car's acceleration.

Answer

7. (b) 1.5 m/s^2 [E]

SUMMARY   *Newton's Third Law of Motion*

- Newton's third law of motion, which analyzes the force on two objects, states that for every action force there is a reaction force equal in magnitude but opposite in direction.

- Action–reaction pairs of forces are applied in many situations, including a person walking, a helicopter flying, and a rocket being launched into space.

▶ *Section 1.8 Questions*

Understanding Concepts

1. Go back to your answer to question 1 in the chapter introduction on page 6. Update the list of forces.

2. Explain, in writing and with a system diagram of the action–reaction pairs of forces, each of the following events:
   (a) A rocket accelerates in the vacuum of space.
   (b) A truck accelerates forward on a dry road.

3. (a) A certain string breaks when the tension force in it reaches 200 N. If two students pull on opposite ends of the string each with a force of 150 N, will the string break? Explain your answer.
   (b) Draw a system diagram of the situation in (a) showing all the action–reaction pairs of forces.

4. A helicopter of mass 6.4×10^3 kg is hovering above a launch pad.
   (a) Draw a system diagram showing all the action–reaction pairs of forces as the helicopter is hovering.
   (b) Draw an FBD of the helicopter as it is hovering, and calculate the upward force that balances gravity.
   (c) If the upward force is now increased to 7.5×10^4 N [up], what acceleration does the helicopter undergo? (Hint: Draw an FBD to show the new net force.)

Applying Inquiry Skills

5. Explain how you would safely demonstrate Newton's third law of motion to students in an elementary class using toys.

Making Connections

6. Explain why a turbo jet engine would not work in space.

7. An "ion propulsion system" is a proposed method of space travel. It uses ejected charged particles, or ions (**Figure 8**). Research this system and describe how it relates to the third law of motion.

Figure 8
Once it is in space, the centre part of this space probe ejects ions, propelling it forward.

 www.science.nelson.com

8. (a) What does a bathroom scale measure? Draw an FBD to help you justify your answer.
   (b) Why do you think the scale is calibrated in kilograms? What assumption is the manufacturer making when calibrating it in kilograms?

Figure 1
To reduce the friction between skis and snow, skiers choose a wax that is designed for certain conditions. At temperatures of about −10 °C, for example, skiers use a wax that is more slippery than the wax used at higher temperatures.

coefficient of friction the ratio of the magnitude of the force of friction between two surfaces to the magnitude of the normal force between the surfaces; symbol μ

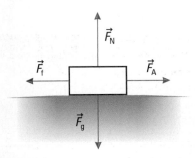

Figure 2
When an object experiences an applied force on a horizontal surface, there are four forces involved. The magnitudes of two of these forces, F_f and F_N, are used to calculate the coefficient of friction.

coefficient of kinetic friction the ratio of the magnitude of the force of kinetic friction to the magnitude of the normal force; symbol μ_K

coefficient of static friction the ratio of the magnitude of the maximum force of static friction to the magnitude of the normal force; symbol μ_S

Friction was introduced in section 1.4; here we look at it in greater detail. Most friction forces are complex because friction is affected by many factors. The friction force between a moving object and a surface depends on the materials of the object and the surface; the size, shape, and speed of the moving object; and perhaps temperature, as well as other factors. For example, the friction between skis and snow changes when the temperature and moisture content change (**Figure 1**), so different waxes are used for different conditions.

Coefficients of Friction

The **coefficient of friction** is the ratio of the magnitude of the force of friction between two surfaces to the magnitude of the normal force between the surfaces. We use the Greek letter mu, μ, to represent this ratio. Thus,

$$\mu = \frac{F_f}{F_N},$$

where F_f is the magnitude of the force of friction, in newtons;
F_N is the magnitude of the normal force, in newtons; and
μ is the coefficient of friction. (It has no units because it is a ratio of forces.)

Rearranging this equation gives the equation for the force of friction:

$$F_f = \mu F_N \quad \text{(see **Figure 2**).}$$

In most situations, the force needed to start an object moving from rest is greater than the force needed to keep it going at a constant velocity. This means that the maximum static friction is slightly greater than the kinetic friction. Thus, the coefficients of friction for these situations are different. To account for the difference, we have two coefficients of friction:

- The **coefficient of kinetic friction** is the ratio of the magnitude of the kinetic friction to the magnitude of the normal force; $\mu_K = \dfrac{F_K}{F_N}$

- The **coefficient of static friction** is the ratio of the magnitude of the maximum static friction to the magnitude of the normal force. This "maximum" occurs just when the stationary object begins to move; $\mu_S = \dfrac{F_S}{F_N}$

The coefficients of friction for various surfaces can only be determined experimentally. Even with careful control of the variables, results are often inconsistent. For example, consider measuring the kinetic friction of steel on ice by using skates on an ice rink. With sharp skate blades on clean, smooth ice, the coefficient of kinetic friction may be 0.010. However, using different skates on slightly rougher ice, the coefficient of kinetic friction may be 0.014.

Table 1 lists typical coefficients of kinetic friction and static friction for sets of common materials in contact; they are based on experimental results.

Table 1 Approximate Coefficients of Kinetic Friction and Static Friction

Materials in Contact	Coefficients of Friction		Materials in Contact	Coefficients of Friction	
	Kinetic	Static		Kinetic	Static
wood on wood, dry	0.2	0.25–0.5	rubber on other solid surfaces	1	1–4
waxed hickory on dry snow	0.18	0.22	rubber on concrete, dry	1.0	1.1
steel on wood	0.25	0.45	rubber on concrete, wet	0.97	1.0
steel on steel, dry	0.41	0.60	aluminum on steel	0.47	0.61
steel on steel, greasy	0.05	0.12	leather on oak, dry	< 0.01	< 0.01
steel on ice	0.010	0.10	lubricated ball bearings	0.5	0.6
rubber on asphalt, dry	1.0	1.1	Teflon on steel	0.00–0.04	0.041
rubber on asphalt, wet	0.95	1.0	Teflon on Teflon	0.04	0.04
rubber on ice	0.005	?	cartilage on synovial fluid	0.003	0.01

▶ **SAMPLE** problem **1**

Calculating the Coefficient of Static Friction

In the horizontal starting area for a bobsled race, four athletes, with a combined mass of 295 kg, need a horizontal force of 41 N [forward] to get the 315-kg sled to start moving. Calculate the coefficient of static friction.

Solution
The normal force is equal in magnitude to the weight of the sled. Also, since the applied force of 41 N [forward] is just sufficient to get the sled moving, the maximum force of static friction must be 41 N [backward]. Thus, omitting directions,

$F_N = mg$
   $= (315 \text{ kg})(9.8 \text{ N/kg})$
$F_N = 3.1 \times 10^3 \text{ N}$
$F_S = 41 \text{ N}$
$\mu_S = ?$

$$\mu_S = \frac{F_S}{F_N}$$

$$= \frac{41 \text{ N}}{3.1 \times 10^3 \text{ N}}$$

$\mu_S = 0.013$

The coefficient of static friction is 0.013. Notice that the combined mass of the athletes was not needed in this case.

LEARNING TIP

Determining the Normal Force
In Sample Problems 1 and 2, we apply the fact that the normal force is equal in magnitude to the weight of the object. You can show this in an FBD of the object. However, it is only true for objects on a horizontal surface with horizontal applied forces acting on them.

Calculating the Force of Kinetic Friction

A truck's brakes are applied so hard that the truck goes into a skid on the dry asphalt road. If the truck and its contents have a mass of 4.2×10^3 kg, calculate the magnitude of the force of kinetic friction on the truck.

Solution

$m = 4.2 \times 10^3$ kg

$g = 9.8$ N/kg

$\mu_K = 1.0$ (from **Table 1**, page 53, for rubber on dry asphalt)

$F_K = ?$

$F_N = mg$

$$F_K = \mu_K F_N$$
$$= \mu_K mg$$
$$= (1.0)(4.2 \times 10^3 \text{ kg})(9.8 \text{ N/kg})$$
$$F_K = 4.1 \times 10^4 \text{ N}$$

The magnitude of the force of kinetic friction is 4.1×10^4 N (in the direction opposite to the truck's initial motion).

▶ **Practice**

Understanding Concepts

1. Provide an example that shows that the coefficient of static friction tends to be greater than the coefficient of kinetic friction for two surfaces in contact.

2. As a car accelerates forward, the action force is the force of the driving tires backward against the road, while the reaction force is the force of the road forward on the tires. Are these forces static friction or kinetic friction? Explain your answer.

3. Calculate the appropriate coefficient of friction in each of the following:
   (a) It takes 59 N of horizontal force to get a 22-kg suitcase to just start to move across a floor.
   (b) A horizontal force of 54 N keeps the suitcase in (a) moving at a constant velocity.

4. A 73-kg hockey player glides across the ice with steel blades. Calculate the magnitude of the force of kinetic friction acting on the skater. (See **Table 1**.)

5. A 1.5×10^3-kg car moving along a concrete road has its brakes locked but skids to a smooth stop. Calculate the magnitude of the force of kinetic friction on (a) a dry road and (b) a wet road.

Applying Inquiry Skills

6. **Table 2** gives the data from an experiment in which a box containing different masses is pulled along the same floor at a constant speed.
   (a) Plot a graph of F_K (vertical axis) versus F_N.
   (b) Calculate the slope of the line. State what the slope represents.

Making Connections

7. Use the data in **Table 1** to verify that driving on an icy highway is much more dangerous than driving on a wet one.

Answers

3. (a) $\mu_S = 0.27$
   (b) $\mu_K = 0.25$
4. 7.2 N
5. (a) 1.5×10^4 N
   (b) 1.4×10^4 N
6. (b) 0.22

Table 2   Data for Question 6

F_N (N)	F_K (N)
0.0	0.0
10.0	2.2
20.0	4.3
30.0	6.7
40.0	8.8

SUMMARY   *Friction and the Coefficients of Friction*

- The coefficient of friction, μ, is the ratio of the magnitude of the force of friction to the magnitude of the normal force between two surfaces in contact.

- The coefficient of kinetic friction is given by $\mu_K = \dfrac{F_K}{F_N}$, and the coefficient of static friction is given by $\mu_S = \dfrac{F_S}{F_N}$.

- Coefficients of friction are found experimentally, and for any set of surfaces, μ_K tends to be less than μ_S.

▶ Section 1.9 Questions

Understanding Concepts

1. State the SI unit (if any) used to measure
   (a) the force of kinetic friction
   (b) the coefficient of kinetic friction

2. Calculate the coefficient of kinetic friction in each of the following:
   (a) A horizontal force of magnitude 28 N is needed to keep a sleigh moving at a constant velocity. The sleigh has a weight of magnitude 4.8×10^2 N.
   (b) The normal force between a car and the road has a magnitude of 1.2×10^4 N, and the force of kinetic friction as the car skids to a stop has a magnitude of 1.1×10^4 N.

3. A 15-kg wooden table requires an applied horizontal force of magnitude 46 N to push it across the floor at a constant velocity.
   (a) Calculate the magnitude of the normal force acting on the table.
   (b) Determine the coefficient of kinetic friction between the table and the floor.
   (c) Using **Table 1**, page 53, suggest the floor material.

4. A 5.2×10^2-N girl is inline skating when she falls and slides along the floor. The coefficient of kinetic friction between the girl and the floor is 0.12. Calculate the magnitude of the force of friction on the girl.

5. Assume you are on a luge toboggan that has a regulation mass of 22 kg and no brakes. The luge relies partly on friction to slow it down.
   (a) If the coefficient of kinetic friction between the luge and the horizontal icy surface is 0.012, what is the magnitude of the kinetic friction acting on the luge? (Use your own mass.)
   (b) Calculate the magnitude of your average acceleration as the luge slows down.

Applying Inquiry Skills

6. (a) Describe how you would determine the coefficient of static friction between a wooden block (or some other appropriate object) and a lab bench using equipment available in your physics classroom.
   (b) Repeat (a) for the coefficient of kinetic friction.
   (c) With your teacher's permission, try your methods and compare your answers in (a) and (b).

Making Connections

7. What do you think is meant by the term *aquaplaning*? Is it the same as *hydroplaning*? What should drivers understand about these concepts? Research the concepts, and see if you are right. Explain your answer.

 www.science.nelson.com

Measuring the Coefficients of Friction

Do all shoes have the same coefficients of friction? No, because different shoes are used on different surfaces. For example, there is a good reason why leather-soled shoes are not recommended for use on gymnasium floors. New shoes may also have different coefficients of friction from heavily used shoes. In the first part of this investigation you will use a ramp to determine the coefficients of static and kinetic friction of running shoes and other objects made of a variety of materials. In the second part you will experiment with other variables to see the effect, if any, on the coefficients of friction.

Question

How are the coefficients of friction affected by the state of the object (at rest or in motion), the materials in contact, the mass of the object, and the type of friction (rolling or sliding)?

Prediction

(a) Predict an answer to the Question.

Materials

For each group of three or four students:
wooden board(s) at least 1.2 m in length (if possible, use one board of unfinished wood and another of finished wood)
metre stick
objects that can slide or roll along the board (e.g., wooden block, steel block, plastic block, book, leather shoe, running shoe, dynamics cart)
mass balance
various metal masses (e.g., 200 g and 500 g)
protractor (optional)

LEARNING TIP

The Tangent Ratio
Recall that for a right-angled triangle with an angle θ, the tangent, or "tan," of the angle is $\tan \theta = \dfrac{\text{opposite}}{\text{adjacent}}$.
Therefore, if you simply measure the angle that the ramp makes to the horizontal, you can use the tan function on your calculator to find the slope of the ramp. When using your calculator for this purpose, be sure that the angle setting is on degrees (deg, not rad or grad).

Procedure
Part A

1. The board is used as a ramp in this investigation. Place the first object to be tested near the top of the ramp, as shown in **Figure 1**. Gradually raise the end of the ramp until the object just begins to slide down. Try this several times to be sure of the best position. Hold the ramp steady, and use the metre stick to determine the slope of the ramp $\left(\text{slope} = \dfrac{\text{rise}}{\text{run}} \right)$. This value is the coefficient of static friction. Record your data and calculations.

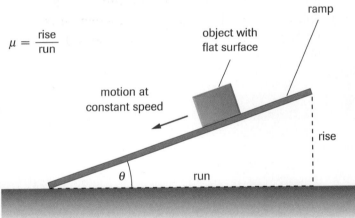

Figure 1
Measuring the coefficient of friction. Tan θ will also give the slope. See the Learning Tip.

2. Repeat step 1 with the same object, but this time, after the object starts moving, adjust the ramp until the object moves with constant velocity down the ramp. In this case, the slope of the ramp yields the coefficient of kinetic friction. Record your data and calculations.

3. Repeat steps 1 and 2 for other objects in order to determine the coefficients of friction for different materials.

Part B

4. Design and perform your own procedure to determine whether increasing the mass of the object has any effect on the coefficients of friction. Describe your procedure and record your data and calculations.

5. Repeat step 4 for any other factor you intend to test, such as how rolling friction compares to sliding friction. (Hint: You can use a dynamics cart both upright and inverted on a ramp.)

Analysis

(b) Which material had the greatest friction with the board? the least?

(c) How did the coefficients of friction you found in this investigation compare with the coefficients listed for the same materials in **Table 1**, page 53?

(d) Answer the Question.

(e) Summarize your findings in your lab report.

Evaluation

(f) Describe the most likely sources of error in your investigation. How might these sources of error be reduced?

The wheels on skateboards and inline skates are designed to have very low friction. Inside the wheels are small steel balls, called ball bearings, that are sealed in a bearing unit (**Figure 1**). Bearings reduce friction between the wheels and the axles

Figure 1

(a) The wheels on skateboards and inline skates use ball bearings to help reduce friction.

(b) The two surfaces are smooth, so the ball bearings move smoothly on the central surface, which allows the wheels to turn with greatly reduced friction.

(a)

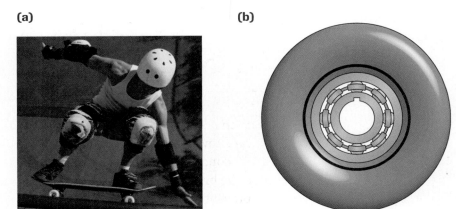

(b)

Ball bearings are just one example of devices that reduce unwanted friction. If you were responsible for designing an artificial limb, like the hand shown in **Figure 2**, you would want to minimize the friction between the moving parts. There are many other situations where we would like to reduce friction:

Figure 2
Artificial hands are designed to operate with as little friction as possible. How much friction do you feel within your hand when you wrap your fingers around a pen?

- Car manufacturers try to maximize the efficiency of engines by minimizing the friction in the moving parts. They do this by using materials with low friction, making surfaces smooth, and lubricating the surfaces with grease or oil.

- A layer of air between a hovercraft (**Figure 3**) and the water reduces fluid friction in a manner similar to the use of air pucks and the linear air track used in a physics laboratory.

- A human joint is lubricated by synovial fluid between the layers of cartilage lining the joint (**Figure 4**). The amount of lubrication provided

Figure 3
This hovercraft carries cars across the English Channel.

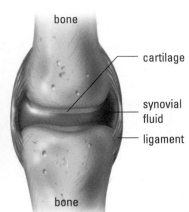

Figure 4
A typical joint in the human body

by the synovial fluid increases when a person moves, giving an excellent example of the efficiency of the human body. In fact, our lubrication system works so well, it is difficult for technologists to design artificial joints to function as well.

On the other hand, scientists and engineers often search for ways to increase friction. Without friction, driving on highways and running on playing fields would not only be dangerous but impossible! Friction is needed in the design of roads, bridges, automobile brake systems, automobile tires, athletic shoes, and surfaces of playing fields.

Cars rely on friction for stopping. Most new cars have disk brakes, especially on the front wheels. The brakes on bicycles operate in a similar way. When the brakes are applied, a piston pushes the brake pads toward a rotating disk called the rotor (**Figure 5**). The rotor is attached to the hub of the wheel, so exerting friction on the rotor causes the wheel to slow down.

Slamming on the brakes in older cars often results in skidding and losing control of the car's direction. Many new cars now feature antilock brakes, which help control stopping under extreme conditions. With a computer-controlled antilock system, the friction on the brakes of individual wheels is adjusted 20 times or more per second to prevent the wheels from locking. This makes the stopping distance less than it would be if the car were skidding.

> ### ▶ Practice
>
> **Understanding Concepts**
>
> 1. List examples that were not given in the text of situations where (a) increased friction and (b) decreased friction would be an advantage.
>
> 2. Explain each of the following statements, taking into consideration the force of friction:
>   (a) Friction is needed to open a closed door that has a doorknob.
>   (b) A highway sign reads "Reduce speed on wet pavement."
>   (c) Screw nails are useful for holding pieces of wood tightly together.
>
> 3. Describe how bicycle handbrakes and automobile disk brakes are similar.
>
> **Applying Inquiry Skills**
>
> 4. Describe an experiment you could use to determine the coefficient of kinetic friction of bicycle tires on asphalt when undergoing a safe, controlled skid.
>
> **Making Connections**
>
> 5. Complete this statement: "Friction is important to transportation engineers because. ..."

Case Study Car Tires

Car tires need friction to provide traction under all conditions. For example, when it is raining, a film of water accumulates between the tires and the road, reducing friction. Tire treads are designed to disperse that water so that friction is maintained (**Figure 6**).

DID YOU KNOW ?

Teflon

When low friction is desired in frying pans, nonstick coatings such as Teflon may be used. Two research scientists created the chemical by chance in 1938, but significant uses were not discovered until 20 years later. Teflon does not stick to any materials, so the process used to make it stick onto a frying pan surface is unique: The Teflon is blasted into tiny holes in the surface of the pan where the material sticks well enough for use.

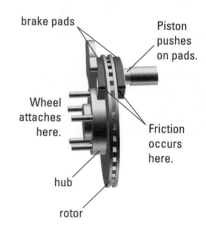

Figure 5
Basic operation of disk brakes

Many car owners leave the same tires on their cars in all four seasons. These tires, called all-season radial tires, provide a compromise between the shallow treads useful in dry conditions and the deeper treads desired in snowy conditions. To increase friction and improve traction in snowy or icy conditions, winter tires require deeper treads. Some even have small metal studs. (These studs are illegal in Ontario but are used in certain conditions in other provinces.)

Technicians and engineers are constantly trying to improve the design features of tires to maximize their friction. However, proper care is important too. Tires are most effective and safest when they are properly maintained. If the air pressure in a tire is not correct, the treads wear out much more quickly and unevenly. Also, when the pressure is too low, the tires heat up. Overheating increases at high speeds and in hot weather. When a tire becomes hot, the layers of the tire (**Figure 7**) can separate, leading to a crash and perhaps injury or death.

Figure 6
Tire treads have thin cuts, called sipes, to gather water on a wet road. This water is then pushed by the zigzag channels out behind the moving car.

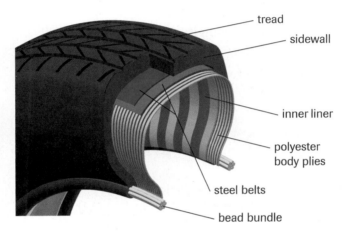

Figure 7
The typical tread of an all-season radial tire is between 7 mm and 8 mm deep. Beneath the rubber treads are various layers of materials that add strength and safety to the tire.

▶ Practice

Understanding Concepts

6. Tire X has treads that are 8 mm deep, and tire Y has treads of the same design that are only 3 mm deep. Which tire provides greater friction as a car accelerates on a wet highway? Explain your answer.

7. Draw three cross-section diagrams (each in the shape of a U) to show the shapes of tires with an interior air pressure that is (a) exactly what the manufacturer recommends, (b) too high, and (c) too low. Use your diagrams to explain why over-inflation and under-inflation cause extra wear and tear on the tires.

Making Connections

8. Do you think that it is better to check tire pressure before the car is driven or after the car has travelled several kilometres? Explain your answer.

▶ **TRY THIS** activity   *Tire Tread Ratings*

Tires are rated according to the degree of traction they provide. The ratings are indicated on the sidewalls of the tires. Find out about traction ratings (AA, A, B, and C), treadwear ratings, and temperature ratings (A, B, and C). Check the labelling on the tires of several cars, and describe how safe you think these tires are. Follow the links for *Nelson College Physics 12.*

 GO   www.science.nelson.com

SUMMARY   *Controlling Friction*

- Friction can be either an advantage or a disadvantage, depending on the situation.

- Unwanted friction can be reduced by changing sliding features to rolling ones and by using bearings and lubrication.

- Technological advances have helped to improve transportation safety in many ways, for example, in the design of tires that can be used in all seasons.

▶ **Section 1.11 Questions**

Understanding Concepts

1. An aluminum surface looks very smooth, but a microscopic view shows otherwise. Describe what the micrograph in **Figure 8** reveals about friction.

Applying Inquiry Skills

2. Exercise bikes have a control that allows the rider to adjust the amount of friction in the wheel. Describe how you would perform an experiment to determine the relationship between the setting on the control and the minimum force needed to move the bike's pedals.

Making Connections

3. Is friction an advantage or a disadvantage when you tie a knot in a string? Explain your answer.

4. Using your understanding of friction, explain how you would solve each of the following problems:
   (a) A refrigerator door squeaks when it is opened or closed.
   (b) A small throw rug at the entrance to a home slides easily on the hardwood floor.
   (c) A picture frame hung on a wall falls down because the nail holding it slips out of its hole.

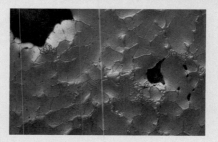

Figure 8
This micrograph of a polished aluminum surface was taken at a magnification of 50×.

5. Friction may be a help in some situations and a hindrance in others. Describe two ways it might help and two ways it can be a hindrance. For those situations where friction is helpful, what ways can be used to increase the friction?

6. Graphite powder can be used as a "dry" lubricant to reduce friction. Research the use of this material and other dry lubricants. Describe specific uses.

Key Understandings and Skills

1.1 Kinematics: The Study of Motion

- **Kinematics**, the study of motion, is important background information needed for **dynamics**, the study of the causes of motion.
- **Average speed** and **average velocity** can be found using the equations $v_{av} = \dfrac{d}{t}$ and $\vec{v}_{av} = \dfrac{\Delta \vec{d}}{\Delta t}$, respectively.
- Plotting graphs and determining slopes and areas on graphs are important skills in physics.

1.2 Activity: Calibrating a Ticker-Tape Timer

- A ticker-tape timer can be calibrated experimentally by determining its frequency and period of vibration.

1.3 Constant Acceleration and Acceleration Due to Gravity

- **Acceleration** is the rate of change of velocity. **Average acceleration** can be found using $\vec{a}_{av} = \dfrac{\Delta \vec{v}}{\Delta t} = \dfrac{\vec{v}_f - \vec{v}_i}{\Delta t}$ or, for constant acceleration, finding the slope of the line on a velocity–time graph.
- The **acceleration due to gravity** at Earth's surface is, on average, $\vec{g} = 9.8$ m/s^2 [down].

1.4 Forces and Free-Body Diagrams

- **Force** is a push or a pull; its SI unit is the **newton**.
- Examples of forces are **gravity**, **normal force**, **friction**, and **tension**.
- **System diagrams** and **free-body diagrams** are useful when analyzing forces.

1.5 Inertia and Newton's First Law of Motion

- The **net force** on an object is the vector sum of all the forces acting on the object.
- **Newton's first law of motion,** also called the law of **inertia**, states that if the net force on an object is zero, the object will maintain its state of rest or constant velocity.

1.6 Investigation: What Factors Affect Acceleration?

- A controlled experiment can be used to determine the relationship between an object's acceleration and the net force on the object and the object's mass.

1.7 Newton's Second Law of Motion and Weight

- **Newton's second law of motion** in equation form, $\vec{a} = \dfrac{\vec{F}_{net}}{m}$ or $\vec{F}_{net} = m\vec{a}$, can be used to define the newton.
- An object's **weight** is the force of gravity on the object; it is the product of the **mass** and the gravitational constant: $\vec{F}_g = m\vec{g}$.

1.8 Newton's Third Law of Motion

- **Newton's third law of motion,** also called the action–reaction law, states that for every action force there is a reaction force equal in magnitude but opposite in direction.
- The third law is applied in many situations, for example, swimming and operating jet engines and rocket engines.

1.9 Friction and the Coefficients of Friction

- The **coefficient of friction** for two objects in contact with each other is the ratio $\dfrac{F_f}{F_N}$.
- The kinetic friction and the maximum static friction are used to determine the **coefficient of kinetic friction** and the **coefficient of static friction**, respectively.

1.10 Investigation: Measuring the Coefficients of Friction

- The coefficients of static and kinetic friction can be found experimentally.

1.11 Controlling Friction

- Physics principles can be applied to increase friction where it is needed, such as on wet roads.
- If desired, friction can be reduced by various means, for example, using lubrication and bearings.

Problems You Can Solve

1.1

- Distinguish between and give examples of scalar and vector quantities.
- Distinguish between average speed and average velocity.
- Use equations and graphs to determine average speed, average velocity, and displacement.

1.2

- Determine the frequency and period of a vibrating timer and determine the percent error of the measurements given the accepted values.

1.3

- Determine the average acceleration of an object given the initial velocity, final velocity, and time interval for the change in velocity.

1.4

- Identify contact and action-at-a-distance forces.
- Draw the system diagram and free-body diagram for an object, given the forces acting on the object.

1.5

- Given the forces acting on an object, draw the free-body diagram and determine the net force.
- State Newton's first law of motion, and apply it to mechanical systems.

1.6

- Determine experimentally how the acceleration of an object depends on the net force applied to the object and the mass of the object.

1.7

- State Newton's second law of motion in words and equation form.
- Given any two of acceleration, net force, and mass, determine the third quantity.
- Determine an object's weight, given its mass and the gravitational strength.

1.8

- Identify action–reaction pairs of forces.
- State Newton's third law of motion, and apply it to mechanical systems.

1.9

- Define the coefficient of friction, and calculate it given the force of friction and the normal force. (This applies to both kinetic friction and static friction.)

1.10

- Experiment with several different objects on a ramp to determine the coefficients of static and kinetic friction.

1.11

- Describe advantages and disadvantages of friction in various situations, and describe how friction is increased or decreased in those situations.

Key Terms

1.1
kinematics
dynamics
scalar quantity
vector quantity
position
displacement
instantaneous speed

average speed
instantaneous velocity
average velocity

1.3
accelerated motion
acceleration
constant acceleration

average acceleration
free fall

1.4
force
force of gravity
normal force
friction

static friction
kinetic friction
air resistance
tension
applied force
newton
system diagram
free-body diagram

1.5
inertia
net force
Newton's first law of
  motion

1.7
Newton's second law of
  motion
newton
weight
mass

1.8
Newton's third law of
  motion

1.9
coefficient of friction
coefficient of kinetic
  friction
coefficient of static
  friction

Key Equations

1.1

- $\Delta \vec{d} = \vec{d}_f - \vec{d}_i$

- $v_{av} = \dfrac{d}{t}$

- $\vec{v}_{av} = \dfrac{\Delta \vec{d}}{\Delta t}$

1.3

- $\vec{a}_{av} = \dfrac{\Delta \vec{v}}{\Delta t} = \dfrac{\vec{v}_f - \vec{v}_i}{\Delta t}$

- $\vec{g} = 9.8 \text{ m/s}^2 \text{ [down]}$

1.7

- $\vec{a} = \dfrac{\vec{F}_{net}}{m}$ or $\vec{F}_{net} = m\vec{a}$

- $\vec{F}_g = m\vec{g}$

1.9

- $\mu = \dfrac{F_f}{F_N}$

- $\mu_K = \dfrac{F_K}{F_N}$

- $\mu_S = \dfrac{F_S}{F_N}$

▸ **MAKE** *a summary*

A dynamics cart (**Figure 1**) can be used to
demonstrate many of the concepts related to motion
and forces. For example, a cart can be inverted on a
ramp to study coefficients of friction, and one or two
carts can be used to demonstrate Newton's laws of
motion. Carts can also be used to illustrate system
diagrams, free-body diagrams, mass, weight, free fall,
etc. Draw several different diagrams of carts in ways
that summarize as many of the key expectations, key
words, and key equations as possible from this
chapter.

Figure 1
A dynamics cart can be used to illustrate many
concepts in the topics of motion and forces.

Write the numbers 1 to 8 in your notebook. Indicate beside each number whether the corresponding statement is true (T) or false (F). If it is false, write a corrected version.

1. Speed can be measured in any of these units: m/s, km/s, km/h, mm/s.

2. A satellite orbiting Earth at a steady 31 000 km/h is an example of constant velocity.

3. If you multiply metres per second by seconds, the product is metres per second squared.

4. A pen experiencing free fall in your classroom has an acceleration of 9.8 m/s² [down].

5. The acceleration of a train increases when the net force on the train increases.

6. One newton is the magnitude of the force needed to give a 1-kg object an acceleration of magnitude 1 m/s².

7. It is impossible for an object to be travelling eastward while experiencing a westward net force.

8. One SI unit of weight is the kilogram.

Write the numbers 9 to 14 in your notebook. Beside each number, write the letter corresponding to the best choice.

9. According to the graph in **Figure 1,** the order of the speeds from fastest to slowest is
   (a) B, C, A, D   (c) C, B, A, D
   (b) D, C, B, A   (d) A, B, C, D

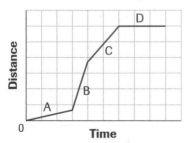

Figure 1

10. Which of the following can be a unit of acceleration?
    (a) (km/h)/h   (c) (km/h)/s
    (b) (m/s)/s   (d) all of these

11. You throw a tennis ball vertically upward. After the ball leaves your hand, the correct list of the force(s) acting on the ball is
    (a) a force from the throw and the downward force of gravity
    (b) a force from the throw, a force exerted by the air, and the downward force of gravity
    (c) a force exerted by the air and the downward force of gravity
    (d) the downward force of gravity

12. The coefficients of friction between a stationary box of mass 9.5 kg and the horizontal floor beneath it are $\mu_S = 0.65$ and $\mu_K = 0.49$. The magnitude of the minimum horizontal force needed to just get the box moving is
    (a) 61 N   (c) 6.2 N
    (b) 46 N   (d) 4.7 N

13. **Figure** 2 shows a rubber stopper suspended from the inside of a moving bus. This situation can be caused by
    (a) an eastward velocity and an eastward acceleration only
    (b) an eastward velocity and a westward acceleration only
    (c) a westward velocity and a westward acceleration only
    (d) either an eastward velocity and an eastward acceleration or a westward velocity and an eastward acceleration

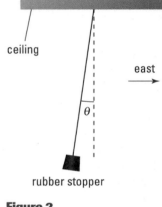

Figure 2

14. A child is standing on the floor. If the action force is the force of gravity of Earth pulling down on the child, then the reaction force is
    (a) the force of the floor pushing up on the child
    (b) the force of gravity of the child pulling up on Earth
    (c) the force of the child pushing down on the floor
    (d) a net force causing the child to accelerate

Understanding Concepts

1. Name three scalar quantities and three vector quantities.

2. State what is represented by each of the following:
   (a) the slope on a position–time graph
   (b) the area on a velocity–time graph
   (c) the slope on a velocity–time graph

3. In the Canadian Grand Prix auto race, the drivers travel a total distance of 304.3 km in 69 laps around the track.
   (a) If the fastest lap time is 84.12 s, what is the average speed for this lap? Express your answer in metres per second and kilometres per hour.
   (b) Assuming that the race starts and ends at the same position, what is the average velocity for the entire race?

4. Can an object have a northward velocity while experiencing a southward acceleration? Explain your answer.

5. A bus, travelling initially at 18 m/s [N], brakes smoothly and comes to a stop in 3.5 s. Calculate the average acceleration of the bus.

6. A ball is thrown vertically upward and is not affected by air resistance. Draw an FBD of the ball and state its acceleration
   (a) after it has left the thrower's hand and is travelling upward
   (b) at the instant it reaches the top of its flight
   (c) on its way down

7. A hovercraft is travelling at a constant velocity eastward. Draw an FBD of the hovercraft. What is the net force on the hovercraft?

8. Use two principles of physics from this chapter to explain the danger in travelling too quickly around a curve on an icy highway.

9. The scale used to draw the forces in **Figure 1** is 1.0 cm = 2.0 N.
   (a) Calculate the net force acting on the cart.
   (b) If the cart's mass is 1.2 kg, what is its acceleration?

Figure 1

10. One of the world's most powerful jumpers is the flea. For a brief instant, a flea is estimated to accelerate with a magnitude of 1.0×10^3 m/s^2. Calculate the magnitude of the net force a 6.0×10^{-7}-kg flea needs to produce this acceleration.

11. Calculate the magnitude of the weight acting on a car of mass 1500 kg.

12. The normal force between a snowmobile and the snow has a magnitude of 2200 N; the horizontal force needed to get the snowmobile to just start moving has a magnitude of 140 N. Calculate the coefficient of static friction is this case.

13. A 3.5-kg computer printer is pushed at a constant velocity across a desk with a horizontal force. The coefficient of kinetic friction between the printer and the desk is 0.36.
    (a) Calculate the magnitude of the normal force acting on the printer.
    (b) Calculate the magnitude of the kinetic friction acting on the printer.
    (c) Draw an FBD of the printer in this situation.

14. A demolition worker places a ramp from a window on the second floor of an old building to a trolley bin below, as shown in **Figure 2**. The chunks of material slide down the ramp at a constant velocity. Use the information in the scale diagram to determine the coefficient of kinetic friction between the material and the ramp.

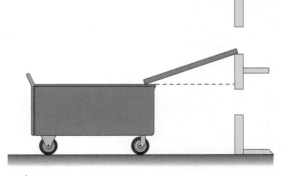

scale 1.0 cm = 1.0 m

Figure 2

15. Describe ways of reducing sliding friction in devices such as car engines.

16. If you get to your feet in a stationary canoe and move toward the front, which way does the canoe move? Explain why.

17. A box containing a new stereo system has a mass of 15 kg. A student pushes the box along the floor with a horizontal force of 58 N [W], and there is a force of kinetic friction of 12 N [E] acting on the box.
    (a) Draw an FBD of the box as it is being pushed.
    (b) Calculate the net force acting on the box.
    (c) Calculate the acceleration of the box.
    (d) Determine the coefficient of kinetic friction between the box and the floor.

18. Using Newton's third law of motion, explain how the operation of the propeller in a boat motor causes the boat to move forward.

19. A tractor pulls forward on a moving plow with a force of magnitude 2.5×10^4 N, which is just large enough to overcome kinetic friction.
    (a) What are the action and reaction forces between the tractor and the plow?
    (b) Are these action and reaction forces equal in magnitude but opposite in direction? Can there be any acceleration? Explain your answer.

20. Go back to your answers to the questions in the Try This Activity in the Chapter 1 introduction. How would you change your answers?

Applying Inquiry Skills

21. Sketch a graph to show how the first variable in the list below depends on the second one.
    (a) position versus time (for an object moving at a constant velocity)
    (b) force versus mass (for an experiment that tests Newton's second law of motion)

22. The results of a motion experiment are shown in **Table 1**.

Table 1 Data for Question 22

Time (s)	0.0	0.10	0.20	0.30	0.40	0.50
Position (cm [E])	12	19	25	31	38	45

    (a) Plot a position–time graph of the motion.
    (b) Determine the slope of the line of best fit on the graph.
    (c) Plot a velocity–time graph of the motion.

23. State at least one possible source of error when using a spring scale to determine the force needed to support an object.

Making Connections

24. Several world records in sporting and other events are featured in questions and sample problems in this chapter. Research the record times for various events of interest to you. Calculate and compare the average speeds or accelerations of these events.

25. Explain these warnings found on the visor of a new automobile:
    "Children 12 and under should be seated in the rear seat."
    "Never seat a rearward-facing child in the front seat."

26. Using concepts from this chapter, explain why signs like the one in **Figure 3** have helped to improve highway safety.

Figure 3

27. What tire tread design is appropriate for mountain bikes? Explain why, using principles you have learned in this chapter.

28. Research the Internet and other sources to find out how buildings in earthquake-prone areas use a bearing support system. Describe, using an illustration, what you discover.

 www.science.nelson.com

29. Research the evolution of skateboards from clay wheels to today's urethane wheels. Describe in a paragraph what you discover.

 www.science.nelson.com

2 Machines

In this chapter, you will be able to

- investigate and apply quantitatively the relationships among torque, force, and displacement in simple machines
- state the law of the lever and apply it quantitatively in a variety of situations for three classes of levers
- explain the operation, applications, and mechanical advantage of simple machines
- define and describe the concepts and units related to torque and mechanical advantage
- determine the mechanical advantage of a variety of compound machines and biomechanical systems
- construct a simple or compound machine and determine its mechanical advantage
- describe the role of machines in everyday life
- analyze natural and technological systems that use the principles of simple machines, and explain their function and structure
- analyze the operation of a bicycle

Getting Started

When you grasp your pen and lift it upward, your arm is acting as a simple machine called a *lever*. To form compound machines, like the artificial arm in **Figure 1**, simple machines are linked together. Artificial limbs share some of the features found in bicycles, robots, cranes, window blinds, food processors, snowplows, and numerous other devices made of one or more simple machines.

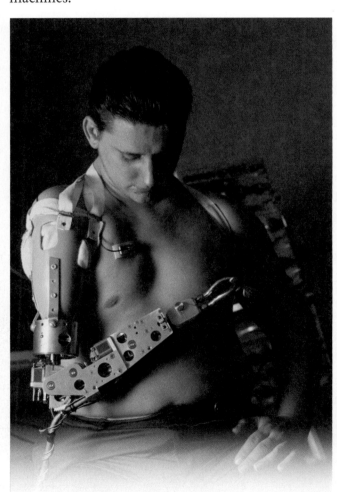

Figure 1
This artificial arm is a compound machine made of several simple machines. It is connected to the user's nerves so it can be controlled by impulses from the brain.

The lever is one of several simple machines used in the design of compound machines. In this chapter, you will study and perform experiments related to simple and compound machines. You will also discover how the principles of machines are applied in devices in the home and in industry.

As you study machines and mechanical systems, think about how you would design a machine to perform a simple task, such as raising a small object from the floor to the top of your desk. You will then be prepared for the Unit 1 Performance Task.

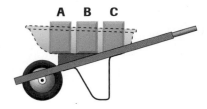

Figure 2
To move the wheelbarrow, you would first apply an upward force near the end of the handle.

REFLECT on your learning

1. A steel block can be placed in the wheelbarrow in **Figure 2** at position A, B, or C. Which position would allow you to most easily raise the handle end of the wheelbarrow? Why?

2. In **Figure 3**, ramps D and E lead to the same level above the ground. Would it be easier to push a wheelchair up ramp D or ramp E? Explain your choice.

3. Explain why each of the following operates on the basis of at least one simple machine. (If you can, name the simple machine or machines involved.)
   (a) a rotating water faucet
   (b) a screwdriver
   (c) scissors
   (d) a can opener

4. When a wrench is used to take a nut off a wheel, do you think the turning effect depends on
   (a) the magnitude of the force you apply to the wrench?
   (b) the distance between the nut and the point at which you apply the force to the wrench (assuming you apply a constant force)?

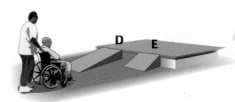

Figure 3
If you were using a wheelchair, which ramp would you use?

▶ TRY THIS activity   *Breaking Pencils*

Predict whether it is easier to break (by hand) a short pencil or a long pencil (**Figure 4**).

(a) Write your prediction, including a reason.

   Observe as a volunteer demonstrates the effort required in each case.

(b) Use diagrams to explain what you observed.

 The person breaking the two pencils should wear safety goggles and protective gloves.

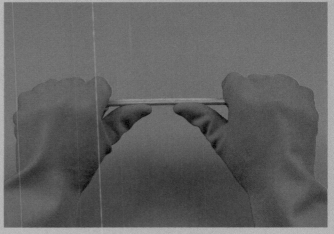

Figure 4
Breaking a pencil

Although many modern machines are large and complex, they usually operate on the basic principles of simple machines, principles that have been used for centuries (**Figure 1**). In museums we often see early tools and weapons whose sharp edges are examples of a simple machine (a wedge). The early Egyptians used ramps to move blocks when building pyramids; ramps are also an example of a simple machine (inclined planes). Even today, the ancient technology of pulleys and levers is often used to raise water out of rivers and wells.

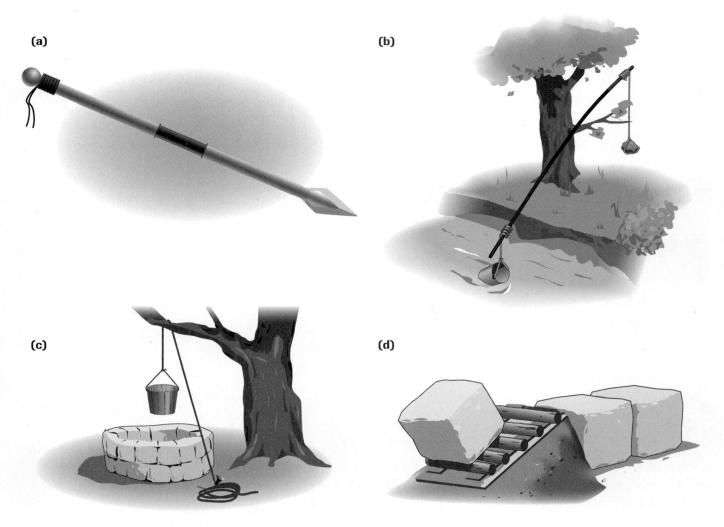

(a)

(b)

(c)

(d)

Figure 1
(a) The wedge at the end of the spear applies a large force to split the hide of a deer or other food source.
(b) The lever helps raise objects such as containers of water.
(c) In this pulley system, the bucket of water is raised when the rope is pulled down.
(d) The inclined plane reduces the force needed to move a stone block to a higher level.

A **machine** is a device that helps us perform tasks. It is designed to achieve at least one of five main functions:

- change energy from one form into another (e.g., a hydroelectric generator changes the energy of falling water into electrical energy);

- transfer forces (e.g., a car transmission transfers the force from the motor to the wheels);

- change the direction of a force (e.g., to raise a flag up a pole, we pull down on a rope attached to a pulley at the top of the pole);

- change the magnitude of a force (e.g., a system of pulleys helps a mechanic exert a small force to hoist a heavy engine out of a car);

- change distance or speed (e.g., the outer circumference of the rear wheel of a bicycle moves farther and faster than the circumference of the sprocket wheel attached to the pedals).

Simple machines can be split into two families that share characteristics: the *lever family of machines* and the *inclined family of machines*. The lever family consists of the lever, the pulley, the wheel and axle, and gears. The inclined family consists of the inclined plane, the wedge, and the screw. As you study details of these machines, think of what the members of each family have in common.

machine a device that helps us perform tasks by achieving at least one of the five main functions of machines

> ▶ *Practice*

Understanding Concepts

1. State the main function or functions of each of the following machines:
   (a) an axe
   (b) a tennis racket
   (c) a round doorknob
   (d) a screwdriver

2. Give two or three examples, not found in the text, of simple machines in the kitchen that increase the force you apply.

The Lever Family of Machines

A **lever** is a rigid bar that can rotate freely around a support called a **fulcrum**. A seesaw (or teeter-totter) is a typical example of a lever. An **effort force**, F_E, is a force applied to one part of a lever to move a load at another part; the load exerts a **load force**, F_L. (You learned in Chapter 1 that force is a vector, but in this chapter we are considering only the magnitudes of the forces when we analyze their effects. Therefore, the symbols for effort force and load force are written without the arrows.)

Two other variables are measured on levers: the perpendicular distance from the fulcrum to the effort force, the **effort arm**, symbol d_E, and the perpendicular distance from the fulcrum to the load force, the **load arm**, symbol d_L.

lever a rigid bar that can rotate freely around a fulcrum

fulcrum a support around which a lever can rotate or pivot

effort force a force applied to one part of a lever to move a load at another part; symbol F_E

load force the force exerted by the load on a lever; symbol F_L

effort arm the perpendicular distance from the fulcrum to the effort force, symbol d_E

load arm the perpendicular distance from the fulcrum to the load force, symbol d_L

first-class lever a lever with the fulcrum between the load and the effort force

second-class lever a lever with the load between the fulcrum and the effort force

third-class lever a lever with the effort force exerted between the fulcrum and the load

Levers are divided into three classes, depending on the positions of the load, the effort force, and the fulcrum:

- In a **first-class lever** the fulcrum is between the load and the effort force (**Figure 2(a)**).
- In a **second-class lever** the load is between the fulcrum and the effort force (**Figure 2(b)**).
- In a **third-class lever** the effort force is exerted between the fulcrum and the load (**Figure 2(c)**).

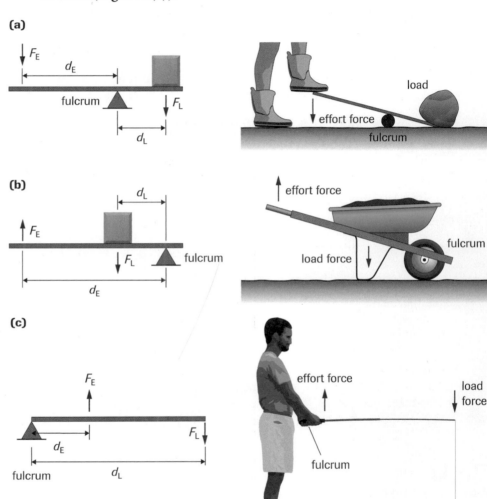

CAREER CONNECTION

Employment opportunities for personal trainers—who require a solid understanding of biomechanics—are varied: health clubs, rehabilitation centres and hospitals, community centres, and in sports.

 www.science.nelson.com

Figure 2
(a) First-class lever
(b) Second-class lever
(c) Third-class lever

biomechanical system a system of a living body

Biomechanical systems, which are systems involving a living body, can also be understood as simple machines. An example is the human forearm (**Figure 3**), which works as a third-class lever. The action of the forearm can be analyzed and then closely copied to create robotic arms. Other examples of biomechanical systems are your lower legs, animal jaws, bird beaks, and the claws of crayfish and hawks. ✛▮

Simple machines that rotate around a fulcrum, such as a *pulley,* are considered part of the lever family of machines. A **pulley** is a wheel with a grooved rim in which a rope runs. The wheel rotates around a central fulcrum. **Figure 4** illustrates two single-pulley arrangements.

A **wheel and axle** is a large-diameter, rigid, circular disk (the wheel) connected to a small-diameter, rigid rod (the axle). Both the wheel and the axle rotate around a fulcrum, so this machine is also part of the lever family (**Figure 5**).

Gears are toothed wheels of different diameters linked together to increase or decrease speed or to change direction. A pair of gears resembles a wheel and axle because each gear rotates around a fulcrum. Gears can be linked together directly at their rigid teeth, or they can be linked with a chain, as on a bicycle (**Figure 6**).

pulley a wheel with a grooved rim in which a rope or cable can run

wheel and axle a large-diameter, rigid disk connected to a small-diameter rigid rod

gears toothed wheels of different diameters linked together to change the speed or direction moved

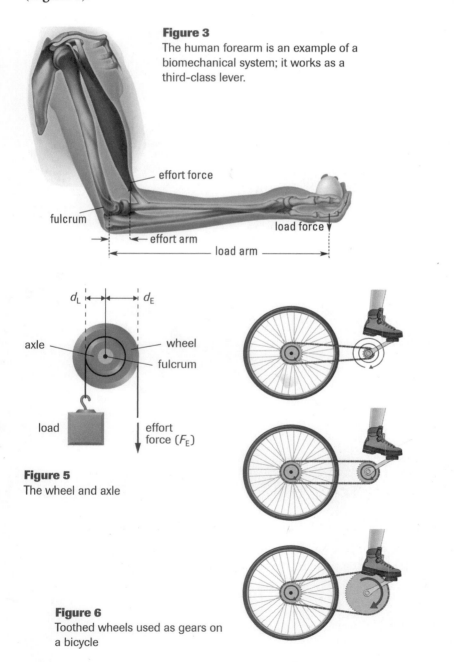

Figure 3
The human forearm is an example of a biomechanical system; it works as a third-class lever.

effort force

fulcrum

effort arm

load force

load arm

d_L d_E

axle

wheel

fulcrum

load

effort force (F_E)

Figure 5
The wheel and axle

Figure 6
Toothed wheels used as gears on a bicycle

Figure 4
(a) Single fixed pulley
(b) Single movable pulley. Notice the location of the fulcrum.

(a)

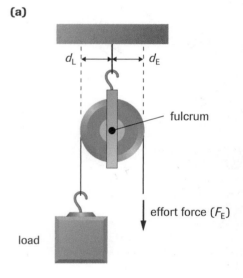

d_L d_E

fulcrum

effort force (F_E)

load

(b)

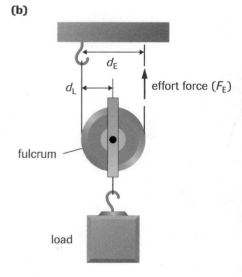

d_E

d_L

effort force (F_E)

fulcrum

load

Gears in a Car

In a car, a gearbox contains a set of toothed gear wheels arranged to cause the wheels to rotate faster or slower. Cars usually have one reverse gear and three or four forward gears.

Understanding Concepts

3. State the class of lever each device shown in **Figure 7** belongs to, and draw a diagram showing the fulcrum, load, and effort force.

4. To which class of lever does each pulley in **Figure 4**, page 73, belong?

5. To which class of lever does the wheel and axle in **Figure 5**, page 73, belong?

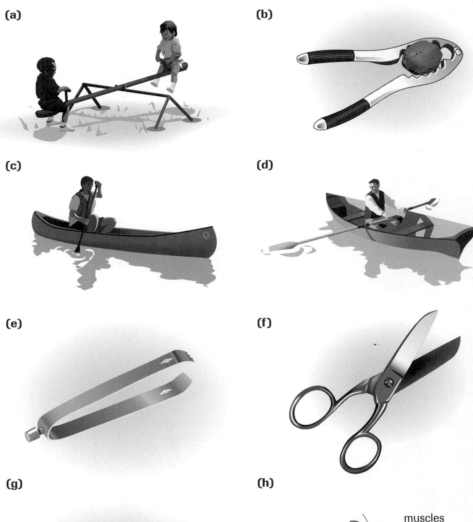

(a)

(b)

(c)

(d)

(e)

(f)

(g)

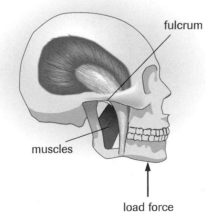

fulcrum

muscles

load force

(h)

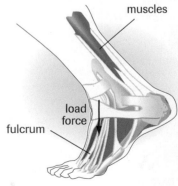

muscles

load force

fulcrum

Figure 7

(a) Seesaw

(b) Nutcracker

(c) Paddling a canoe (Assume the fulcrum is at the top end of the paddle.)

(d) Rowing a boat (Assume the fulcrum is at the point at which the oar is attached to the boat.)

(e) Sugar tongs

(f) Scissors

(g) Human jaw

(h) Human foot

The Inclined Plane Family of Machines

The inclined plane and related machines are shown in **Figure 8**. An **inclined plane** is a ramp that increases the load that can be moved by an effort force. The force required to move an object up a ramp is less than the force needed to lift the object vertically upward. For example, think of how much easier it is to walk up a stairway than to climb up a ladder. A **wedge** is a double inclined plane that increases the applied force, for example, in an axe blow. A **screw** is an inclined plane wrapped around a central shaft. It too increases the applied force.

inclined plane a ramp that increases the load that can be raised by an effort force

wedge a double inclined plane that increases the applied or effort force

screw an inclined plane wrapped around a central shaft that increases the applied or effort force

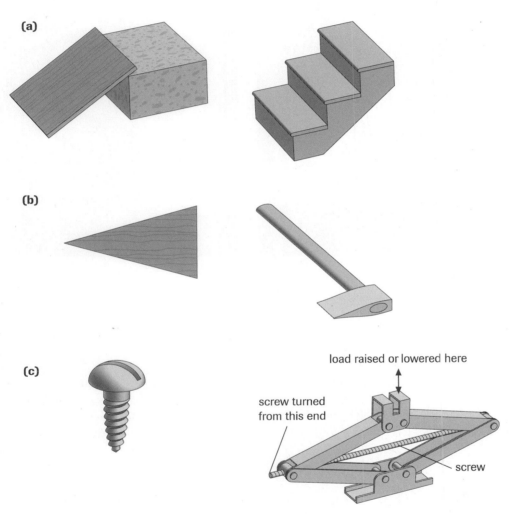

(a)

(b)

(c)

load raised or lowered here

screw turned from this end

screw

Figure 8
(a) The inclined plane and stairs
(b) The wedge and an axe
(c) The screw and a car jack

Notice in **Figure 8(c)** that the screw is one of the two simple machines of the car jack. The handle and central shaft act as a wheel and axle. Thus, the car jack is an example of a *compound machine*. A **compound machine** is a machine made of two or more simple machines.

compound machine a machine made of two or more simple machines

Liquid Wedges

In recent years, large chunks of Antarctica's ice shelves have broken off. Scientists have discovered that a wedge effect causes the ice to break. At higher summer temperatures caused by global warming, surface ice melts. Some of the resulting water seeps into cracks in the ice. This water, acting as a "liquid wedge," somewhat like an axe, puts extra pressure on the ice. The water cracks the ice further and pushes the chunks of ice apart. Some of the chunks pushed away from the shelves are huge; one from the Larsen Ice Shelf was 200 m thick and covered a surface area half the size of Prince Edward Island.

▶ **Practice**

Understanding Concepts

6. State which simple machine forms the basis of each of the following:
   (a) a doorstop, with the pointed end placed between the floor and the bottom of the door
   (b) an escalator (**Figure 9**)
   (c) a mountain highway (**Figure 10**)
   (d) a hand-held pencil sharpener

Figure 9

Figure 10

7. Explain how the parts of the bicycle shown in **Figure 6**, page 73, form a compound machine made up of gears and another simple machine.

Making Connections

8. You have been asked to design a two-level bridge over a river. One level is for trains and the other is for cars, trucks, and buses. Which would you place in the upper level? (Hint: Think of the forces required to go up an inclined plane.) Explain your answer. Draw a diagram to illustrate your design.

SUMMARY   *Simple Machines*

- A machine helps us perform tasks by performing one or more of five main functions.
- The lever family of machines consists of the lever, the pulley, the wheel and axle, and gears.
- The inclined plane family of machines consists of the inclined plane, the wedge, and the screw.
- Compound machines and biomechanical systems can be analyzed in terms of simple machines.

▶ *Section 2.1 Questions*

Understanding Concepts

1. Name the simple machine(s) associated with each of the following:
   (a) a water slide
   (b) a triple beam balance
   (c) a tricycle
   (d) a letter opener
   (e) the screw cap on a water bottle
   (f) a skateboard

2. (a) Explain why a wheel and axle system is considered a member of the lever family of machines.
   (b) Explain why a screw is considered a member of the inclined plane family of machines.

3. How are gears similar to a wheel and axle? How are they different?

4. For a wheel and axle system, on which part would you apply the effort force in order to
   (a) increase the force?
   (b) increase the distance or speed?

5. State the class of lever of each biomechanical system in **Figure 11**.

6. (a) When an axe is used as a wedge to split a log, how does the direction of the load force compare to the direction of the effort force? Draw a sketch to illustrate your answer.
   (b) Axes used to split logs are heavier than axes used to chop down trees. Explain why.

Applying Inquiry Skills

7. Cut a sheet of paper diagonally in half to create two right-angled triangles. Wrap one of the triangles around a pencil in a way that illustrates one type of simple machine. Explain which machine it is and why.

Making Connections

8. Before the wheel and axle system was invented, people used logs to move large stones and other heavy objects. Explain how straws or pencils can be used to model this ancient form of the wheel.

9. Choose one of the following topics to research, and write a summary of what you discover. Relate your findings to simple machines.
   (a) Early inhabitants of Easter Island (an island in the South Pacific, called Rapa Nui locally) used simple machines to move huge volcanic rocks. The inhabitants then carved these rocks into figures of heads.
   (b) An estimated 4500 years ago, inhabitants of southern England transported huge stones to build stone circles such as the famous Stonehenge. Scientists are trying to recreate the trip using only simple machines.

GO www.science.nelson.com

(a)

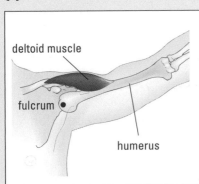

deltoid muscle

fulcrum

humerus

(b)

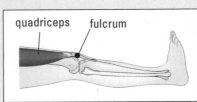

quadriceps   fulcrum

Figure 11
(a) The deltoid muscle pulls on the humerus (the bone extending from the shoulder to the elbow) to pivot it around the shoulder joint.
(b) The quadriceps are muscles that pull on the lower leg to pivot it around the knee.

Forces on Levers

Anyone who has tried to remove the lid of a paint can using fingers only soon realizes it is impossible. As depicted in **Figure 1**, the task becomes much easier if we use a rigid device, for example, a screwdriver, to pry the lid off. In this case, the screwdriver is acting as a lever.

In this investigation, you will measure effort and load forces, and effort and load arms with levers that are stationary, or balanced. You will then look for any pattern(s) when you calculate the following products:

effort force × effort arm ($F_E d_E$)

load force × load arm ($F_L d_L$)

If the effort and load forces on a lever act vertically, the force of gravity needs to be considered in the calculations. To eliminate the effect of gravity, the load and effort forces will be applied horizontally in the parts of the investigation involving levers.

If spring scales are used to measure forces, they should be checked for proper adjustment by pulling one against the other to see if their readings are equal. Also, the scales should be adjusted to read zero under no tension when holding them horizontally (in steps 2 to 4). The scales should be readjusted to read zero under no tension when holding them vertically (in step 5).

If a computer spreadsheet is available, you can use it to design a data table for this investigation.

Question

For any member of the lever family that is stationary, how does the product $F_E d_E$ compare to the product $F_L d_L$?

Prediction

(a) Predict whether the products in the Question will be equal or whether one will be larger. Give a reason for your prediction.

Inquiry Skills

○ Questioning   ● Conducting   ● Evaluating
● Predicting   ● Recording   ● Communicating
● Planning   ● Analyzing   ○ Synthesizing

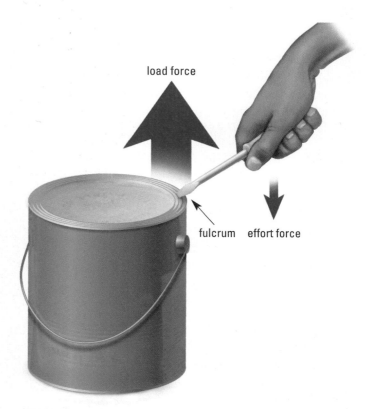

load force

fulcrum   effort force

Figure 1
A small effort force exerted by the hand produces a large load force on the lid. Notice that the effort arm is much greater than the load arm.

Materials

For each group of three or four students:

metre stick with a hole drilled at the centre (for steps 2 to 4)

metric ruler (for step 5)

two force scales or force sensors (each to 10 N or more)

board

nail (pounded partway into the board, as shown in **Figure 2** on page 80)

single fixed pulley

single movable pulley

wheel and axle

For each student:
goggles (if spring scales are used)

Procedure

1. Set up a table of data based on **Table 1**. Read the rest of the Procedure steps and fill in as much of the table as possible without taking measurements. For step 5, you will have to decide which distance measurements to make. Ask your teacher to approve your table.

2. Set up the apparatus as shown in **Figure 2** on page 80. The lever should be able to rotate freely around the nail. With the load force at the 10-cm mark and the effort force at the 70-cm mark, exert a 4.0-N load force. Measure the effort force required to keep the lever balanced. Be sure both forces are perpendicular to the metre stick. Enter your readings in your data table, and calculate the products required.

3. Change the variables according to the data in **Table 1** for this step. Complete the measurements and calculations.

4. Complete the measurements and calculations for the situations in **Figure 3** on page 80. For the lever with two loads, add the product $F_L d_L$ for the two loads to obtain the total product.

5. Perform the measurements and calculations related to the setups shown in **Figure 4** on page 80. Complete your data table.

Analysis

(b) What is meant by the word "balanced" in the term "balanced lever"?

(c) Describe the pattern(s) you observe in the calculations you made.

(d) Write concluding statements to summarize your answer to the original Question.

Evaluation

(e) Describe the main sources of error in the measurements in this investigation.

Table 1 Data Table for Investigation 2.2

Procedure Number	2	3	3	4(a)	4(b)	4(c)	5(a)	5(b)	5(c)
Class of lever	1st	?	?	?	?	?	?	?	?
F_L (N)	4.0	8.0	6.0	10.0	4.0	3.0/4.0	10.0	10.0	10.0
Load position (cm mark)	10	30	25	30	10	10/30	?	?	?
d_L (m)	0.40	?	?	?	?	?	?	?	?
F_E (N)	?	?	?	?	?	?	?	?	?
Effort position (cm mark)	70	90	90	10	30	90	?	?	?
d_E (m)	0.20	?	?	?	?	?	?	?	?
$F_L d_L$ (N·m)	?	?	?	?	?	?	?	?	?
$F_E d_E$ (N·m)	?	?	?	?	?	?	?	?	?

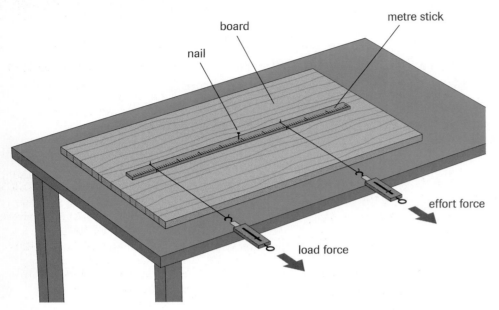

Figure 2
Setup for step 2

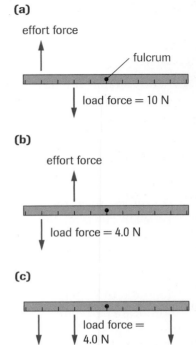

(a)

effort force

fulcrum

load force = 10 N

(b)

effort force

load force = 4.0 N

(c)

load force =
4.0 N

load force =
3.0 N

effort force

Figure 3
For step 4
(a) Second-class lever
(b) Third-class lever
(c) First-class lever with two loads

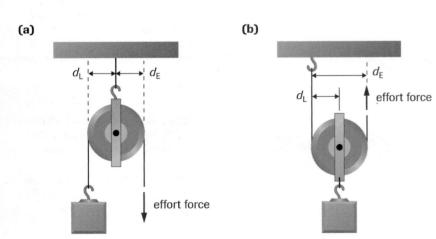

(a)

d_L d_E

effort force

load force = 10 N

(b)

d_E

d_L effort force

load force = 10 N

(c)

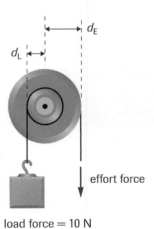

d_E

d_L

effort force

load force = 10 N

Figure 4
For step 5
(a) Single fixed pulley
(b) Single movable pulley
(c) Wheel and axle

When a force or set of forces causes a rigid body to rotate, we say a *torque* has been applied. A **torque** is a turning effect on a rigid object around an axis. If the rigid object is a lever, the axis is the fulcrum. Every time you apply a force to open or close a door, you are producing a torque on the door. With a door, the hinges act as the fulcrum. A small force applied far from the hinges can produce the same torque as a large force applied closer to the hinges. This is illustrated in **Figure 1**.

A good way to get a "feel" for torque is to try to push a door at various distances from the hinges. You will notice that as the distance from the hinge to the effort force increases, the size of the effort force needed to cause the same torque decreases. But you will also discover something else. Our natural tendency is to apply the effort force at an angle of 90° to the door. We have learned from experience that this causes the largest turning effect (i.e., torque) on the door. If you try applying the same effort force at various angles to the same point on the door, you will see that the torque is reduced. (In this text, we will consider only situations in which the effort force is perpendicular to the rigid object.)

We can conclude from observing what happens when we push the door that the amount of torque produced depends on two main factors. One factor is the magnitude of the force (F) applied to the rigid object. The other is the distance (d) between the force and the axis or fulcrum. Using the symbol T for the magnitude of the torque, the following statements hold true:

T increases as F increases (i.e., $T \propto F$)

T increases as d increases (i.e., $T \propto d$)

(a)

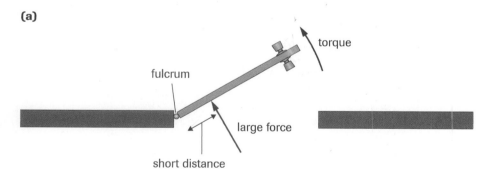

fulcrum · torque · large force · short distance

(b)

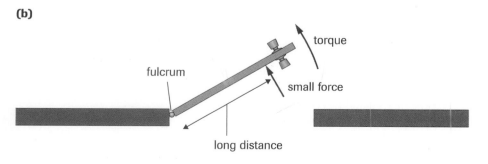

fulcrum · torque · small force · long distance

torque the turning effect caused by a force on a rigid object around an axis or a fulcrum, symbol T; it is measured in newton-metres, or N•m; it can also be called a "moment of force"

LEARNING TIP

Rigid Objects

A *rigid object* (also called a *rigid body*) is solid and strong enough to withstand the forces applied to it. It does not deform or break under the applied forces. An example of a rigid object is a seesaw. A seesaw will not usually break under the weight of two people.

DID YOU KNOW?

Torque in Skating and Ballet

In pairs figure skating and ballet, one partner can help the other partner spin by producing two torques. One torque is caused by pushing on one side, and the other by pulling on the other side.

Figure 1

A large force applied at a short distance from the fulcrum **(a)** can cause the same torque as a small force applied at a longer distance **(b)**.

Thus,   torque = force × distance

or $T = Fd$ (with F perpendicular to the rigid object)

Using SI units, force is measured in newtons (N) and distance is measured in metres (m), so torque is measured in newton-metres (N·m).

A simple activity to "feel" torque is to try to raise a 2.0-kg mass a few centimetres off the desk in the following way: With a clamp, secure an S-shaped hook to the top end of a support stand. Attach the 2.0-kg mass to the hook. While sitting on a chair holding the stand horizontally, and with your arm straight (no bent elbow!), try to lift the mass with the hook (**Figure 2**). Explain why this task is not easy to do.

Figure 2
Why is this difficult to do?

▶ **SAMPLE** problem **1**

Calculating Torque

Calculate the magnitude of the torque on the wrench shown in **Figure 3**.

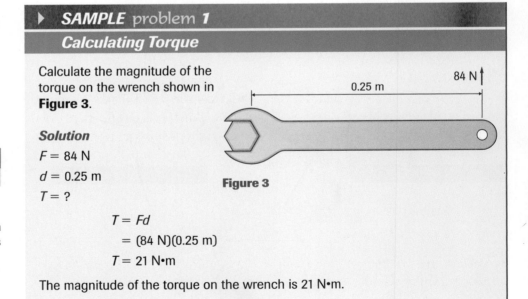

0.25 m   84 N

Figure 3

Solution

$F = 84$ N

$d = 0.25$ m

$T = ?$

$T = Fd$

$\quad = (84 \text{ N})(0.25 \text{ m})$

$T = 21$ N·m

The magnitude of the torque on the wrench is 21 N·m.

▶ **Practice**

Understanding Concepts

1. List the conditions needed for a lever to be considered a rigid object.

2. In **Figure 1**, page 81, what class of lever is represented in (a) and (b)?

3. A mechanic applies a force of magnitude 540 N perpendicular to a wrench to loosen a nut. Calculate the magnitude of the torque if the distance from the applied force to the nut is (a) 0.30 m and (b) 0.50 m.

4. A person applies a force of magnitude 150 N at the hinge side of a door, which means that d is zero. What is the magnitude of the torque produced on the door?

5. Rearrange the equation for torque to solve for (a) d and (b) F.

6. A cyclist is signalling with one hand and turning the handlebar to turn a corner with the other. The cyclist applies a force of magnitude 4.5 N at 90° to the handlebar, which produces a torque of magnitude 0.99 N•m on the handlebar. Calculate the distance from the fulcrum to the applied force.

7. A corral gate is partly open. A cow hits the gate with her head at a distance of 3.6 m from the hinges. If the torque produced by the cow has a magnitude of 540 N•m, what magnitude of force did the cow apply to the gate? (Assume that the force is perpendicular to the gate.)

Applying Inquiry Skills

8. Did any of the calculations you performed in Investigation 2.2 involve torque? Explain your answer.

Making Connections

9. If a wrench is unable to loosen a wheel nut, how can you use a hollow pipe to increase the torque provided by the wrench? Illustrate your answer with a sketch.

Answers

3. (a) 1.6×10^2 N•m
   (b) 2.7×10^2 N•m

6. 0.22 m

7. 1.5×10^2 N

Torques on Levers

Two torques can be calculated for a lever: the effort torque (T_E) and the load torque (T_L). The associated distances are the effort distance, or effort arm (d_E), and the load distance, or load arm (d_L). The corresponding equations are

$$\text{effort torque} = \text{effort force} \times \text{effort arm},$$

or $$T_E = F_E d_E$$

$$\text{load torque} = \text{load force} \times \text{load arm},$$

or $$T_L = F_L d_L$$

Remember that torque is measured in newton-metres (N•m). In each case, the force is perpendicular to the lever, which allows us to deal with magnitudes only, thus avoiding vector signs.

Calculating Effort Torque

A camper is using a large plank as a first-class lever to move a rock. The effort force has a magnitude of 4.5×10^2 N, and the distance from the fulcrum to the effort force is 2.2 m. What is the magnitude of the effort torque produced? (Ignore the mass of the plank.)

Solution

$F_E = 4.5 \times 10^2$ N

$d_E = 2.2$ m

$T_E = ?$

$$T_E = F_E d_E$$
$$= (4.5 \times 10^2 \text{ N})(2.2 \text{ m})$$
$$T_E = 9.9 \times 10^2 \text{ N·m}$$

The magnitude of the effort torque produced is 9.9×10^2 N·m.

▶ **Practice**

Understanding Concepts

10. Calculate the magnitude of the effort torque or the load torque in each case in **Figure 4**.

11. Estimate the magnitude of the maximum effort torque you could produce on a wheel nut using a tire wrench that is 50 cm long. (Hint: Do you think your maximum effort force could be greater than your own weight?)

Answers

10. (a) 1.1×10^2 N·m
    (b) 6.8 N·m

(a)

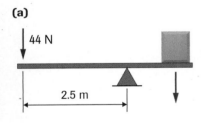

(b)

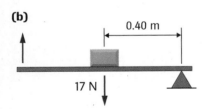

Figure 4
For question 10

Static Equilibrium of Levers

The word *static* means at rest. A rigid object that is in static equilibrium is at rest in two ways. First, it is not moving in any direction. Second, and more important for levers, it is not rotating. This brings us to the *law of the lever*.

Law of the Lever
When a lever is in static equilibrium, the magnitude of the effort torque equals the magnitude of the load torque.

This law can be written in equation form as follows:

effort torque = load torque

effort force × effort arm = load force × load arm

$$F_E d_E = F_L d_L$$

In this equation for the law of the lever, only the magnitudes of the quantities are considered. This eliminates the need to consider positive and negative signs.

Any one of the four variables in the law of the lever equation can be found by rearranging the equation. For example, to find the effort force, F_E, the equation is

$$F_E = \frac{F_L d_L}{d_E}$$

> ▶ **SAMPLE** problem **3**
> ### *Applying the Law of the Lever*

A camper wants to mount a trailer on blocks for the winter. One corner of the trailer is lifted by applying an effort force using a 3.00-m steel bar (**Figure 5**). Determine the magnitude of the effort force required. (Ignore the mass of the bar.)

Solution

$F_L = 1.8 \times 10^3$ N

$d_L = 0.45$ m

$d_E = 3.00$ m $- 0.45$ m $= 2.55$ m

$F_E = ?$

$$F_E = \frac{F_L d_L}{d_E}$$

$$= \frac{(1.8 \times 10^3 \text{ N})(0.45 \text{ m})}{2.55 \text{ m}}$$

$$F_E = 3.2 \times 10^2 \text{ N}$$

The magnitude of the effort force is 3.2×10^2 N.

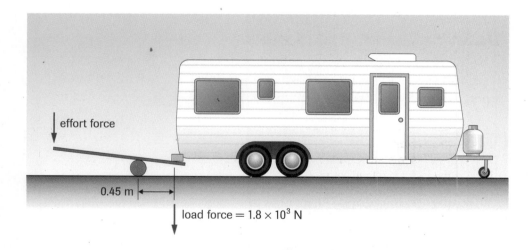

effort force

0.45 m

load force $= 1.8 \times 10^3$ N

Figure 5
For Sample Problem 3

DID YOU KNOW ?

Using Torque to Find an Object's Balance Point

In the Chapter 1 Getting Started *Try This Activity* on page 7, you balanced a metre stick on two fingers and slid your fingers until they met, just beneath the balance point. In that case, it was the centre of the stick. This experiment works because the clockwise and counterclockwise torques around the balance point are equal in magnitude. However, if you tie a light mass to one end of the stick and repeat the experiment, you will find a new balance point. You can do the same with a broom, a golf club, a baseball bat, a support stand, etc. You can apply this principle when designing a mobile in which branches of different masses are suspended from a single string or thread. (If you try this, be sure to start from the bottom of the mobile and work toward the top.)

For any rigid object, the law of the lever can be stated in more general terms. Rather than using the terms "load" and "effort," consider the direction of possible rotation. Thus, when any rigid object is in static equilibrium, *the clockwise torque is balanced by the counterclockwise torque.* Using symbols, the general condition for static equilibrium is

$$T_{CW} = T_{CCW}$$

where T_{CW} is the magnitude of the clockwise torque on an object around the fulcrum, and T_{CCW} is the magnitude of the counterclockwise torque on an object around the fulcrum. This is illustrated for a first-class lever in **Figure 6**.

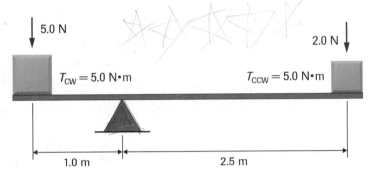

Figure 6
When a lever is in static equilibrium, the magnitude of the clockwise torque equals the magnitude of the counterclockwise torque.

Answers

12. 96 N

14. (a) 0.75 m
    (b) 3.2 × 10² N
    (c) 0.10 m

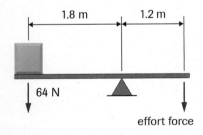

Figure 7
For question 12

> ### ▶ Practice

Understanding Concepts

12. Calculate the effort force for the situation shown in **Figure 7**.

13. Rearrange the equation for the law of the lever to solve for (a) d_E, (b) F_L, and (c) d_L.

14. (a) Find d_E given $F_E = 15$ N, $F_L = 75$ N, and $d_L = 0.15$ m.
    (b) Find F_L given $F_E = 64$ N, $d_E = 3.5$ m, and $d_L = 0.70$ m.
    (c) Find d_L given $F_E = 32$ N, $F_L = 640$ N, and $d_E = 2.0$ m.

SUMMARY *Torque and Levers*

- Torque (T) is a turning effect on a rigid object around a fulcrum; it is measured in newton-metres (N•m); $T = Fd$.

- The magnitudes of the effort torque and load torque can be found for a lever using the equations $T_E = F_E d_E$ and $T_L = F_L d_L$.

- When a lever is in static equilibrium, the effort torque is equal in magnitude to the load torque. This is the law of the lever, also expressed as an equation: $F_E d_E = F_L d_L$.

Section 2.3 Questions

Understanding Concepts

1. Revisit your diagram and explanation of the breaking pencils in the Getting Started *Try This* Activity. Rewrite your explanation.

2. Calculate the magnitude of the effort torque or the load torque for each case in **Figure 8**.

   (a)

   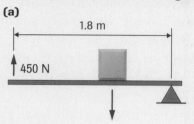

   (b)

   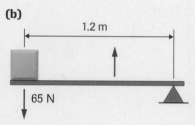

   Figure 8
   For question 2

3. Calculate the magnitude of the effort force needed to produce an effort torque of magnitude 24 N•m at a distance of 0.25 m from the fulcrum of a first-class lever.

4. (a) How far from the fulcrum of a second-class lever is a load force of magnitude 1.6×10^3 N located if the magnitude of the load torque is 1.2×10^3 N•m?
   (b) Calculate the mass of the load. (Hint: Apply the equation for weight from Chapter 1.)

5. Calculate the effort force for the situation shown in **Figure 9**.

   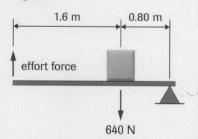

   Figure 9
   For question 5

6. A student uses an effort force on a lever to balance a load of mass 5.0 kg. The load arm is 84 cm and the effort arm is 24 cm.
   (a) Calculate the magnitude of the load force.
   (b) Calculate the magnitude of the effort force.

7. A homeowner wants to raise one end of an upright piano in order to place coasters under the wheels. It takes 1.6×10^3 N to raise one end of the piano, but the maximum force the homeowner can apply is 4.0×10^2 N. To accomplish the task, she places a fulcrum 0.30 m from the piano and uses a strong board as a first-class lever to raise the piano. (Ignore the mass of the board.)
   (a) Draw a sketch of this situation showing the fulcrum, the lever, and the distances and forces involved.
   (b) Calculate the minimum length of the effort arm required to lift the piano.
   (c) Find the total length of the board.
   (d) If she applied her effort farther from the fulcrum, would the effort force increase or decrease? Explain your answer.

8. Each oar in a rowboat is held 0.40 m from the oarlock (**Figure 10**). The oars exert a force on the water at an average distance of 1.4 m from the oarlock. If a rower can exert a force of magnitude 350 N with each arm, what is the total force exerted by the oars on the water? (For this situation, assume that the fulcrum is at the oarlock.)

   oarlock

   Figure 10
   For question 8

9. The effort force applied to the tweezers in **Figure 11** has a magnitude of 12 N. If the magnitude of the force exerted on a load is 8.0 N, determine the
   (a) distance from the fulcrum to the load
   (b) distance from the effort force to the load

   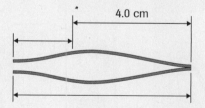

   Figure 11
   For question 9

Applying Inquiry Skills

10. Design a seesaw that is balanced when a child sits at one end and a parent sits at the other end. Using estimated values, show that the torques are equal in magnitude.

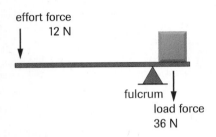

Figure 1
Using a first-class lever to gain a
force advantage

Machines help us perform tasks. In doing so, they give us an advantage. For example, if you use a lever to lift a heavy object (**Figure 1**), the output force (or load force) achieved can be greater than the input force (or effort force) required. In **Figure 1**, the lever provides a force advantage. As another example, an inclined plane reduces the effort force needed to raise a load to a higher level. In each case, the advantage can be calculated.

To determine the force advantage of a machine, we find the ratio of the load force to the effort force, $\frac{F_L}{F_E}$. In **Figure 1**, that ratio is $\frac{36 \text{ N}}{12 \text{ N}} = 3.0$. This means that the load force, F_L, is 3.0 times greater than the effort force, F_E. Rearranging the equation for the law of the lever, we have

$$F_E d_E = F_L d_L$$

$$\frac{d_E}{d_L} = \frac{F_L}{F_E}$$

We now have two ratios in this equation: the force ratio, $\frac{F_L}{F_E}$, and the distance ratio, $\frac{d_E}{d_L}$. In a machine in static equilibrium or in an ideal or perfect machine, these ratios are equal. However, most machines have some friction due to moving parts, so they are not ideal. Thus, we distinguish the *ideal mechanical advantage* from the *actual mechanical advantage*. The

**actual mechanical advantage
(AMA)** the ratio of the load force to
the effort force for a machine

**ideal mechanical advantage
(IMA)** the ratio of the effort arm (or
effort distance) to the load arm (or
load distance) for a machine

actual mechanical advantage (AMA) of a machine is the force ratio $\frac{F_L}{F_E}$. The **ideal mechanical advantage (IMA)** of a machine is the distance ratio $\frac{d_E}{d_L}$. The distance ratio can be modified to find the IMA of certain machines. See **Table 1**. Note that in both the force and the distance ratios, the units cancel, so AMA and IMA have no units.

(a)　　**(b)**　　**(c)**　　**(d)**

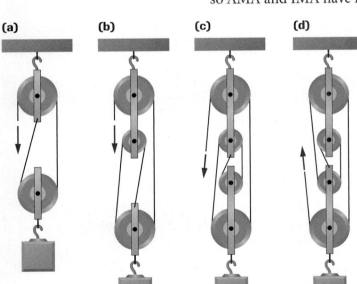

Table 1　IMA of Other Machines

Machine	IMA
wheel and axle	the ratio of the radii, $\frac{r_E}{r_L}$
a set of gears	the ratio of the teeth count, $\frac{N_E}{N_L}$
a set of pulleys	the number of support strands (See **Figure 2**.)
an inclined plane	the ratio $\dfrac{\text{length of inclined plane}}{\text{height}}$

Figure 2
In any pulley system, a support strand is
one that pulls upward.
(a) IMA = 2　　**(c)** IMA = 4
(b) IMA = 3　　**(d)** IMA = 5

> ▶ **SAMPLE** problem **1**
>
> *Calculating AMA and IMA*

In an acrobatic demonstration, one person jumps onto the end of a plank (lever) (**Figure 3**). This creates a large effort force of magnitude 9.2×10^2 N at the end of the board at a distance of 1.7 m from the fulcrum. A smaller person (a load force of 4.6×10^2 N located 3.1 m from the fulcrum) moves a larger distance and rises high enough to perform acrobatic moves. Calculate

(a) the AMA of the board

(b) the IMA of the board

Solution

(a) $F_E = 9.2 \times 10^2$ N

$F_L = 4.6 \times 10^2$ N

AMA = ?

$$AMA = \frac{F_L}{F_E}$$

$$= \frac{4.6 \times 10^2 \text{ N}}{9.2 \times 10^2 \text{ N}}$$

$$AMA = 0.50$$

The AMA of the board is 0.50.

(b) $d_E = 1.7$ m

$d_L = 3.1$ m

IMA = ?

$$IMA = \frac{d_E}{d_L}$$

$$= \frac{1.7 \text{ m}}{3.1 \text{ m}}$$

$$IMA = 0.55$$

The IMA of the board is 0.55.

LEARNING TIP

Static and Moving Components
It is important to distinguish between static and moving situations in machines. When a machine is in static equilibrium, the law of the lever applies, so $\frac{d_E}{d_L} = \frac{F_L}{F_E}$. But when the components of the machine move in order to accomplish their task, the effort force increases to account for friction, so $\frac{d_E}{d_L} > \frac{F_L}{F_E}$, and the IMA is greater than the AMA.

Figure 3
A large effort force results in a small load force, but during the time both are in contact with the board, the load moves farther than the effort force.

In Sample Problem 1, the AMA is 0.50, or less than one, but in **Figure 1**, page 88, the AMA is 3.0, or greater than one. When the AMA is less than one, F_L is reduced but d_L is increased. When the AMA is greater than one, F_L is increased but d_L is decreased. However, d_E must be large to achieve that force advantage. The force advantage can be greater than one, less than one, or in some cases equal to one. These concepts are summarized in **Table 2**.

Table 2 Changing AMAs of a First-Class Lever

AMA	Load Force	Load Distance	Example
$\dfrac{F_L}{F_E} = 1$	load force = effort force	load distance and effort distance are approximately equal	
$\dfrac{F_L}{F_E} > 1$	larger load force than effort force	smaller load distance than effort distance	
$\dfrac{F_L}{F_E} < 1$	smaller load force than effort force	larger load distance than effort distance	

Biomechanical Systems

Up to this point, the calculations of torque and AMA have been kept as simple as possible by ignoring the force of gravity on the machine in use. In Investigation 2.2, the lever was placed horizontally so that its weight did not affect the torques. Also, in sample problems and questions, the mass of the lever was ignored. In some situations, however, the mass of the lever is an important consideration in the torques and the AMA. One such case is the biomechanical system of a human forearm used to hold a load.

Figure 4 shows two diagrams of a human forearm: one with no load and one with a 4.0-kg load. Even with no load, the triceps must exert a tension force in order to hold the forearm in the horizontal position. In the sample problems that follow, you will see that the AMA of the forearm with the load is less than the AMA without the load.

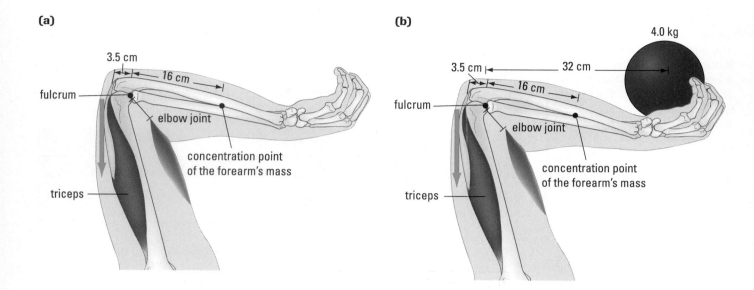

(a)

3.5 cm

16 cm

fulcrum

elbow joint

concentration point
of the forearm's mass

triceps

(b)

4.0 kg

3.5 cm   32 cm

16 cm

fulcrum

elbow joint

concentration point
of the forearm's mass

triceps

Figure 4
(a) A forearm with no load
(b) A forearm with a 4.0-kg load

▶ **SAMPLE** problem **2**

The AMA of an Outstretched Arm

Assume that the forearm in **Figure 4(a)** has a mass of 1.5 kg concentrated at the point shown. The effort force exerted by the triceps to hold the forearm steady has a magnitude of 67 N. Calculate

(a) the magnitude of the load force of the forearm due to its own mass and

(b) the AMA of the arm.

Solution

(a) $m = 1.5$ kg

$g = 9.8$ N/kg

$F_L = ?$

$$F_L = mg$$

$$= (1.5 \text{ kg})\left(9.8 \, \frac{\text{N}}{\text{kg}}\right)$$

$$F_L = 14.7 \text{ N}$$

The magnitude of the load force caused by the forearm's own mass is 15 N.

(b) $F_L = 14.7$ N

$F_E = 67$ N

AMA = ?

$$\text{AMA} = \frac{F_L}{F_E}$$

$$= \frac{14.7 \text{ N}}{67 \text{ N}}$$

$$\text{AMA} = 0.22$$

The AMA of the arm is 0.22.

When the mass of the forearm and the mass of the load are both considered, we apply the law of the lever by considering counterclockwise and clockwise torques. In **Figure 4(b)**, the total clockwise torque is the sum of the torques caused by the forearm and the load. This torque is balanced by the counterclockwise torque caused by the effort force.

▶ **SAMPLE** problem 3

The AMA of an Arm Holding a Load

The same arm is holding a 4.0-kg load, as shown in **Figure 4(b)**, page 91. Calculate

(a) the magnitude of the load force due to the load

(b) the magnitude of the effort force exerted by the triceps to hold the arm in a static position

(c) the AMA of the arm

Solution

(a) $m = 4.0$ kg

$g = 9.8$ N/kg

$F_L = ?$

$$F_L = mg$$
$$= (4.0 \text{ kg})\left(9.8 \frac{\text{N}}{\text{kg}}\right)$$
$$F_L = 39.2 \text{ N}$$

The magnitude of the load force of the load is 39 N.

(b) F_L (forearm) = 14.7 N d_L (forearm) = 16 cm = 0.16 m

F_L (load) = 39.2 N d_L (load) = 32 cm = 0.32 m

$d_E = 3.5$ cm = 0.035 m

$F_E = ?$

$$T_{\text{CCW (total)}} = T_{\text{CW (total)}}$$
$$F_E d_E = (F_L d_L)_{\text{forearm}} + (F_L d_L)_{\text{load}}$$
$$F_E = \frac{(F_L d_L)_{\text{forearm}} + (F_L d_L)_{\text{load}}}{d_E}$$
$$= \frac{(14.7 \text{ N})(0.16 \text{ m}) + (39.2 \text{ N})(0.32 \text{ m})}{0.035 \text{ m}}$$
$$F_E = 4.256 \times 10^2 \text{ N}$$

The magnitude of the effort force is 4.3×10^2 N.

(c) $F_L = 14.7\,N + 39.2\,N = 53.9\,N$
$F_E = 4.256 \times 10^2\,N$
AMA = ?

$$AMA = \frac{F_L}{F_E}$$

$$= \frac{53.9\,N}{4.256 \times 10^2\,N}$$

$$AMA = 0.13$$

The AMA of the arm is 0.13.

▶ *Practice*

Understanding Concepts

1. Calculate the AMA in each of the following cases:
   (a) In a pulley system, an effort force of magnitude 17 N raises a load force of magnitude 32 N.
   (b) To turn a truck's steering wheel, an effort force of magnitude 2.9 N on the wheel creates a load force on the steering column of magnitude 15 N.

2. Calculate the IMA in each of the following cases:
   (a) Assume that the load arm in **Figure 1**, page 88, is 0.35 m and the effort arm is 1.3 m.
   (b) In raising a flag 6.2 m up a pole, the effort force is moved 6.2 m downward.

3. A child pulls a friend on a toboggan up a small hill that is 9.8 m long and 1.2 m higher at the top than at the bottom. The effort force, applied parallel to the slope of the hill, has a magnitude of 65 N, and the load force has a magnitude of 4.8×10^2 N. Calculate the AMA and IMA in this situation.

4. Explain why the AMA of a machine is generally less than its IMA.

5. Is the IMA always less than one for a third-class lever but always greater than one for a second-class lever? Use diagrams to explain your answer.

6. Using the ratios in **Table 1**, page 88, calculate the IMA in each of the following:
   (a) The radius of the smallest sprocket on the rear wheel of a bicycle is 2.5 cm, and the radius of the rear wheel is 0.30 m.
   (b) There are 42 teeth on the largest pedal gear of a bicycle and 14 teeth on the smallest gear on the rear wheel.
   (c) A load is hung from a single pulley, and one end of the cord raising the pulley is attached to the ceiling. (For an example of this arrangement, see **Figure 4(b)**, page 73, in section 2.1.)

7. A person is holding a 17-N carton in the hand of an outstretched forearm held horizontally. The effort distance is 3.4 cm, and the forearm's mass can be assumed to be concentrated at a point 15 cm from the fulcrum. The load is 31 cm from the fulcrum. The magnitude of the weight of the forearm is 14 N. Calculate
   (a) the magnitude of the effort force required to hold the arm steady
   (b) the AMA of the arm

Answers

1. (a) 1.9
   (b) 5.2
2. (a) 3.7
   (b) 1.0
3. 7.4; 8.2
6. (a) 12
   (b) 3.0
   (c) 2
7. (a) 2.2×10^2 N
   (b) 0.14

Efficiency of Machines

As you learned in Chapter 1, in many situations friction is undesirable, and reducing friction improves a machine's efficiency. Knowing the efficiency of a machine tells us how productive it is. Like mechanical advantage, this efficiency can be measured. The **percent efficiency**, which is the ratio of the actual mechanical advantage to the ideal mechanical advantage, is expressed as a percentage:

percent efficiency the ratio of the AMA to the IMA of a machine, expressed as a percentage

$$\% \text{ eff} = \frac{\text{AMA}}{\text{IMA}} \times 100\%$$

As you can predict, machines that have a large amount of friction have a low percent efficiency.

▶ **SAMPLE** problem **4**

Calculating Percent Efficiency

A 14-N cart is pulled 1.2 m up a ramp with an effort force of magnitude 5.0 N parallel to the ramp, raising the cart 0.40 m above its initial level. Calculate

(a) the IMA

(b) the AMA

(c) the percent efficiency of the ramp

Solution

(a) $d_E = 1.2$ m

$d_L = 0.40$ m

IMA $= ?$

$$\text{IMA} = \frac{d_E}{d_L}$$

$$= \frac{1.2 \text{ m}}{0.40 \text{ m}}$$

$$\text{IMA} = 3.0$$

The IMA of the ramp is 3.0. (Notice that the ratio $\frac{d_E}{d_L}$ is the same as the ratio $\frac{\text{length of ramp}}{\text{height}}$.)

(b) $F_L = 14$ N

$F_E = 5.0$ N

AMA $= ?$

$$\text{AMA} = \frac{F_L}{F_E}$$

$$= \frac{14 \text{ N}}{5.0 \text{ N}}$$

$$\text{AMA} = 2.8$$

The AMA of the ramp is 2.8.

(c) Using the calculations in (a) and (b):

$$\% \text{ eff} = \frac{\text{AMA}}{\text{IMA}} \times 100\%$$

$$= \frac{2.8}{3.0} \times 100\%$$

$$\% \text{ eff} = 93\%$$

The percent efficiency of the ramp is 93%.

▶ *Practice*

Understanding Concepts

8. State the main reason that machines do not have an efficiency of 100%.

9. Calculate the percent efficiency in each of the following:
   (a) The distance ratio of a lever is 3.6 and the force ratio is 3.1.
   (b) The AMA of a wheel and axle is 6.0 and the IMA is 7.0.

10. A pulley system is used to raise a shipping container from the dock to a cargo ship. An effort force of magnitude 6.8×10^4 N is required to lift a load of mass 3.5×10^4 kg. The effort force moves 54 m while the load is raised 8.7 m.
   (a) Calculate the magnitude of the load force.
   (b) Calculate the IMA, the AMA, and the percent efficiency of the pulley system.

Answers

9. (a) 86%
   (b) 86%

10. (a) 3.4×10^5 N
    (b) 6.2; 5.0; 81%

Applying Inquiry Skills

11. Assume that you are attempting to determine the efficiency of a ramp that is used to raise a box. You investigate by pushing the box up the ramp. What could you do to improve the ramp's efficiency?

Making Connections

12. What happens to the percent efficiency of a bicycle when the moving components rust? What can be done to restore the percent efficiency to the highest possible level?

13. When you use your arm to lift a load of appropriate size, the efficiency of your arm is close to 100%. Why is this biomechanical system so efficient?

SUMMARY   *Mechanical Advantage and Efficiency*

- Machines have an actual mechanical advantage (or force ratio, $\frac{F_L}{F_E}$) and an ideal mechanical advantage (or distance ratio, $\frac{d_E}{d_L}$).

- In general, the AMA is less than the IMA because friction causes machines to be less than 100% efficient.

- Biomechanical systems can be analyzed and their AMA determined.

- The percent efficiency of a machine is determined by the equation $\% \text{ eff} = \frac{\text{AMA}}{\text{IMA}} \times 100\%$.

Understanding Concepts

1. When the smooth fishing line on a fishing rod breaks, the owner replaces it with a rough string.
    (a) When used to reel in fish, how does the IMA of the new version of this "machine" compare with the previous version? Explain your answer.
    (b) Repeat (a) for the AMA.

2. Two screwdrivers are identical, except that the handle of one is double the diameter of the handle of the other. Compare the ideal mechanical advantages of the two screwdrivers.

3. Copy **Table 3** into your notebook and make the calculations needed to complete it.

4. (a) Repeat Practice question 7, page 93, but assume that the magnitude of the load force of the carton is double (i.e., 34 N).
    (b) Based on your answers in (a) and Practice question 7, state what happens to the AMA of a horizontal arm as the mass held in the hand increases.

5. A person in a wheelchair (total weight 6.4×10^2 N) is pushed 9.4 m up a ramp with an effort force of magnitude 1.4×10^2 N parallel to the ramp. The wheelchair is raised 1.8 m above its initial level. Calculate (a) the IMA, (b) the AMA, (c) the percent efficiency of the ramp, and (d) the mass of the person and wheelchair.

6. A rotary cheese grater (**Figure 5**) has a handle that rotates with a radius of 74 mm. It is attached rigidly to a grating drum with a radius of 19 mm. An effort force of magnitude 11 N is just big enough to grate a piece of cheese that exerts a load force of 23 N.
    (a) Determine the IMA of the grater.
    (b) Determine the AMA of the grater.
    (c) Calculate the percent efficiency of the grater.
    (d) How would your answer in (c) change if the cheese were harder to grate?

7. **Figure 6** shows two pulley systems raising the same size loads.
    (a) Which one requires a greater effort to raise the mass? Why?

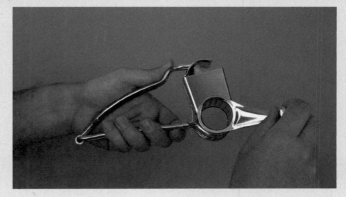

Figure 5
A cheese grater

(b) Which one requires a greater effort distance to move the mass the same amount? Why?

Applying Inquiry Skills

8. How would you experimentally determine the AMA, the IMA, and the percent efficiency of the pulley system shown in **Figure 6(a)**?

Making Connections

9. A new gasoline-powered lawnmower has a percent efficiency of 25%. After frequent use, the mower gives off fumes and odours as it overheats.
    (a) What has happened to the percent efficiency of the mower? What causes the change?
    (b) How can the mower's efficiency be restored to the original value?

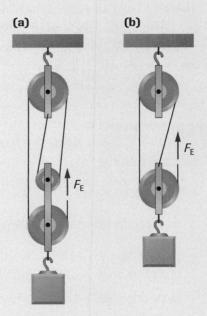

Figure 6

Table 3 Data for Question 3

Question	F_E (N)	d_E (m)	F_L (N)	d_L (m)	IMA	AMA	% eff
(a)	13	0.84	39	0.21	?	?	?
(b)	96	0.60	19	2.5	?	?	?
(c)	?	0.72	35	?	6.0	5.0	?
(d)	2.0×10^2	?	?	0.25	14	13	?

Mechanical Advantage and Efficiency of Machines

This investigation will reinforce the concepts and equations related to the IMA, the AMA, and efficiencies of machines. It will also allow you to design an efficient machine that performs a specific task.

Inquiry Skills

○ Questioning   ● Conducting   ● Evaluating
● Predicting   ● Recording   ● Communicating
● Planning   ● Analyzing   ○ Synthesizing

Questions

What difficulties must be overcome to determine the IMA, AMA, and percent efficiency of various machines?

What design for a machine works well to raise a mass from the floor to the top of the lab bench?

Prediction

(a) Predict an answer to each Question.

Experimental Design

In looking at the data table and diagrams for this investigation, you will see many different machines you can experiment with and the calculations you will make in Part A of the investigation. This part is open ended; you may test machines not shown in the diagrams. Also, if a computer spreadsheet is available, you can use it to create your own data table in Part A.

In Part B, you will design, test, and modify a machine that will raise a mass from the floor to the top of the lab bench. You may change the task to make it more challenging. Get your teacher's approval first.

For some measurements in this activity, you must invert the force scale. This will cause an error in the measurement. You can correct for this error by zeroing the scale when it is inverted.

As you perform Part A of the investigation, think about how you will design the apparatus for Part B.

Materials

For each group of three or four students:
metre stick with a hole drilled at the centre
board
nail
metric ruler
two force scales or force sensors (to 10 N or more)
1.0-kg mass
pulley systems
wheel and axle
inclined plane
cart
block of wood
jackscrew
various apparatus for Part B, including a 200-g mass

For each student:
goggles (if spring scales are used)

Procedure

Part A

1. Set up a table for data using the titles in **Table 1**. The first row in the table has been started for you.

2. Set up the apparatus as shown in **Figure 1(a)**. Exert a 10-N load force at the 30-cm mark and determine the effort force at the 90-cm mark to balance the lever. Enter the data in the first row of your data table, and complete the calculations for that row.

Table 1   Observations and Calculations for Investigation 2.5

Machine	Details	F_L (N)	F_E (N)	d_L (m)	d_E (m)	IMA	AMA	% eff
lever	1st class	10	?	0.20	0.40	?	?	?
?	?	?	?	?	?	?	?	?

(a) First-class lever

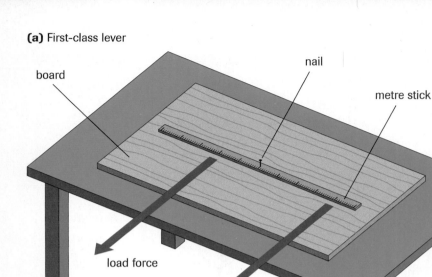

board

nail

metre stick

load force

effort force

(b) Second-class lever

effort force

load

(c) Third-class lever

effort force

load

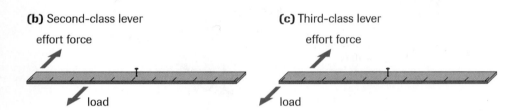

Figure 1
For steps 2 and 3

3. Try other positions of the load force and effort force for a first-class lever. Then take measurements and make calculations for the other classes of lever, as illustrated in **Figure 1(b)** and **(c)**. Record the data in your table.

4. Record measurements and calculations for the machines illustrated in **Figure 2**.

5. Design and experiment with setups other than those suggested here. Complete your data table.

Part B

6. Design a machine with two or more simple machines to raise a 200-g mass from the floor and place it gently onto the top of the lab bench. Get your teacher to approve your design.

7. Build the machine, and test and modify it.

8. Determine the machine's IMA, AMA, and percent efficiency.

Analysis

(b) Describe any difficulties you had in determining the IMA, AMA, and percent efficiency of the machines you explored.

(c) How did your design in Part B compare with the designs of other groups?

Evaluation

(d) Comment on the accuracy of your predictions.

(e) If you could start again, how would you change your design in Part B?

(a) Pulleys

(i) Single fixed pulley   (ii) Single movable pulley   (iii) Pulley system   (iv) Pulley system   (v) Pulley system

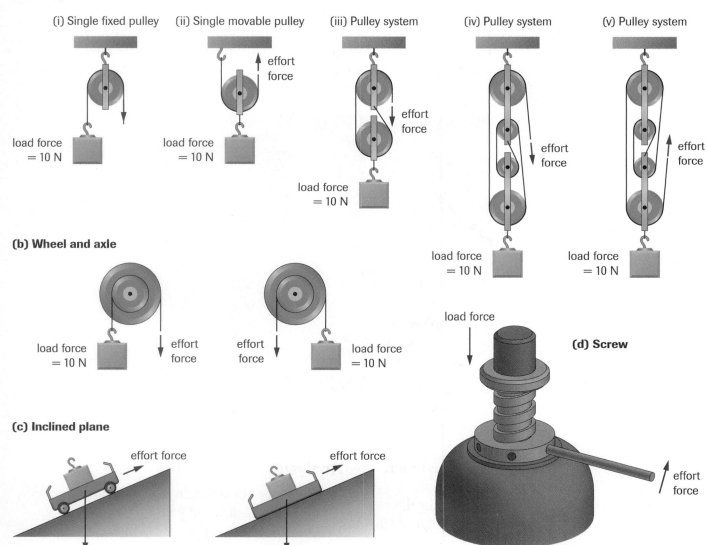

(b) Wheel and axle

(c) Inclined plane

(d) Screw

Figure 2
For step 4

As you learned earlier, a compound machine is a machine made of two or more simple machines. Compound machines can be as basic and inexpensive as a pepper grinder, or they can be as complex and expensive as the robotic Canadarm2 on the *International Space Station* (**Figure 1**).

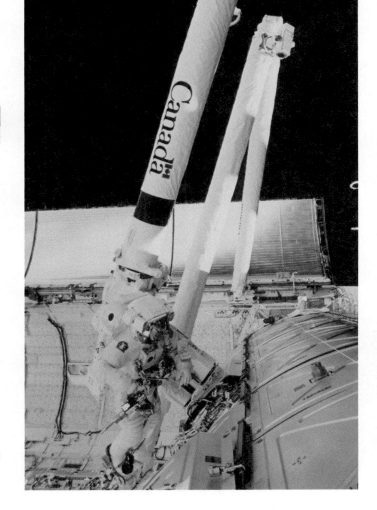

Figure 1
The Canadarm2 robotic arm has several simple machines linked together with great precision, all computer controlled. The arm is used for construction and maintenance of the space station.

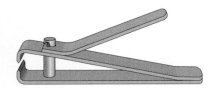

Figure 2
The nail clipper has a second-class lever and a third-class lever. Which is which?

Domestic Machines

Around your home there are many examples of domestic machines. For example, a nail clipper applies the principles of the lever and the inclined plane families of machines. As you can see in **Figure 2**, there are two lever systems: the handle and the cutting blades. The cutting blades are wedges, or inclined planes. Other examples of simple and compound machines around the home are can openers, faucets, window shades, coffee grinders, grandfather clocks, pianos, bathroom scales, exercise machines, clamps, bicycles, and eggbeaters.

An eggbeater has a wheel and axle for the handle connected to a large gear. This gear is connected to small gears that cause a second set of axles to rotate the beaters. The beaters move several times faster than the handle, and they rotate in the horizontal plane while the handle rotates in the vertical plane.

▶ **Practice**

Understanding Concepts

1. What design feature allows the beaters of an eggbeater (**Figure 3**) to rotate in opposite directions?

2. How is a salad spinner (**Figure 4**) similar to an eggbeater? How is it different?

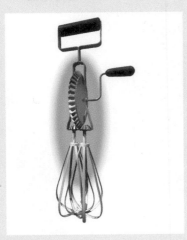

Figure 3
A typical hand-operated eggbeater

Figure 4
A salad spinner

▶ **TRY THIS** activity   *Domestic Machines*

Choose at least one compound machine from each of the following categories: tools, exercise and sports machines, and household appliances. For each machine chosen, identify the simple machines that make up the compound machine, and explain the function of each machine.

Industrial Machines

Large, compound machines are used in many industries, for example, mining and construction. A backhoe (**Figure 5(a)** on page 102) digs holes for the foundations of buildings and moves soil, sand, and gravel. A crane (**Figure 5(b)** on page 102) moves heavy components into position for constructing high-rise buildings, bridges, and other structures.

Pulleys can be arranged in various ways, depending on the application. Many industrial machines, including the crane, incorporate at least one pulley design in their operation. Two common examples are the *block and tackle* and the *chain hoist*.

block and tackle a system of two sets of pulleys and one cable, with the upper set fixed and the load attached to the lower movable set

A **block and tackle** is a compact system of pulleys designed to raise heavy loads (**Figure 6**). The system has one cable or rope wound around two separate sets of pulleys. The pulleys of each set rotate freely on the same axle. The upper set is fixed to a support, and the lower set is attached to the load. Pulling the rope raises the lower set of pulleys, and thus raises the load. The IMA of the block and tackle is equal to the total number of support strands of the cable. This type of machine is used in large cranes as well as in robotic devices.

(a)

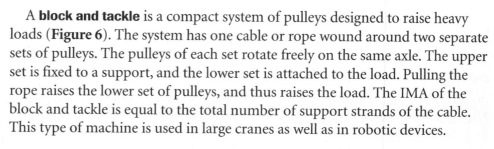

cab   boom   stick

hydraulic ram

bucket

(b)

pulley system

winch

counterweights

boom

hydraulic ram

outriggers

effort force

load force

Figure 6
In this simplified model of a block and tackle, a downward effort force on the rope causes the movable lower pulley to be raised toward the upper fixed pulley.

Figure 5
(a) A typical backhoe
(b) A typical crane (The load is not shown; it would be hanging from the hook.)

A **chain hoist** consists of an endless chain looped around three pulleys. The upper two pulleys have teeth and are joined together, while the lower pulley is free to move up and down with the load suspended from it. The IMA of this machine is the ratio of the radius of the larger fixed pulley to the radius of the smaller fixed pulley. The IMA can also be determined based on the number of teeth. For example, if the larger fixed pulley has 20 teeth and the smaller one has 10 teeth, the IMA is 2. A chain hoist system is used to lift engines and other components that are too heavy to lift by hand.

chain hoist a system of two fixed pulleys and one movable pulley linked by an endless chain; pulling on the chain causes a load attached to the movable pulley to be raised or lowered

---

▶ *Practice*

Understanding Concepts

3. (a) What is the IMA of the block and tackle in **Figure 6**, page 102?
   (b) Suppose the block and tackle is designed so that the effort force is exerted upward rather than downward. What is the IMA?

4. (a) What is the IMA of the chain hoist in **Figure 7**?
   (b) How can the load be lowered?
   (c) How does the distance moved by the effort force compare to the distance moved by the load?

Making Connections

5. A winch is a crank connected to a rotating wheel or axle. At the front end of a boat trailer, a winch is used to pull a boat out of the water. This winch has a safety ratchet.
   (a) How does this winch provide a force advantage?
   (b) Why is the ratchet used?

Figure 7
When the large upper wheel of this chain hoist rotates counter-clockwise, it pulls more chain than the small upper wheel does. This effect raises the lower movable wheel a small distance.

Answers

3. (a) 2
   (b) 3
4. (a) 1.8
   (c) $d_E = 1.8d_L$

DID YOU *KNOW* ?

Winches

The basic design of winches hasn't changed much since Roman times. The Romans used winches to haul pails of water out of wells. In medieval times, large cathedrals were built with the aid of winches, which were used to raise and position heavy blocks. Today, winches are used in fishing rods and in hoists to load small boats onto trailers.

▶ *TRY THIS* activity   *Industrial Machines*

Research the role of compound machines in one of these fields: industry, manufacturing, agriculture, the space program, medical applications, sports or leisure activities, and entertainment (including musical instruments and stage productions). Choose a compound machine, such as the Canadarm2 or an upright piano, and in a short report identify the simple machines that comprise it. Explain the function or functions of each simple machine. Use diagrams in your explanation. In your report, record your source(s) of information.

- Simple machines are linked together to form compound machines used in numerous domestic and industrial applications.

▶ *Section 2.6 Questions*

Understanding Concepts

1. Name at least one domestic machine that applies the principles of
   (a) the lever and the wheel and axle
   (b) the pulley and the lever
   (c) the pulley and gears
   (d) the wheel and axle and gears
   (e) the wedge and the wheel and axle
   (f) the screw and the wheel and axle

2. Name at least one industrial machine that applies the principles of
   (a) the lever and the wheel and axle
   (b) the pulley and the lever
   (c) the pulley and gears
   (d) the wheel and axle and gears

3. Choose a compound machine in the kitchen, such as a coffee grinder, a pepper grinder, or a food processor, and describe the simple machines it uses. Draw a sketch to show how the device works.

Applying Inquiry Skills

4. (a) How would you experimentally determine the AMA of the machines in **Figures 6** and **7**, pages 102–103?
   (b) Would you expect the AMA to be greater than, equal to, or less than the IMA? Explain why.

Making Connections

5. Some of the largest machines in the world are used for industrial purposes, for example, a bulldozer, a front-end loader, a tower crane, a bucket-wheel excavator, a hydraulic shovel excavator, a dragline excavator, and a forklift truck.
   (a) Choose one machine and research its structure and use. Identify the principles of simple machines that it uses.
   (b) What specific tasks require the machine to be the size that it is?

2.7 Investigation

Inquiry Skills

○ Questioning	● Conducting	● Evaluating
● Predicting	● Recording	● Communicating
● Planning	● Analyzing	○ Synthesizing

The Bicycle

The bicycle is an excellent machine to analyze, not only because it is so common, but also because it applies most of the principles of simple machines. In the late 1800s, the "penny-farthing" bicycle (**Figure 1(a)**) was used. It can still be seen in parades and museums today. By the beginning of the 1900s, however, the "safety bicycle" became popular (**Figure 1(b)**). It had many of the same features as today's more high-tech bikes (**Figure 1(c)**, **(d)**).

(a)

(b)

(c)

(d)

Figure 1
(a) The penny-farthing bike. Can you explain the name of this bike? (Hint: In Canada today, it might be called a "loonie-dime bike.")
(b) A bicycle from about 100 years ago
(c) A road bike
(d) A mountain bike

In order to perform observations and make calculations for a bicycle, you need to be familiar with bicycle terminology. **Figure 2** shows one type of bicycle with its main components labelled.

Questions

How does a multi-speed bike apply the principles of simple machines?

What mathematical calculations related to the use of machines can be made?

How do the features of different types of modern bikes compare?

Prediction

(a) Based on your own experience, predict an answer to each Question.

Experimental Design

Because there are many different bicycles to choose from, the procedure steps provide guideline instructions only. Thus, you will be expected to design the procedure yourself and carry out the observations and calculations on your own in some cases. You will also be required to design your own data tables to summarize the measurements and calculations.

To answer the third question, it would be most interesting to have available a mountain bike, a touring bike, and a racing bike.

Materials

For each group of three or four students:
at least one multi-speed bike
metre stick
metric ruler
two force scales (calibrated in newtons)

Procedure

1. Find as many examples of simple machines as possible on the bicycle you are investigating. For each machine you find, describe the machine and/or draw a sketch of how it operates. For each member of the lever family, indicate the class of lever. Where possible, include the main function(s) of each simple machine.

2. Determine all the gear ratios possible for the bicycle. Summarize your findings in a table.

Figure 2
Main components of a typical bicycle

3. For the lowest and highest gears available, determine the IMA. (Hint: You will need to take measurements to determine the distance or speed ratio of the rear wheel rotation to the pedal rotation.)

4. For the gears used in step 3, determine the AMA. To do this, you can tie a piece of string around the tire, as shown in **Figure 3**, so the force can be as close to the wheel circumference as possible. In addition, if the force sensor pulls on a wheel spoke, the spoke may break. (Be sure that the forces you apply to the pedal and the drive wheel are parallel to the circumference of the circle of motion.) Design a data table and record your measurements and calculations.

Figure 3
Determining the AMA

5. Find features on the bicycle that relate to friction and other forces. (For example, look for bearings that reduce friction, brakes that increase friction, and the way in which tension in the chain is kept at a sufficient strength.) Describe your findings.

6. Compare the features of two or three different types of bikes. Create a table, diagram, detailed description, or other instrument for summarizing the main differences.

Analysis

(b) Name the parts of a bicycle to which a torque can be applied.

(c) What is the range of gear ratios for the bike you analyzed? Under what conditions would you use the highest gear ratio? the lowest gear ratio?

(d) How does the maximum IMA compare to the maximum AMA? Explain any difference.

(e) How does the maximum IMA compare to the minimum IMA for the bike you investigated? Which types of bikes usually have the greatest difference between these two values? Explain your answer.

(f) If you were climbing a steep hill on a mountain bike, would you use a low mechanical advantage or a high one? Why?

(g) Compare and contrast bearings that are sealed and bearings that are not sealed.

(h) State, with a reason, which type of bike has
   (i) thin tires
   (ii) clipless pedals
   (iii) disc brakes
   (iv) flexible suspension
   (v) the lightest weight possible

(i) Complete a formal report of your investigation, including a summary of your answers to the Questions.

Evaluation

(j) Describe any difficulties you had with the measurements in this investigation. What did you do to minimize those difficulties?

(k) How good were your predictions?

Synthesis

(l) Why do you think that the first bikes that were constructed along the same lines as today's bikes were called "safety bicycles"?

(m) When determining gear ratios, which is easier to measure: the ratio of the number of teeth or the ratio of diameters? Explain your answer.

(n) How would you calculate the theoretical maximum speed for a bicycle using the data you found in this investigation? Use typical numbers to show a sample calculation.

Key Understandings

2.1 Simple Machines

- A **simple machine** can be classified into one of two families: the lever family (the **lever, pulley, wheel and axle,** and **gears**) and the inclined plane family (the **inclined plane, wedge,** and **screw**).
- Any machine in the lever family can be further classified as a **first-class lever, second-class lever,** or **third-class lever,** depending on the position of the **fulcrum** relative to the **effort force** and the **load force**.
- Simple machines can perform a variety of functions; they form the basis of **compound machines** and **biomechanical systems**.

2.2 Investigation: Forces on Levers

- Measurements and calculations on a balanced lever reveal distinct patterns to the products $F_E d_E$ and $F_L d_L$.

2.3 Torque and Levers

- **Torque** is a turning effect on a rigid object around a fulcrum. It can be calculated using the equation $T = Fd$, assuming F is perpendicular to the rigid object.
- The **law of the lever** applies to any lever in static equilibrium. In equation form, it states $F_E d_E = F_L d_L$.

2.4 Mechanical Advantage and Efficiency

- The **actual mechanical advantage** of a machine is the force ratio $\dfrac{F_L}{F_E}$, and the **ideal mechanical advantage** is the distance ratio $\dfrac{d_E}{d_L}$.
- The **percent efficiency** of a machine is the ratio
$$\% \text{ eff} = \frac{\text{AMA}}{\text{IMA}} \times 100\%.$$

2.5 Investigation: Mechanical Advantage and Efficiency

- The ideal mechanical advantage, actual mechanical advantage, and percent efficiency of simple machines can be determined experimentally.

2.6 Domestic and Industrial Machines

- Complex machines are used in numerous domestic and industrial applications.

2.7 Investigation: The Bicycle

- The bicycle is an excellent device to use to investigate the properties and variables related to machines, torque, ideal mechanical advantage, and actual mechanical advantage.

Key Terms

2.1

machine
lever
fulcrum
effort force
load force
effort arm
load arm
first-class lever
second-class lever

third-class lever
biomechanical system
pulley
wheel and axle
gears
inclined plane
wedge
screw
compound machine

2.3

torque
law of the lever

2.4

actual mechanical
  advantage (AMA)
ideal mechanical
  advantage (IMA)
percent efficiency

2.6

block and tackle
chain hoist

Key Equations

2.3

- $T = Fd$
- $T_E = F_E d_E$
- $T_L = F_L d_L$
- $F_E d_E = F_L d_L$

2.4

- $AMA = \dfrac{F_L}{F_E}$
- $IMA = \dfrac{d_L}{d_E}$
- $\% \ eff = \dfrac{AMA}{IMA} \times 100\%$

Problems You Can Solve

2.1

- Identify, describe, and illustrate applications of simple machines in the lever and inclined plane families.
- For a lever in any of the three classes, locate the fulcrum, load force, effort force, load arm, and effort arm, and identify examples of the lever.

2.2

- For a balanced lever, state how the product $F_E d_E$ compares to the product $F_L d_L$.

2.3

- Given any two of torque, force, and distance, determine the third quantity. (This applies to both load torque and effort torque.)
- State the law of the lever, and apply it to practical situations.

2.4

- Calculate the ideal mechanical advantage, actual mechanical advantage, and percent efficiency of simple machines.

2.5

- Determine experimentally the ideal mechanical advantage, actual mechanical advantage, and percent efficiency of a variety of simple machines.
- Design a simple or compound machine to perform a specific task.

2.6

- Describe the role of machines in domestic life and industry.

2.7

- Analyze experimentally the machine systems of a bicycle to determine input and output forces and the ideal mechanical advantage and actual mechanical advantage of the machines that make up the bicycle.

▶ MAKE a summary

Create a concept map to summarize this chapter. Begin by placing "machines" and the corresponding definition in the middle of the page. Then add the main concepts presented in the chapter, as shown in **Figure 1**. Add more branches to the map stemming from the main concepts, and include definitions, equations, examples, and diagrams. Finally, link the ideas together to show you understand the relationships among the concepts.

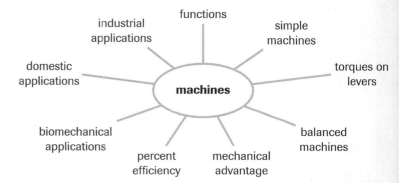

Figure 1
The start of a concept map

Write the numbers 1 to 8 in your notebook. Indicate beside each number whether the corresponding statement is true (T) or false (F). If it is false, write a corrected version.

1. When you use your thumb and index finger to pull up your socks, you are using a first-class lever.

2. For a third-class lever, the load force always exceeds the effort force.

3. To make the toy top in **Figure 1** spin, you spin the axle, creating a speed advantage for the wheel.

Figure 1

4. Torque and ideal mechanical advantage are measured in newton-metres.

5. If the effort distance for a machine part is greater than the load distance, then the machine provides a force advantage.

6. Scissors are an example of a second-class lever combined with an inclined plane.

7. In the equation for torque, the force F is perpendicular to the rigid object.

8. To obtain a force advantage for a wheel and axle, the effort force must be applied to the wheel.

Write the numbers 9 to 15 in your notebook. Beside each number, write the letter corresponding to the best choice.

9. When arm-wrestling, your arm is acting as
   (a) a first-class lever
   (b) a second-class lever
   (c) a third-class lever
   (d) none of these because the arm cannot be a lever

10. The exercise handgrip in **Figure 2** is

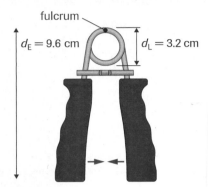

Figure 2

   (a) a first-class lever
   (b) a second-class lever
   (c) a third-class lever
   (d) either a second-class lever or a third-class lever, depending on where the fulcrum is considered to be

11. The IMA of the handgrip in **Figure 2** is
    (a) 3   (b) 1/3   (c) 4   (d) 1/4

12. For any machine, as the friction of the moving parts increases,
    (a) the AMA and the % eff decrease
    (b) the AMA and the % eff increase
    (c) the AMA increases but the % eff decreases
    (d) the AMA decreases but the % eff increases

13. As the angle of a wheelchair ramp changes from 10° to 20°,
    (a) the effort force falls and the AMA rises
    (b) both the effort force and the AMA rise
    (c) both the effort force and the AMA fall
    (d) the effort force rises and the AMA falls

14. When using a hammer to remove a nail from a board,
    (a) the hammer acts as a third-class lever with an AMA greater than one
    (b) the hammer acts as a second-class lever with an AMA greater than one
    (c) the hammer acts as a first-class lever with an AMA greater than one
    (d) none of the above

15. For a bicycle with seven rear-wheel gears and three pedal gears, the number of gear ratios possible is
    (a) $7 + 3 = 10$ (c) $7 \div 3 = 7/3$
    (b) $7 - 3 = 4$ (d) $7 \times 3 = 21$

Understanding Concepts

1. State the main function and class of lever for each machine in **Figure 1**.

   (a)

   (b)

   (c)

   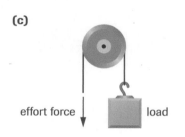

   (d)

   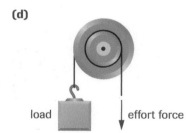

   Figure 1

2. (a) State two examples of a third-class lever in the human body.
   (b) State an example of a second-class lever in the human body.

3. Name all the simple machines that make up each item in **Figure 2**. For any members of the lever family, state the class.

4. What is the angle between a rigid object and the force applied to it that yields torque using the equation $T = Fd$?

5. Explain why cars have smaller steering wheels than large buses and trucks.

6. A first-class lever, 2.8 m long, has a load force of magnitude 6.8×10^2 N located 1.2 m from the fulcrum.
   (a) Draw a diagram of the lever, showing the fulcrum, forces, and distances involved.
   (b) Calculate the magnitude of the effort force at the end of the lever needed to balance the load.

7. A wheelbarrow has a 95-kg load located 0.60 m from the fulcrum. An effort force of magnitude 5.2×10^2 N is needed to lift the handles of the wheelbarrow.
   (a) Calculate the magnitude of the load force.
   (b) Calculate the distance from the effort force to the load.

(a)

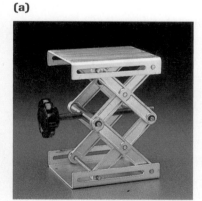

(b)

(c)

Figure 2
(a) Lab jack
(b) Locking pliers
(c) Hamster dragster

8. In a student's arm, acting as a lever, the distance from the fulcrum to the muscle is 4.0 cm, and the distance from the fulcrum to the hand is 31.5 cm.
   (a) What class of lever is the arm?
   (b) If an effort force of magnitude 1.5×10^2 N is required to support a particular load, what is the magnitude of the load force?
   (c) Calculate the mass of the load.

9. The pull tab used to open a typical pop can is an example of a lever.
   (a) Draw a diagram of a pull tab system, and label the fulcrum, effort force, and load force. What class of lever is the opener?
   (b) Use measured values to estimate the IMA of the opener.

10. In **Figure 3**, a tension force in the triceps holds the arm in a static position.

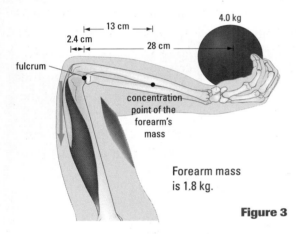

Forearm mass is 1.8 kg.

Figure 3

   (a) What class of lever is illustrated?
   (b) Determine the effort force required to hold the shot in the position shown.
   (c) Calculate the AMA of the arm.

11. An emergency crew is using a plank as a first-class lever to raise one side of a car off the road. The fulcrum is a large block of wood. The magnitude of the load force is 5200 N and that of the effort force is 650 N; the load arm and effort arm are 0.40 m and 3.6 m, respectively. Calculate the
   (a) AMA
   (b) IMA
   (c) percent efficiency of the plank

12. The wheel and axle in **Figure 4** has four wheels, any of which can act as either a wheel or an axle. The wheels are linked together rigidly and can rotate with little friction on the central rod. The wheel diameters are 35 mm, 50 mm, 65 mm, and 105 mm.

Figure 4

   (a) To two significant digits, determine the IMA of these combinations of load and effort forces:
      • load force on the 35-mm wheel; effort force on the 105-mm wheel
      • load force on the 65-mm wheel; effort force on the 50-mm wheel
      • the combination that gives the lowest IMA
   (b) For each situation in (a), name the class of lever involved.
   (c) For each situation in (a), if the effort force moves 14 cm, how far does the load force move?

Applying Inquiry Skills

13. Calculate an approximate value for the effort torque you apply when you sharpen a pencil with a crank pencil sharpener. To do this, you can measure the effort arm and either measure or estimate the effort force required. When is the required torque minimum? maximum?

14. Use two broomsticks and about 5 m of strong cord to set up the arrangement shown in **Figure 5**. The sticks should be parallel and about 40 cm apart.

Figure 5

   (a) Predict what will happen if two strong students in the class hold the sticks firmly apart, as indicated by the coloured arrows, while a third student exerts an effort force, as shown by the black arrow.
   (b) Try the demonstration. Explain the results. (Hint: Relate the situation to a system of pulleys.)

15. **Figure 6** shows a mechanical model of the Sun–Earth–Moon system, with the planet Venus closest to the Sun.
   (a) What simple machines make up this model?
   (b) What measurements related to the ideal mechanical advantage of the machine could you make?

Making Connections

16. A zipper has three wedges. Only a small effort force is required to pull the tab up or down, pushing the teeth of the zipper together or pulling them apart. Without the wedges, pulling the teeth apart is extremely difficult.
   (a) Research the invention of the zipper.
   (b) Explain how the three wedges provide the mechanical advantage in this "machine."

   GO www.science.nelson.com

17. Safety in the workplace is an important issue, especially when working with machines that have exposed moving parts. Think of a career or workplace location that involves potentially dangerous machines. (Examples are the school shop, an exercise gym, a manufacturing assembly line, an auto repair or maintenance shop, construction, demolition, renovation, tool-and-die making, landscaping, etc.) Choose one career or location, and create a poster showing safety rules you would recommend.

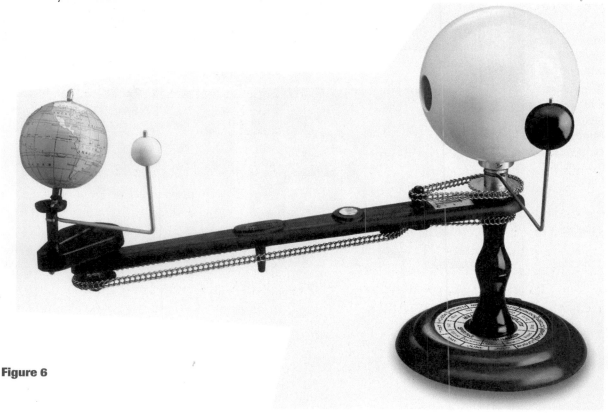

Figure 6

Building a Machine

Machines can help people who have lost or broken a limb; the machines can help them perform exercises during rehabilitation or tasks for daily living. Some machines are designed for general use in hospitals, care-giving facilities, gymnasiums, the home, or the workplace or for transportation of people with disabilities (**Figure 1**).

▶ **Criteria**

Process

- Draw up detailed plans and safety considerations for the design, tests, and modifications of the machine or model.
- Choose appropriate research tools, such as books, magazines, and the Internet (especially for Option 2).
- Choose appropriate materials to construct the machine or model.
- Appropriately and safely carry out the construction, tests, and modifications of the machine or model.
- Analyze the process (as described in the Analysis).
- Evaluate the task (as described in the Evaluation).

Product

- Demonstrate an understanding of the relevant physics principles, laws, and equations.
- Prepare a suitable research summary (Option 2).
- Submit a report containing the design plans for the machine or model, as well as test results and calculations of the AMA, IMA, and percent efficiency.
- Use terms, symbols, equations, and SI metric units correctly.
- Demonstrate that the final product works as explained.

(a)

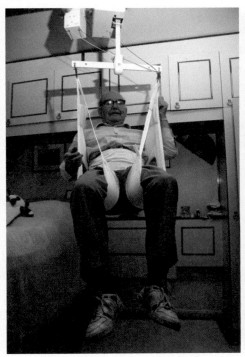

(b)

(c)

(d)

Figure 1

One problem with machines that have a general design is that they are not suitable for everyone's needs. For example, a machine may need to be modified to accommodate a smaller or larger person. Or a patient may have special needs that require adapted machines or perhaps even a new design. So there are many opportunities to design and build a machine that performs an important task.

In this Performance Task, you have two options. In Option 1, you design, build, test, and analyze a compound machine that accomplishes a specific task. In Option 2, you research the details of the design and operation of a compound machine, and then build a working model to analyze its operation. Both options apply the principles you learned in Chapters 1 and 2. The machine or model you build will involve forces, including friction, applied effort, and load forces and will feature at least two simple machines linked together. You will be able to analyze the machine quantitatively by determining its actual mechanical advantage (AMA), ideal mechanical advantage (IMA), and percent efficiency.

The Task

Option 1: A Lifting Device

Your task is to design, build, test, modify, and analyze a compound machine that can raise and lower a known mass; the machine should have a high actual mechanical advantage. Your design will depend on the suggestions made by your teacher. Before starting the task, your group should decide on the object to be lifted (e.g., a 10-kg bag of potatoes or a 200-g mass), the criteria chosen to evaluate the machine, and the necessary safety precautions. For instance, the IMA and the percent efficiencies of the machines made by various groups can be compared.

Option 2: A Model Machine

Your task is to research the design and operation of a compound machine that is used for a specific purpose, such as helping a person with a temporary or permanent disability. You will then create a working model of the machine so that it has the same ideal mechanical advantage as the full-size machine. To build the model, you can use materials that are inexpensive and easily and safely assembled (e.g., thick cardboard, if used wisely, would be an appropriate material). You will then design tests to determine the actual mechanical advantage and percent efficiency of your model.

Analysis

(a) What physics principles apply to the design and use of your machine?

(b) How can you judge if your machine or model was successful?

(c) How can the machine you designed or researched be used? What tasks can it perform?

(d) What careers are related to the manufacture and use of the machine?

(e) What safety precautions did you follow in building and testing your machine or model?

(f) After first testing your machine or model, how did you modify it to improve it?

(g) How could the process you used in this task be applied in business or industry?

(h) List problems you had while building the machine or model, and explain how you solved them.

Evaluation

(i) How does your machine or model compare with the designs of other groups? (Some criteria to consider are friction, IMA, percent efficiency, usefulness, appearance, and the wise use of materials.)

(j) Evaluate the tools you used in constructing the machine (Option 1), or evaluate the resources you used in your research (Option 2).

(k) If you did this task again, how would you modify the process to obtain a better final product?

1. Write the letters (a) to (e) in your notebook. Beside each letter, write the corresponding letter from the position–time graph in **Figure 1**.
   (a) zero velocity
   (b) decreasing velocity
   (c) increasing velocity
   (d) fast constant velocity
   (e) slow constant velocity

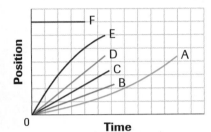

Figure 1

2. Write the letters (a) to (h) in your notebook. Beside each letter, write the word or phrase that corresponds to each of the following:
   (a) the motion of a stone dropped vertically downward from your hand
   (b) the force parallel to the road that allows a car to accelerate
   (c) the force within your flexed muscles
   (d) the law of inertia
   (e) the pivot point of a lever
   (f) the product of the effort force and effort arm of a third-class lever
   (g) an inclined plane wrapped around a central shaft
   (h) the ratio of the load force to the effort force of a machine

3. Write the letters (a) to (f) in your notebook. Beside each letter, write the letter from A to I that corresponds to each of the following terms:
   (a) torque
   (b) second-class lever
   (c) weight
   (d) third law of motion
   (e) mass
   (f) normal force

   A. considers objects at rest or moving at constant velocity
   B. the law relating force, mass, and acceleration
   C. considers action and reaction forces
   D. a turning effect on a rigid object around a fulcrum
   E. a lever with the load between the fulcrum and the effort force
   F. the quantity of matter in an object
   G. the force of gravity on an object
   H. a lever with the effort force between the load and the fulcrum
   I. the force between surfaces perpendicular to friction

Write the numbers 4 to 11 in your notebook. Indicate beside each number whether the corresponding statement is true (T) or false (F). If it is false, write a corrected version.

4. The slope of a straight line on a velocity–time graph indicates the average acceleration of the motion.

5. The acceleration of a 2.0-kg ball toward the ground is greater than the acceleration of a 1.0-kg ball.

6. When a ball is rising upward after you toss it vertically, the net force on the ball is equal to the force of gravity on the ball.

7. Static friction is always greater than kinetic friction.

8. To increase the force of a machine, the distance from the fulcrum must be decreased.

9. The torque on a rigid object is greatest when the applied force is parallel to the rigid object.

10. In any machine that experiences friction, the ideal mechanical advantage exceeds the actual mechanical advantage.

11. A machine with a high percent efficiency has a high amount of kinetic friction.

Write the numbers 12 to 21 in your notebook. Beside each number, write the letter corresponding to the best answer for the question.

12. If the frequency of a certain vibration is 100 Hz, then the period of the vibration is
    (a) 100 s   (b) 0.1 s
    (c) 0.01 s   (d) 0.001 s

13. A pop can is resting on a table. If Earth's force of gravity on the can is the action force, the reaction force is a(n)
    (a) upward normal force exerted by the table on the can
    (b) downward gravitational force exerted by Earth on the table
    (c) downward normal force exerted by the can on the table
    (d) upward force of gravity by the can on Earth

14. If your head is initially bent downward, and then you raise it slowly, your neck is acting as a
    (a) first-class lever
    (b) second-class lever
    (c) third-class lever
    (d) none of the above because the neck is not a lever

15. If a lever has an actual mechanical advantage of 0.5, then it can be
    (a) a second-class lever only
    (b) either a first-class lever or a third-class lever
    (c) a third-class lever only
    (d) a first-class lever only

16. If a pair of gears has 12 teeth where the load is attached and 6 teeth where the effort force is applied, then the
    (a) IMA is 2
    (b) IMA is 0.5
    (c) AMA is 2
    (d) AMA is 0.5

17. If the effort force is applied to the small axle of a wheel and axle,
    (a) the load force is decreased, but the load distance is increased
    (b) both the load force and the load distance are increased
    (c) the load force is increased, but the load distance is decreased
    (d) both the load force and the load distance are decreased

18. The IMAs of the two pulley systems in **Figure 2** are, respectively,
    (a) 4, 4   (b) 5, 4
    (c) 4, 5   (d) 5, 5

(a)   (b)

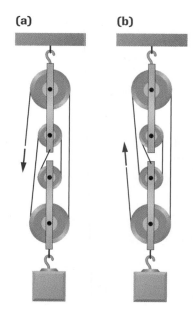

Figure 2

19. On a multi-speed bike, you choose a pedal gear with 45 teeth and a rear-wheel gear with 15 teeth. You are likely
    (a) riding up an incline into the wind
    (b) riding down an incline with a tail wind
    (c) moving slowly while pedalling quickly
    (d) any of the above because the choice of gears depends only on the strength of the rider's legs

20. If you were on one end of a seesaw that is between 2 m and 3 m long, the load torque caused by your body would be closest to
    (a) 10 N•m   (b) 100 N•m
    (c) 1000 N•m   (d) 10 000 N•m

21. The simple machines that make up the can opener (**Figure 3**) are
    (a) lever, wedge, inclined plane, gears
    (b) lever, wedge, wheel and axle, gears
    (c) pulley, wheel and axle, lever
    (d) gears, pulleys, lever

Figure 3

Understanding Concepts

1. At a certain instant, the period of rotation of the rotating ride in **Figure 1** is 1.3 s. Calculate the frequency of rotation at that instant.

Figure 1

2. Calculate the average speed of a migrating bird that travels 1100 km in a 24-h day.

3. Explain how a velocity–time graph can be used to determine (a) displacement and (b) acceleration.

4. A car travelling initially at 42 km/h on the on-ramp of an expressway accelerates in a straight line to 105 km/h in 26 s. Calculate the magnitude of the average acceleration in kilometres per hour per second.

5. A 1.2×10^4-kg truck, travelling initially at 21 m/s [S], brakes smoothly and comes to a stop in 14 s.
   (a) Calculate the average acceleration of the truck.
   (b) Calculate the net force required to bring the truck to a stop.
   (c) Draw an FBD of the truck as it is braking, and state the cause of the net force.

6. Explain, in terms of Newton's first law of motion, why we should wear seat belts.

7. Use physics principles to explain why people take very short steps on slippery surfaces.

8. Explain why the use of airbags is an application of acceleration and Newton's laws of motion.

9. Using the equation $F = ma$, where F is in newtons and a is in metres per second squared,
   (a) isolate mass, m
   (b) use unit cancellation to show that mass is in the unit of kilograms

10. State the net force acting on a 1.0-kg object when it is
    (a) at rest
    (b) moving at a constant velocity
    (c) falling freely with no air resistance

11. **Figure 2** shows an experiment in which a 0.75-kg glider moves without friction on an air track. The tension in the right-hand string is 9.8 N, and the tension in the left-hand string is 6.9 N.

Figure 2

    (a) Draw an FBD of the glider at the instant shown.
    (b) Calculate the glider's acceleration.

12. In a ramp experiment, a brick slides at a constant velocity down a wooden board. A student finds that the height of one end of the board above the other end is 0.80 m, and the horizontal "run" of the inclined plane is 2.3 m. Determine the coefficient of kinetic friction between the surfaces.

13. If you rotate the plastic wheels of a child's toy wagon, you notice fairly loud sounds. But if you rotate the front wheel of a bicycle, you notice very little sound. Give reasons for the difference. (Relate your answer to friction and percent efficiency.)

14. (a) Why do gymnasts and weightlifters put powder on their hands before performing their exercises?
    (b) What other athletes use the same type of powder?

15. A skater, initially at rest, pushes on the boards with an average force of 2.5×10^2 N [W]. The skater's velocity after 0.85 s is 3.5 m/s [E].
    (a) Determine the skater's average acceleration during this short time interval.
    (b) Assuming friction on the ice is zero, what is the net force on the skater? What causes this net force?
    (c) Determine the skater's mass.

16. Name the simple machine(s) in each biomechanical system in **Figure 3**. For any members of the lever family, include the class.

17. Draw a diagram of the type of nutcracker that is a lever.
    (a) What class of lever is the nutcracker?
    (b) Use estimated measurements to determine an approximate value of the IMA of the nutcracker.

(a)

(b)

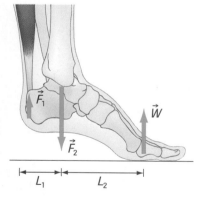

(c)

Figure 3

18. The distance between the effort force and fulcrum of a wheelbarrow is 1.5 m. An effort force of magnitude 1.3×10^2 N can support a load force of magnitude 3.4×10^2 N.
    (a) Calculate the distance between the load and the fulcrum.
    (b) Calculate the mass of the load.

19. To compare modern variations on the pulley with ancient ones, a student rigs up two simple pulleys to raise a 1.0-kg mass a distance of 75 cm. An effort force of magnitude 9.9 N is required to raise the mass using a nearly frictionless modern pulley. The effort force needed when a tree branch is used as a pulley has a magnitude of 14.4 N. In both cases, the effort distance moved is 75 cm.
    (a) Determine the IMA, AMA, and percent efficiency of the modern pulley.
    (b) Repeat (a) for the tree-branch pulley.
    (c) Which is more efficient? Explain why.

Applying Inquiry Skills

20. Explain why it is wise to perform more than one trial when using a motion sensor to observe the motion of a cart or a moving student.

21. Describe how you would demonstrate to a grade 8 student that the acceleration due to gravity does not depend on the mass of an object.

22. Use the velocity–time graph in **Figure 4** to determine the
    (a) instantaneous velocity at 0.40 s and 0.80 s
    (b) average acceleration between 0.0 s and 0.60 s
    (c) average acceleration between 0.60 s and 1.40 s

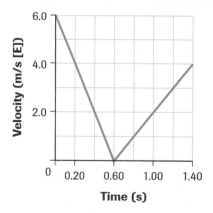

Figure 4

23. Draw a diagram to help you describe how you could determine the coefficient of kinetic friction between your calculator and the cover of this textbook. Try it, and compare your answer to the answers found by other students.

24. Design an experiment in which you use pliers, a wrench, or a similar tool to determine the torque required to loosen or tighten a nut.

Making Connections

25. Ramps are used to make entrances to buildings wheelchair accessible. Use physics principles to explain why these ramps must be kept free of snow and ice in the winter.

26. **Figure 5** shows a tool called a level.
    (a) Describe how the level can be used to draw lines on a wall that are either exactly horizontal or exactly vertical.
    (b) Research the use of lasers in modern levels. Describe what you discover.
    (c) Research how simple levels (e.g., water levels) were used before modern times. Describe what you find out.

Figure 5

27. Explain why the topic of acceleration has more applications today than in previous centuries. Consider acceleration in transportation, the space program, and amusement parks.

28. List advantages and disadvantages of using seat belts in school buses. Be sure to include physics principles in your answer.

29. **Figure 6** shows a cart carrying a battery-powered fan. The cart has a slot into which a card can be inserted.
    (a) If the card is absent, explain what will happen when the fan is turned on.
    (b) Explain what will happen if the card is inserted.
    (c) Some cartoons show the cartoon character blowing on a sail to make a sailboat move. Is this good physics? Relate your answer to your answer in (b).

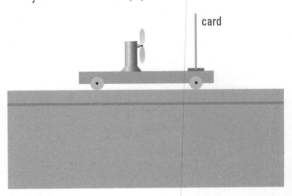

Figure 6

30. The soles of different types of modern curling shoes are designed to have different coefficients of friction, depending on the shoe's function on the ice.
    (a) State what you think the following terms mean: slider sole; gripper sole; half-sole slider; perimeter slider.
    (b) Would you expect all players to wear a "slider" on the same foot? Explain your answer.
    (c) Research curling shoes to check your answers in (a) and (b) above. What other design features of modern curling shoes help the players control friction?

 www.science.nelson.com

31. (a) List three advantages and three disadvantages of using machines in our society.
    (b) A trip to Mars is planned, to set up a colony for humans. You have been chosen to select the machines allowed on the trip. What are the first five machines you would choose? Justify your choices.

32. Each of the features in **Figure 7** is used on or by the astronauts on the *International Space Station.* Using the concepts you studied in Chapters 1 and 2, explain each one.

33. Some wrenches are designed so that when a preset torque has been applied, they rotate without tightening the nut any further.
    (a) When using this type of wrench, how can you tell that the preset maximum torque has been reached?
    (b) What is the advantage of this design?

34. Skateboarding applies principles of torques and forces. For example, the manoeuvre called the "ollie" is a jump that allows the skater to move through the air while the skateboard appears to be attached to the skater's feet (**Figure 8**). Research this manoeuvre on the Internet. Describe how the torques and forces cause the motion observed.

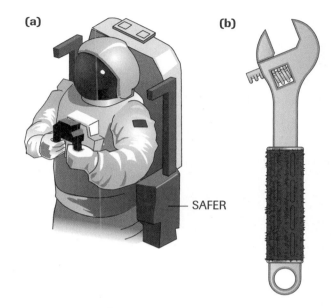
(a)

(b)

SAFER

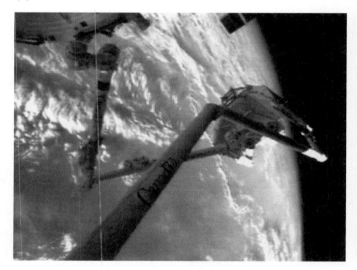
(c)

Figure 7
(a) The SAFER (Simplified Aid for EVA Rescue) system is a mini-jetpack, to be used in emergencies, if the astronaut moves away from the station.
(b) An adjustable wrench (the handle is large and rough)
(c) Canadarm2 is the space station's robotic arm. It can move end over end around the exterior by locking one end into one of many special fixtures, then detaching the other end and pivoting it forward. Some liken the movement to that of an inchworm. Its mass is 1800 kg, but it can move loads up to 100 000 kg.

Figure 8
Partway through an ollie manoeuvre

 www.science.nelson.com

Energy Transformations

Energy transformations can be observed in a light bulb, with electrical energy changing into light energy and thermal energy. But to produce the electrical energy in the first place, energy transformations must occur at a source. The photograph on these pages shows one example of a renewable energy source that Canada is taking advantage of: wind energy.

In this unit, you will study forms of energy, work, energy sources, energy transformations, efficiency, and the operation of energy-transforming devices. You will investigate devices involving energy sources, energy transformations, and energy losses and assess their efficiency. This will lead to the Unit Performance Task in which you can design and build an energy-transforming device. You will also analyze various energy-transforming technologies and evaluate sources of renewable energy.

Unit 2 is divided into two chapters. Chapter 3 deals with energy, work, and energy transformations, and Chapter 4 applies these concepts to the study of power and efficiency.

What you learn about the energy transformations you examine in this unit will be useful as you consider a variety of careers.

▶ Overall Expectations

In this unit, you will be able to

- demonstrate an understanding of work, forms of energy, energy sources, energy transformations, energy losses, efficiency, and the operation of energy-transforming devices

- construct devices that involve energy sources, energy transformations, and energy losses and assess their efficiency

- describe and analyze the operation of various technologies based on energy transfers and transformations, and evaluate the potential of energy-transformation technologies that use sources of renewable energy

- identify and describe science- and technology-based careers related to energy transformations

► **Prerequisites**

Concepts

- energy
- forms of energy
- energy transformations or changes
- conduction, convection, and radiation
- efficiency

Skills

- cost–benefit analysis
- draw and analyze graphs
- convert percent values to decimal numbers and vice versa
- apply appropriate safety precautions required in a laboratory environment
- write lab reports for investigations

Knowledge and Understanding

1. Refer to **Figure 1**.
   (a) List all the forms of energy you can see in the drawing.
   (b) List several examples from the drawing that show how energy changes from one form to another. (For example, in a TV, electrical energy changes into sound energy, light energy, and thermal energy.)

Figure 1

2. (a) Draw a model of the atom; label as many components as you can.
   (b) Which particle is responsible for the electric current in wire conductors?
   (c) Is the particle named in (b) positively charged, negatively charged, or neutral?

3. Heat can be transferred by conduction, convection, and radiation. Copy **Figure 2** into your notebook, and use symbols (e.g., particles and waves) to show how these three methods are illustrated in the drawing.

water

metal pot

Figure 2

Inquiry and Communication

4. You are asked to do a cost-benefit analysis of operating a radio using a battery as compared with using an electrical outlet. You are to include economic, social, and environmental impacts. Describe what is meant by a "cost-benefit analysis" using the radio example in your answer. Show that you understand the meaning of costs, benefits, economic impact, social impact, and environmental impact.

Math Skills

5. Sketch an x-y graph to show how the y variable depends on the x variable in each of the following cases:
   (a) If x doubles, y doubles (i.e., $y \propto x$, a direct, or linear, variation).
   (b) If x increases by a factor of 2, y increases by a factor of 2^2, or 4 (i.e., $y \propto x^2$, a quadratic variation).
   (c) If x increases by a factor of 2, y decreases by a factor of 2 (i.e., $y \propto \dfrac{1}{x}$, an inverse variation).

6. A machine has an efficiency of 75%. Express the efficiency as a decimal number.

Technical Skills and Safety

7. Assume you are performing an investigation to determine how the stretch of an extension spring depends on the mass hung vertically from it. What safety precautions would you follow for this investigation?

Making Connections

8. Describe what you know about the generation of electrical energy in the part of the province in which you live. (A diagram will help your description.)

Energy and Energy Transformations

In this chapter, you will be able to

- define and describe the concepts and units related to energy, forms of energy, and work
- describe and compare various energy transformations
- describe, with the aid of diagrams, the operation of energy-transforming devices
- design a device that uses at least four energy transformations to complete a task
- in an experiment, calculate the percent efficiency of a variety of springs
- describe and analyze examples of technologies based on various combinations of energy transformations
- describe the benefits and drawbacks of a source of renewable energy

Getting Started

This chapter is about energy, forms of energy, how energy is changed, or transformed, from one form to another, and sources of energy. When you turn the ignition key in a car, the gas is *ignited,* causing a small explosion. Energy from this small explosion is changed into other forms of energy. For example, one form moves the pistons and another we feel as heat. This is why engines that have been running for a while are hot. The motion energy of the pistons is transferred through other moving parts to the wheels. Car designers must take into consideration that energy does not disappear (**Figure 1**). It simply changes forms.

The operation of a car is just one example of energy changes that you will study in this chapter. You will also explore how energy relates to the work done by machines.

Figure 1
A car engine is designed to change the energy stored in fuel into energy of motion. The process also produces high temperatures.

Figure 2

💡 *REFLECT* on your learning ▼

1. Some springs are compression springs; others are extension springs. Explain the difference, giving two or three examples of each.

2. To get to a building entrance that is above street level, you can walk up a set of stairs or take a ramp (**Figure 2**). What is the advantage of each method in terms of (a) the force required and (b) the energy required?

3. What is the difference between heat and temperature?

4. What is meant by "the law of conservation of energy"?

▶ *TRY THIS* activity *Analyzing a Toy's Action*

The toy in **Figure 3** operates on two principles: changes in forms of energy and the force of friction. Of course, it only operates on these principles if you discover how to make it work. Good luck!

1. Obtain the toy from your teacher and look carefully at its design. Try to predict how you can make it move in its most interesting way. Test your prediction using trial and error until you master the operation of the toy.

2. Have your teacher inspect your attempted motion and tell you whether you are ready to proceed with the next step.

3. Try to explain the operation of the toy.

Figure 3
Can you figure out how to operate this toy?

Without light and other radiant energies from the Sun, life as we know it would not exist. With these energies, plants can grow, and the oceans and atmosphere can maintain temperature ranges that support life. A simple definition of **energy** is the capacity to do work or to accomplish a task. Thus, when you think of energy, think of the work or task involved. For example, the energy from the burning fuel in a car's engine allows the engine to do the work of moving the car.

energy the capacity to do work or to accomplish a task

Forms of Energy

The various forms of energy are classified as follows:

- Atoms combine to form many different kinds of molecules, involving various amounts of energy. In chemical reactions, new molecules are formed and *chemical potential energy* is released or absorbed (**Figure 1**).

- *Sound energy* is produced by vibrations; the energy travels by waves through a material to the receiver (**Figure 1**).

- Visible light and other forms of *radiant energy* belong to the electromagnetic spectrum (**Figure 2**). Components of the electromagnetic spectrum have characteristics of waves, such as wavelengths, frequencies, and energies; they travel in a vacuum at the speed of light (3.00×10^8 m/s). (The electromagnetic spectrum will be covered in more detail in section 10.1.)

- The nucleus of every atom has stored energy. This *nuclear energy* can be transformed in nuclear reactions called *fission* and *fusion* (**Figure 2**).

LEARNING TIP

Heat
Note that heat is not listed as a form of energy. Heat is actually a transfer of energy. The definition of heat and the methods of heat transfer are discussed in section 3.6.

Figure 1
When fireworks explode, they release chemical potential energy. Some of that energy is changed into sound energy. Besides sound, what other forms of energy are apparent in this photo?

Figure 2
The Sun emits radiant energy in the form of infrared radiation, visible light, ultraviolet radiation, and other forms. Nuclear fusion reactions in the Sun's interior release energy that is transformed into these types of energy.

- Electrons in an electric circuit possess *electrical energy*. The electrons can transfer energy to the components of the circuit (**Figure 3**).

- The atoms and molecules of a substance possess *thermal energy*. The more rapidly the atoms and molecules move, the greater their total thermal energy (**Figure 3**).

- A raised object has stored *gravitational potential energy* due to its position above some reference level (**Figure 4**).

- Every moving object has *kinetic energy*, or energy of motion (**Figure 4**).

- *Elastic potential energy* is stored in objects that are stretched or compressed (**Figure 4**).

Figure 3
Electrical energy delivered to the stove heats the water in the pot. Thermal energy in the boiling water transfers to the pasta and cooks it.

> ▶ *Practice*

Understanding Concepts

1. Give three examples of devices that use energy available today that your grandparents did not have when they were your age. Describe your dependence on those devices.

2. Name at least one form of energy associated with each object in italics:
   (a) A *bonfire* roasts a marshmallow.
   (b) A *baseball* smashes a window.
   (c) A *solar collector* heats water in a swimming pool.
   (d) A stretched *rubber band* is used to launch a rolled-up T-shirt toward the fans during intermission at a hockey game.
   (e) The *siren* of an ambulance warns of an emergency.

Figure 4
At the highest position above the trampoline, this athlete has the greatest amount of gravitational potential energy. This energy gradually changes to kinetic energy as her downward speed increases. Then that energy changes into elastic potential energy in the trampoline, which helps her bounce back up.

Energy Transformations

The forms of energy listed above are able to change from one to another; this change is called an **energy transformation**. For example, in a microwave oven, electrical energy transforms into radiant energy (microwaves), which then transforms into thermal energy, which cooks the food.

We can summarize these changes using an *energy-transformation equation*. For the microwave oven example, the equation is

electrical energy → radiant energy → thermal energy

The oven is an example of an **energy-transformation technology**, which is a device used to transform energy for a specific purpose. You can find these devices in homes (e.g., an oven or a clothes dryer), transportation (e.g., a car engine or an air bag), industry (e.g., a construction crane or an oil-well drill), and entertainment (e.g., a TV or a child's spinning top).

energy transformation the change of energy from one form to another

energy-transformation technology a device used to transform energy for a specific purpose

Typical Energy-Transformation Equations

Describe the energy transformations, and write the energy-transformation equations for the following:

(a) a battery-powered portable flashlight

(b) an electric drill with a rechargeable battery

Solution

(a) The battery has chemical potential energy that is transformed into electrical energy when the flashlight is turned on. The electrical energy is then transformed into radiant (or light) energy and thermal energy as the bulb becomes hot. The equation is

chemical potential energy → electrical energy → radiant energy + thermal energy

(b) Electrical energy is used to recharge the battery, which then has chemical potential energy. When the drill is turned on, the chemical potential energy in the battery is transformed into electrical energy. This electrical energy is transformed into kinetic energy as the drill bit rotates, as well as sound energy and some thermal energy as the bit and its surroundings warm up. The equation is

electrical energy → chemical potential energy → electrical energy → kinetic energy + sound energy + thermal energy

---

▶ **Practice**

Understanding Concepts

3. Write the energy-transformation equation for each of the following examples:
   (a) Fireworks explode.
   (b) An arrow is shot horizontally off a bow and flies through the air.
   (c) A paved driveway becomes hot on a clear, sunny day.
   (d) A camper raises an axe to chop a chunk of wood.
   (e) A lawnmower with a gasoline engine cuts a lawn.

4. Make up at least three energy transformations where sound energy is the final product. Each transformation must involve a different type of energy. Write the energy-transformation equation for each example.

▶ **TRY THIS** activity **Analyzing Energy-Transforming Technologies**

Choose one of the energy-transformation technologies listed below. Research the design and operation of the device, and create a poster or model to explain what you discover. Include the energy-transformation equation.

- shock absorber
- air bag in a car
- Mars Rover landing system
- wind-up toy
- solar-powered toy car (**Figure 5(a)**)

- demonstration wind-powered generator (**Figure 5(b)**)
- electric motor used to operate a toy train or similar device
- hand-held electric generator (**Figure 5(c)**)
- demonstration electric motor/generator (**Figure 5(d)**)
- mousetrap-powered toy car
- your own choice of an appropriate technology (e.g., take apart a broken toy and analyze the energy transformations involved)

(a)

(b)

Figure 5

(a) A solar-powered toy car

(b) A demonstration wind-powered generator

(c) A hand-held electric generator

(d) A demonstration motor/generator

(c)

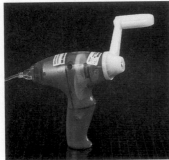

(d)

| SUMMARY | *Energy Forms and Transformations* |

- Energy, the capacity to do work or to accomplish a task, exists in different forms, for example, thermal energy and kinetic energy.

- In an energy transformation, energy changes from one form into another. The transformation can be described using an energy-transformation equation.

Figure 6

Section 3.1 Questions

Understanding Concepts

1. Rub your hands together vigorously. Write an energy-transformation equation to describe what happens.

2. Using an energy-transformation equation, show how energy is transformed for each of the following:
   (a) A hotdog is grilled on an outdoor barbecue.
   (b) A truck is accelerating on a highway.
   (c) A child jumps on a trampoline.
   (d) A portable CD player operates with a rechargeable battery.
   (e) An incandescent light bulb is switched on.
   (f) A waterfall (**Figure 6**). (Hint: Without the Sun, waterfalls would not be possible.)

3. Give an example (not yet given in this text) of each of the following energy transformations:
   (a) electrical energy → thermal energy
   (b) kinetic energy → sound energy
   (c) gravitational potential energy → elastic potential energy
   (d) chemical potential energy → thermal energy → kinetic energy
   (e) electrical energy → kinetic energy + thermal energy + sound energy

Making Connections

4. Describe at least one energy transformation that a person in each of the following careers would observe or experience regularly:
   (a) the fast-food industry
   (b) food preservation
   (c) heating and air conditioning
   (d) forensic science
   (e) firefighting

Figure 1
An off-road dump truck

work in physics, the amount of energy transferred to an object by a force applied over a distance

joule (J) the SI unit of work; it is also the SI unit of energy

The off-road dump truck in **Figure 1** can hold 325 t (325 000 kg) of gravel. To load the gravel into the truck, a large force must be applied to the gravel to raise it more than 9.0 m to the top of the truck. In this case, the force applied over the distance of 9.0 m does work to overcome the force of gravity.

The term *work* has a specific meaning in physics: **Work** is the energy transferred to an object by a force applied over a measured distance. As the force or the distance increases, the work also increases. This relationship is expressed in the equation for work:

$$W = F\Delta d$$

where W is the work done on the object,
F is the magnitude of the applied force in the direction of the displacement, and
Δd is the magnitude of the displacement.

Notice that this equation applies if the applied force and the displacement are *in the same direction*. (This allows us to omit the vector signs, even though force and displacement are vector quantities.) Work is a scalar quantity; it has magnitude but no direction.

Because force is measured in newtons (N) and displacement in metres (m), work is measured in newton-metres (N•m). The newton-metre is called the **joule** (J).

> ### SAMPLE problem 1
> #### Calculating Work
>
> A store employee exerts a horizontal applied force of magnitude 44 N on a set of carts (**Figure 2**). How much work is done by the employee when pulling the carts 15 m? Express the answer in joules and kilojoules.
>
> **Solution**
> $F = 44$ N
> $\Delta d = 15$ m
> $W = ?$
>
> $W = F\Delta d$
> $\quad = (44\ \text{N})(15\ \text{m})$
> $W = 6.6 \times 10^2$ J
>
> The work done by the employee in pulling the carts is 6.6×10^2 J, or 0.66 kJ.
>
>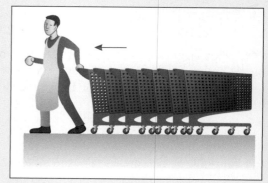
>
> **Figure 2**
> The force and displacement are parallel.

Negative Work

In Sample Problem 1, the applied force and the displacement are in the same direction, so the work done by the force is positive. If the force is opposite in direction to the displacement, however, the work done is negative. Consider a situation similar to that in the sample problem, but this time a second employee exerts a horizontal force of magnitude 14 N on the carts in the opposite direction to the 44-N force. Since the force is in the opposite direction to the displacement, the work done by the second employee on the carts is negative:

$$W = -F\Delta d$$
$$= -(14 \text{ N})(15 \text{ m})$$
$$W = -2.1 \times 10^2 \text{ J}$$

Now, the total work done by the two employees on the carts is the sum of the positive and negative values:

work done by the 44-N force:	$+6.6 \times 10^2$ J
work done by the 14-N force:	-2.1×10^2 J
total work done by the forces on the carts:	$+4.5 \times 10^2$ J

Negative work also occurs with kinetic friction because the force of kinetic friction always acts in a direction opposite to the direction of motion of the object, that is, opposite in direction to the displacement. The equation in this case is

$$W = -F_K \Delta d$$

where W is the work done by the force of kinetic friction,
  F_K is the magnitude of the force of kinetic friction, and
  Δd is the magnitude of the displacement.

▶ **SAMPLE** problem 2

Calculating Work by Kinetic Friction

A toboggan carrying two children (total mass of 85 kg) reaches its maximum speed at the bottom of a hill. It then glides to a stop in 21 m along a horizontal surface. The coefficient of kinetic friction between the toboggan and the snowy surface is 0.11.

(a) Draw a system diagram and an FBD of the toboggan when it is gliding on the horizontal surface.

(b) Calculate the magnitude of the force of kinetic friction acting on the toboggan.

(c) Calculate the work done by the force of kinetic friction on the toboggan.

LEARNING *TIP*

The Joule
The joule is named after James Prescott Joule (1818–89), an English physicist who studied heat and electrical energy. It is a derived SI unit, so it can be expressed in terms of the base units of metres, kilograms, and seconds. Recall that the newton, expressed in base units, is $N = (kg)(m/s^2)$. Thus, the joule is

$$J = N \cdot m$$
$$= (kg)\left(\frac{m}{s^2}\right)(m)$$
$$J = \frac{kg \cdot m^2}{s^2}$$

The newton-metre is called a joule when the force and displacement are parallel. When the force and displacement are perpendicular, as you studied in Chapter 2, the quantity calculated is called torque, but its unit remains the newton-metre.

Solution

(a) The system diagram and FBD are shown in **Figure 3**.

(a)

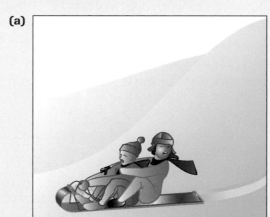

(b)

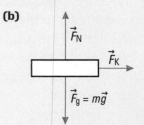

Figure 3
(a) The system diagram
(b) The FBD

(b) $m = 85$ kg

$g = 9.8$ N/kg

$\mu_K = 0.11$

$F_K = ?$

$$F_K = \mu_K F_N$$
$$= \mu_K F_g$$
$$= \mu_K mg$$
$$= (0.11)(85 \text{ kg})(9.8 \text{ N/kg})$$
$$F_K = 92 \text{ N}$$

The magnitude of the force of kinetic friction is 92 N.

(c) $\Delta d = 21$ m

$W = ?$

$$W = -F_K \Delta d$$
$$= -(92 \text{ N})(21 \text{ m})$$
$$W = -1.9 \times 10^3 \text{ J}$$

The work done by the force of kinetic friction on the toboggan is -1.9×10^3 J, or -1.9 kJ.

thermal energy the energy possessed by the atoms and molecules of a substance

What happens to the work done by friction? The answer is simple but important: The work changes to **thermal energy**, which is the energy produced as the result of the atoms and molecules of the two substances in contact rubbing together. In most cases, friction causes waste energy. This thermal energy is observed as an increase in temperature. (In question 1 of the section 3.1 questions, you felt the thermal effects of friction simply by rubbing your hands together vigorously.)

▶ **Practice**

Understanding Concepts

1. A farmer applies a constant horizontal force of magnitude 21 N on a wagon and moves it a horizontal distance of 3.2 m. Calculate the work done by the farmer on the wagon.

2. Rearrange the equation $W = F\Delta d$ to solve for (a) F and (b) Δd.

3. A truck does 3.2 kJ of work pulling horizontally on a car to move it 1.8 m horizontally in the direction of the force. Calculate the magnitude of the force.

4. A store clerk moves a 4.4-kg box of soap at a constant velocity along a shelf by pushing it with a horizontal force of magnitude 8.1 N. The clerk does 5.9 J of work on the box.
   (a) How far did the box move?
   (b) What was the magnitude of the force of kinetic friction during the push?
   (c) How much work was done by the force of kinetic friction on the box?

5. A student pushes a 0.85-kg textbook across a cafeteria table toward a friend. As soon as the student stops pushing, the book slows down, coming to a stop after moving 65 cm horizontally. The coefficient of kinetic friction between the surfaces in contact is 0.38.
   (a) Draw a system diagram and an FBD of the book as it slows down, and calculate the magnitude of all the forces in the diagram.
   (b) Calculate the work done on the book by the friction force between the book and the table.

6. (a) Calculate the area under the line on the graph in **Figure 4**.
   (b) State what that area represents. (Hint: Look at the units of the area calculation.)
   (c) Describe a situation that this graph might represent.

Applying Inquiry Skills

7. **Table 1** lists the results of an experiment to determine the work done by a student in pushing a desk a short distance across a floor using a horizontal force.
   (a) Plot a graph of the work done by the student on the desk (vertical axis) versus the distance moved, and draw the line of best fit.
   (b) Calculate the slope of the line of best fit.
   (c) State what the slope represents. (Hint: Consider the units of the slope calculation.)

Answers

1. 67 J

3. 1.8×10^3 N

4. (a) 0.73 m

5. (a) $F_g = 8.3$ N; $F_N = 8.3$ N; $F_K = 3.2$ N
   (b) −2.1 J

6. (a) 12 J

7. (b) 55 N

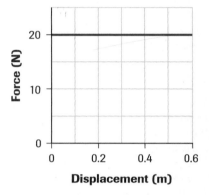

Figure 4
You can analyze the units on this force–displacement graph to determine what the area calculation represents. (Only magnitudes are considered.)

Table 1 Data for Question 7

Distance (m)	Work (J)
0	0
0.20	11
0.40	23
0.60	32
0.80	44

Work Done in Raising Objects

To lift an object to a higher position, an upward force must be applied against the downward force of gravity acting on the object. If the force applied and the displacement are both vertically upward and no acceleration occurs, the work done by the upward force is positive; it is calculated by $W = F\Delta d$. The force in this case is equal in magnitude to the weight of the object or the force of gravity on the object, $F = mg$.

Calculating Work in Raising Objects

A bag of groceries of mass 8.1 kg is raised vertically at a slow, constant velocity from the floor to a countertop, for a distance of 92 cm. Calculate

(a) the force needed to raise the bag of groceries at a constant velocity

(b) the work done on the bag of groceries by the upward force

Solution

(a) $m = 8.1$ kg

$g = 9.8$ N/kg

$F = ?$

$$F = mg$$
$$= (8.1 \text{ kg})(9.8 \text{ N/kg})$$
$$F = 79 \text{ N}$$

The force needed is 79 N.

(b) $\Delta d = 92$ cm $= 0.92$ m

$W = ?$

$$W = F\Delta d$$
$$= (79 \text{ N})(0.92 \text{ m})$$
$$W = 73 \text{ J}$$

The work done on the bag by the upward force is 73 J.

LEARNING TIP

Applying Newton's Laws
Newton's laws of motion help us understand Sample Problem 3. Any upward force greater in magnitude than 79 N would move the mass upward (Newton's second law). Once the mass starts moving upward, a force of magnitude 79 N will keep it moving at a constant velocity (Newton's first law). The force needed to raise an object without acceleration (i.e., at a constant velocity) is equal in magnitude to the force of gravity acting on the object.

▶ **Practice**

Understanding Concepts

8. A 150-g book is lifted from the floor to a shelf 2.0 m above. Calculate
   (a) the force needed to lift the book without acceleration
   (b) the work done by the force on the book to lift it up to the shelf

9. A world-champion weightlifter does 5.0×10^3 J of work in raising a weight from the floor to a height of 2.0 m. Calculate
   (a) the average force exerted to lift the weight
   (b) the mass of the weight

10. An electric forklift truck is capable of doing 4.0×10^5 J of work on a 4.5×10^3-kg load. To what height can the truck lift the load?

Answers

8. (a) 1.5 N
   (b) 2.9 J
9. (a) 2.5×10^3 N
   (b) 2.6×10^2 kg
10. 9.1 m

Zero Work

Sometimes an object experiences a force, a displacement, or both, yet no work is done on the object. For example, if you are holding a box on your shoulder, you are exerting an upward force on the box. But the box is not moving, so the displacement is zero, and the work done on the box, $W = F\Delta d$, is also zero.

In another example, a puck on an air table experiences negligible friction while sliding a certain displacement. There is no force parallel to the displacement, so the work done on the puck is zero. (Of course, initial work was done on the puck to start it sliding.)

Finally, consider the force exerted by the figure skater in **Figure 5**, who glides along the ice while holding his partner above his head. The partner experiences both a vertical force and a horizontal displacement. However, in this case the displacement is perpendicular (not parallel) to the force, so no work is done on the partner by the skater as they glide. Of course, work was done to lift the partner vertically to the height shown.

Figure 5
If the applied force and the displacement are perpendicular, no work is done by the applied force.

> ▶ **Practice**

Understanding Concepts

11. A student pushes against a large tree with a force of magnitude 250 N, but the tree does not move. How much work has the student done on the tree?

12. A 500-kg asteroid is travelling through space at 100 m/s. If it travels for 25 years at a constant velocity, how much work is done on the asteroid? (Disregard the force of gravity.)

13. A nurse holding a 3.0-kg newborn baby at a height of 1.2 m above the floor carries the baby 15 m at constant velocity along a hospital corridor. How much work has the nurse done on the baby?

14. Based on the above questions, write general conclusions about when the work done on an object is a positive quantity, a negative quantity, or zero.

Work and Springs

You discovered in Practice question 6 that the area under the line on a force–displacement graph (**Figure 4**, page 135) is equal to the work done. In that case, the force was constant. But what if the force changes as the displacement changes? This is the case when you stretch an elastic material, like a spring or a rubber band. (Recall what stretching a rubber band feels like: The farther you stretch it, the more difficult it becomes to stretch.) The area under the line of the force–displacement graph still yields the work done.

If we stretched a typical spring and then graphed the force applied to it against the stretch experienced by the spring, the line would resemble the graph in **Figure 6**. The area under the line is a triangle whose area, calculated using the equation $A = \dfrac{bh}{2}$, yields the work done by the force used to stretch the spring by an amount Δx. The slope of the line, found by applying the equation $\text{slope} = \dfrac{\text{rise}}{\text{run}}$ or $k = \dfrac{\Delta F}{\Delta x}$, represents a quantity that physicists call the *force constant* of the spring, k. The force constant represents the stiffness of the spring. In Activity 3.3, you can discover how some springs compare to the one represented in this graph.

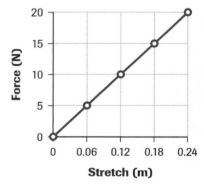

Figure 6
In this force–stretch graph for a spring, the shape of the area under the line is a triangle, and it represents the work done in stretching the spring. The slope of the line yields the force constant of the spring.

> ▶ **Practice**

Understanding Concepts

15. Calculate the work done in stretching the spring represented in the graph in **Figure 6** after it has stretched (a) 0.12 m and (b) 0.24 m.

Answers

15. (a) 0.60 J
    (b) 2.4 J

SUMMARY · Work

- Work is the energy transferred to an object by an applied force over a distance.
- If the force and displacement are in the same direction, the work done by the force, $W = F\Delta d$, is a positive value. If the force and displacement are in opposite directions, the work done is a negative value. If the force and displacement are perpendicular to each other, the work done by the force is zero.
- Work is a scalar quantity measured in joules (J).
- The work done by kinetic friction on a moving object is negative and is changed into thermal energy; $W = -F_K\Delta d$.
- The area under the line on a force–displacement graph equals the work done by the force. For a constant force, the area is a rectangle; for a force that varies directly with the stretch, as in a typical spring, the area is a triangle.

► Section 3.2 Questions

Understanding Concepts

1. An average horizontal force of magnitude 32 N is exerted on a chair on the floor. If the chair moves 7.8 m along the floor, how much work does the force do on the chair?

2. An elevator lifts you upward without acceleration a distance of 36 m. How much work does the elevator do on you against the force of gravity to lift you this far? (You need to know your own mass.)

3. An off-road dump truck, like the one shown in **Figure 1**, page 132, can hold 325 t of gravel (1 t = 1000 kg). How much work must be done on a new load of gravel to raise it an average of 9.2 m into the truck? Express your answer in joules and megajoules.

4. A camper does 7.4×10^2 J of work in lifting a pail of water 3.4 m vertically up a well at a constant speed.
   (a) Calculate the force exerted by the camper on the pail of water.
   (b) Calculate the mass of the water in the pail.

5. The driver of a 1300-kg car suddenly slams on the brakes, causing the car to skid forward on the road. The coefficient of kinetic friction between the tires and the road is 0.97, and the car comes to a stop after travelling 27 m horizontally. Calculate the work done by the force of friction on the car during the skid.

6. For the equation $W = F\Delta d$, describe when the equation yields
   (a) a positive value of work
   (b) a negative value of work
   (c) a zero value of work

7. A puck of mass 0.16 kg is sliding along the ice when it reaches a rough section where the coefficient of kinetic friction is 0.37. If −2.8 J of work is done on the puck to bring it to rest, how far does the puck slide before stopping?

8. Explain why the equation $W = F\Delta d$ is not used to determine the work done by an applied force to stretch a spring.

Applying Inquiry Skills

9. The graph in **Figure 7** was generated by a computer connected to a force sensor that collected data several times per second as a wooden block was pulled with a horizontal force across a desk.
   (a) Estimate the work done by the force on the block. Show your calculations.
   (b) Describe the sources of error when using a force sensor in this type of investigation.

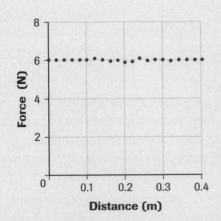

Figure 7
For question 9. Only magnitudes are considered.

10. **Table 2** gives the magnitudes of the data from an experiment in which weights are hung from a vertical spring.
    (a) Plot a force–stretch graph of the data, and draw the line of best fit.
    (b) From the graph, determine the force constant of the spring.
    (c) From the graph, calculate the work done in stretching the spring 0.20 m.
    (d) From the graph, calculate the work done in stretching the spring 0.40 m.
    (e) How much work was done in stretching the spring from 0.20 m to 0.40 m?
    (f) Explain why more work was done in stretching the spring from 0.20 m to 0.40 m than from 0.00 m to 0.20 m.

Table 2   Data for Question 10

Stretch (m)	Force (N)
0	0
0.10	1.6
0.20	3.0
0.30	4.4
0.40	6.0

3.3 Activity

Comparing Springs

In this activity, you will compare the stiffnesses and determine the force constants of different springs. If possible, label the springs and save them, as well as your data, for Investigation 3.5.

Materials

For each group of three or four students:
support stand with clamp to hold the springs
clamp to secure the support stand to the lab bench
string
3 extension springs of different stiffnesses
mass set (masses from 50 g to 200 g for a sensitive
   spring; 500 g to 2000 g for a stiff spring)
metre stick

For each student:
graph paper   safety goggles

 **Do not allow the springs to overstretch.
Make sure that the support is secure so that
a mass cannot tip it over. Use the clamp.
Wear safety goggles.**

Procedure

1. Set up a table like **Table 1**. Complete column 2 by determining the magnitude of the weight of each mass.

2. Put on the safety goggles. Obtain the three springs, and pull gently on each one to decide which one is least stiff. Using string, suspend this spring from the clamp attached to the stand (**Figure 1**). Measure the initial length of the spring when no mass is attached. Then attach the first mass and determine the final length of the spring. Repeat by suspending more masses from the spring. Record the values in your data table.

3. Repeat steps 1 and 2 with the other springs.

Analysis

(a) Make the calculations to determine the values of Δx in metres. Enter the values in your data table.

(b) Plot, on a single graph, the magnitude of the force (F) applied to the spring versus the stretch (Δx) of the spring, for each spring. In each case, draw the line of best fit, starting at the origin of the graph.

(c) Calculate the slopes of the lines on your graph in (b) and compare them. What does each slope represent? Enter the values in your data table.

(d) Calculate and compare the total area under each line on your graph in (b). What does each area represent? Enter the values in your data table.

(e) Use your graph to determine the stretch of each spring under an applied force of magnitude 1.5 N.

(f) Use your graph to determine the work done in stretching each spring from zero to 0.030 m.

Evaluation

(g) Describe the major sources of error in performing this activity.

Synthesis

(h) How would a person who designs ropes for bungee jumping apply the principles discovered in this activity?

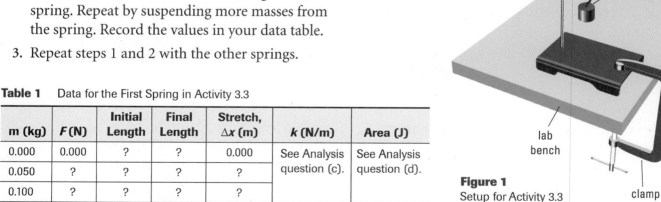

Figure 1
Setup for Activity 3.3

Table 1   Data for the First Spring in Activity 3.3

m (kg)	F (N)	Initial Length	Final Length	Stretch, Δx (m)	k (N/m)	Area (J)
0.000	0.000	?	?	0.000	See Analysis question (c).	See Analysis question (d).
0.050	?	?	?	?		
0.100	?	?	?	?		

Old buildings are often torn down to make way for new ones. One way is through chemical explosions using dynamite. Another much slower way is to use a wrecking ball (**Figure 1**). What energy transformations allow a relatively small wrecking ball to destroy a building?

To raise the wrecking ball, work is done by a machine on the ball; the ball gains what is called *potential energy* as it rises. The energy is called *potential* because it is stored and not used until later, in this case, when the ball is released. This potential energy arises because the force of gravity pulls downward on the ball. The type of energy that an object possesses because of its position above some level is called **gravitational potential energy**, E_g. This potential energy can be used to do work on an object at a lower level.

When the ball is released, it falls. As the ball falls, its gravitational potential energy is gradually transformed into *kinetic energy* as its speed increases. Energy due to the motion of an object is called **kinetic energy**, E_K. ("Kinetic," like the word "kinematics," stems from the Greek word *kinema*, which means motion.)

Gravitational Potential Energy

Suppose you are erecting a tent and using a hammer to pound the tent pegs into the ground, as in **Figure 2**. To lift the hammer a height Δh, you must transfer energy to it, based on the equation for work, $W = F\Delta d$. Here, F is the magnitude of the force required to lift the hammer from the ground without acceleration; it is equal in magnitude to the hammer's weight, which is mg. The transferred energy, or work, equals the hammer's gravitational potential energy above a reference level, such as the ground. That is, $E_g = F\Delta h$, where E_g is the gravitational potential energy of the hammer raised to a height Δh above the original level. Since $F = mg$, we can now write the common equation for gravitational potential energy:

$E_g = F\Delta h = mg\Delta h$, where $g = 9.8$ N/kg

In SI, energy is measured in joules, mass in kilograms, and height (or displacement) in metres.

What we usually need to know is the potential energy *relative* to a particular **reference level**, the level to which the object may fall, such as the ground. Then the Δh in the potential energy equation is the height h of the object above the reference level. Thus, the equation for the gravitational potential energy of an object relative to a reference level is

$E_g = mgh$

When answering questions about relative potential energy, it is important to state the reference level. For example, when a tent peg is hammered, the hammer has a greater potential energy relative to the ground than it has relative to the top of the peg.

Figure 1
Several principles of mechanics are applied in the demolition of this structure.

gravitational potential energy the type of energy possessed by an object because of its position above a reference level; symbol E_g; it is a scalar quantity, measured in joules (J)

kinetic energy the type of energy due to an object's motion; symbol E_K; it is a scalar quantity, measured in joules (J)

reference level the level to which a raised object may fall

Figure 2

> **SAMPLE** problem **1**

Calculating Gravitational Potential Energy

In the sport of pole vaulting, the jumper's point of mass concentration, called the centre of mass, must clear the pole. Assume that a 59-kg jumper must raise the centre of mass from 1.1 m off the ground to 4.6 m off the ground. What is the jumper's gravitational potential energy at the top of the bar relative to the point at which the jumper started to jump?

Solution

The height of the jumper's centre of mass above the reference level is
4.6 m − 1.1 m = 3.5 m.

$$m = 59 \text{ kg}$$
$$g = 9.8 \text{ N/kg}$$
$$E_g = ?$$

$$E_g = mgh$$
$$= (59 \text{ kg})(9.8 \text{ N/kg})(3.5 \text{ m})$$
$$E_g = 2.0 \times 10^3 \text{ J}$$

The jumper's gravitational potential energy relative to the lower position is 2.0×10^3 J.

> **Practice**

Understanding Concepts

Answers

1. (a) 0.0 J
   (b) 2.8 J
4. 1.7×10^3 m
5. (a) 1.5×10^4 kg
6. (a) 1.2×10^2 kg
   (b) 3.2 N/kg or 3.2 m/s²

1. A 0.45-kg book is resting on a desktop 0.64 m high. Calculate the book's gravitational potential energy relative to (a) the desktop and (b) the floor.

2. Estimate your own gravitational potential energy relative to the lowest floor in your school when you are standing at the top of the stairs of the highest floor.

3. Rearrange the equation $E_g = mgh$ to solve for (a) m, (b) g, and (c) h.

4. The elevation at the base of a ski hill is 350 m above sea level. A ski lift raises a skier (total mass is 72 kg, including equipment) to the top of the hill. If the skier's gravitational potential energy relative to the base of the hill is now 9.2×10^5 J, what is the elevation at the top of the hill?

5. The spiral shaft in a grain auger raises grain from a farmer's truck into a storage bin (**Figure 3**). Assume that the auger does 6.2×10^5 J of work on a certain amount of grain to raise it 4.2 m from the truck to the top of the bin.
   (a) What is the total mass of the grain moved? Ignore friction.
   (b) What simple machine is used in the auger?

6. An astronaut, with a total weight on Earth of 1.2×10^3 N, including the space suit, is about to jump down from a space capsule that has just landed safely on planet X. The drop to the surface of planet X is 2.8 m, and the astronaut's gravitational potential energy relative to the surface is 1.1×10^3 J.
   (a) Calculate the mass of the astronaut wearing the space suit.
   (b) Calculate the magnitude of the acceleration due to gravity (g) on planet X.

Figure 3
A grain auger

Kinetic Energy

An object's kinetic energy depends on two factors: the object's mass and its speed. The kinetic energy increases in direct proportion to the mass, and it increases in proportion to the square of the speed. The equation relating these factors is

$$E_K = \frac{mv^2}{2}$$

where E_K is the kinetic energy measured in joules (J),
m is the mass in kilograms (kg), and
v is the speed in metres per second (m/s).

▶ **SAMPLE** problem **2**
Calculating Kinetic Energy

Find the kinetic energy of a 6.0-kg bowling ball rolling at 5.0 m/s.

Solution
$m = 6.0$ kg
$v = 5.0$ m/s
$E_K = ?$

$$E_K = \frac{mv^2}{2}$$

$$= \frac{(6.0 \text{ kg})(5.0 \text{ m/s})^2}{2}$$

$$E_K = 75 \text{ kg} \cdot \frac{m^2}{s^2} = 75 \text{ J}$$

The kinetic energy of the bowling ball is 75 J.

Figure 4
The ostrich has powerful legs that allow it to run fast.

▶ **Practice**

Understanding Concepts

7. Calculate the kinetic energy in each of the following:
   (a) During a shot put, a 7.2-kg shot leaves an athlete's hand at a speed of 12 m/s.
   (b) A 140-kg ostrich (**Figure 4**) is running at 14 m/s.

8. Using the equation $E_K = \frac{mv^2}{2}$, write an equation to solve for (a) m and (b) v.

9. A softball travelling at 34 m/s has a kinetic energy of 98 J. Calculate its mass.

10. A 97-g cup falls from a kitchen shelf and shatters on the ceramic tile floor. Assume that the maximum kinetic energy obtained by the cup is 2.6 J and that air resistance is negligible.
    (a) Calculate the cup's maximum speed.
    (b) What happened to the 2.6 J of kinetic energy after the crash?

Answers

7. (a) 5.2×10^2 J
   (b) 1.4×10^4 J

9. 0.17 kg

10. (a) 7.3 m/s

Answers

12. (c) 16 times
    (d) 25 times
13. (a) $E_K = F\Delta d$
    (b) $E_K = ma\Delta d$
    (c) $E_K = m\left(\dfrac{v_f - v_i}{\Delta t}\right)\Delta d$
    (d) $E_K = m(v_f - v_i)\dfrac{(v_f + v_i)}{2}$
    (e) $E_K = \dfrac{mv_f^2}{2}$

LEARNING TIP

Square Root Solutions
When the equation for kinetic energy is rewritten to solve for the speed, the solution is a square root, which can be either positive of negative; for example,
$\sqrt{4.0 \dfrac{m^2}{s^2}} = \pm 2.0 \dfrac{m}{s}$. Each situation should be examined to decide whether the solution is positive or negative. In most cases in this text, the solution is positive.

Applying Inquiry Skills

11. Create an experiment in which you measure your own speed when you are running as fast as you can—safely—for 50 m to 100 m. Use your speed and mass to determine your own kinetic energy.

12. A 1.0-kg object accelerates from a speed of 0.0 m/s to a speed of 5.0 m/s.
    (a) Determine the object's kinetic energy at each of these speeds: 0.0 m/s, 1.0 m/s, 2.0 m/s, 3.0 m/s, 4.0 m/s, and 5.0 m/s. Organize the values of kinetic energies and speeds in a table.
    (b) Draw a graph of the kinetic energy (vertical axis) as a function of the speed.
    (c) By what factor does the kinetic energy increase when the speed increases by a factor of 4 (from 1.0 m/s to 4.0 m/s)?
    (d) By what factor does the kinetic energy increase when the speed increases by a factor of 5 (from 1.0 m/s to 5.0 m/s)?
    (e) What is the increase in kinetic energy from a speed of 1.0 m/s to a speed of 2.0 m/s? Compare that to the increase in kinetic energy when the speed goes from 4.0 m/s to 5.0 m/s.

Extension

13. You can derive the equation for kinetic energy yourself from concepts you learned in Chapters 1 and 2.
    (a) Start with the equation for work, $W = F\Delta d$; substitute the symbol for kinetic energy, E_K, for W.
    (b) From Newton's second law of motion, $F = ma$; replace F with ma.
    (c) From Chapter 1, $a = \dfrac{v_f - v_i}{\Delta t}$. Drop in the new value for a.
    (d) Also, from Chapter 1, $v_{av} = \dfrac{\Delta d}{\Delta t}$, or $\dfrac{v_f + v_i}{2} = \dfrac{\Delta d}{\Delta t}$; replace $\dfrac{\Delta d}{\Delta t}$ with the new value.
    (e) Consider an object starting from rest; $v_i = 0$. Drop in this value and simplify the final equation.

Mechanical Energy

The sum of gravitational potential energy and kinetic energy is called **mechanical energy**. For example, assume that a ball, initially at rest, is held above the floor. Assume its initial gravitational potential energy is 1.00 J. This value remains constant as the ball falls. Ignoring the effects of air resistance,

mechanical energy the sum of gravitational potential energy and kinetic energy; it is a scalar quantity, measured in joules (J)

	$E_{mechanical}$ =	E_K +	E_g
at the top position:	1.00 J =	0.00 J +	1.00 J
one-quarter of the way to the floor:	1.00 J =	0.25 J +	0.75 J
halfway to the floor:	1.00 J =	0.50 J +	0.50 J
three-quarters of the way to the floor:	1.00 J =	0.75 J +	0.25 J
just before landing	1.00 J =	1.00 J +	0.00 J

The example of the wrecking ball in **Figure 1**, page 141, illustrates a useful application of mechanical energy. An object is raised to a position above a reference level. When released, the force of gravity causes the object to accelerate. The object gains speed and thus kinetic energy, allowing it to crash into the wall and do work on the wall—demolishing it. Another example is a roller coaster at its highest position. At that point, its gravitational potential energy is maximum. The force of gravity causes the coaster to accelerate downward, giving it enough speed and, thus, kinetic energy, to travel around the track.

Figure 5 shows two more applications of mechanical energy. In **Figure 5(a)**, the hammer of a pile driver is about to be lifted by a motor high above the pile (the long column). This lifting does work on the hammer. The hammer will then have gravitational potential energy, which changes into kinetic energy as it falls. This kinetic energy then does work on the pile, driving the pile into the ground. The pile will act as a support for a high-rise building.

In **Figure 5(b)**, water stored in a dammed river has gravitational potential energy relative to the base of the dam. At hydroelectric generating stations, this gravitational potential energy is transformed into the kinetic energy of the falling water at the bottom of the falls. The water is directed through turbines that are connected to electric generators. The generators rotate and transform their kinetic energy into electrical energy.

> ▶ **Practice**

Understanding Concepts

14. Write the energy-transformation equation for each of the following situations:
    (a) A fuel-powered engine pumps water from a lake to a tower at a higher level.
    (b) A hammer is used to pound in a tent peg. (Start with the energy stored in the food eaten by the camper.)

SUMMARY   *Gravitational Potential Energy and Kinetic Energy*

- Gravitational potential energy, which is energy possessed by an object because of its position above a reference level, is given by the equation $E_g = mgh$.
- Kinetic energy, which is energy of motion, is given by the equation $E_K = \dfrac{mv^2}{2}$.
- Mechanical energy is the sum of an object's gravitational potential energy and kinetic energy.

DID YOU *KNOW* ?

Smart Birds Apply Mechanical Energy

Some animals take advantage of mechanical energy. One example is the bearded vulture (or lammergeier), the largest of all vultures. This bird, found in South Africa, can digest bones. Often its food consists of bones picked clean by other animals. When a bone is too large to crack, the vulture carries it to a great height and drops it onto a rock, so that the bone shatters. Then the bird circles down to scoop up pieces of bone and marrow.

GO www.science.nelson.com

(a)

(b)

Figure 5
(a) A pile driver
(b) Damming a river to produce electrical energy

Understanding Concepts

1. Explain why a roller coaster is called a "gravity ride."

2. In April 1981, Arnold Boldt of Saskatchewan set a world high-jump record for disabled athletes in Rome, Italy, jumping to a height of 2.04 m. (At the age of three, following an accident, Arnold had his right leg amputated above the knee.) Calculate Arnold's gravitational potential energy relative to the ground. (Assume that his mass was 68 kg at the time of the jump.) Express your answer in joules and kilojoules.

Figure 6
Arnold Boldt's World and Paralympic High Jump record remains unbroken.

3. A hockey puck has a gravitational potential energy of 2.3 J when it is held by a referee at a height of 1.4 m above the rink surface. Calculate the mass of the puck.

4. A 636-g basketball has a gravitational potential energy of 19 J at the basket. How high is the ball from the floor?

5. Calculate your own kinetic energy when you are running at a speed of 5.5 m/s.

6. At what speed would a 1200-kg car be moving to have a kinetic energy of 2.0×10^5 J?

Applying Inquiry Skills

7. Use your graphing skills to show the relationship between each set of variables listed below. (You can review graphing relationships in Appendix A1 at the back of the text.)
   (a) gravitational potential energy; acceleration due to gravity (*g*)
   (b) kinetic energy; mass of the object (at a constant speed)
   (c) kinetic energy; speed of the object (constant mass)

Making Connections

8. (a) By what factor does the kinetic energy of a car increase when its speed doubles? triples?
   (b) What happens to the kinetic energy if the car crashes?
   (c) As a driver education tool, make up a cartoon that relates speed, higher energy, and the extent of collision damage.

3.5 Investigation

Inquiry Skills

○ Questioning   ● Conducting   ● Evaluating
● Predicting   ● Recording   ● Communicating
○ Planning   ● Analyzing   ● Synthesizing

Energy in Springs

If a rubber band is stretched, the work done on it becomes elastic potential energy ($E_{elastic}$). If the rubber band is then launched vertically upward, it has maximum kinetic energy (E_K) just after the launch. As it rises, it slows down, losing kinetic energy and gaining gravitational potential energy (E_g). At its highest position (h) above the launch level, $E_g = mgh$. Using symbols, the energy-transformation equation is

$$W \rightarrow E_{elastic} \rightarrow E_K + E_g$$

An extension spring acts in the same way as a rubber band. In this investigation, you will determine the elastic potential energy and gravitational potential energy of a spring launched vertically upward from a reference level (**Figure 1**).

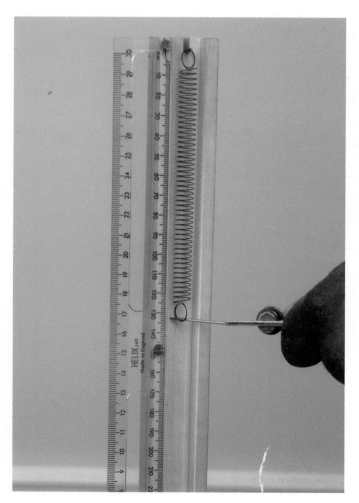

Figure 1
Launching an extension spring vertically upward

You will also find the efficiency of the launched spring using the following ratio, which is similar to the percent efficiency ratio from Chapter 2:

$$\% \text{ eff} = \frac{E_g}{E_{elastic}} \times 100\%,$$

where E_g is the gravitational potential energy of the spring at the top of the motion, and

$E_{elastic}$ is the elastic potential energy of the spring at the launch position, which is equal to the work done in stretching the spring.

If you saved the springs and data from Activity 3.3, you can use the data for Procedure steps 1 and 2 and Analysis (b).

For this investigation, you need a launch pad to launch the springs vertically upward. If a launch pad is not available, you can design, build, and test one with your teacher's guidance.

Question

When a spring is launched vertically upward, is all of its elastic potential energy transformed into gravitational potential energy at the top of its flight?

Prediction

(a) Predict an answer to the Question, giving reasons.

Materials

For each group of three or four students:

For Part A

3 extension springs of different stiffnesses
metric ruler
safety goggles
support stand with clamp on which to support the springs
clamp to secure the support stand to the lab bench
string
mass set (with masses from 50 g to 200 g for a sensitive spring, or 500 g to 2000 g for a stiff spring)

For Parts B and C
launch pad (or the materials for building the launch pad)
metre stick
clamp to secure the launch pad

For each student
graph paper

 Wear safety goggles.
Do not allow the springs to become overstretched.

Procedure

Part A

1. Put on the safety goggles. Using the appropriate mass set and the metric ruler, make the measurements and do the analysis needed to determine the work done in stretching each spring by an amount that does not overstretch the spring. (Refer to Activity 3.3 to review how to do this. If you saved the springs and data from that activity, you can use them here.) Set up a data table to summarize the data.

2. Repeat step 1 using the other springs.

Part B

3. Test the launch pad to be sure it is in safe working order. The launch pad should allow the springs to be launched vertically upward by a measurable amount. Make sure that the support is secure so that an applied force cannot tip it over. Use the clamp. Also, be sure the design allows the spring to be caught after it begins falling downward.

Part C

4. With your safety goggles on, practise launching the most sensitive spring from the launch pad vertically upward. Be sure the spring does not travel more than halfway to the ceiling. Design a way to stretch the spring the exact amount of the final stretch you used in step 1. After the spring is launched, determine the height it reaches above the release point. Repeat the measurement of the change of position at least three or four times, and determine an average value of this vertical displacement.

5. Repeat step 4 using the other springs.

Analysis

(b) For each of the springs, plot, on a single graph, the magnitude of the force (F) applied to the spring versus the stretch (Δx) of the spring. In each case, draw the line of best fit, starting at the origin of the graph. Calculate and compare the total areas under the lines on your graph. What does each area represent? Enter the values in your data table.

(c) Calculate the maximum gravitational potential energy for each spring after its launch.

(d) For each spring, compare the value in (c) to the value found in (b). Account for any differences.

(e) Calculate the percent efficiency of each spring.

(f) Complete a formal report of your investigation, including a summary of your answer to the original Question.

Evaluation

(g) Describe any difficulties you had with the measurements in this investigation. What did you do to minimize those difficulties?

(h) How good was your prediction?

Synthesis

(i) What measurements and calculations would you need to make in order to determine the speed with which the spring left the launch pad? Explain your reasoning.

Thermal energy and heat play significant roles in our lives: Thermostats control furnaces, winds are generated by the uneven heating of Earth's surface and atmosphere, and the weather influences the clothes we wear. In addition, much of the energy we consume when we eat is eventually transformed into thermal energy.

Thermal energy and *heat* are not exactly the same thing, and *temperature* is different from both. Recall that thermal energy is the total kinetic energy and potential energy of the atoms or molecules of a substance. It depends on the mass, temperature, nature, and state of the substance. **Heat** is a measure of the energy *transferred* from a warm body to a cooler one; it can be thought of as a process. **Temperature** is a measure of the average kinetic energy of the atoms or molecules of a substance. It increases if the motion of the particles increases.

Consider, for example, two samples of water: 100 g at 50 °C and 500 g at 50 °C (**Figure 1(a)**). The samples have the same temperature, but the bigger sample contains more thermal energy simply because there's more of it. If these samples were mixed, no heat would transfer between them because they are at the same temperature.

Now consider two more samples of water: 500 g at 50 °C and 500 g at 90 °C (**Figure 1(b)**). The masses are the same but the warmer sample has more thermal energy because the motion—that is, the average kinetic energy—of the molecules is greater at a higher temperature. If these two samples were mixed, heat would transfer from the 90 °C sample to the 50 °C sample.

heat a measure of the energy transferred from a warm body to a cooler body because of a difference in temperature

temperature a measure of the average kinetic energy of the atoms or molecules of a substance

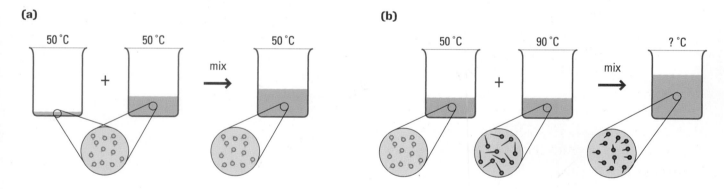

(a)   **(b)**

Figure 1
(a) When two samples have the same average kinetic energy, the temperatures are the same, and no heat will flow if the samples are mixed.
(b) When the samples have the same mass but different temperatures, the one with the higher temperature has both higher average kinetic energy and higher thermal energy. If they are mixed, heat will flow from the warmer sample to the cooler sample.

Methods of Heat Transfer

There are three ways that heat can transfer from a warmer body to a cooler body: *conduction, convection,* and *radiation.*

The process by which the collision of atoms and electrons transfers heat through a material or between two materials in contact is called **conduction**. For example, a metal rod comprises billions of vibrating atoms. When one end of the rod is heated (**Figure 2(a)**), the atoms at the heated end gain kinetic energy and vibrate more quickly. They collide with adjacent atoms, causing them to vibrate more quickly (**Figure 2(b)**). This action continues along the rod from the hot end to the cool end. (The action resembles a row of standing dominoes falling after the first domino is knocked over.) Metals are the best heat conductors because they have electrons that vibrate more freely than those of other substances. (Metals are also good electric conductors for the same reason.) Conduction occurs much less in solids such as concrete, brick, and glass and only slightly in liquids and gases.

conduction the process by which the collision of atoms and electrons transfers heat through a material or between two materials in contact

(a)

(b)

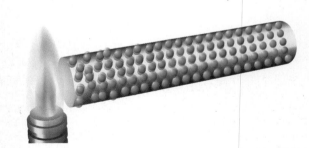

Figure 2
(a) The metal rod must be hot before it can be bent into the desired shape. Heat from the fire is transferred through the metal by conduction.
(b) Heat is conducted through the metal rod by the collisions of atoms.

The second method of heat transfer, **convection**, is through a circulating path of fluid particles. (The fluid can be a gas, such as air, or a liquid, such as water.) The circulating path is called a *convection current*. The particles of the fluid actually move, carrying energy with them. Consider, for example, a room with an electric heater (without a fan) on one wall (**Figure 3**). When the heater is turned on, the air particles near the heater gain thermal energy and move faster, colliding with each other. As they collide, they spread out. As they spread out, the heated air becomes less dense than the surrounding cooler air. The warmer air rises and is replaced with the denser, cooler air, forming the convection current. This current distributes energy throughout the room.

Both conduction and convection require particles to transfer energy. However, we know that energy can also transfer through a vacuum because we receive energy from the Sun. This is a third method of heat transfer, one that requires no particles. **Radiation** is the process in which energy is transferred by means of electromagnetic waves. Examples of these waves are visible light, microwaves, radio waves, radar, X rays, and infrared radiation.

Infrared radiation is the dominant form of radiation from objects at everyday temperatures. (See the drawing of the electromagnetic spectrum in section 10.1.) Heat emitted from an object in the form of infrared radiation can be detected by an infrared camera and recorded on an image called a *thermograph* (**Figure 4**). For example, a cancerous tumour is slightly warmer than its surroundings, so it shows as a shaded region in a thermograph. Some new cars are equipped with infrared detectors that allow night drivers to "see" objects such as a deer or a jogger about four times farther away than their headlights allow.

convection the process of transferring heat by a circulating path of fluid particles

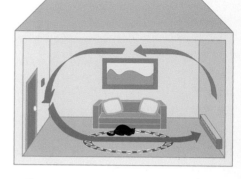

Figure 3
The thermal energy from the electric heater starts a convection current in the room.

radiation the process in which energy is transferred by means of electromagnetic waves

Figure 4
This is an infrared photograph of a farmhouse in Ireland. The darkest colours indicate the highest temperatures.

▶ **TRY THIS** activity   *Effects of Heat Transfer*

Predict what you will observe if the demonstrations illustrated in **Figure 5** are set up. Observe the demonstrations, and explain what happens by applying the concepts of conduction, convection, and/or radiation. (To explain the action of the radiometer, you should also apply at least one of Newton's laws of motion.)

 Follow strict safety procedures whenever an open flame is used.

(a)

(b)

Figure 5
(a) In this apparatus, a candle flame heats air at one location while smoke is introduced at another location.
(b) This device, called a *radiometer*, has much of the air evacuated from inside. It operates in the presence of a bright light source or a heat lamp.

Understanding Concepts

4. Explain each of the following in terms of the concepts in this section:
    (a) Curling irons and clothes irons have plastic handles, even if other parts are metal.
    (b) High-quality cooking pots are often made with thick copper bottoms.
    (c) Inserting a metal skewer into a potato before baking it will decrease the required baking time.
    (d) Smoke in a fireplace rises up the chimney.
    (e) Campers stay warm sitting near the hot embers of a campfire.

5. Discuss whether the following statement is true or false: In heat conduction, energy is transferred, but the particles themselves are not transferred.

6. (a) Explain what happens to the density of a substance when it is heated.
    (b) Describe how convection currents are created using this principle.

7. In what way does heat radiation differ from conduction and convection?

Making Connections

8. If air were a good conductor, you would feel cool even on a day when the air temperature is 25 °C. Explain why.

9. Explain your answer to each question.
    (a) Would it be better to place an electric heater near the floor or the ceiling of a room?
    (b) Would it be better to place an air conditioning vent near the floor or ceiling?
    (c) Why are fans placed near the ceilings of tall rooms?

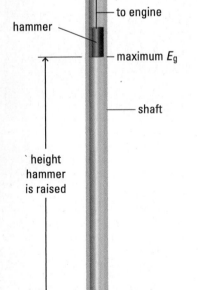

Figure 6
The design of a pile driver

Conservation of Energy

As you have learned, energy can change from one form into another. Scientists say that when any such change occurs, energy is conserved. In other words, no energy disappears and no new energy suddenly appears; it simply changes forms. This is expressed as the *law of conservation of energy*.

Law of Conservation of Energy
When energy changes from one form to another, no energy is created or destroyed.

The law of conservation of energy applies to all energy transformations. In ideal situations, just as with ideal machines, no energy is lost to friction. However, in actual situations and machines, some energy is usually needed to overcome friction. This results in the production of waste thermal energy and, sometimes, sound energy. Thus, energy-transformation equations are more complete when they include thermal energy in particular, and perhaps other forms, such as sound energy. This is illustrated in the operation of a pile driver (**Figure 6**) in which the overall goal of the energy transformations is the work done on the pile. The energy-transformation equation in this example is

$$E_{chem} \rightarrow W_1 + E_{therm} \rightarrow E_g \rightarrow E_K + E_{therm} \rightarrow W_2 + E_{therm} + E_{sound}$$

where E_{chem} is the chemical potential energy released in the burning of the fuel in the engine,

W_1 is the work done by the force on the hammer to raise it to its highest position,

E_{therm} is the thermal energy resulting from heat losses in the engine and from friction forces, as the hammer is raised and lowered and the pile is pushed into the ground,

E_g is the maximum gravitational potential energy of the hammer at its top position,

E_K is the maximum kinetic energy of the hammer just before it hits the pile,

W_2 is the work done by the hammer as it pushes downward on the pile,

E_{sound} is the sound energy resulting from operating the engine, from the collision of the hammer and the pile, etc.

CAREER CONNECTION

Graduates in a mechanical engineering technologist program from an Ontario community college design heavy equipment such as construction and agricultural machinery, then, if desired, can move into sales or technical writing.

GO www.science.nelson.com

▶ Practice

Understanding Concepts

10. Draw a sketch of the first two hills of a roller-coaster ride, with the coaster on its way toward the top of the first hill. Using labels on the diagram, apply the law of conservation of energy to describe the energy changes that occur during the operation of the roller coaster. Include the energy-transformation equation.

11. A ball is dropped vertically from a height of 1.5 m and bounces back to a height of 1.2 m. Does this violate the law of conservation of energy? Explain your answer.

12. A 0.20-kg ball is held at rest 2.2 m above the ground, and then it is dropped. Apply the law of conservation of energy to determine the ball's speed after it has fallen (a) 1.1 m and (b) 2.2 m. Ignore air resistance. (Hint: As the ball falls, what type of energy does its gravitational potential energy change into?)

Answers

12. (a) 4.6 m/s
    (b) 6.6 m/s

▶ TRY THIS activity | Design a Device Using Energy Transformations

In a group, design a device that uses at least four functional energy transformations to complete a task. One idea is a device that sounds an alarm when a locker door is opened by someone other than its owner. Another is a device that rings a bell when someone hits a compression spring hard enough. Describe the operation of the device, and include an energy-transformation equation. (Save your ideas for the Unit 2 Performance Task where you may be able to build and test your design.)

SUMMARY Thermal Energy and Heat

- It is important to distinguish between thermal energy, heat, and temperature.
- Heat transfer can occur through conduction, convection, and radiation.
- The law of conservation of energy states that when energy changes from one form into another, no energy is lost.
- The law of conservation of energy applies to all energy transformations, such as operating machines and hydroelectric generating stations.

▶ Section 3.6 Questions

Understanding Concepts

1. Distinguish between heat and thermal energy.

2. One morning, you walk barefoot across a rug and onto a tile floor. The rug and the floor must be at the same temperature, yet the floor feels much colder. Explain why.

3. What is the most likely method of heat transfer through (a) a metal, (b) a vacuum, and (c) a liquid?

4. Hang gliders and birds of prey ride convection currents called thermals. Describe the conditions that cause thermals.

5. Use the law of conservation of energy to describe the energy changes that occur when a wrecking ball is used to demolish a building. Assume that the machine controlling the ball has a fuel-powered engine. Include the energy-transformation equation.

6. A student places the toy described in the *Try This* Activity in the introduction to this chapter on the desk so that its curved side rests on the desk. The student does work on the toy by pushing down on one end of it.
   (a) Starting with the input work, state the energy transformations that occur until the toy comes to rest.
   (b) Relate the energy transformations in (a) to the law of conservation of energy.
   (c) How did your attempted explanation in the activity compare to your answers in (a) and (b) above?

7. A skier with a mass of 65.0 kg, including equipment, starts from rest and accelerates down a slope. The slope is 27.0 m higher at the top than at the bottom. The work done on the skier by the kinetic friction is -1.20×10^4 J.

(a) Determine the skier's gravitational potential energy at the top of the slope relative to the bottom.
(b) Determine the skier's kinetic energy at the bottom of the slope. State which law you are applying in your calculation.
(c) Calculate the skier's speed at the bottom of the slope.

Applying Inquiry Skills

8. You are given a ball that bounces well and three different surfaces on which to bounce it: a hard floor, a thin piece of cardboard on the floor, and a thick piece of cardboard on the floor.
   (a) Predict how the ball's bounce on the three surfaces will compare.
   (b) List the steps you would take to test your prediction in (a). Include safety considerations.
   (c) With your teacher's approval, either carry out your procedure in (b) or use a computer simulation to compare the bounces of a ball on different surfaces. Relate what you observe to the law of conservation of energy.

9. Describe how you would set up a demonstration to show
   (a) convection in water
   (b) convection in air
   (c) conduction in a solid

Making Connections

10. At the bottom of a cliff, police discover a car that skidded off the road. How could a forensic scientist use infrared photography to determine approximately how long ago the mishap occurred?

Canadians are among the highest per capita consumers of energy in the world, and our use of energy continues to grow. Currently, we rely heavily on energy from fossil fuels, such as coal, oil, and natural gas. Energy-transformation technologies transform this energy into other forms of energy, such as electrical energy and the kinetic energy of moving vehicles. As the world's population grows and our sources of fossil fuels are used up, we must find new sources of energy and more efficient energy-transforming technologies.

An original source of energy, called an **energy resource**, is a raw material, obtained from nature, that can be used to do work. A resource is called **renewable** if it renews itself in a normal human lifetime. All other resources are called **nonrenewable**.

Figure 1 illustrates how much of Canada's energy comes from renewable and nonrenewable sources. Approximately 11% of our energy consumption comes from hydraulic energy (waterfalls, for example). This resource is renewable. But most of the remaining energy comes from nonrenewable resources: oil, natural gas, and coal, which are fossil fuels, and uranium (nuclear energy).

energy resource a raw material, obtained from nature, that can be used to do work; also called an energy source

renewable resource an energy resource that renews itself in a normal human lifetime

nonrenewable resource an energy resource that does not renew itself in a normal human lifetime

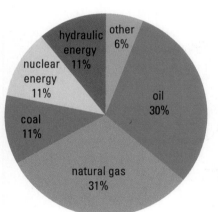

Figure 1
Canada's sources of energy. (Note: These percentages are approximate.)

Nonrenewable Energy Resources

Energy from fossil fuels begins as radiant energy from the Sun; this is absorbed by plants. The plants use the energy to manufacture carbohydrates, which store energy. Most of that energy is used during the lifetime of the plants, but some remains after the plants die. If the plants become buried, they do not decay; rather, they are compressed into new forms of organic material. Their energy (chemical potential energy) can be extracted later. However, it takes millions of years for plant life to become useful fuel. Once we have consumed the fossil fuels currently available, they will be gone forever.

Fossil fuels are recovered from the ground in raw form. Before they can be used, they must undergo transformations in an energy-transforming device.

When burned directly, fuel products can be used to operate engines of cars and other vehicles and to heat buildings. The fuel products can also generate electrical energy (**Figure 2**).

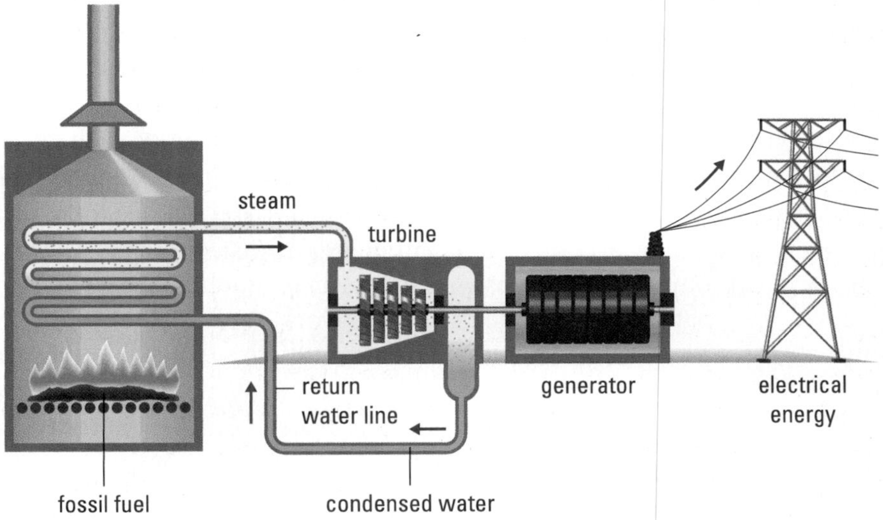

Figure 2
This fossil fuel generating station is a common energy-transforming device. Chemical energy stored in the fuel changes to thermal energy. The thermal energy boils water, which changes to steam. The steam, under pressure, forces the turbine to spin. The generator, connected to the turbine, changes the kinetic energy of spinning into electrical energy.

LEARNING TIP

Electric Generators and Motors
The way in which rotating generators produce electrical energy is described in more advanced physics texts. The main idea is that windings of wires inside the generator are forced to spin while surrounded by magnetic forces. In other words, in a generator, mechanical energy is transformed into electrical energy as the magnetic and electric forces interact. An electric motor looks similar to a generator, and it, too, has wire windings surrounded by magnetic forces. However, in an electric motor the electrical energy is transformed into mechanical energy as the electric and magnetic forces interact. These interacting forces cause the motor to spin.

SAMPLE problem 1

Generating Electrical Energy

Starting with energy from the Sun, write the basic energy-transformation equation for the fossil fuel generating station in **Figure 2**.

Solution
The basic energy-transformation equation is

$$E_{rad} \rightarrow E_{chem} \rightarrow E_{therm} \rightarrow E_K \rightarrow E_{elec}$$

where E_{rad} is the radiant energy from the Sun, or solar energy,
E_{chem} is the chemical potential energy stored in the plant deposits,
E_{therm} is the thermal energy released during the burning of the fuel and transferred to the water to produce steam,
E_K is the kinetic energy of the turbines as the moving steam causes them to rotate, and
E_{elec} is the electrical energy produced in the rotating generator.

This equation is called "basic" because it does not show the waste thermal energy at each stage.

Fossil fuels have both disadvantages and advantages. They are very convenient, but they are costly to mine and deliver to the user, and their availability is limited. They cause environmental problems because of pollution and the release of certain gases (e.g., carbon dioxide), which contribute to global warming. Of the fossil fuels available, natural gas produces the least amount of pollution because it burns more cleanly than oil and much more cleanly than coal.

A source of energy for producing electrical energy that is not a fossil fuel is uranium, which uses a process called **nuclear fission**. In this process, the nucleus of an atom is split, either naturally or by being bombarded with other particles that make it split. In either case, it releases energy. If enough uranium nuclei are split, the energy released from the controlled fission reaction can be used to heat water and produce steam. The steam then operates generators. Thus, for electrical energy production, uranium serves the same function as fossil fuels.

Canadians have designed and built several nuclear reactors, called CANDU reactors. In a CANDU reactor, the uranium fuel is pressed into pellets and placed in long metal tubes sealed at the ends. Several of these tubes are assembled into bundles, which are arranged horizontally in the reactor. Each reactor contains about 5000 of these bundles. As the uranium undergoes fission, the energy released heats the liquid coolant surrounding the bundles. The coolant, a type of water called heavy water, circulates under pressure to heat the ordinary water in a boiler. Steam from the boiler water moves through and drives the turbines to produce electrical energy (**Figure 3**). Fission reactors are very expensive to set up, and the nuclear waste produced is a concern. However, the emissions to the atmosphere are low, and uranium is readily available in Canada.

nuclear fission a nuclear reaction in which the nucleus of an atom is split

DID YOU KNOW ?

Refuelling a CANDU Reactor
The horizontal arrangement of the fuel bundles in a CANDU reactor allows the bundles to be replaced without shutting down the reactor to refuel. Most other reactor designs require a shutdown of about one week for refuelling, which disrupts the supply of electricity. A shutdown may also affect the local ecology in lakes or rivers: If the cooling water is taken from a river or lake, it is returned there, at a warmer temperature. Species that have adapted over the years to these warmer waters are suddenly faced with cooler temperatures, which they may not tolerate. Other species were likely displaced by warmer water when the reactor was first built.

GO www.science.nelson.com

Figure 3
The basic operation of a CANDU generating station. CANDU means that this fission reactor is **CAN**adian in design, uses **D**euterium oxide (heavy water) to control the rate of the nuclear reaction, and uses ordinary **U**ranium as its fuel. (Some other types of reactors use enriched uranium as their fuel.)

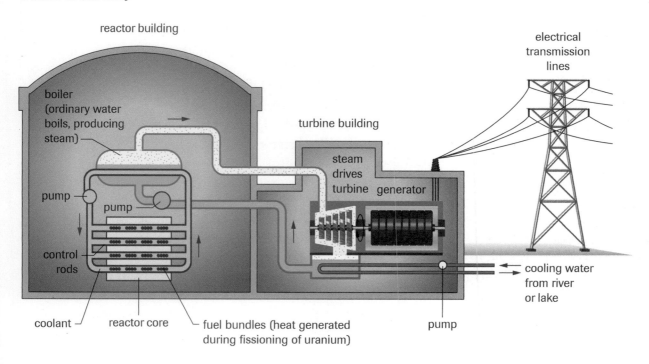

solar energy radiant energy from the Sun

solar cells electronic devices that transform light energy into electrical energy directly; also called photovoltaic cells, photoelectric cells, or simply photocells

passive solar heating the design and building of a structure to best take advantage of solar energy at all times of the year

active solar heating an energy-transformation technology that absorbs solar energy and converts it into thermal energy, and distributes it where needed in the structure

---

▶ Practice

Understanding Concepts

1. Name the energy resources and energy-transforming devices described so far in this section.

2. Write the energy-transformation equations for the following energy-transforming devices:
   (a) a CANDU electric generating station
   (b) a propane-driven car (Propane is a fossil fuel in the form of a gas.)

3. (a) List advantages of natural gas over other types of fuels.
   (b) List advantages and disadvantages of using nuclear fission rather than fossil fuels to produce electrical energy.

Making Connections

4. An area known as the "oil sands" in northern Alberta has vast amounts of oil mixed with sand. Research this fossil fuel resource. List advantages and disadvantages of extracting this electrical energy.

 www.science.nelson.com

Renewable Energy Resources

Earth's supplies of fossil fuels and uranium are limited. Some believe that oil reserves will not last many more decades. Therefore, we must develop devices to transform the energy from renewable energy resources into energy we can use.

Many renewable energy resources are briefly described here. Some are used today, and others are still being researched. As you read each description, note whether the resource is available locally or whether it must be transported from another location.

Solar energy, radiant energy from the Sun, can be used in **solar cells**, or *photovoltaic cells*. These electronic devices transform light energy into electrical energy directly. The materials used to make solar cells are the same as the materials used to make transistors and computer chips. Currently, solar cells are popular in portable devices, such as solar-powered calculators and watches. They are also very useful in remote areas, such as cottages, weather stations, and isolated communities (**Figure 4(a)**). This method of generating electrical energy can be used in a single dwelling (**Figure 4(b)**) or a central power generating station (**Figure 4(c)**).

Solar energy can also be used to heat buildings and other structures directly. However, as illustrated in **Figure 5**, the Sun's rays arrive at different angles, depending on the season. A design that takes best advantage of the Sun's energy at all times of the year is called **passive solar heating**.

In a structure with **active solar heating**, solar collectors absorb the Sun's energy and convert it into thermal energy that can be used elsewhere in the structure. For instance, solar energy can heat water in a circulating system used to heat a home (**Figure 6**). Active solar heating is much more expensive to install than passive solar heating.

(a)

(b)

(c)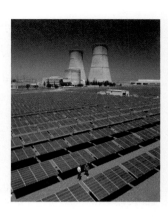

Figure 4
(a) Solar cell arrays form the two solar cell panels that provide electrical energy for this nomad in a remote location in Mongolia.
(b) Solar panels cover the entire south-facing roof of this home.
(c) This "solar farm" uses numerous solar panels to generate electrical energy, which will be distributed to customers.

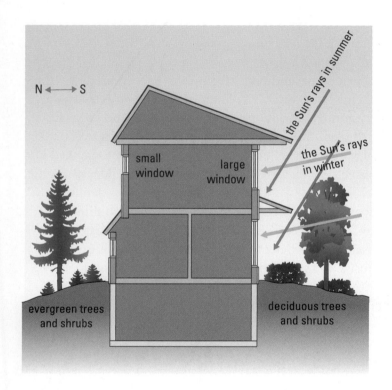

Figure 5
A home with passive solar heating. Other features that assist solar heating may include carpets that absorb light energy in winter and window shutters that are closed at night to prevent heat loss.

(a)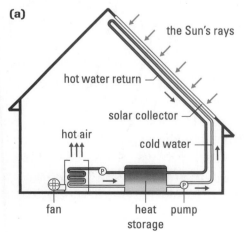

(b)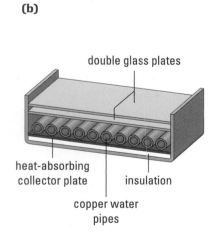

Figure 6
(a) An active solar heating system
(b) An example of solar collector design. Solar energy heats the heat-absorbing plate, which heats the water flowing through the pipes.

Solar energy is abundant, nonpolluting, and available locally. Although devices that use solar energy are expensive to construct, they are fairly inexpensive to operate. However, we must find ways to store the energy when sunlight is not available.

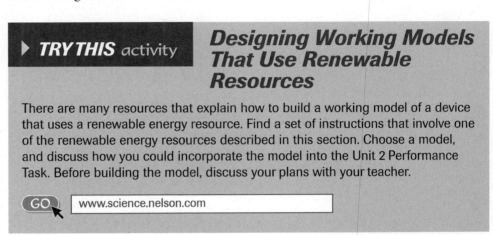

► **TRY THIS** activity

Designing Working Models That Use Renewable Resources

There are many resources that explain how to build a working model of a device that uses a renewable energy resource. Find a set of instructions that involve one of the renewable energy resources described in this section. Choose a model, and discuss how you could incorporate the model into the Unit 2 Performance Task. Before building the model, discuss your plans with your teacher.

GO   www.science.nelson.com

hydraulic energy energy generated by harnessing the gravitational potential energy of water

Hydraulic energy, which comes indirectly from solar energy, harnesses the gravitational potential energy of water. The Sun's radiant energy strikes water on Earth. The water evaporates, rises, condenses into clouds, and then falls as rain. The rain gathers in rivers and lakes. At the top of a dam or waterfall, the water has gravitational potential energy. That energy can then be changed into another form, such as electrical energy. Although this resource does not cause pollution, it does affect the ecology of the area where the river is dammed. In addition, the electrical energy produced at the generating station must be transmitted long distances to the consumers. As you will see in Chapter 4, long-distance transmission is inefficient. The generating station is expensive to set up but less expensive to operate than a station that uses fossil fuel or nuclear fission.

wind energy energy generated by harnessing the kinetic energy of wind

Wind energy, obtained indirectly from solar energy, is a possible energy source in those areas of Canada where it is windy throughout the year. Wind generators can change the kinetic energy of the wind into clean, nonpolluting electrical energy or into energy for pumping water. **Figure 7** shows a set of horizontal axis wind turbines, which transform wind energy into electrical energy. Wind energy can be used locally or transmitted to consumers. However, wind is not consistent and the turbines are noisy.

tidal energy energy generated by the rising and falling of ocean tides

Tidal energy is a potential energy resource in regions where ocean tides are large (**Figure 8**). It is one of the few resources that do not result from the Sun's energy. Tides are produced by the gravitational forces of the Moon and, to a lesser extent, the Sun, on Earth. To obtain electrical energy from tidal action, a dam must first be built across the mouth of a river that empties into the ocean. The gates of the dam are opened when the tide rolls in. The moving water spins turbines that produce electrical energy. When the tide stops rising, the gates are closed until low tide. Then the gates are opened, and once again the trapped water rushes out past the turbines, producing electrical energy. A major advantage of this system is that it does not produce air pollution or

Figure 7
This wind farm in Alberta has several horizontal generators.

Figure 8
The Annapolis Tidal Generating Station, Nova Scotia, is North America's first tidal generating station. It is linked to the Bay of Fundy system, whose tides are more than 15 m, among the highest in the world. Other possible sites are at Ungava Bay in northern Quebec, Frobisher Bay and Cumberland Sound on Baffin Island, and Jervis and Sechelt Inlets near Vancouver.

thermal pollution. However there are disadvantages: It does not produce a consistent supply of electrical energy for peak times; also, the construction of the dam may change the local ecology. Finally, tidal generating stations require turbine systems that allow the water to flow in two directions. A technology different from other turbine installations is required.

Biomass energy is the chemical potential energy stored in plants and in animal waste. Again, this energy comes indirectly from the Sun. Wood is commonly burned as a source of biomass energy, not only in home fireplaces and woodstoves, but also in large industries that burn the leftover products of the forestry industry. Of course, wood is a renewable resource only if new trees are planted. Advantages of this resource are that it uses products that otherwise might be discarded, and it can be used locally. Its disadvantages are the same as those of fossil fuels: the creation of pollution and unwanted gases in the atmosphere.

Geothermal energy is thermal energy or heat taken from beneath Earth's surface. The principle source of this heat is radioactive decay deep within Earth, mainly from elements such as uranium and potassium. This enormous resource increases Earth's subsurface temperature an average of 25 °C with each kilometre of depth. However, the heat generated in areas that were once volcanic is easier to harness. The rocks in these areas stay hot for thousands of years. As a result, any groundwater that seeps down becomes heated, forming hot springs and geysers. The hot water can be used directly to heat homes and generate electrical energy (**Figure 9**). In Canada, geothermal energy is plentiful only in a small number of places, such as the former volcanic regions of British Columbia and Yukon, as well as in the sedimentary basin in the Prairie provinces. Geothermal activity can also be used for energy in Iceland,

biomass energy the chemical potential energy stored in plants and in animal waste

geothermal energy thermal energy or heat taken from beneath Earth's surface

Figure 9
New Zealand has several active geothermal areas, some of which are used to generate electrical energy. The steam field shown here provides energy to a local power station.

Energy and Energy Transformations **161**

nuclear fusion the process in which the nuclei of the atoms of light elements, such as hydrogen, join together at extremely high temperatures and densities to become larger nuclei, releasing energy in the process

California, and New Zealand. The only major disadvantage of geothermal energy is that it must be used close to the source.

Nuclear fusion is the process in which the nuclei of the atoms of light elements join together at extremely high temperatures and densities to become larger nuclei. (Notice that this process differs from nuclear *fission*, in which the nuclei of heavy elements split apart.) With each fusion reaction, some mass is lost because it changes into energy. Fusion is the energy source for the Sun and stars.

Hydrogen, one of the most abundant substances on Earth, is used, in certain forms, as a fuel to operate fusion reactors. Nuclear fusion has at least two advantages: There is a potentially limitless supply of fuel from the world's oceans. It also produces much less radioactive waste than nuclear fission, which means that it is safer for the environment.

Two main problems must be overcome before fusion can be an efficient source of electrical energy: First, temperatures as high as hundreds of millions of degrees are needed to begin the fusion reaction. Second, the reacting materials must be confined so that fusion may continue. Researchers are investigating the use of magnetic fields and lasers to solve both problems.

fuel cell a device that changes chemical potential energy directly into electrical energy

Hydrogen can also be used in an energy-transformation technology called a hydrogen **fuel cell** (**Figure 10**). This technology attracted much attention as a result of its use in the space program. In a fuel cell, the chemical potential energy of the fuel, usually hydrogen gas, is changed directly into electrical energy. (Thus, the fuel cell can also be called an "electrochemical cell.") The hydrogen combines chemically with oxygen in the presence of a third chemical called a *catalyst*. The result is the production of water and an electric current. The hydrogen fuel cell has a number of advantages. Because the chemical energy turns directly into electrical energy, the fuel cell is much more efficient than electric generating stations. It can operate at a relatively low temperature, so it emits fewer pollutants. Furthermore, it has few moving

(a)

(b)

(c)

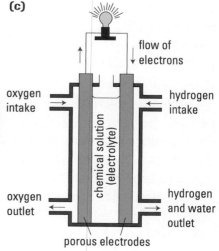

flow of electrons

oxygen intake

hydrogen intake

chemical solution (electrolyte)

oxygen outlet

hydrogen and water outlet

porous electrodes

Figure 10
(a) Stacked fuel cells
(b) This transit bus uses fuel cell technology.
(c) The basic design of a fuel cell

parts, so it is quiet and easy to maintain. A major disadvantage is that energy is needed to obtain the hydrogen from water. If fossil fuels are burned for this purpose, then pollution and the use of nonrenewable energy sources remain problems.

The atmosphere can also be used as a source of heat. An electric **heat pump**, for example, can cool a home in summer and heat it in winter. It works on the following principles: A substance called a *refrigerant* circulates in the heat pump system. The refrigerant flows in one direction in summer and the opposite direction in winter. When a substance changes from a liquid to a vapour, energy is absorbed (i.e., evaporation requires heat). The refrigerant evaporates inside the home in summer, absorbing heat. However, when a substance changes from a vapour to a liquid, energy is given off (i.e., condensation releases heat) (**Figure 11**). Therefore, the refrigerant evaporates outside in winter, again absorbing heat.

Heat pumps are relatively expensive to set up, and they only work above a certain temperature. As a result, a backup heating system is needed in most parts of Canada. However, they help save energy in the long term. The refrigerant used should not be toxic or harmful to the environment.

heat pump a device that uses evaporation and condensation to heat a home in winter and cool it in summer

DID YOU *KNOW*❓

Gasohol
Many new energy ideas depend on forms of biomass other than wood. One proposal is to burn trash to produce heat. Another is to capture the gas emitted from decaying matter in garbage dumps. A third is the fermentation of sugar molecules in grain by bacteria to produce methane and ethanol (grain alcohol). A mixture of one part alcohol in nine parts gasoline can be used to run automobile engines. This mixture, called gasohol, is used in various parts of Canada. It is widely used in Brazil, where fermented sugar cane, not grain, is the source of alcohol.

 www.science.nelson.com

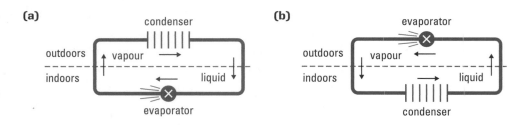

Figure 11
(a) In summer, evaporation occurs indoors to absorb heat; condensation occurs outdoors to give away the heat.
(b) In winter, evaporation occurs outdoors to absorb heat; condensation occurs indoors to give away the heat.

DID YOU *KNOW*❓

The Ground as a Source of Heat
The principle of the operation of a heat pump (air-to-air heat extraction) can be applied to a system in which the outside source of heat is the ground, rather than the air. A closed-loop piping system is buried underground, where the temperature is relatively constant. This type of system requires less energy to operate than a heat pump that uses the atmosphere. Therefore it is more efficient.

 www.science.nelson.com

▶ ***Practice***

Understanding Concepts

5. (a) List as many renewable energy resources that originate in the Sun's radiant energy as you can.
   (b) Which renewable energy resources do not originate in the Sun's radiant energy?
   (c) Classify each of the resources you named in (a) and (b) as either available locally or available somewhere from which the energy would have to be transported.

6. Starting with the Sun, list the energy transformations that occur when cooking a roast in an electric oven. Write the energy-transformation equation. Assume that the electrical energy comes from a hydraulic generating system.

7. List the energy resources that match each of the following criteria:
   (a) available locally
   (b) produces no pollution
   (c) available only at certain locations
   (d) significantly affects the local ecology

8. Draw a diagram to show how a hydrogen fuel cell can be used to operate a heater in a barn.

Making Connections

9. The hydrogen fuel cell is gaining importance as an energy-transforming technology. Research this technology, and answer the following questions:
   (a) Research some uses of the hydrogen fuel cell, and describe the advantages and disadvantages you discover in each case.
   (b) List some careers linked to the development and use of hydrogen fuel cells.

SUMMARY | *Nonrenewable and Renewable Energy Resources*

- Energy-transformation devices change energy from a resource into a usable form; for example, a fossil fuel generating station changes chemical potential energy into electrical energy.

- Nonrenewable energy resources include uranium and all fossil fuels, such as coal, oil, and natural gas.

- Renewable energy resources include solar, hydraulic, wind, tidal, and geothermal energies, as well as nuclear fusion, biomass, and the atmosphere.

- All energy resources have advantages and disadvantages.

▸ Section 3.7 Questions

Understanding Concepts

1. List two nonrenewable energy resources and five renewable energy resources.

2. Write the energy-transformation equation for each of the following resources used to produce electrical energy:
   (a) hydraulic energy
   (b) the Sun
   (c) biomass
   (d) nuclear fission

3. Draw two sketches to show the difference between generating electrical energy using a fossil fuel resource and using hydraulic energy.

Applying Inquiry Skills

4. (a) Make up three or four survey questions that will help you judge how much the average citizen knows and cares about energy resources and energy-transforming devices in Canada.

(b) Ask several people who are not in your physics class to participate in your survey. Describe what you discover.

Making Connections

5. List some harmful effects to the environment of the following types of electrical generating stations:
   (a) coal-fired
   (b) hydraulic
   (c) tidal
   (d) nuclear fission

6. List some reasons why renewable energy resources are not encouraged more throughout Canada. Show that you understand the arguments some people give for not using certain renewable resources.

7. Which alternative energy resource described in this section is most likely to be developed in your area? Explain why, showing that you understand the advantages of using the resource.

8. A major international project called Iter aims to develop nuclear fusion as a clean and plentiful resource for the production of electrical energy. Research this project on the Internet, paying special attention to Canada's contribution, and prepare a summary that addresses the following questions:
   (a) What countries are involved in the project?
   (b) What are the scientific contributions Canadians have made and intend to make to the project?
   (c) Iter's design choice for its fusion power plant is called *tokomak*. Describe the basic operation of this design. (If possible, test the operation of a tokomak on the Internet to see what happens when you change the variables involved.)

    www.science.nelson.com

9. Stuart Energy is a world-leading energy company based in Mississauga, Ontario. One of the company's main products is a hydrogen fuel cell system that takes advantage of renewable energy sources or electrical energy during low-cost periods to produce hydrogen. The system is illustrated in **Figure 12**.
   (a) Starting with solar energy, write the energy-transformation equation for this system that results in electrical energy consumed in a home.
   (b) Repeat (a) starting with wind energy, and use an internal combustion engine as the output.
   (c) Explain how the Stuart Energy system overcomes the problem of an inconsistent supply of input energy from renewable sources, such as wind and solar energy.

    www.science.nelson.com

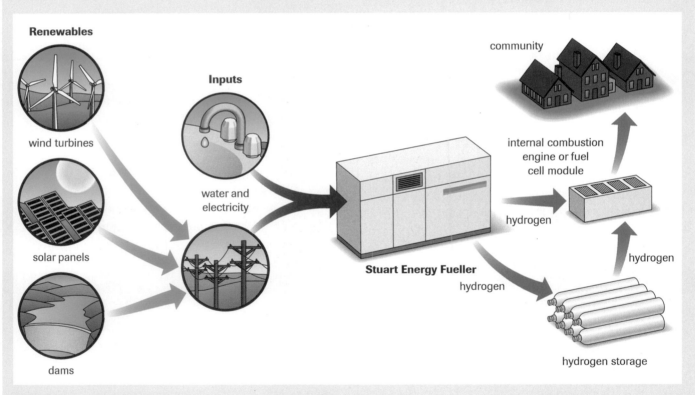

Figure 12
Stuart Energy's hydrogen fuel cell technology. (Figure based on material provided by Stuart Energy.)

Key Understandings and Skills

3.1 Energy Forms and Transformations

- **Energy** is the capacity to do work; it exists in different forms, such as kinetic energy and gravitational potential energy.
- In an **energy transformation**, energy changes from one form into another.

3.2 Work

- The **work**, or energy transferred to an object by an applied force over a distance, is found using the equation $W = F\Delta d$, where F and d are parallel.
- The SI unit of work is the **joule** (J).
- The work done by kinetic friction, $W = -F_K\Delta d$, is transformed into **thermal energy**.

3.3 Activity: Comparing Springs

- On a force–stretch graph of a spring, the slope equals the force constant in newtons per metre (N/m), and the area under the line equals the work done in stretching the spring.

3.4 Gravitational Potential Energy and Kinetic Energy

- **Gravitational potential energy** is the energy possessed by an object because of its position above a **reference level**; it is found using the equation $E_g = mgh$.

- **Kinetic energy** is energy of motion; it is found using the equation $E_k = \dfrac{mv^2}{2}$.
- **Mechanical energy** is the sum of an object's gravitational potential energy and kinetic energy.

3.5 Investigation: Energy in Springs

- The efficiency of a spring launched vertically upward is the ratio of the output energy ($E_g = mgh$) to the input energy (the work done in stretching the spring).

3.6 Thermal Energy and Heat

- **Heat** is energy transferred; **temperature** is the average kinetic energy of the particles of a substance.
- Heat transfers by **conduction**, **convection**, and **radiation**.
- The **law of conservation of energy** states that when energy changes from one form into another, no energy is lost; it applies to all energy transformations.

3.7 Nonrenewable and Renewable Energy Resources

- A **renewable energy resource** can be replaced within a human lifetime, and a **nonrenewable resource** cannot.

Key Terms

3.1
energy
energy transformation
energy-transformation
  technology

3.2
work
joule (J)
thermal energy

3.4
gravitational potential
  energy
kinetic energy
reference level
mechanical energy

3.6
heat
temperature
conduction

convection
radiation
law of conservation of
  energy

3.7
energy resource
renewable resource
nonrenewable resource
nuclear fission
solar energy

solar cells
passive solar heating
active solar heating
hydraulic energy
wind energy
tidal energy
biomass energy
geothermal energy
nuclear fusion
fuel cell
heat pump

Key Equations

3.2
- $W = F\Delta d$
- $W = -F_K\Delta d$

3.4
- $E_g = mgh$
- $E_K = \dfrac{mv^2}{2}$

Problems You Can Solve

3.1

- Recognize the common forms of energy.
- Describe examples of energy transformations, and write the corresponding energy-transformation equations.

3.2

- Given any two of work, force, and displacement, determine the third quantity.
- Describe situations in which negative work and zero work can be done on an object.

3.3

- Describe how to experimentally determine the force constant of a spring and the work done in stretching the spring.

3.4

- Given any three of gravitational potential energy, mass, gravitational strength, and height, determine the fourth quantity.
- Given any two of kinetic energy, mass, and speed, find the third quantity.

3.5

- Describe how you would experimentally determine the efficiency of a spring launched vertically upward from a launching pad.

3.6

- Describe how heat can transfer through solids, liquids, gases, and a vacuum.
- Apply the law of conservation of energy to explain the operation of a variety of devices, such as a gravity ride at an amusement park.

3.7

- Describe the advantages and disadvantages of common nonrenewable and renewable energy resources.
- Use diagrams to explain the operation of energy-transformation technologies.

▶ *MAKE* a summary

A small cottage is located in a remote area close to a river with a small waterfall, as illustrated in **Figure 1**. Make a larger diagram of the cottage that shows how you would

- operate a water pump
- generate electrical energy for lighting and for operating appliances
- provide heat for the cottage on cold days and nights

Include solar cells, passive solar heating, an active solar heating system, hydraulic energy, wind energy, and biomass energy. In your diagram, include as many concepts, key words, and equations (including energy-transformation equations) from this chapter as you can.

Figure 1
Many of the concepts in Chapter 3 can be applied to provide this cottage's energy needs.

Write the numbers 1 to 10 in your notebook. Indicate beside each number whether the corresponding statement is true (T) or false (F). If it is false, write a corrected version.

1. Sound is a form of energy that travels by waves through a material such as air.

2. The type of energy stored in a stretched or compressed spring is called elastic potential energy.

3. When a student pushes on a wall with a force of magnitude 250 N, the work done on the wall is zero.

4. The gravitational potential energy of a 4-kg hammer is the same as the gravitational potential energy of a 2-kg hammer as long as the height above the floor is the same for each hammer.

5. A golf ball loses the same amount of gravitational potential energy as it gains in kinetic energy as it falls from your hand to the ground.

6. The kinetic energy of an object depends only on the object's speed.

7. The slope of the line on the graph in **Figure 1** represents the force constant of the spring.

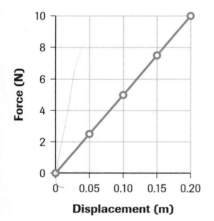

Figure 1

8. The transfer of energy from one part of an object to another is called temperature.

9. In a convection current, heated molecules become more dense.

10. The biomass energy from burning wood is a nonrenewable resource because trees must be cut down to obtain the wood.

Write the numbers 11 to 17 in your notebook. Beside each number, write the letter corresponding to the best choice.

11. Mechanical energy is the sum of
    (a) kinetic energy and gravitational potential energy
    (b) sound energy and chemical potential energy
    (c) thermal energy and electrical energy
    (d) all of the above

12. Scientifically, a person's weight is measured in
    (a) newtons   (c) grams
    (b) kilograms   (d) m/s^2

13. The area under the line on the graph in **Figure 1** is equal to
    (a) 50 N/m   (c) 2.0 J
    (b) 0.020 m/N   (d) 1.0 J

14. A spring has a spring constant of 50 N/m. (Assume two significant digits for this question.) In order to increase the length of the spring from 3.0 m to 5.0 m, the extra force needed would be
    (a) 250 N   (c) 100 N
    (b) 150 N   (d) 50 N

15. In explaining matter and energy, it is true that
    (a) matter is made of atoms and molecules that are constantly in motion
    (b) as objects become hotter, their particles move faster
    (c) thermal energy transfers from one part of an object to another by means of collisions
    (d) all of the above

16. When a heat pump is used in the winter, evaporation occurs
    (a) outdoors to release heat
    (b) indoors to absorb heat
    (c) outdoors to absorb heat
    (d) indoors to release heat

17. Renewable energy resources include
    (a) solar energy, natural gas, and biomass energy
    (b) nuclear fission, natural gas, and solar energy
    (c) wind energy, tidal energy, and biomass energy
    (d) natural gas, tidal energy, and hydraulic energy

An interactive version of the quiz is available online.
GO www.science.nelson.com

NEL

Understanding Concepts

1. Name and give an example of four forms of energy that have influenced your life today.

2. Write out the energy-transformation equation that summarizes the operation of a car as described in the Chapter 3 introduction.

3. **Figure 1** shows a person wearing shoes with built-in springs. Describe the energy transformations that occur when these shoes are used, and write the energy-transformation equation.

Figure 1
These "Flying Shoes" help to absorb shocks to knees, ankles, and heels. (Invented in Japan, they are also called Pyon-pyon or hop-hop shoes.)

4. In physics, what is the relationship between energy and work?

5. Explain when the work done by a force is negative.

6. A black bear's greatest enemy is the grizzly bear. To escape a grizzly attack, a black bear does what its enemy cannot do: It climbs a tree whose trunk has a small diameter. If the mass of the black bear is 140 kg, calculate
   (a) the magnitude of the force of gravity on the black bear
   (b) the work done by the black bear in climbing 18 m up a tree

7. A golf ball is given 114 J of energy by a golf club that exerts a force over a distance of 4.4 cm while the club and the ball are in contact.
   (a) Calculate the magnitude of the average force exerted by the club on the ball.
   (b) If the ball's mass is 47 g, find the magnitude of its average acceleration while the force is applied.

8. A cyclist (total mass of 68 kg, including the bike), travelling initially at 6.5 m/s, slams on the brakes and skids to a stop in 9.4 m. The coefficient of kinetic friction between the tires and the trail is 0.93. Calculate
   (a) the magnitude of the force of kinetic friction
   (b) the work done by the force of friction during the skid

9. A hoist in an automobile service centre raises a 1600-kg car 1.8 m off the floor.
   (a) Calculate the car's gravitational potential energy relative to the floor.
   (b) How much work did the hoist do on the car to raise it? Ignore friction.

10. A roast of beef in a refrigerator's freezer compartment has a potential energy of 35 J relative to the floor. If the roast is 1.7 m above the floor, what is the mass of the roast?

11. A 55-kg diver has 1.62 kJ of gravitational potential energy relative to the water when standing on the edge of a diving board. How high is the board above the water?

12. A group of skiers is travelling in a car at 97 km/h along a highway. A pair of ski boots having a total mass of 2.8 kg has been placed on the shelf of the rear window.
    (a) Calculate the speed of the car in metres per second.
    (b) Calculate the kinetic energy of the pair of boots.
    (c) Explain what happens to that energy if the driver suddenly stops the car.

13. What happens to an object's kinetic energy when its speed increases by a factor of 4?

14. A 51-kg cyclist on an 11-kg bicycle speeds up from 5.0 m/s to 10.0 m/s.
    (a) Calculate the total kinetic energy before the acceleration.
    (b) Calculate the total kinetic energy after the acceleration.
    (c) Calculate the work done to increase the kinetic energy of the cyclist.
    (d) Is it more work to speed up from 0 m/s to 5.0 m/s than from 5.0 m/s to 10.0 m/s? Explain your answer.

15. A discus travelling at 18 m/s has 260 J of kinetic energy. Find the mass of the discus.

16. An interesting and practical feature of the Montreal subway system, called the Métro, is that, in some cases, the level of the station is higher than the level of the adjacent tunnel (**Figure 2**). Explain the advantages of this design, taking into consideration concepts such as force, acceleration, work, gravitational potential energy, and kinetic energy.

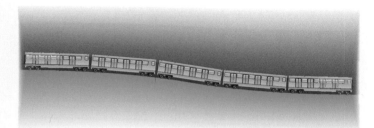

Figure 2

17. In a diagram, summarize the energy transformations that occur when an electric motor is used to operate a roller coaster at an amusement park.

18. State which method of heat transfer
    (a) does not require particles
    (b) works because particles collide with their neighbours
    (c) travels at the speed of light
    (d) works when particles circulate in a path

19. Compare the motion of particles in heat conduction and heat convection. (Diagrams may make the comparison easier.)

20. It is possible to heat a cold kitchen by opening the oven door, but it is not possible to cool a hot kitchen by opening the refrigerator door. Explain why, taking into consideration the law of conservation of energy.

21. Explain why it is impossible to have a motor that is 100% efficient.

22. Two balls have the same mass and diameter. One ball is made of wood and the other is a table-tennis ball partly filled with sand. Both are rolled along a desk. The wooden ball rolls along nicely, but the table-tennis ball stops after a few centimetres. What happened to its kinetic energy? Explain your answer.

23. A child of mass 33 kg slides down a slide 2.2 m high relative to the bottom of the slide. The child's speed at the bottom of the slide is 5.1 m/s.
    (a) Calculate the child's gravitational potential energy at the top of the slide.
    (b) Calculate the child's kinetic energy at the bottom of the slide.
    (c) Does the fact that the values in (a) and (b) are different violate the law of conservation of energy? Explain your answer.

24. The male world record for the pole vault is about 5.8 m. To achieve this feat, an athlete must raise his centre of mass about 4.6 m off the ground to clear the bar.
    (a) How fast must the vaulter run before takeoff to be able to get over the bar? (Assume that all his kinetic energy changes to potential energy at the top of the jump.)
    (b) How does this speed compare to your own maximum running speed?

25. Contrast and compare the use of tidal energy with the use of fossil fuels for generating electrical energy. Use diagrams in your answer.

26. State one main advantage and one main disadvantage of using each of the following energy resources:
    (a) wind energy   (d) biomass energy
    (b) geothermal energy (e) uranium
    (c) coal   (f) hydraulic energy

Applying Inquiry Skills

27. You have been hired by a manufacturer to test three new elastic materials, the best of which will be used to make trampolines. You decide to use a strip of each material to launch a ball vertically upward from the sturdy arrangement shown in **Figure 3**.
    (a) What measurements would you need to make to determine the work done in stretching the material before the launch?
    (b) What measurements would you need to make to determine the maximum gravitational potential energy of the ball?

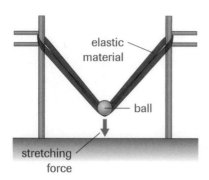

Figure 3

(c) How would you use the results of (a) and (b) to determine the percent efficiency of the elastic strip?

(d) What other tests would you conduct before recommending the material to use for the trampoline?

(e) What safety precautions would you follow in this investigation?

28. In the *Try This* Activity in section 3.6, page 153, one group submitted the design shown in **Figure 4**.

(a) What "task" did the group have in mind?

(b) What forms of energy are involved in the operation of the device in the diagram?

(c) Write the energy-transformation equation for this device, starting with the work done in raising the hammer.

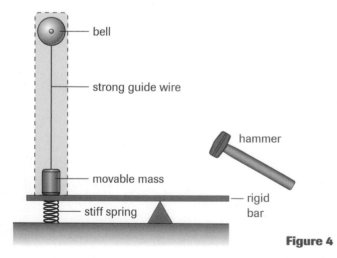

Figure 4

Making Connections

29. The sign in **Figure 5** warns drivers of a danger resulting from cold weather.

(a) Explain why the road surface on a bridge is likely to freeze before a regular road surface nearby. (Hint: Think of conduction and the structure and location of the bridge.)

(b) What methods would you suggest to reduce this danger?

Figure 5

30. One solution to question 29(b) is a design used on a bridge in Switzerland. In summer, coils leading from the bridge transfer heat to holes drilled in nearby rock. The rock holds the heat well, and in winter the heat transfers back slowly to the coils.

(a) Sketch two diagrams to show the heat flow, first in summer and then in winter.

(b) Describe the advantages of this system. (Consider economic and environmental issues, as well as safety.)

31. Suppose you are a planner for a Canadian electrical utility company. You wish to build a hydroelectric generating station. Describe briefly the main factors you would have to consider in selecting a site for the station.

4 Power and Efficiency

- define and describe the concepts and units related to power and efficiency

- analyze and describe, using energy flow diagrams, the relationships among and efficiencies of various sources of energy

- determine the power and efficiency of energy transformations in some common devices

- analyze the operation and efficiencies of common heat engines

- in an experiment, determine the efficiency of an electric motor and a mechanical device

- analyze the benefits and drawbacks of a source of renewable energy

Getting Started

Figure 1 shows examples of various types of light bulbs. Different types of bulbs exhibit different qualities. For example, incandescent bulbs feel hot when you touch them. Generally, they are inefficient in their use of energy. High-power incandescent bulbs, for example, a 200-W floodlight, are especially inefficient. The fluorescent bulbs used in schools and office buildings feel cooler when you touch them. They are also more efficient than incandescent bulbs. The small fluorescent bulbs available for household and desk lamps also provide bright light in a more efficient way.

In this chapter, you will compare the power and efficiency of light bulbs, as well as other devices. To do this, you will learn how to calculate efficiencies. You can calculate the efficiency of something as simple as the bounce of a ball or as complex as a gasoline engine that powers a car. You will also use diagrams to illustrate the flow of energy in common devices such as light bulbs, heating systems, and cars. Finally, you will have the opportunity to evaluate the benefits and drawbacks of a renewable energy resource.

💡 REFLECT on your learning

1. Light bulbs are labelled 20 W, 40 W, 60 W, and so on.
   (a) What does the W stand for?
   (b) What quantity is measured in this unit?

2. When you are alone on the stairs, you take 10 s to climb them. When other people are using the stairs, you take 20 s to climb the same stairway.
   (a) How does the work you do in the two cases compare?
   (b) How does the power you generate in the two cases compare?
   (c) Do you think the 10-s climb is more efficient than the 20-s climb? Explain your answer.

3. Some light bulbs operate at higher temperatures than others.
   (a) Which type of bulb tends to be cooler?
   (b) How do you think the temperature of the bulb relates to its efficiency? Explain your answer.

4. Electrical energy can be produced close to your community or far away. Which method, if either, do you think is more efficient? Why?

5. Which ball will bounce higher when dropped from rest from the same height: an overinflated basketball or an underinflated basketball? Explain your answer.

6. Describe how the following groups of people can promote the use of renewable energy resources to produce electrical energy:
   (a) federal politicians   (c) scientists and engineers
   (b) business executives   (d) consumers (including you)

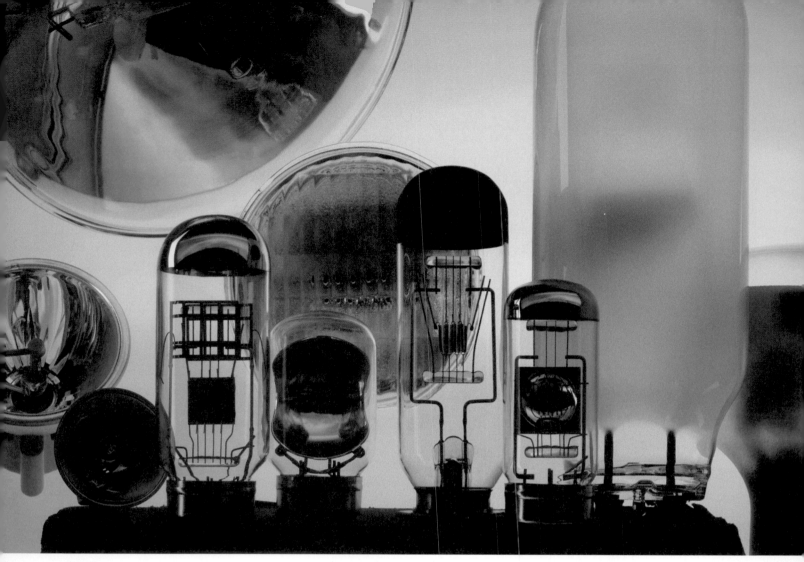

Figure 1
Light bulbs come in a variety of shapes, sizes, and efficiencies.

In this activity, you will predict the outcome when ball A and ball B are dropped from the same height (**Figure 2**). Before any class discussion, your teacher will pass ball A around the class, followed by ball B. Do *not* bounce them. When you obtain each ball, try to decide whether ball A will bounce higher than, lower than, or the same height as ball B.

 Stand at a distance from others when bouncing balls. Be ready to control a rebound.

(a) Write your prediction, and give a reason for your choice.

(b) Drop the two balls to test the prediction. Assess your prediction and reasoning.

(c) Link your observations to the efficiency of energy transfer, wasted energy, and the law of conservation of energy.

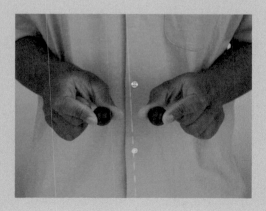

Figure 2

Figure 1
The CN Tower

power the rate of doing work or transforming energy, or the rate at which energy is supplied; symbol P; it is a scalar quantity, measured in watts

watt the SI unit of power; symbol W; 1 W is equivalent to 1 J/s

Figure 2
James Watt (1736–1819), a Scottish physicist and the inventor of the first practical steam engine

Two students of equal mass run up the stairs of Toronto's CN Tower (**Figure 1**), to raise money for charity. The students climb the same vertical displacement, 342 m up the stairs, one in 24 min and the other in 36 min. How does the work done by each student against gravity compare? It is the same because they have the same mass. (Recall that the work done equals the gravitational potential energy at the top relative to the bottom, $E_g = mgh$.) But the times are different, so some other factor must make a difference. That other factor is called *power*. The person who does the work in a shorter time generates more power.

Power (P) is the rate of doing work or transforming energy. Thus,

$$\text{power} = \frac{\text{work}}{\text{time interval}} \quad \text{or} \quad \text{power} = \frac{\text{energy transformed}}{\text{time interval}}$$

Using symbols,

$$P = \frac{W}{\Delta t} \quad \text{or} \quad P = \frac{\Delta E}{\Delta t}$$

Work and energy are measured in joules, and time is measured in seconds. Therefore, power is measured in joules per second (J/s). This SI unit is called the **watt** (W), in honour of James Watt (**Figure 2**). Watts and kilowatts are commonly used to measure the power used by electrical appliances while megawatts are often used to measure the power of electric generating stations.

> ▶ **SAMPLE** problem 1
> ### Power

A cyclist transforms 2.7×10^4 J of energy in 3.0 min. What is the cyclist's power? Express the answer in watts and kilowatts.

Solution

$\Delta E = 2.7 \times 10^4$ J

$\Delta t = 3.0$ min $= 180$ s $= 1.8 \times 10^2$ s

$P = ?$

$$P = \frac{\Delta E}{\Delta t}$$

$$= \frac{2.7 \times 10^4 \text{ J}}{1.8 \times 10^2 \text{ s}}$$

$$P = 1.5 \times 10^2 \text{ W}$$

The cyclist's power is 1.5×10^2 W. Since 1000 W = 1 kW, in kilowatts the power is 0.15 kW.

> **SAMPLE** problem **2**

Power in Climbing Upward

A 52-kg student climbs 3.0 m up a ladder in 4.7 s. Calculate

(a) the student's gravitational potential energy at the top of the ladder relative to the bottom

(b) the student's power for the climb

Solution

(a) $m = 52$ kg

$g = 9.8$ N/kg

$h = 3.0$ m

$E_g = ?$

$$E_g = mgh$$
$$= (52 \text{ kg})(9.8 \text{ N/kg})(3.0 \text{ m})$$
$$E_g = 1528.8 \text{ J, or } 1.5 \times 10^3 \text{ J}$$

The gravitational potential energy is 1.5×10^3 J.

(b) The work done in climbing the ladder is equal to the gravitational potential energy at the top of the ladder. Thus, $W = 1.5288 \times 10^3$ J, the unrounded value in (a).

$\Delta t = 4.7$ s

$P = ?$

$$P = \frac{W}{\Delta t}$$
$$= \frac{1.5288 \times 10^3 \text{ J}}{4.7 \text{ s}}$$
$$P = 3.3 \times 10^2 \text{ W}$$

The student's power is 3.3×10^2 W.

> ▶ **Practice**

Understanding Concepts

1. An electric clock uses 150 J every minute. Assuming two significant digits, calculate the clock's power rating.

2. A fully outfitted mountain climber, complete with camping equipment, has a mass of 85 kg. In exactly 1.5 h, she climbs from an elevation of 2700 m to 3800 m. Calculate
   (a) the climber's gravitational potential energy at the top position relative to the bottom position
   (b) the climber's power, in watts and kilowatts

3. Rearrange the equation $P = \dfrac{\Delta E}{\Delta t}$ to solve for (a) ΔE and (b) Δt.

4. How much electrical energy does a stereo system, rated at 85 W, use each minute? Express your answer in joules and kilojoules.

DID YOU KNOW ?

Horsepower
The "horsepower" (symbol hp) is a unit of power in the imperial system of measurement. It is equal to 746 W, which is the power output of an average working horse. The unit was first used by James Watt when he compared his steam engines to the rate at which a horse could pull coal out of a mine.

GO www.science.nelson.com

LEARNING TIP

The Watt in Base Units
Like the newton, the joule, and other derived units, the watt can be expressed in terms of base SI units:

$$1 \text{ W} = 1 \frac{\text{J}}{\text{s}}$$
$$= 1 \text{ J} \times \frac{1}{\text{s}}$$
$$= 1 \text{ N} \cdot \text{m} \times \frac{1}{\text{s}}$$
$$= 1 \text{ kg} \cdot \frac{\text{m}}{\text{s}^2} \cdot \text{m} \times \frac{1}{\text{s}}$$
$$1 \text{ W} = 1 \text{ kg} \cdot \frac{\text{m}^2}{\text{s}^3}$$

Answers

1. 2.5 W

2. (a) 9.2×10^5 J
   (b) 1.7×10^2 W; 0.17 kW

4. 5.1×10^3 J; 5.1 kJ

5. With favourable winds, the power of one of the world's largest wind generators is 3.2 MW, or 3.2×10^6 W. How long would it take such a generator to produce 3.5×10^7 J, the amount of energy needed to cook a large turkey in an oven?

Making Connections

6. Research the life and scientific contributions of James Watt. Give a reason why the unit for power is named in his honour.

 www.science.nelson.com

▶ **TRY THIS** activity   *Student Power*

You can design and carry out activities to determine the power a student generates during simple exertions such as climbing stairs. (Consider only upward motion when determining the power generated by these activities.) You might think of activities other than those suggested here.

1. Determine the power you or another student generates while walking up a set of stairs. Safety considerations are important here. Wear running shoes and walk at your normal pace. (For interest, calculate the power in horsepower using the relationship 1 hp = 746 W.)

2. Design an experiment to determine the power generated by a student when performing a variety of activities at a safe, comfortable pace. Examples are lifting weights in an exercise room, climbing a rope in the gymnasium, lifting books, doing pushups, digging in the garden, and shovelling snow. Have your teacher approve your activity and method, and then carry out the experiment.

 Students with health problems should not participate in this activity. Running shoes or other traction shoes are essential for option 1. Any student who experiences dizziness or discomfort during an activity should stop immediately. Warm-up exercises are strongly recommended to prevent injury.

SUMMARY   *Power*

- Power is the rate of doing work or transforming energy; it is found using the equation $P = \dfrac{W}{\Delta t}$ or $P = \dfrac{\Delta E}{\Delta t}$.
- Power is a scalar quantity measured in watts (W).

▶ *Section 4.1 Questions*

Understanding Concepts

1. Copy **Table 1** into your notebook, and complete the missing information. Show the equation used to solve each part.

Table 1 For Question 1

	Power, P	Energy Transformed, ΔE	Time Interval, Δt
(a)	?	63 J	1.5 s
(b)	64 W	?	15 min
(c)	12 W	1.8×10^4 J	?
(d)	?	8.2 MJ	2.0 h
(e)	25 kW	?	24 min

2. A 5.1×10^2-kg prize-winning pumpkin is raised from the ground to a truck platform, a distance of 1.1 m, by several volunteers. The action takes 3.2 s. Calculate
   (a) the work required to raise the pumpkin
   (b) the power of the lift

3. A 62-kg student does 55 pushups in 45 s. With each pushup, the student must lift an average of 68% of the body mass a height of 41 cm off the floor.
   (a) Calculate the work the student does against gravity for each pushup. Assume that work is done only when pushing upward.
   (b) Calculate the total work done against gravity in 45 s.
   (c) Calculate the power generated for this period.
   (d) Explain why the student does not have to lift 100% of the body mass with each pushup.

4. (a) Determine how long, in seconds and minutes, it would take to consume 5.0 MJ of energy using a hair dryer rated at 1500 W.
   (b) How many times could you dry your hair using the dryer described in (a)?

5. The largest bulldozer ever built had a mass of about 1.5×10^5 kg and a blade 7.4 m wide. Its two engines had a total maximum power output of about 1.7 MW. How much work (in megajoules) could this machine do each hour?

Applying Inquiry Skills

6. Students in a physics class were asked to design a simple machine and determine its power in raising a 2.0-kg mass. (Simple machines were described in Chapter 2.) One group designed a crank system, based on the wheel and axle, as illustrated in **Figure 3**.
   (a) What laboratory equipment do the students need to make the measurements needed to determine the power?
   (b) Use typical values to calculate this machine's power.
   (c) How could the students maximize the power of the machine?

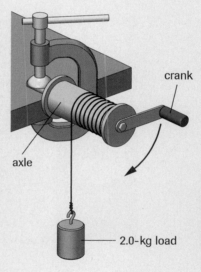

crank

axle

2.0-kg load

Figure 3
As the crank of this simple machine is turned, the rope attached to the axle raises the load.

The purpose of an incandescent light bulb, like the one shown in **Figure 1(a)**, is to provide light energy. Unfortunately, it also produces a lot of thermal energy at the same time. In fact, only about 5% of the electrical energy delivered to the bulb is transformed into light energy; 95% becomes waste thermal energy. Therefore, we say that this incandescent light bulb is only 5% efficient. This is illustrated in the energy flow diagram in **Figure 1(b)**.

efficiency (eff) a decimal number or percentage that rates how well a device transforms energy; calculated as the ratio of the useful energy output of a device to the energy input required to operate the device

Efficiency is the ratio of the useful energy output of a device to the energy input required to operate the device. It can be expressed as a decimal number or as a percentage. The efficiency and percent efficiency can be calculated as follows:

$$\text{efficiency} = \frac{\text{useful energy output}}{\text{energy input}} \quad \text{and \% efficiency} = \frac{\text{useful energy output}}{\text{energy input}} \times 100\%$$

Using symbols, these equations are

$$\text{eff} = \frac{E_{out}}{E_{in}} \quad \text{and} \quad \% \text{ eff} = \frac{E_{out}}{E_{in}} \times 100\%$$

where eff is the efficiency,
E_{out} is the useful energy output, and
E_{in} is the energy input.

(a)

(b)

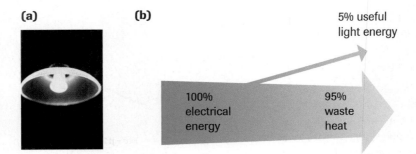

Figure 1
(a) An incandescent light bulb gets quite hot when in use.
(b) The energy flow diagram of an incandescent light bulb

▶ **SAMPLE** problem 1

The Efficiency of a Pulley System

A helicopter crew fighting a forest fire uses a pulley system to raise a large bucket of water from a lake. The energy used to operate the pulley system is 3.6×10^4 J, and the energy output in raising the water is 3.2×10^4 J.

(a) What is the efficiency of the pulley system? Express the answer as a decimal number and a percentage.

(b) Determine the percentage of the energy input "wasted."

(c) Show the wasted energy on an energy flow diagram.

Solution

(a) $E_{out} = 3.2 \times 10^4$ J

$E_{in} = 3.6 \times 10^4$ J

eff = ?

$$\text{eff} = \frac{E_{out}}{E_{in}}$$

$$= \frac{3.2 \times 10^4 \text{ J}}{3.6 \times 10^4 \text{ J}}$$

$$\text{eff} = 0.89$$

The efficiency of the pulley system is 0.89, or 89%.

(b) The percentage of the energy input wasted is 100% − 89% = 11%.

(c) The wasted energy (3.6×10^4 J − 3.2×10^4 J = 4×10^3 J) is transformed into thermal energy and sound energy, as shown in **Figure 2**.

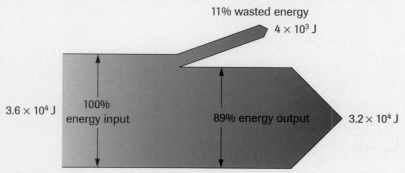

Figure 2
The energy flow diagram of the pulley system

▶ SAMPLE problem 2

The Efficiency of an Inclined Plane

A family sets up an inclined plane as shown in **Figure 3** to slide a 350-kg piano onto the back of a truck. The back of the truck is 81 cm above the ground, and the inclined plane is 3.0 m long. An average force of magnitude 1.5×10^3 N is needed to slide the piano up the ramp.

(a) Determine the work done by the family in loading the piano.

(b) Calculate the efficiency of the inclined plane used to load the piano.

(c) Draw an energy flow diagram for the situation.

Solution

(a) $F = 1.5 \times 10^3$ N

$\Delta d = 3.0$ m

$W = ?$

$$W = F\Delta d$$

$$= (1.5 \times 10^3 \text{ N})(3.0 \text{ m})$$

$$W = 4.5 \times 10^3 \text{ J}$$

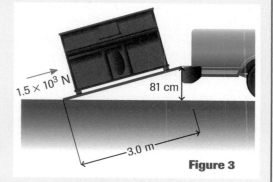

Figure 3

The work done by the family in loading the piano is 4.5×10^3 J.

LEARNING TIP

Efficiency of Machines
You learned in Chapter 2 that the efficiency (or percent efficiency) of a machine can be found using the ratio of the actual mechanical advantage (AMA) to the ideal mechanical advantage (IMA). To show that this is the same as the equation presented here, we can apply the equation for work or energy transferred, $W = F\Delta d$:

$$\text{eff} = \frac{\text{AMA}}{\text{IMA}}$$

$$= \frac{\left(\dfrac{F_L}{F_E}\right)}{\left(\dfrac{\Delta d_E}{\Delta d_L}\right)}$$

$$= \left(\frac{F_L}{F_E}\right)\left(\frac{\Delta d_L}{\Delta d_E}\right)$$

$$= \frac{F_L \Delta d_L}{F_E \Delta d_E} \left(\text{or } \frac{\text{energy output}}{\text{energy input}}\right)$$

$$\text{eff} = \frac{E_{out}}{E_{in}}$$

The concepts of work and energy show why some efficiencies are high and some are low.

(b) The energy input is equal to the work done in pushing the piano up the inclined plane. The useful energy output is the increase in gravitational potential energy of the piano going from the ground to the higher level.

$E_{in} = W = 4.5 \times 10^3$ J

$m = 350$ kg

$g = 9.8$ N/kg

$h = 81$ cm $= 0.81$ m

$E_{out} = ?$

eff $= ?$

The energy output is

$$E_{out} = mgh$$
$$= (350 \text{ kg})(9.8 \text{ N/kg})(0.81 \text{ m})$$
$$E_{out} = 2.8 \times 10^3 \text{ J}$$

The efficiency is

$$\text{eff} = \frac{E_{out}}{E_{in}}$$
$$= \frac{2.8 \times 10^3 \text{ J}}{4.5 \times 10^3 \text{ J}}$$
$$\text{eff} = 0.62$$

The efficiency of the inclined plane is 0.62, or 62%.

(c) The energy flow diagram for this situation is shown in **Figure 4**. The "wasted" energy has been transformed into thermal energy and sound energy as a result of friction.

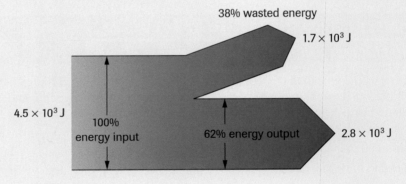

Figure 4

Efficiency and percent efficiency can also be found using power. Using symbols, the equations are

$$\text{eff} = \frac{P_{out}}{P_{in}} \quad \text{or} \quad \% \text{ eff} = \frac{P_{out}}{P_{in}} \times 100\%$$

where   eff is the efficiency,
       P_{out} is the useful power output, and
       P_{in} is the power input.

> ▶ **SAMPLE** problem **3**
> ### The Efficiency of a Motorcycle

A motorcycle has a gasoline engine with a power rating of 2.6×10^4 W. The engine does 3.4×10^5 J of work at full power in moving the motorcycle over a period of 85 s. Calculate

(a) the power output of the engine

(b) the percent efficiency of the motorcycle

Solution

(a) $W = 3.4 \times 10^5$ J

    $\Delta t = 85$ s

    $P_{out} = ?$

$$P_{out} = \frac{W}{\Delta t}$$

$$= \frac{3.4 \times 10^5 \text{ J}}{85 \text{ s}}$$

$$P_{out} = 4.0 \times 10^3 \text{ W}$$

The power output of the engine is 4.0×10^3 W.

(b) $P_{out} = 4.0 \times 10^3$ W

    $P_{in} = 2.6 \times 10^4$ W

    $\% \text{ eff} = ?$

$$\% \text{ eff} = \frac{P_{out}}{P_{in}} \times 100\%$$

$$= \frac{4.0 \times 10^3 \text{ W}}{2.6 \times 10^4 \text{ W}} \times 100\%$$

$$\% \text{ eff} = 15\%$$

The percent efficiency is 15%.

LEARNING TIP

Using Power to Find Efficiency
The reason that power can be used to find efficiency is that the output power and input power take the same time interval, Δt, so the Δt values cancel out:

$$\text{eff} = \frac{P_{out}}{P_{in}}$$

$$= \frac{\left(\dfrac{E_{out}}{\Delta t}\right)}{\left(\dfrac{E_{in}}{\Delta t}\right)}$$

$$= \left(\frac{E_{out}}{\Delta t}\right)\left(\frac{\Delta t}{E_{in}}\right)$$

$$\text{eff} = \frac{E_{out}}{E_{in}}$$

Answers

1. (a) 0.25, or 25%
   (b) 75%
2. (a) 5.8×10^4 J
   (b) 0.39, or 39%
3. (a) 0.86, or 86%
4. (a) 0.85 J
   (b) 0.81 J
   (c) 0.95, or 95%
6. 1.3 MJ

> ## ▶ Practice

Understanding Concepts

1. (a) Determine the efficiency of a certain swimmer if 4.4 MJ of energy input (originally from food) is required to provide 1.1 MJ of energy output during a swim.
   (b) What percentage of the energy input is not used for swimming?

2. A fuel-powered loader raises a 2.2×10^3-kg load from the ground to a loading platform, which is 2.7 m above the ground. The loader consumes 1.5×10^5 J of energy from the fuel while raising the load.
   (a) Determine the work done by the loader on the load.
   (b) Calculate the efficiency of the loader.
   (c) Draw an energy flow diagram for this situation.

3. A stereo system consumes power at a rate of 85 W while producing sound at a power of 73 W.
   (a) Determine the efficiency of the transformation.
   (b) Draw an energy flow diagram for this situation.

4. During an energy-efficiency test, a student pulls a toy car up a ramp that is 2.5 m long. The student exerts a force of magnitude 0.34 N parallel to the ramp. The toy has a mass of 75 g, and its height increases by 1.1 m.
   (a) Calculate the work done in pulling the car up the ramp.
   (b) Calculate the energy output in raising the car 1.1 m.
   (c) Calculate the ramp's efficiency.
   (d) Explain why the ramp's efficiency is so much higher than the efficiency of the inclined plane in Sample Problem 2, page 179.

5. Rewrite the equation $\text{eff} = \dfrac{E_{out}}{E_{in}}$ to solve for (a) E_{out} and (b) E_{in}.

6. How much energy does an electric kettle transfer to the water it holds if the kettle has an efficiency of 0.94 and consumes 1.4 MJ of electrical energy?

Case Study The Efficiencies of Heat Engines ▼

heat engine an engine that transforms heat from burning fuels into the kinetic energy of the moving parts of a machine

Many machine engines burn fuel to produce motion. A **heat engine** transforms heat from burning fuel into the kinetic energy of the moving parts of the machine. Examples of heat engines are the following:

- gasoline engines in cars
- diesel engines in trucks, buses, and locomotives
- gasoline engines in lawn mowers, snow blowers, leaf blowers, chain saws, motorcycles, and outboard motors
- rocket engines
- jet engines
- steam turbines used to produce electrical energy at electric generating stations and drive the propellers of most large ships

◆ CAREER CONNECTION

An efficiency administrator with the federal government designs energy programs and workshops to promote energy-efficient fuels, strategies, and driving techniques to save companies with large fleets of vehicles energy and, therefore, money.

 www.science.nelson.com

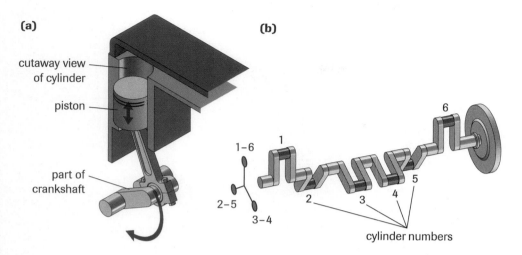

(a)

cutaway view
of cylinder

piston

part of
crankshaft

(b)

6

1–6   1

2–5

3–4

2   3   4   5

cylinder numbers

Figure 5
(a) The piston's reciprocating (up
and down) motion is changed
to rotational motion by a
crankshaft.
(b) This is the basic design of the
crankshaft of an in-line
6-cylinder engine.

Heat engines are very common, and most of them use fossil fuels. Therefore, it is important to maximize their efficiencies, especially because their efficiencies are relatively low. This case study looks at how different types of heat engines operate. The gasoline and diesel engines described below are *reciprocating engines*. The part that reciprocates (or moves back and forth, or up and down) is the piston, located in the cylinder. The reciprocating motion is changed into rotational motion by a crankshaft, which, as you will see below, starts the process of moving the car or other device (see **Figure 5**).

Four-Stroke Engines

The engine most commonly used in cars is the four-stroke gasoline engine. (Some cars and many trucks use four-stroke diesel engines.) **Figure 6** on page 184 illustrates the four strokes for one piston in the gasoline engine. (The diagram does not show the camshaft, which controls the opening and closing of the valves, or the engine's cooling system.)

Car engines can have 4, 6, 8, or even 12 cylinders. As seen in **Figure 6**, the piston in each cylinder is connected to the crankshaft. The crankshaft converts the up-and-down motion of the pistons into circular motion. It is connected to the flywheel at the front of the engine. This heavy wheel helps keep the pistons moving during the compression stroke. The crankshaft is also connected to the transmission gears, and then to the drive shaft, and finally to the differential. The differential connects to the drive axle, causing the wheels to turn. The pistons go through their stroke cycles out of step with each other so that the engine can operate smoothly.

The efficiency of this type of engine is typically between 25% and 35%. As a result, between 75% and 65% of the energy available from the fuel is wasted. Most of the waste energy is thermal energy and sound energy. Because a car's engine has many moving parts that operate in a complex way, it must be well maintained to have the highest possible efficiency.

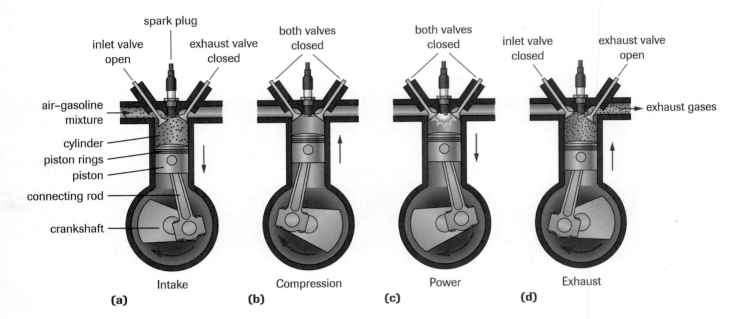

Figure 6

(a) On the *intake stroke,* the piston moves downward, reducing the pressure in the cylinder. The inlet valve opens, allowing an air–gasoline mixture from the carburetor to enter the cylinder. (An engine with a fuel-injection system works slightly differently.)

(b) On the *compression stroke,* both valves are closed and the piston moves upward, compressing the air–gasoline mixture.

(c) On the *power stroke,* an electric spark is created in the spark plug. The air–gasoline mixture explodes, forcing the piston downward.

(d) During the *exhaust stroke,* the exhaust valve opens and the piston rises, pushing the exhaust gases out of the cylinder.

Two-Stroke Engines

A two-stroke engine has ports rather than valves. The piston position controls the opening and closing of the ports. This design allows one cycle of operation to be complete after two strokes rather than four. Two-stroke engines can be either gasoline-powered or diesel fuel-powered. The two-stroke gasoline-powered engine is commonly used in small machines, such as lawnmowers.

Figure 7 shows the operation of the two-stroke gasoline engine. In (**a**), the piston is near the top of the cylinder, blocking the transfer and exhaust ports. The inlet port is open, which allows the air–gasoline mixture to enter the crankcase, which houses the crankshaft. The compressed air–gasoline mixture, above the piston from the previous cycle, is ignited by a spark from the spark plug. The explosion forces the piston downward. In (**b**), the piston is near the bottom of the stroke. The intake port is closed, but the exhaust and transfer ports are open. The new mixture of gases moves up to replace the exhaust gases. The piston is designed to prevent the new gases from mixing with the exhaust gases. Even so, some mixing does occur, which is a disadvantage of this type of engine. The crankshaft of a two-stroke engine may be connected to gears in order to drive blades, wheels, or other components of a machine. Two-stroke gasoline-powered engines have efficiencies between 25% and 35%.

The two-stroke diesel engine is commonly used in heavy machinery and large ships. A diesel engine has valves but no spark plugs. With this

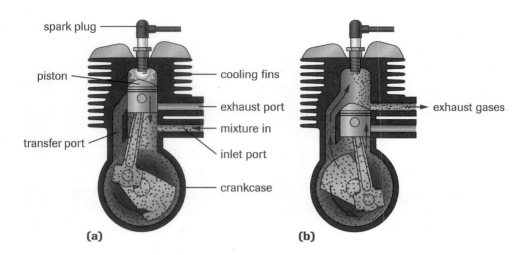

spark plug

piston

cooling fins

exhaust port

mixture in

transfer port

inlet port

crankcase

(a)

(b)

exhaust gases

Figure 7
The basic operation of a two-stroke gasoline engine. The engine shown is air-cooled: It requires cooling fins that are exposed to the air outside the engine.

arrangement, the piston can complete one cycle in two strokes. In **Figure 8(a)**, the piston is moving downward in the cylinder near the bottom of the stroke. The exhaust valve has just closed and the intake valve has just opened, allowing air to enter the cylinder. Then the piston moves upward, compressing the air to a high pressure. The high pressure causes the air to reach a high temperature just as the fuel is injected. The fuel ignites automatically, driving the piston downward. (If you compress the air in a bicycle pump, you can feel the air getting warmer). A diesel engine for a truck is shown in **Figure 8(b)**.

Diesel engines operate at relatively high temperatures so that they can ignite the air–fuel mixture without a spark plug. The high temperatures are needed for another reason as well: Diesel fuel, which is denser and oilier than gasoline, has a high boiling point, higher than water. Diesel engine efficiency can be as high as 40%, which means that diesel engines are more efficient than gasoline engines. One reason for the higher efficiency is that diesel fuel yields more energy per litre than gasoline. However, maximum efficiency is achieved only with a well-maintained engine.

DID YOU KNOW ?

Glow Plugs
Diesel engines can be hard to start because there are no spark plugs to ignite the fuel. Cold weather makes the problem worse. To overcome this problem, many diesel engines have a glow plug in each cylinder. Before the engine is started, electric current from the battery heats the plugs. This in turn heats the air–fuel mixture, making ignition easier.

GO www.science.nelson.com

(a)

fuel

inlet valve

exhaust valve

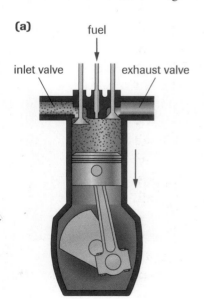

(b)

Figure 8
(a) The cylinder design of a two-stroke diesel engine
(b) A diesel engine used in some large trucks

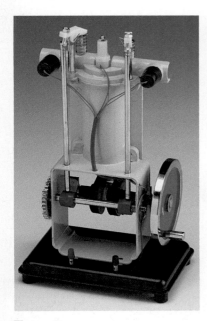

Figure 9
A model of a gasoline engine

Answers

7. (a) 33%
   (b) 28%
   (c) 39%

▶ **TRY THIS** activity   *A Gasoline Engine Model*

Use a model of a gasoline engine, like the one shown in **Figure 9**, or a computer simulation to observe the action of a heat engine. What can you observe from the model or simulation that was not discussed in this section?

▶ **Practice**

Understanding Concepts

7. Determine the percent efficiency of each of the following:
   (a) A four-stroke gasoline engine consumes 9.3 MJ of energy in doing 3.1 MJ of work on the crankshaft.
   (b) The power input provided by the fuel of a two-stroke gasoline engine is 6.0 kW, and the power output to the crankshaft is 1.7 kW.
   (c) A 3.4×10^5-W diesel engine on a fire truck does 8.0×10^6 J of work on the engine's crankshaft each minute.

8. Name the four strokes in the operation of a gasoline engine in a car.

9. Summarize the design features that allow a lawnmower engine to complete a cycle in two strokes rather than four.

10. Explain how a two-stroke diesel engine differs from a two-stroke gasoline engine.

11. Starting with chemical potential energy, draw an energy flow diagram for a well-maintained diesel engine.

12. The apparatus shown in **Figure 10** is a "fire piston." A piston with a sturdy handle safely compresses the air in the strong glass tube, causing a small piece of cotton in the tube to ignite. Describe how this demonstration relates to one of the heat engines presented in this section.

Making Connections

13. State your opinion of our dependence on heat engines, taking into account
    (a) the availability of fossil fuels
    (b) the efficiency of heat engines
    (c) noise pollution
    (d) air pollution

14. Some municipalities prohibit car owners from leaving their engines idling.
    (a) What are the advantages of this bylaw?
    (b) Are there disadvantages to this bylaw? Explain your answer.

Figure 10
Compressing air to produce heat

SUMMARY *Efficiency*

- The efficiency of an energy-transformation technology can be found using the equations

$$\text{eff} = \frac{E_{out}}{E_{in}} \text{ or } \% \text{ eff} = \frac{E_{out}}{E_{in}} \times 100\%, \text{ and eff} = \frac{P_{out}}{P_{in}} \text{ or } \% \text{ eff} = \frac{P_{out}}{P_{in}} \times 100\%$$

- Almost all energy-transformation technologies operate at efficiencies less than 100%. Much energy is transformed into waste thermal energy.

- Heat engines, such as gasoline and diesel engines used in automobiles and trucks, have relatively low efficiencies.

▶ Section 4.2 Questions

Understanding Concepts

1. Copy **Table 1** into your notebook, and complete the missing information. Show the equation used to solve each part.

2. An electric can opener consumes 2.2 kJ of electrical energy as it does 1.5 kJ of work while opening a can.
   (a) Calculate the can opener's efficiency.
   (b) Draw the energy flow diagram for this situation.

3. A 3.5×10^2-W electric sander has an output power of 2.2×10^2 W. Calculate its efficiency.

4. A 55-W incandescent bulb provides light energy at an efficiency of 5.1%.
   (a) Calculate the amount of electrical energy consumed each minute.
   (b) Calculate the amount of light energy output the bulb provides each minute.
   (c) Draw the energy flow diagram for the bulb.

5. A hospital volunteer pushes a patient in a wheelchair (total mass of 62 kg) up a ramp with a force of magnitude 88 N parallel to the ramp. The ramp is 8.5 m long, and it is 1.2 m higher at the top than at the bottom.

   (a) Calculate the input work done by the volunteer in pushing the patient up the ramp.
   (b) Calculate the output work accomplished by raising the patient to the higher level.
   (c) Determine the efficiency of the ramp.
   (d) What factors explain the high efficiency in this case?

6. The valves of a certain diesel engine are worn so that gases under pressure leak past them. Does this affect the efficiency of the engine? Explain your answer.

7. **Figure 11** on page 188 shows a device called Hero's engine, named after a scientist who lived in ancient Egypt.
   (a) Use the information in the diagram to describe how the device is forced to spin. Include an energy-transformation equation.
   (b) Which of Newton's laws of motion explains why the device spins? (Newton's laws were discussed in Chapter 1.)
   (c) Would you classify this device as a heat engine? Explain your answer.
   (d) Would you expect this device to have a high or low efficiency? Why?

Table 1 For Question 1

	Efficiency	Energy Output, E_{out}	Energy Input, E_{in}	% of E_{in} Wasted
(a)	?	63 J	95 J	?
(b)	?	0.65 MJ	13 MJ	?
(c)	0.25	?	1.8×10^4 J	?
(d)	?	44 kJ	?	16%

(a)

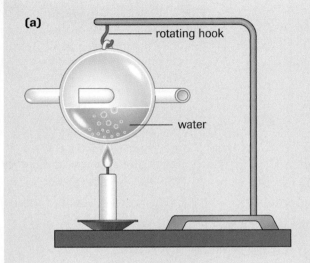

rotating hook

water

(b)

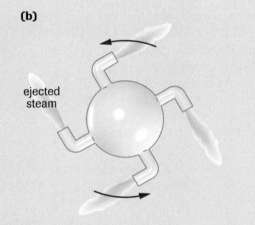

ejected
steam

Figure 11
(a) Side view of Hero's engine
(b) Top view of Hero's engine

Applying Inquiry Skills

8. A single, fixed pulley is used to raise a bucket of material from the ground to a scaffold several metres above the ground. (Pulleys were described, along with other simple machines, in section 2.1.)
   (a) Draw a system diagram of the situation, and identify the forces and distances involved in operating the pulley.
   (b) List the measurements and calculations that can be used to determine the efficiency of the pulley.
   (c) What main sources of measurement error must be overcome to obtain accurate results?

Making Connections

9. Research one of the types of heat engines listed at the beginning of the case study (page 182). Using diagrams, describe changes in design that have helped improve the efficiency of the engine.

10. A "kinetic watch" is not battery-operated, and it does not need to be wound.
    (a) What source of energy does it use?
    (b) Is this method of energy transformation good or bad for the environment? Explain your answer.

 www.science.nelson.com

11. Ontario has a Drive Clean program. Research this program, and describe its advantages.

 www.science.nelson.com

4.3 Investigation

Determining Efficiencies

In this investigation, you will determine the efficiency of an electrical energy-transformation technology and a spring-driven mechanical technology using the equations

$$\text{eff} = \frac{E_{out}}{E_{in}} \quad \text{and} \quad \text{eff} = \frac{P_{out}}{P_{in}}$$

For example, you can determine the efficiency of an electric motor used to raise a load of known mass (**Figure 1(a)**). To calculate the power input, P_{in}, first measure the voltage, V, in volts and the current, I, in amps supplied to the motor as the load is being raised (**Figure 1(b)**). Then take the product of the two measurements: $P_{in} = VI$. The power output, P_{out}, can be found by measuring the time it takes the motor to raise the mass a specific height. As in Sample Problem 3 in section 3.2, page 136, you will need to determine the work done in raising the mass.

Part A of the investigation is described in detail for you, so all you need to do is follow the instructions carefully and then make the calculations. In Part B, you apply the inquiry skills you developed in Chapters 1 and 3 to determine the energy input of a spring mechanism and the speed (and then the kinetic energy) of a cart that is moved by the spring.

 Voltmeters and ammeters should be used only under the supervision of your teacher.

Inquiry Skills

○ Questioning	● Conducting	● Evaluating
● Predicting	● Recording	● Communicating
● Planning	● Analyzing	● Synthesizing

You can find the efficiency of a spring-driven mechanical device by applying the equations for energy. When you use a dynamics cart and compression spring, the elastic potential energy of the spring provides the energy input; the energy output is the kinetic energy of the moving cart (**Figure 2**).

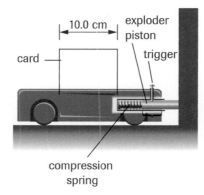

Figure 2
Elastic potential energy transforms into kinetic energy as the cart accelerates away from the barrier.

Questions

What measurements and calculations must be made to determine the efficiency of an electrical energy-transformation technology and a mechanical energy-transformation technology?

How does the efficiency of an electrical energy-transformation technology compare with the efficiency of a spring-driven mechanical energy-transformation technology?

(a)

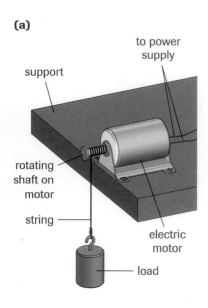

(b)

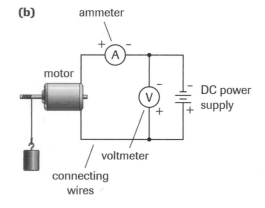

Figure 1
(a) A small electric motor raises a load.
(b) Using a voltmeter and ammeter to measure the voltage and current

Predictions

(a) For the electric motor described above, state what you must know or measure to determine the power input and power output.

(b) For the mechanical technology described above, state what you must know or measure to determine the energy input and energy output.

(c) Predict an answer to the second Question above.

Materials

For each student:
safety goggles
graph paper

For each group of three or four students:

Part A
small electric motor
power supply for the motor
voltmeter
ammeter
connecting wires
clamp, to attach the motor to the bench
mass (e.g., 200 g)
piece of string (long enough to raise the mass from the floor to the level of the motor)
masking tape (if required)
metre stick
stopwatch

Part B
dynamics cart with an exploding spring
force sensor
centimetre ruler
mass scale (to determine the mass of the cart)
method to determine the speed of the cart (e.g., a ticker-tape timer, a motion detector, or a photogate and related apparatus)
10.0-cm card mounted vertically on the cart (if the photogate is used)
barrier (against which the cart is launched)

Procedure
Part A

1. Use the clamp to secure the motor safely to the lab bench. Refer to **Figure 1** on page 189 for the setup.

2. Tie or tape one end of the string to the rotating shaft of the motor, and attach the mass to the other end so it is just above the floor. Align the metre stick vertically beside the string so you can determine the distance the mass is raised.

3. Put on your safety goggles. Use the stopwatch to determine the time interval in which the motor raises the mass a specific distance, such as 0.60 m. As the mass is being raised, record the voltage and current readings on the voltmeter and ammeter. Repeat the trial three or four times to determine an accurate average value of the time interval, voltage, and current.

 Wear your safety goggles in case the string breaks.

Ensure that the electrical equipment is in good condition and connected correctly.

Keep hands away from the rotating shaft when the motor is operating.

Stop the motor before the mass reaches the rotating shaft.

Part B

4. Refer to **Figure 2** on page 189 for the setup for this part. Create a list of measurements, calculations, and safety considerations you intend to follow in order to determine
   (i) the energy input of the compressed spring in the dynamics cart (Hint: Refer to the materials list as well as to Activity 3.3, page 140, and Investigation 3.5, page 147.)
   (ii) the energy output (i.e., the kinetic energy) of the cart just after the spring has been exploded against the barrier
   Have your teacher approve your list before proceeding.

5. Carry out the measurements and calculations you listed in step 4. Repeat two or three times to ensure accuracy.

Analysis

(d) Write an energy-transformation equation for each energy-transformation technology you investigated.

(e) Using the data in Part A, determine the output power, input power, and efficiency of the electric motor. Draw an energy flow diagram for the motor.

(f) Using the data in Part B, determine the output energy, input energy, and efficiency of the mechanical energy-transformation device. Draw an energy flow diagram for the situation.

(g) Compare the efficiencies of the technologies you investigated.

(h) Compare the efficiencies you found with the values obtained by other groups.

(i) Summarize the measurements and calculations as set out in the first Question in this investigation.

Evaluating

(j) Describe any difficulties you had with the measurements in this investigation. What did you do to minimize those difficulties?

(k) How does your prediction regarding the second Question compare with your answer for (g) above?

Synthesis

(l) Describe the features of an energy-transformation technology that has
(i) a high efficiency
(ii) a low efficiency

An electric hair dryer in good condition is approximately 95% efficient (**Figure 1(a)**). In other words, about 95% of the electrical input energy has been transformed into useful thermal energy in the heating coil (to heat the air) and kinetic energy of the molecules of moving air (driven by the fan). But you may know from experience that if the movement of the air becomes blocked by lint and hair, the dryer overheats. It may shut down if it gets too hot. The dryer's efficiency has been reduced because more of the electrical input energy now heats the dryer itself, and less input energy is available to heat the air molecules (**Figure 1(b)**).

A hair dryer is a typical example of a device whose efficiency can vary significantly. Almost all devices that have low efficiency waste some input energy as unwanted heat, which is waste thermal energy. But energy can be wasted in other ways. In this section, we look at the efficiencies of several energy-transformation technologies and some ways in which those efficiencies might be improved.

Table 1 lists typical percent efficiencies of several energy resources and energy-transformation technologies. (A thorough discussion of energy resources, including their benefits and drawbacks, can be reviewed from section 3.7.) Note that most of the efficiencies given in **Table 1** cover a single technology. However, as you will see below, a fair comparison of efficiencies is much more complex and must include several energy transformations. For example, although the efficiency of generating electrical energy at a generating station that uses fossil fuel is about 40%, a great deal of additional energy is required to extract the petroleum, refine it, and deliver it to the station. This reduces the overall efficiency.

CAREER CONNECTION

A career of increasing importance is an energy auditor. Energy auditors implement strategies to ensure the efficiency of a building—commercial and residential—in many areas, for example, energy, air and water quality, and waste management.

GO www.science.nelson.com

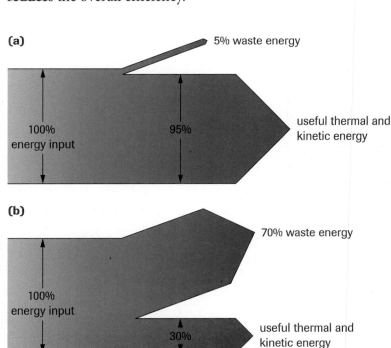

Figure 1
(a) The hair dryer operating with a high efficiency
(b) The hair dryer operating with a reduced efficiency and increased waste energy

Table 1   Typical Efficiencies of Energy-Transformation Technologies

Technology	% Efficiency	Comments
Nonrenewable energy resources		
fossil fuel electric generating station	40%	This is Canada's most common source of electrical energy. Oil and gas supplies are limited, and using them adds to pollution and global warming. Coal is more plentiful, but it also contributes more pollution.
nuclear fission reactor	30%	Ontario has some very large CANDU reactors. They are expensive to build and maintain, but they provide large quantities of electrical energy. Storage of nuclear waste poses hazards to the environment.
Renewable energy resources		
solar cells	10%–15%	These cells operate at low efficiency because they only use the higher-energy radiation from the Sun. Commercial solar cells are gradually improving in efficiency. Some experimental cells with new materials are reaching 35% efficiency.
passive solar heating	varies	This method of using solar energy can easily be included in new construction or added to existing buildings.
active solar heating	varies	Installation costs are high at present but will decrease as the technology becomes more common.
hydraulic generating station	95%	Canada has already constructed many dams along rivers to divert the water flow through generators. Large hydraulic sites are often far from the consumer.
wind generator	45%–55%	The efficiency of a wind turbine depends on how well the wind turbine "stops the wind" that hits it. The air molecules continue to move with a reduced kinetic energy after striking the blades, giving an efficiency of about 50%. Approximately 10 000 wind turbines would be needed to replace one fossil fuel generating station. Wind generators are becoming more common as fossil fuel prices rise.
tidal generator	95%	Tides undergo two complete cycles of rising and falling in just over a day, so a large amount of gravitational potential energy could be transformed into electrical energy along Canada's ocean coasts. However the setup cost is high.
biomass generator	5%	This resource has a very low efficiency when used to produce electrical energy; however, it uses material that would otherwise be burned or discarded.
geothermal generator	15%–20%	This resource is used primarily in Iceland, New Zealand, and the United States but may become more common in parts of western Canada in the future.
nuclear fusion reactor	20%–50%	This may become an important resource in the future. Its efficiency is uncertain, and initial costs are very high.
hydrogen fuel cell	40%	This resource is currently being developed and tested.
Energy-transformation technologies		
fan-driven electric heater	95%	This is similar to a fan-driven electric hair dryer.
transporting oil	97%	The fuel must be transported from the source to the consumer.
large electric motor	95%	Little thermal energy results from this transformation.
automobile	10%–16%	This efficiency rating includes energy transformations in all parts of the car, from the engine to the transmission to the wheels.
home gas furnace	85%	Natural gas burns cleanly, so its efficiency for heating is high.
fluorescent light	20%	Long fluorescent tubes have fairly high installation costs; the small fluorescent lamps are more convenient.
incandescent light	5%	*Incandescent* means "at a high temperature," which underscores the low efficiency in producing light energy.

Note: All values are approximate. They depend on several factors not described here.

(a)

(b)

Figure 2
(a) Low-voltage transmission lines
(b) High-voltage transmission lines

The efficiencies listed in **Table 1** do not include the additional energy needed to transmit the electrical energy produced by each of these devices to the consumer. Generating stations are usually far from consumers. Thus, the electrical energy must be transmitted large distances using transmission lines. **Figure 2** shows two examples of transmission lines: **(a)** low-voltage lines and **(b)** high-voltage lines. The high-voltage lines allow the electric charges to flow much more easily, so the lines do not heat up as much as low-voltage lines. Thus, the high-voltage lines waste a smaller portion of the energy and have higher efficiency in transmitting it.

We can illustrate the efficiency of transmission lines using the power efficiency equations: Two different transmission lines transmit electrical energy. Both require a power input of 2.00×10^5 W. The low-voltage lines yield a power output of 1.80×10^5 W; the high-voltage lines yield a power output of 1.99×10^5 W. The percent efficiency of the low-voltage lines is

$$\% \text{ eff} = \frac{P_{out}}{P_{in}} \times 100\%$$

$$= \frac{1.80 \times 10^5 \text{ W}}{2.00 \times 10^5 \text{ W}} \times 100\%$$

$$\% \text{ eff} = 90.0\%$$

The percent efficiency of the high-voltage lines is

$$\% \text{ eff} = \frac{1.99 \times 10^5 \text{ W}}{2.00 \times 10^5 \text{ W}} \times 100\%$$

$$\% \text{ eff} = 99.5\%$$

It is evident that high-voltage transmission lines are much more efficient for transmitting electrical energy than low-voltage lines. However, high-voltage lines require a lot of space, and they are expensive to build. Like so many other technologies, they have both advantages and disadvantages.

We also waste electrical energy in our homes and offices without realizing it: Many electronic appliances draw power 24 h a day, even though they are in use for only a small fraction of that time. These appliances are in "standby mode"; they are standing by, waiting for you to use the remote control or other activator. Examples include computers, TVs, VCRs, CD and DVD players, and garage-door openers. Some appliances draw as high as 15 W or 20 W on standby. Although this is less than a 60-W light bulb, most homes and offices across the country have one or more devices on standby every day, all day long—the total effect is huge. (You can verify this in Practice question 3.)

> **Practice**

Understanding Concepts

1. (a) Rank the following energy resources from the highest to lowest efficiency in generating electrical energy: wind generator; biomass generator; hydraulic generating station; oil-fired generating station; geothermal generator.
   (b) Explain why some resources are much more efficient than others at generating electrical energy.

2. The input power of a transmission line is 2.75×10^5 W, and the output power is 2.74×10^5 W.
   (a) Determine the efficiency of this transmission line.
   (b) Determine the percentage of the input power lost during transmission.
   (c) Is this transmission line high or low voltage? How do you know?

Making Connections

3. Assume there are 32 million TVs in Canada, with average standby power of 8.0 W each. Also assume that the standby feature is used for an average of 22 h each day.
   (a) Calculate the total energy consumed on TV standby each 22 h. Express your answer in joules, megajoules, and terajoules ($1 \, TJ = 10^{12} \, J$).
   (b) To save energy, would you be willing to push the button on the TV itself to turn it on, rather than use the remote control? Discuss your answer.
   (c) If you were paying the hydro bill for your home, how would you feel about paying for this wasted energy?

4. Do you think the Canadian government should require standby power use to be labelled on all new appliances? Justify your answer.

Answers

2. (a) 0.996, or 99.6%
   (b) 0.4%

3. (a) 2.0×10^{13} J; 2.0×10^7 MJ; 2.0×10^1 TJ

Overall Efficiency

You have seen that the efficiency of a fan-driven electric heater is about 95%. The efficiency of an electric baseboard heater with no fan approaches 100%. However, these figures are the percentage efficiency of one energy transformation, from electrical energy to thermal energy, after the electrical energy has already been delivered to the heater. The methods used to produce and deliver the electricity might be less efficient. To compare the use of technologies fairly, we must determine the **overall efficiency** of the use of the technology. Overall efficiency is the accumulated efficiency of all the energy transformations required to operate an energy-transformation technology. In equation form, the overall efficiency is the product of the efficiencies of each transformation:

$$\text{eff}_{overall} = (\text{eff}_1)(\text{eff}_2)(\text{eff}_3) \ldots$$

overall efficiency the total efficiency of all the energy transformations required to operate an energy-transformation technology; obtained by calculating the product of the efficiencies of all stages

▶ **SAMPLE** problem **1**

Overall Efficiency

Determine the overall efficiency of an oil furnace given the following energy transformations and their efficiencies: extracting the oil is 97% efficient; transporting the oil is 97% efficient; operating the oil furnace is 75% efficient.

Solution

$\text{eff}_1 = 0.97$

$\text{eff}_2 = 0.97$

$\text{eff}_3 = 0.75$

$\text{eff}_{overall} = (\text{eff}_1)(\text{eff}_2)(\text{eff}_3)$

$\qquad\qquad = 0.97 \times 0.97 \times 0.75$

$\text{eff}_{overall} = 0.71$

The overall efficiency of the oil furnace is 0.71, or 71%.

LEARNING TIP

Calculating Overall Efficiency
To see why overall efficiency is the product of the individual efficiencies, consider this example: If you received 80% on a test and your friend received 75% of what you did, then your friend's mark is 75% of your mark. The "of" means taking the product, so your friend's mark is 75% of 80% or (0.75)(0.80) = 0.60, which is 60%.

The energy transformations and the overall efficiency of the oil furnace in Sample Problem 1 are illustrated in the energy flow diagram in **Figure 3(a)**. Similar calculations can be used for the energy transformations and overall efficiency of an electric heater. The results are shown in the energy flow diagram in **Figure 3(b)**. Note that the efficiency drops dramatically when the electrical energy is obtained from a coal-fired generating station. Some energy is needed to mine and transport the coal. A large amount of energy is needed to generate the electrical energy and transport it to the customer. At that stage, even though the electric heater itself has an efficiency of 100%, only 34% of the original chemical potential energy in the coal is left. Note that that the overall efficiency of using the oil furnace is higher than the overall efficiency of using the electric heater, even though the oil furnace itself is less efficient than the electric heater.

(a)

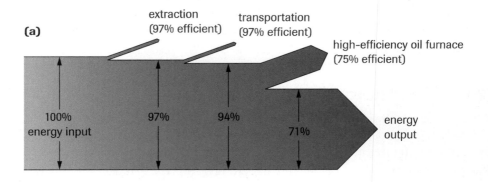

extraction (97% efficient)
transportation (97% efficient)
high-efficiency oil furnace (75% efficient)
100% energy input
97%
94%
71%
energy output

(b)

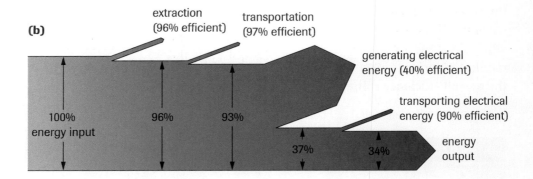

extraction (96% efficient)
transportation (97% efficient)
generating electrical energy (40% efficient)
transporting electrical energy (90% efficient)
100% energy input
96%
93%
37%
34%
energy output

Figure 3
(a) Overall efficiency of an oil furnace
(b) Overall efficiency of an electric heating system

Answer

5. 16%

> ▶ **Practice**

Understanding Concepts

5. Using the following information, draw an energy flow diagram to show the overall efficiency of operating a gasoline-driven car. Then calculate the overall efficiency.
   - The oil extraction is 97% efficient.
   - The oil is refined to produce gasoline with 95% efficiency.
   - The transportation of the fuel to the gasoline station is 94% efficient.
   - The engine is 24% efficient.
   - The transmission is 76% efficient.

Using Energy-Transforming Devices Efficiently

We all rely heavily on electrical energy, but its production creates much pollution and consumes vast amounts of nonrenewable resources. To improve the efficiency of energy production, we could use some of the thermal energy created during electricity generation as an alternative source of energy. This would reduce thermal pollution and provide a source of useful heat at the same time. The process of producing electrical energy or heat from waste thermal energy is called **cogeneration**. Today, cogeneration is used mainly in industrial plants located near generating stations. It may become important in the future as new ways are found to broaden cogeneration so it applies to other forms of renewable energy, such as solar energy.

As you have learned, transferring energy over long distances using huge transmission lines is expensive; it also requires large areas of land. Therefore, another way to improve the efficiency of generating electrical energy is to produce it for **local consumption**. Sources of energy for localized power include renewable resources such as solar, hydraulic, and wind energies, especially when combined with hydrogen fuel cell technology.

cogeneration the process of using waste thermal energy for the production of heat or electrical energy

local consumption the production of electrical energy close to the consumer

▶ *Practice*

Making Connections

6. Could cogeneration and local consumption be implemented to supply some of the energy needs of your school? Explain your answer.

7. Research the Kyoto Accord, and answer the following questions:
   (a) What is the Kyoto Accord, and when did Canada sign it?
   (b) How did the Accord alter energy policy in Canada?
   (c) Has it stimulated "new energy" technologies?
   (d) Are we meeting our goals?
   (e) Have other countries followed through with their goals?

 www.science.nelson.com

Evaluating a Renewable Energy Resource

Canada's population is growing, fossil fuel resources are decreasing, and the costs of electrical energy are increasing. Therefore, we must find ways of generating electrical energy from renewable energy resources. All sources of energy have advantages and disadvantages. To complicate the issue, sources appropriate for one location may be unusable at other locations.

For example, a ranch or farm in an area with regular winds can use a wind turbine connected to an electric generator. However, a home in an area that seldom experiences strong winds would be a poor location for a wind turbine. The wind energy map in **Figure 4(a)** shows which areas of Canada experience strong enough winds to warrant building wind turbines. Note that local conditions may differ from the average values depicted on the map. The wind turbine shown in **Figure 4(b)** is in a windy location in Toronto, Ontario, near Lake Ontario.

To evaluate the effectiveness of a source of energy not used regularly in your area, you can analyze the advantages and disadvantages of one or more sources. Start by researching the topic using a variety of references, such as books, magazines, and the Internet. Determine the benefits of the source of energy as well as the risks to the environment and human health. Then summarize what you discover, suggest how the alternative source might be used in your area, and evaluate your research and your plan.

Understanding the Issue

(a) Describe the difference between an energy-transforming device and a source of energy. Give examples of each, using renewable energy sources available in your area.

(b) Give reasons why we must find alternative sources of energy and energy-transforming devices.

(c) Describe centralized and decentralized generation of electrical energy. Give examples using renewable energy resources.

Take a Stand

Research the advantages and disadvantages of an alternative energy source or energy-transforming device in a specific location, such as your own area. Decide whether you would recommend the resource and/or device, and defend your decision. Create a position paper (such a poster board, an essay, a letter, a web page, a video, or an audio presentation) to summarize your research and stand. Your analysis should include answers to the following questions:

(d) What physics principles apply to the design and/or use of the resource or technology?

(e) What are the economic, social, and environmental impacts of the resource or technology?

(f) What are the benefits of the resource or technology?

(g) What careers are related to the design and/or use of the resource or technology?

Evaluation

(h) Evaluate the usefulness of your reference material.

(a)

(b)

Figure 4
(a) Mean annual wind energy density, 1996–2000. Pale yellow areas are not considered good locations to generate electrical energy by wind. (The units of measurement are W/m^2 at an elevation of 50 m.)
(b) A wind turbine near the lakeshore in Toronto

SUMMARY *Using Energy Efficiently*

- When comparing the efficiency of energy resources and transformation technologies, the overall efficiencies must be calculated.

- Overall efficiency can be calculated using the equation
  $\text{eff}_{\text{overall}} = (\text{eff}_1)(\text{eff}_2)(\text{eff}_3). \ldots$

- Overall efficiencies include the efficiency of the extraction, refining, and transportation of the resource, or transmission of the electrical energy.

- The efficiency of energy use may be improved through renewable energy resources, cogeneration, and local consumption.

▶ Section 4.4 Questions

Understanding Concepts

1. To maintain high efficiency, what type of maintenance do hair dryers and fuel-fired furnaces need?

2. Suggest ways of maximizing the efficiency of lighting your home.

3. (a) Determine the overall efficiency of fluorescent lights in a school using these assumptions:
   - Oil is extracted at an efficiency of 97%.
   - Oil is transported at an efficiency of 97%.
   - Oil is used to generate electrical energy at an efficiency of 39%.
   - Electrical energy is transmitted at an efficiency of 98%.
   - The fluorescent lights operate at an efficiency of 19%.
   (b) Draw an energy flow diagram to illustrate the overall efficiency found in (a).

4. **Table 2** lists the data for four sets of transmission lines for electrical energy. Copy the table into your notebook, and complete the missing information, showing the equation used to solve each part. Then rank the four sets of transmission lines from highest voltage to lowest voltage based on their efficiencies.

Applying Inquiry Skills

5. Two new electric kettles, one made of stainless steel and the other of plastic, are both rated at 1500 W. Describe how you would investigate the two kettles to compare their efficiencies as they heat water. (Numerical values are not required here.)

Making Connections

6. Make up four campaign slogans to promote "Energy-Wise Week" in your area. Choose one of the slogans, and create an effective diagram to accompany it. Trade your diagram with another student and evaluate each other's ideas. (Refer to Appendix A6 for a sample form you can use for peer evaluation.)

7. Should the Canadian government set a maximum standby power for all new appliances and electronic equipment? Explain your answer. What maximum would you recommend?

8. Research the use of light sources not mentioned in this section, such as halogen and xenon lights. Where are they used? How do their efficiencies compare with the efficiencies of incandescent and fluorescent light bulbs?

Table 2 For Question 4

	% Efficiency	Power Output, P_{out}	Power Input, P_{in}
(a)	?	0.198 MW	0.200 MW
(b)	?	1.32 MW	1.44 MW
(c)	95.5%	?	888 kW
(d)	97.4%	914 kW	?

Key Understandings

4.1 Power

- **Power** is the rate of doing work or transforming energy; it is found using the equation $P = \dfrac{W}{\Delta t}$ or $P = \dfrac{\Delta E}{\Delta t}$.
- The SI unit of power is the **watt** (W).

4.2 Efficiency

- The **efficiency** of an energy-transformation technology can be found using the equation $\text{eff} = \dfrac{E_{out}}{E_{in}}$ or $\text{eff} = \dfrac{P_{out}}{P_{in}}$.
- A **heat engine** transforms thermal energy into mechanical energy with a typical efficiency in the range of 25%–40%.

4.3 Investigation: Determining Efficiencies

- The efficiency of an electric motor used to raise a mass is found using the equation $\text{eff} = \dfrac{P_{out}}{P_{in}}$, and the efficiency of a spring-driven mechanism is found using the equation $\text{eff} = \dfrac{E_{out}}{E_{in}}$.

4.4 Using Energy Efficiently

- The **overall efficiency** of an energy-transformation technology is found using the equation $\text{eff}_{overall} = (\text{eff}_1)(\text{eff}_2)(\text{eff}_3)\ldots$.
- The efficiency of energy use may be improved through renewable energy resources, **cogeneration**, and **local consumption**.

Key Terms

4.1
power (P)
watt (W)

4.2
efficiency (eff)
heat engine

4.4
overall efficiency
cogeneration
local consumption

Key Equations

4.1

- $P = \dfrac{W}{\Delta t}$

- $P = \dfrac{\Delta E}{\Delta t}$

4.2

- $\text{eff} = \dfrac{E_{out}}{E_{in}}$

- $\% \text{ eff} = \dfrac{E_{out}}{E_{in}} \times 100\%$

- $\text{eff} = \dfrac{P_{out}}{P_{in}}$

- $\% \text{ eff} = \dfrac{P_{out}}{P_{in}} \times 100\%$

4.4

- $\text{eff}_{overall} = (\text{eff}_1)(\text{eff}_2)(\text{eff}_3) \ldots$

Problems You Can Solve

4.1

- Given one of power, energy transferred, or time interval, determine the other two quantities.
- Apply the SI unit of power, the watt.

4.2

- Determine the efficiency of an energy-transformation device, given the energy output and input, or the power output and input.
- Identify the key parts of a gasoline or diesel engine, and describe the function of each part.
- Describe factors that affect the efficiency of a heat engine.

4.3

- Describe how you can experimentally determine the efficiency of an energy-transformation device.

4.4

- Account for the efficiency of a variety of energy-transformation technologies.
- Determine the overall efficiency of an energy-transformation technology given the efficiencies of the step-by-step transformations.
- Describe ways in which the efficiency of energy use can be improved, and explain why it is wise to do so.

▶ *MAKE* a summary

In the middle of a piece of notepaper, sketch an apartment building or a house in your area. On the left side, show all the ways in which energy can be provided for appliances, heating, and any other uses you can think of. On the right side, show ways in which the energy transmitted to the building can be used. Wherever possible, indicate the power and approximate efficiency of the devices shown. Include at least one energy flow diagram, and use as many concepts, key words, and equations from this chapter as you can.

Write the numbers 1 to 8 in your notebook. Indicate beside each number whether the corresponding statement is true (T) or false (F). If it is false, write a corrected version.

1. The power of a student running up a set of stairs increases if the time taken increases.

2. One watt is equivalent to one joule-second.

3. A 1000-kW generating station produces the same amount of electrical energy per day as a 1.0-MW generating station.

4. In an engine with an efficiency of 25%, the portion of the original energy input that is wasted is 25%.

5. A jet engine is an example of a heat engine.

6. In a two-stroke gasoline engine, input and output ports are used rather than valves.

7. A two-stroke gasoline engine does not require spark plugs because the air–gasoline mixture is highly compressed.

8. Low-voltage electrical-energy transmission lines are more efficient than high-voltage lines.

Write the numbers 9 to 16 in your notebook. Beside each number, write the letter corresponding to the best choice.

9. The time interval required for a 5.0-W portable radio to consume 1.0 kJ of energy is
   (a) 5.0 s
   (b) 2.0×10^2 s
   (c) 1.0×10^3 s
   (d) 5.0×10^3 s

10. In **Figure 1**, the portion of energy wasted is
    (a) 33%
    (b) 67%
    (c) 81%
    (d) 41%

11. In **Figure 1**, the overall efficiency of the technology is
    (a) 33%
    (b) 67%
    (c) 81%
    (d) 41%

12. If the energy input of the technology in **Figure 1** is 8.0 MJ, then the useful output energy is
    (a) 7.6 MJ
    (b) 6.8 MJ
    (c) 3.3 MJ
    (d) 2.6 MJ

13. The technology represented in **Figure 1** is most likely a
    (a) large electric motor
    (b) hydraulic generating station
    (c) home heating system
    (d) fluorescent light bulb

14. An investigation compares the work done in pulling a cart up a ramp (the "pulling work") with the work needed to lift the cart vertically (the "vertical work").
    (a) The pulling work = the vertical work because the cart is raised to the same height.
    (b) The vertical work > the pulling work because the vertical force is greater.
    (c) The pulling work > the vertical work because friction must be overcome.
    (d) The pulling work > the vertical work because the distance moved is greater.

15. In a four-stroke heat engine, the order of strokes after intake is
    (a) exhaust, compression, power
    (b) power, compression, exhaust
    (c) exhaust, power, compression
    (d) compression, power, exhaust

16. Of the following choices, the renewable energy resource most suitable for local consumption in Canada's largest cities is
    (a) tidal energy
    (b) geothermal energy
    (c) hydraulic energy
    (d) solar energy

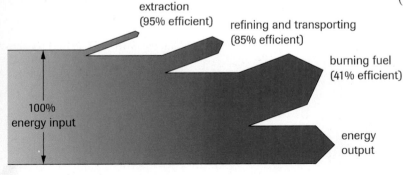

extraction
(95% efficient)

refining and transporting
(85% efficient)

burning fuel
(41% efficient)

100%
energy input

energy
output

Figure 1

202 Chapter 4   An interactive version of the quiz is available online.   NEL

GO   www.science.nelson.com

Understanding Concepts

1. State which of the following quantities, if any, are scalar: power; work; time interval; energy output; percent efficiency.

2. Calculate the power of a light bulb that consumes 54 kJ of electrical energy in 1.0 h.

3. The world's smallest submarine, *The Water Beetle*, is only 3.0 m long. It operates on air cylinders with a total power of 3.0 kW. When searching for submerged wrecks, the submarine can stay under water for 4.0 h. How much energy, in joules and megajoules, does the submarine use in this time?

4. The world's tallest fountain (in Fountain Hills, Arizona) shown in **Figure 1** spews out water that reaches a height of 1.7×10^2 m. Each minute, the water pump ejects 2.6×10^4 kg of water to that height. Calculate
   (a) the total work done by the pump on the water each minute
   (b) the power of the pump in watts and megawatts (Assume 100% efficiency.)

Figure 1

5. Sometimes scientific symbols or terms are used incorrectly in the media. State what is wrong with each of the following examples, and write a corrected version:
   (a) "The CANDU reactor has a power of 1.8 mw."

(b) "The energy of the light bulb is 54 W."
   (c) The power of the wind generator is 3.5 kW/day.
   (d) On a package for a 52-watt bulb, intended to replace a 60-watt bulb: "Save 8 w of energy."

6. Objects A, B, and C in **Figure 2** are pushed up a ramp, which is 4.0 m long and 1.2 m high. The applied forces are parallel to the ramp with magnitudes of 9.8 N on A, 4.9 N on B, and 3.1 N on C.
   (a) Predict how the ramp's efficiency will compare in the three situations. Explain your reasoning.
   (b) Determine the efficiency of the ramp when A is pushed up, then B, then C.
   (c) Evaluate your prediction in (a).

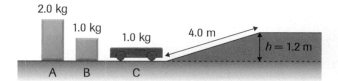

Figure 2

7. One of the world's largest wind energy farms is located at Altamont Pass, California. Its 7300 turbines produce a total power of 34 MW. If the power that reaches the nearly one million homes serviced is 33 MW, what is the efficiency of transmitting the electrical energy?

8. A certain swimmer consumes energy at a rate of about 0.60 kW.
   (a) What is the source of energy consumed?
   (b) How long does it take a long-distance swimmer to consume an amount of energy equivalent to that stored in one kilogram of coal (31 MJ)? Express your answer in seconds and hours.
   (c) How does the swimmer's power compare with that of an "average" horse?

9. **Figure 3** is a force–compression graph for a compression spring in a dynamics cart of mass 1.0 kg. After the spring explodes against a barrier, the cart moves the first 10.0 cm in 0.30 s. Calculate the
   (a) energy input of the system
   (b) energy output
   (c) efficiency of the transformation

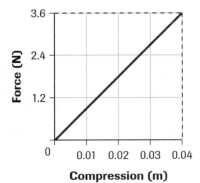

Compression (m)   **Figure 3**

10. State the function of each of the following parts of a four-stroke gasoline engine:
    (a) carburetor
    (b) spark plug
    (c) piston
    (d) flywheel

11. **Figure 4** shows a two-stroke diesel engine during one part of its cycle of operation.
    (a) What is happening during this part of the cycle?
    (b) Redraw the diagram in your notebook to show what happens a short time later, when the piston is moving upward and both valves are closed.

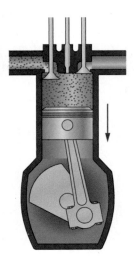

Figure 4

12. Describe the benefits and drawbacks of using high-voltage lines to transmit electrical energy.

13. For the situation described below, calculate the overall efficiency and draw the corresponding energy flow diagram:
    • Petroleum is extracted at 97% efficiency.
    • Petroleum is refined to obtain propane at 95% efficiency.
    • Propane is transferred to a supply depot at 97% efficiency.
    • A propane engine in a city bus operates at 42% efficiency.
    • The transmission and wheels of a bus operate at 51% efficiency.

14. Describe three ways in which electrical energy could be used more efficiently in your school.

15. Give the factors you must consider in order to compare the usefulness of different renewable energy resources in your region.

16. An alternative unit to the joule and megajoule is the kilowatt hour (kW•h), which is still used in many parts of Canada to measure electrical energy. One kilowatt hour is equivalent to one kilowatt of power used for one hour. Prove that 1.0 kW•h = 3.6 MJ.

Applying Inquiry Skills

17. **Figure 5** shows two dynamics carts initially at rest. The one on the left has a spring that can be compressed and then released. This setup can be used to determine the efficiency of the transformation from elastic potential energy to the kinetic energy of the carts.
    (a) Give the measurements you must make to determine the efficiency described.
    (b) Indicate how you would calculate the efficiency.
    (c) Describe how you would keep the sources of error in the measurements to a minimum.

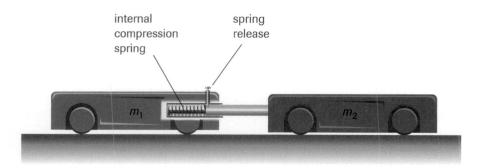

internal
compression
spring

spring
release

m_1

m_2

Figure 5

18. Think of a toy that involves at least one of the energy transformations presented in this chapter or in Chapter 3.
    (a) Describe the toy and the energy transformations that make it work.
    (b) Suggest an original design for a toy that involves at least one energy transformation.

19. The device shown in **Figure 6** can be used to determine the bounce efficiency of spherical objects made from a variety of materials, such as brass, plastic, steel, wood, rubber, and aluminum.
    (a) How would you use the device to determine the bounce efficiency of spheres made of various materials?
    (b) With your teacher's approval, use either this device or another method to determine the bounce efficiency of various spheres.

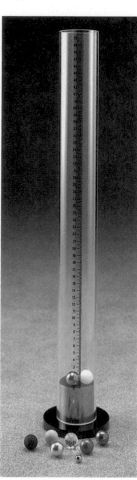

Figure 6

Making Connections

20. (a) List the appliances in your home that usually consume the most electrical energy.
    (b) Describe ways your family could reduce the electrical energy consumed by the appliances you listed in (a).

21. Gasohol is a mixture of gasoline and ethanol (pure grain alcohol) that can be used to operate vehicles. Research the use of gasohol in Canada.
    (a) How is ethanol obtained from corn, wheat, or other crops?
    (b) How does the ethanol you described in (a) compare with the ethanol obtained from cellulose in agricultural wastes, such as straw?
    (c) Which method is more environmentally friendly? Explain your answer.

 www.science.nelson.com

22. Suggest how cars could be more efficiently used.

23. Use the Internet or another resource to research the newest designs in automobile engines and hydrogen fuel cells. Research the materials used in making the new engines or fuel cells, the design features, the efficiencies, and so on. Write a summary of your findings.

PERFORMANCE TASK

Build an Energy-Transformation Technology

This unit is about energy and how it is transformed into various usable forms. Your task is to design and build a device that uses four energy transformations to accomplish a task. Depending on the option you choose, the device can be full-size or a scale model. Some concepts and ideas follow.

The Toyota Prius (**Figure 1**) is the world's first production car to combine a gasoline engine with an electric motor that never needs to be plugged in. The electric motor gains energy and becomes charged whenever the brakes are applied: In an energy transformation, the kinetic energy changes back into stored electrical energy. The concept makes sense—perhaps a physics student had the idea first!

Another idea is to connect a small generator to a bicycle. The generator could have an armature that rotates when it contacts the sidewall of a tire. In this way, the generator could produce electrical energy to operate lights or an attached radio or CD player. The same idea can be applied in a gym (**Figure 2**), where lights and/or music operate only when the exercise bikes or treadmills are being used.

Consider another example of an energy transformation in which waste heat can be made to generate electrical or mechanical energy: On a typical summer day, air in the attics of homes gets very hot. Fans on the roof begin rotating and help circulate the hot air (**Figure 3**). The rapid rotation of these fans can be used to operate a small electric generator or a fan in the living area below. This technology could reduce the amount of energy required to operate an air conditioning unit.

Figure 1
The Toyota Prius

Figure 2
A gym has a variety of pieces of cardiovascular equipment that could be connected to mini-generators to operate lights, radios, or other devices.

Figure 3
A rotating roof fan has kinetic energy that can be changed to other forms of energy.

These are just a few of the ways in which an energy-transformation technology can be combined with wise use of energy. In other words, your design should help improve efficiency. Your design provides a good start for the *Try This* Activity in section 3.6, page 153, or the one in section 3.7, page 160. Many ideas are also available in books and magazines, and on the Internet.

 www.science.nelson.com

Two options are available for this Performance Task. In Option 1, you design, build, test, modify, and analyze an energy-transformation technology that accomplishes a specific task. In Option 2, you research the details of the design and operation of an energy-transformation technology and then build a working model to analyze its operation. Both options apply the principles you learned in Chapters 3 and 4. The technology or model you build will involve energy, energy transformations, and efficiency. For either model, there should be at least four energy transformations.

The Task
Option 1 Build an Energy-Transformation Technology
Your task is to design, build, test, modify, and analyze an energy-transformation technology that uses at least four energy transformations. Your design will depend on the suggestions made by your teacher. Before starting the task, your group should decide on the final objective of the transformations, as well as the criteria to evaluate the technology. For example, the efficiencies of the technologies designed by various groups can be compared.

Option 2 A Model Technology
Your task is to research the design and operation of an energy-transformation technology that increases the efficiency of energy use. You will then create a model of the technology that involves the same energy transformations as the technology you researched. To build the model, you can use materials that are inexpensive and easily assembled (e.g., a 2-L plastic pop or water bottle can be altered to catch the wind and rotate on its axis). You will then describe its operation, including all the energy transformations.

Analysis
(a) What physics principles apply to the design and use of your technology?

(b) How can you judge whether your technology or model was successful?

(c) Who would benefit from the technology you designed or researched?

(d) What careers are related to the manufacture and use of this type of technology?

(e) What safety precautions did you follow in building and testing your technology or model?

(f) After testing your technology or model, how did you modify it to make improvements?

(g) How could the process you used in this task be applied in business or industry?

(h) List problems you had while building the technology or model, and explain how you solved them.

Evaluation
(i) How does your technology or model compare with the designs of other groups?

(j) If you did this task again, how would you modify the process to obtain a better final product?

1. Write the letters (a) to (h) in your notebook. Beside each letter, write the word or phrase that corresponds to each of the following:
   (a) the SI unit of power
   (b) the SI unit of work
   (c) the transfer of energy by means of a circulating path of fluid particles
   (d) "In any energy transformation, energy is neither created nor destroyed."
   (e) the sum of the gravitational potential energy and kinetic energy of a falling object
   (f) a measure of the energy transferred from a hot body to a cooler one
   (g) the ratio of the work to the time interval during which the work is done
   (h) a measure of the average kinetic energy of the atoms or molecules of a substance

   A the quantity that increases in proportion to the square of the object's speed
   B the quantity that depends on the force and the displacement in the direction of the force
   C the process in which a large nucleus splits into small nuclei
   D the process in which small nuclei join to produce a larger nucleus
   E the process of transferring energy by means of a circulating path of fluid particles
   F the process of transferring energy by means of particle collisions
   G the quantity that depends on the height of an object above a reference level
   H uranium
   I cogeneration

2. **Figure 1** is a diagram of a four-cylinder gasoline engine during one of its strokes. Write the letters A to J in your notebook, and then write beside each the name of the part of the engine labelled in the diagram. Also, identify the stroke that is represented.

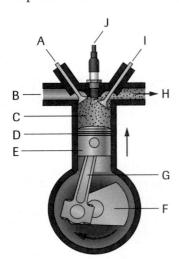

Figure 1

3. Write the letters (a) to (e) in your notebook. Beside each letter, write the letter from A to I that corresponds to each of the following terms:
   (a) gravitational potential energy
   (b) conduction
   (c) nuclear fusion
   (d) nonrenewable resource
   (e) work

Write the numbers 4 to 12 in your notebook. Indicate beside each number whether the corresponding statement is true (T) or false (F). If it is false, write a corrected version.

4. The unit of work is the joule, which is equivalent to a newton-metre.

5. For an ideal extension spring, the force applied to the spring increases directly as the stretch increases.

6. The type of heat transfer that does not require any particles is called radiation.

7. Power is a scalar quantity measured in watts.

8. As the time interval required to perform a given amount of work increases, the power increases.

9. When expressed as a percentage, an efficiency of 0.55 is 55%.

10. Electric motors tend to have much lower efficiencies than gasoline motors.

11. In a four-stroke gasoline engine, the spark jumps across the gap in the compression stroke.

12. The overall efficiency of an energy-transformation technology is generally lower than the efficiency of only one energy transformation.

Write the numbers 13 to 22 in your notebook. Beside each number, write the letter corresponding to the best choice.

13. **Figure 2** shows a graph of the magnitude of the force applied to a bungee cord versus the stretch of the cord. Assuming two significant digits, the area under the line is
    - (a) 50 J
    - (b) 25 J
    - (c) 200 N•m
    - (d) 200 N/m

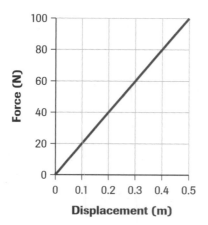

Displacement (m)   **Figure 2**

14. When force is plotted on the *y*-axis and the stretch of a spring on the *x*-axis, the slope of the line represents
    - (a) the elastic potential energy of the spring
    - (b) the force constant of the spring
    - (c) the work done in stretching the spring
    - (d) the stretch in metres per newton of force

15. If the speed of a car doubles, its kinetic energy
    - (a) stays the same
    - (b) becomes 2 times as much
    - (c) becomes 4 times as much
    - (d) becomes half as much

16. A power rating of 1.0 kW is equal to
    - (a) 1000 W
    - (b) 1.0 kJ/s
    - (c) 1.0×10^3 J/s
    - (d) all of these

17. If a 100-W light bulb consumes 1000 J of electrical energy in a time interval Δt, then in the same time interval, a 200-W light bulb consumes
    - (a) 1000 J
    - (b) 500 J
    - (c) 2000 J
    - (d) none of these

18. A ball, initially held at rest at a height of 100 cm above the floor, is released. After bouncing from the floor, it bounces back to a height of 40 cm. The useful output energy and the waste energy, expressed as a percentage of the initial gravitational potential energy, are, respectively,
    - (a) 100% and 0%
    - (b) 40% and 60%
    - (c) 60% and 40%
    - (d) 0% and 100%

19. On the second bounce of the ball in question 18, the ball reaches a height of
    - (a) 40 cm
    - (b) 24 cm
    - (c) 20 cm
    - (d) 16 cm

20. A pulley system operated with an electric motor consumes 5.0 kJ of electrical energy as it does 4.2 kJ of work in raising an object. The efficiency of this technology is
    - (a) 84
    - (b) 0.10
    - (c) 0.19
    - (d) 0.84

21. In an energy transformation, the energy output is 20 J, and the waste thermal energy is 80 J. The percent efficiency of the transformation is
    - (a) 20%
    - (b) 80%
    - (c) 60%
    - (d) 25%

22. An example of cogeneration is
    - (a) operating a nuclear generating plant located beside a natural gas generating plant
    - (b) heating a manufacturing building located beside a fossil fuel generating plant
    - (c) operating two fossil fuel generating plants side by side
    - (d) operating a windmill generator beside an active solar heating system

Understanding Concepts

1. A gravity clock, like the one in **Figure 1**, has a chain that is pulled to raise the metal cylinders to their top position. As the cylinders slowly move downward, the round pendulum swings back and forth and creates ticking sounds, the hands of the clock move around, and chimes ring every 15 min.
   (a) Starting with the work done in pulling the chain, write the energy-transformation equation for the gravity clock.
   (b) Calculate the work done on the chain by a force of magnitude 12 N applied over a distance of 0.80 m.
   (c) After the work calculated in (b) is complete, the output energy in operating the clock is 9.2 J. Find the clock's percent efficiency.
   (d) Does an efficiency of less than 100% violate the law of conservation of energy? Explain your answer.

Figure 1

2. State the SI unit used to measure (a) thermal energy, (b) work, and (c) power.

3. Describe the difference between "energy transferred" and "an energy transformation," giving an example of each.

4. Hydraulic energy-generating facilities, such as the one in **Figure 2**, store large quantities of water behind the dam.
   (a) Write the energy-transformation equation that ends in electrical energy for this facility.
   (b) What evidence is there in the photograph that the electrical energy is transmitted by high-voltage transmission lines?
   (c) Calculate the gravitational potential energy of the water in the lake behind the dam if it contains 3.2×10^{12} kg of water at an average height of 12 m above the level of the turbines.

Figure 2
This is the Revelstoke Dam and Generating Station on the Columbia River in British Columbia.

5. Calculate the kinetic energy in each of the following:
   (a) A Pacific leatherback turtle, the world's largest turtle species, has a mass of 8.6×10^2 kg and is swimming at 1.3 m/s.
   (b) A 7.0×10^3-kg African elephant is running at 7.9 m/s.
   (c) A 19-kg mute swan is flying at 94 km/h.

6. A personal watercraft and its rider have a combined mass of 405 kg and a kinetic energy of 3.11×10^4 J. Determine the speed of the craft.

7. A bullet travelling at a speed of 9.0×10^2 m/s has a kinetic energy of 2.0×10^3 J. Calculate its mass.

8. Assume that the executive jet shown in **Figure 3** has a mass, including passengers and cargo, of 6.7×10^3 kg and is travelling at 6.4×10^2 km/h.
   (a) Determine its speed in metres per second.
   (b) Calculate the jet's kinetic energy in joules and megajoules.

Figure 3
The Bombardier Learjet 45, built by Canadian firm Bombardier Inc., is one of the first executive jets designed entirely on computer.

9. Explain why roller coaster rides always start by going uphill.

10. You drop a hard-boiled egg, initially at rest, from a height of 11 cm onto a countertop in order to crack the shell. The egg's mass is 0.052 kg.
    (a) Calculate the egg's initial gravitational potential energy relative to the countertop.
    (b) What is the egg's final gravitational potential energy just as it hits the countertop?
    (c) Into what form of energy did the gravitational potential energy transform as it fell? How much of this form of energy is there just as the egg is about to hit the countertop?
    (d) Use your answer to (c) to determine the maximum speed of the egg just before it lands on the countertop.

11. A skateboarder is initially at rest at the top edge of a vertical ramp half-pipe with a radius of 2.5 m (**Figure 4**). The total mass of the skateboard and rider is 64 kg. Assume that all the gravitational potential energy at the top of the pipe is transformed into kinetic energy at the bottom.
    (a) Determine the skateboarder's speed at the bottom of the half-pipe.
    (b) If you apply the law of conservation of energy to this type of question, you can solve the problem without needing to know the mass. Show that this is true in this case.

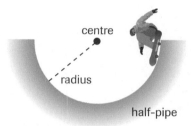

Figure 4

12. Two students are pulling horizontally in opposite directions on a desk, and the desk moves 3.7 m at a constant velocity. One force is 95 N to the right (considered to be the positive direction), and the other force is 68 N to the left. A force of kinetic friction of 29 N acts to the left.
    (a) Draw an FBD of the desk, and state why it must be moving at a constant velocity.
    (b) Determine the work done on the desk by each of the three horizontal forces.
    (c) Determine the total work done on the desk by the three horizontal forces.

13. Explain each of the following in terms of the concept of heat transfer by conduction, convection, or radiation:
    (a) An astronaut on a space walk outside the *International Space Station* is concerned about only one method of heat transfer.
    (b) Which method heats the air in your classroom in winter?
    (c) Weather stripping around doors and windows reduces energy losses in a home.

14. State one advantage and one disadvantage of each of the following energy resources:
    (a) active solar heating
    (b) passive solar heating
    (c) tidal energy
    (d) natural gas

15. (a) Compare the amount of work you do in climbing a vertical rope in 6.0 s with the work done in climbing the same rope in 12 s, if both activities get you 6.0 m above your starting point.
    (b) Repeat (a) for power rather than work.

16. Calculate the power of a light bulb that transforms 1.5×10^4 J of energy per minute.

17. Calculate the amount of energy transformed by a 1.2×10^3-W electric kettle during 5.0 min of operation.

18. What is the final form of energy that renders all energy-transformation technologies less than 100% efficient?

19. Draw a sketch of the parts and the piston of a two-stroke gasoline engine showing how the position of the piston controls which port is open at any particular instant.

20. Every 35 s, a chair lift takes snowboarders to the top of a hill that is 3.6×10^2 m high. The average mass of a snowboarder complete with equipment is 72 kg. Determine the power required to deliver three snowboarders. (Assume the snowboarders join and leave the lift at full speed.)

21. Draw an energy flow diagram to show what happens at a local electrical generating facility that uses biomass energy.

22. Calculate the overall efficiency and draw the corresponding energy flow diagram for the following process:
    • petroleum is extracted at 97% efficiency
    • the petroleum is refined to obtain gasoline at 94% efficiency
    • the gasoline is transferred to a service centre with 96% efficiency
    • the gasoline engine in a car operates at 26% efficiency
    • the transmission and wheels of the car operate at 66% efficiency

Applying Inquiry Skills

23. (a) Use actual measurements to determine how much work you must do on this textbook to raise it from the floor to your desktop.
    (b) Describe sources of error in your measurements.

24. The Bat, a roller coaster at Paramount Canada's Wonderland, takes its riders both forward and backward (**Figure 5**). An electric motor pulls the coaster from the loading platform backward to the top of the starting side. After the coaster is released, it travels down the ramp, along the tracks, through the loops, and partway up the second side. There, the coaster is pulled up to the top by another electric motor before being released for the backward trip.
    (a) Write the energy-transformation equation to show the transformations that occur from the time the coaster starts to be pulled up the first hill to the time it brakes to a stop. (Numerical values are not required.)
    (b) Why is the coaster unable to get to the top of the second side without the aid of a motor?
    (c) Describe what you must know or measure to determine the efficiency of this ride.

Figure 5
In order to travel through the loops, this roller coaster must be released from the highest position of the track.

Making Connections

25. Explain why some roller coaster rides are shut down on very cold or windy days.

26. It is possible to judge a material's ability to conduct heat simply by touching the material.
    (a) Touch the following objects with your fingertips: this page, the cover of this textbook, your chair or desk legs, your desktop, a pencil, a pen, and some cloth. Which materials that you touched are the best heat conductors? How can you tell?
    (b) When testing the temperature of the milk in a baby bottle, we are advised to let a few drops of milk fall on our wrist. Why do you think that the wrist is recommended, rather than fingertips?

27. The label on a cereal box says that the energy content of the 620 g of food in the box is 9.1 MJ. It also states that there are 31 portions in the box.
    (a) Calculate the energy content per portion.
    (b) Assume a student has a power of 9.5×10^2 W while running. How long (in seconds and minutes) must this student run to consume the energy of one portion of this cereal?
    (c) How could you judge whether the number of grams per portion is truly the average size portion?

28. (a) What is a perpetual motion machine?
    (b) Is such a machine possible? Justify your answer. (Include the law of conservation of energy in your reasoning.)

29. The airbag shown in the test car in **Figure 6** is activated after the car experiences a front-end impact. A sensor sends an electric signal to a computer, which sends signals to the airbag's ignition system. The resulting explosion sends gases quickly into the airbag, inflating it in less than a second.
    (a) Write the energy-transformation equation for this situation. (Include sound energy and thermal energy, where appropriate.)

Figure 6

   (b) The crash-test dummy in the photograph is secured by a seat belt. Why is it wise to wear a seat belt? (You can apply such physics principles as the law of conservation of energy and Newton's first law of motion.)

30. The cost of electrical energy is an important issue.
    (a) Is the cost of electrical energy regulated by the government in your province?
    (b) What are two advantages of government regulation of the price of electrical energy to the consumer?
    (c) What are two disadvantages of government regulation of the price of electrical energy to the consumer?

31. Our society has come to rely heavily on energy-transformation technologies, which we often take for granted. However, we soon realize how dependent we are on these technologies when there is an electrical blackout. Describe how your area would be affected by an electrical blackout that lasted from several hours to several days.

32. Describe long-term objectives that governments in Canada should pursue to ensure that we have a plentiful supply of low-pollution energy in the future.

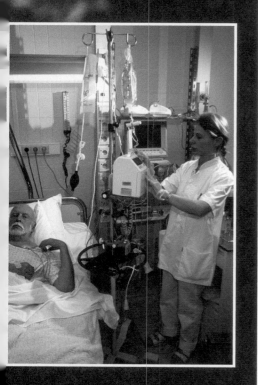

Hydraulic and Pneumatic Systems

Pumping water to fight a fire is an example of a hydraulic system. Pumping air to inflate a bicycle tire is an example of a pneumatic system. Both systems use a fluid, that is, a liquid or a gas. Hydraulic and pneumatic systems can use fluids at rest, for example, a boat floating in water. Or they can use fluids in motion, such as natural gas flowing in a pipeline.

Many modern technologies use hydraulic and pneumatic systems. However, nature applies the same basic principles, which scientists study and then copy. For example, sharks and other sea creatures have streamlined bodies that move easily through water.

In this unit, you will study the scientific principles related to fluids at rest and in motion, and to hydraulic and pneumatic systems. You will design and carry out investigations on these topics. You will also analyze and describe the social and economic consequences of related technologies.

This unit applies many of the concepts and principles presented in Units 1 and 2. It is divided into two chapters: Chapter 5 deals with fluids at rest, and Chapter 6 deals with fluids in motion. In the Unit Performance Task you can design, build, and evaluate a hydraulic or pneumatic system.

What you learn in this unit will be useful as you consider a variety of careers.

▶ Overall Expectations

In this unit, you will be able to

- demonstrate an understanding of the scientific principles related to fluids at rest and in motion, and to hydraulic and pneumatic systems
- design and carry out investigations of fluids at rest and in motion, and of hydraulic and pneumatic systems
- describe and analyze the social and economic consequences of the development of technological applications related to the motion and control of fluids
- identify and describe science- and technology-based careers related to hydraulic and pneumatic systems

ARE YOU READY?

Knowledge and Understanding

1. Is it true that all fluids are liquids? Is it true that all liquids are fluids? Explain your answers.

2. The heart is the main part of one system in the human body, and the lungs are the main part of another system.
   (a) Name the two systems.
   (b) Which system is hydraulic, and which is pneumatic?

3. Three objects, A, B, and C, of equal volume, are placed in water, as shown in **Figure 1**.
   (a) How do the densities of A, B, and C compare with each other?
   (b) How does the density of A compare with the density of water?
   (c) How does the density of C compare with the density of water?
   (d) Are the masses of the three objects the same? Explain your answer.

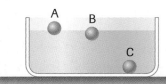

Figure 1

4. (a) Can a gas be easily compressed? Use an example to illustrate your answer.
   (b) Can a liquid be easily compressed? Use an example to illustrate your answer.

5. In general, a liquid expands when it is heated. Use the kinetic molecular theory (also called the particle theory) of matter to explain why. (Use diagrams to help explain your answer.)

Inquiry and Communication

6. You are asked to perform an investigation to determine how the viscosity (or resistance to flow) of honey depends on its temperature.
   (a) Formulate a question for the investigation.
   (b) Write a prediction answering the question.
   (c) List the steps in the procedure, including safety considerations, for your investigation.

Math Skills

7. Solid volume is expressed in cubic units (m^3, dm^3, cm^3, mm^3, etc.), and liquid volume is expressed in capacity units (L, mL, etc.). Solid and liquid volumes can be equated as follows: $1\ L = 1\ dm^3$, and $1\ mL = 1\ cm^3$. Copy **Table 1** into your notebook, and complete it. (More information about volume conversions can be found in Appendix C, **Table 4**.)

Table 1

	Solid Volume		Liquid Capacity	
	dm³	cm³	L	mL
(a)	?	?	2.0	?
(b)	?	555	?	?
(c)	?	?	?	305

8. Given the equation $D = \dfrac{m}{V}$,
   (a) Write out the equation in words.
   (b) Rewrite the equation to solve for m and then V.
   (c) Copy **Table 2** into your notebook and complete it.

9. (a) Calculate the slope of the line on the graph in **Figure 2**.
   (b) What does the slope represent? (Look at the units.)

Table 2

	m	V	D
(a)	26 g	2.0 cm³	?
(b)	25 g	75 L	?
(c)	?	7.0 mL	$12\dfrac{g}{mL}$
(d)	3.6 kg³	?	$1.2 \times 10^3 \dfrac{kg}{m^3}$

Technical Skills and Safety

10. You are asked to design a model of a hydraulic or pneumatic system, such as a car's brake system. Describe how you would apply your research skills to determine the structure and operation of the system before you design the model.

11. Describe at least one major safety concern when using hydraulic or pneumatic systems in each of the following occupations:
    (a) a mechanic working at an automotive service centre
    (b) a maintenance worker at a milk processing factory
    (c) a nurse
    (d) a firefighter
    (e) a high school teacher in a technical course, such as woodworking or auto shop

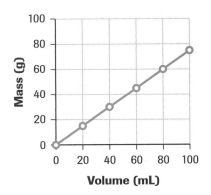

Figure 2

Making Connections

12. A hot-air balloon (**Figure 3**) is observed rising slowly after the on-board burner has been fired.
    (a) What happens to the air in the balloon as it is heated?
    (b) As the balloon rises, does the atmospheric pressure surrounding it change? Explain your answer.
    (c) Under what conditions is this activity dangerous?

13. Identify three industries that require knowledge of hydraulic and pneumatic systems.

Figure 3

Fluid Statics

- define and describe the concepts and units related to fluids and to hydraulic and pneumatic systems

- identify factors that affect the pressure in static fluids, and see how experimental values compare to theoretical values

- state Pascal's principle, explain its application in the transmission of forces in enclosed liquid systems, and experimentally verify it

- describe common components used in hydraulic and pneumatic systems

- analyze quantitatively and experimentally variables such as force, area, pressure, work, power, and time in hydraulic and pneumatic systems

- using circuit symbols, draw simple hydraulic and pneumatic circuits

- design and construct a hydraulic or pneumatic system

- describe and analyze examples of the historical development of fluid systems

- identify and analyze the social and economic consequences of the use of robotic systems

- identify and evaluate the impact of applications of hydraulic and pneumatic systems in everyday life

Getting Started

The Jaws of Life pry and cut parts of vehicles and other structures to free trapped occupants (**Figure 1**). This tool, which has saved many lives, is a hydraulic system. A hydraulic system uses liquids under pressure to provide the force needed to perform an action, in this case, prying and cutting. A nonflammable liquid is required for the Jaws of Life because they are often used near flying sparks or open flames.

In this chapter, you will study the properties of fluids at rest, or *fluid statics*. You will learn how pressure and forces are transmitted in hydraulic systems and in pneumatic systems (systems that use gas under pressure to provide the force), and how the liquids and gases in these systems are controlled with pumps, valves, and other components. You will also discover how the properties of fluids at rest are applied in many fields, such as medicine, manufacturing, transportation, and construction.

Figure 1
A hydraulic system, such as the Jaws of Life, uses a liquid under pressure.

1. Liquids and gases are both called fluids.
   (a) Name three liquids and three gases.
   (b) What properties do liquids and gases have in common?
   (c) How are liquids and gases different? Use examples in your answer.

2. A weather forecaster reports that the atmospheric pressure is 100.2 kPa.
   (a) What causes atmospheric pressure?
   (b) What does the unit kPa represent in words?
   (c) If you were at the top of a mountain, would you expect the atmospheric pressure to be higher than, lower than, or the same as the atmospheric pressure in the valley below? Use a sketch and the particle theory of matter to explain why.

3. What causes your ears to pop when going up or down in an airplane?

4. **Figure 2** shows air being drawn out of a straw.
   (a) How does the air pressure compare at the locations marked A, B, and C in the diagram? Explain your answer.
   (b) Redraw the diagram in your notebook, and add information to it to explain how a straw works.

5. How does a siphon work?

6. Is the pipeline system that distributes natural gas at high pressure to populated regions of Canada a hydraulic system or a pneumatic system? Give a reason for your answer.

7. Describe how the use of robotic devices has influenced each of the following:
   (a) the employment of workers in industry
   (b) worker safety in the field of manufacturing
   (c) worker safety in the field of hazardous waste disposal

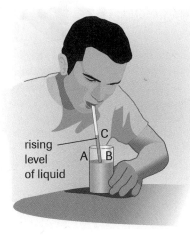

rising
level
of liquid

C
A B

Figure 2

Comparing Force and Pressure

Place your left forearm flat on a desk or table. With the palm of your right hand, press down firmly on your left forearm near your elbow (**Figure 3(a)**). Now try to exert the same force downward using only the tip of the index finger of your right hand (**Figure 3(b)**).

(a) Describe what you observed.

(b) How does the surface area of your fingertip compare to the surface area of the palm of your hand? Use your answer to explain what you observed.

(c) Compare the pressure exerted by a flat shoe heel with the pressure exerted by a high heel.

Figure 3

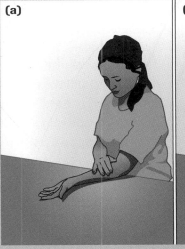
(a)

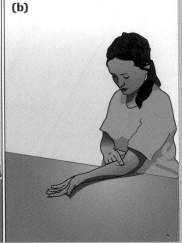
(b)

(a) Push with your palm.   **(b)** Push with your finger.

fluid a substance that flows and takes the shape of its container; liquids and gases are both fluids

A **fluid** is a substance that flows and takes the shape of its container. Liquids and gases are both fluids. We can easily show that water conforms to the definition of a fluid by pouring water from one container into another. We can do the same with air, which is a gas: A sealed flask filled with air is held under water; when the seal is removed, it is evident that the air occupies the entire flask (**Figure 1(a)**). When we tilt the flask, the air "pours" out into the beaker (**Figure 1(b)**).

(a)  **(b)** inverted container initially filled with water

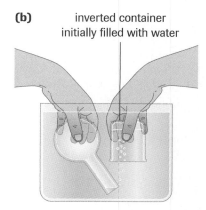

Figure 1
(a) Air takes the shape of its container. Because air is invisible, we cannot easily observe this property. However, we can observe it when the flask, with the air sealed inside, is inverted in water and the seal is removed.
(b) Air can be "poured" from one container into another. Why does the gas (air) flow upward in this case?

In most cases, it is easy to determine whether a substance is a fluid at room temperature. However, there are exceptions: Butter is solid when cold, but at room temperature (about 20 °C) it is neither a rigid solid nor a readily flowing liquid. When heated, butter is a liquid.

Sand, sugar, and other similar solids can be poured, so why are they not called fluids? These solids can be formed into irregular piles, whereas liquids and gases cannot. When piled high above the top of a container, the poured solids do *not* take the shape of the container.

hydraulics the science and technology of the mechanical properties of liquids

The science and technology of the mechanical properties of water and other liquids is called **hydraulics** (the prefix *hydro* comes from the Greek word for water or liquid). Similarly, the science and technology of the mechanical properties of air and other gases is called **pneumatics** (the prefix *pneu* comes from the Greek word for wind or breath). If the fluids are at rest, the study is called **fluid statics**, which is the topic of this chapter.

pneumatics the science and technology of the mechanical properties of gases

fluid statics the study of fluids at rest

hydraulic system a mechanical system that operates using a liquid under pressure

The fields of hydraulics and pneumatics include the study of fluids that are not enclosed, such as water in a lake and air in the atmosphere. They also include the study of fluids that are under pressure in systems that are enclosed. A **hydraulic system** is a mechanical system that operates using a liquid under pressure. Often the liquid is water, but, depending on the system, it might be oil, blood, or a vaccine. An example of a simple hydraulic system is a syringe used to give an injection (**Figure 2**). A **pneumatic system** is a mechanical system that operates using a gas under pressure. Often the gas is air (**Figure 3**), but it could also be another gas or mixtures of gases, for example, hydrogen, helium, oxygen, natural gas, and acetylene.

pneumatic system a mechanical system that operates using a gas under pressure

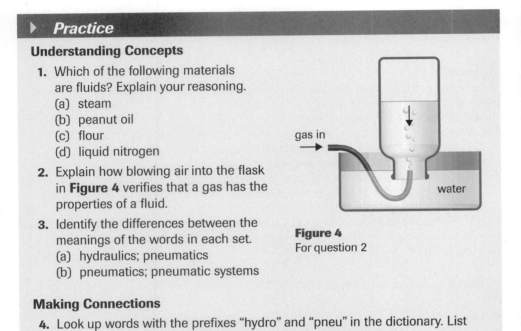

> ▶ *Practice*

Understanding Concepts

1. Which of the following materials are fluids? Explain your reasoning.
   (a) steam
   (b) peanut oil
   (c) flour
   (d) liquid nitrogen

2. Explain how blowing air into the flask in **Figure 4** verifies that a gas has the properties of a fluid.

3. Identify the differences between the meanings of the words in each set.
   (a) hydraulics; pneumatics
   (b) pneumatics; pneumatic systems

Making Connections

4. Look up words with the prefixes "hydro" and "pneu" in the dictionary. List four words you know that begin with each of these prefixes.

gas in →

water

Figure 4
For question 2

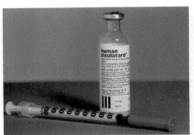

Figure 2
A syringe is an example of a simple hydraulic system.

Figure 3
This air-powered tool, which hammers concrete to break it apart, is a pneumatic system.

Density and Compressibility

Before we consider specific properties of fluids, review the kinetic molecular theory of matter by studying **Figure 5**. This theory will help you visualize the properties of matter.

Two properties of matter are particularly important to the study of fluid statics: *density* and *compressibility*. **Density** is a substance's mass per unit volume. The term "per unit volume" means that we must compare the masses of equal volumes of substances. For example, a cubic metre (1.0 m^3) of brass has a mass of about 8.5×10^3 kg; a cubic metre of water has a mass of 1.0×10^3 kg. For the same volume (1.0 m^3), brass has 8.5 times more mass than water.

The SI unit for density is kilograms per cubic metre (kg/m^3). Other metric units that are commonly used to express density are grams per litre (g/L) and grams per millilitre (g/mL) for liquids, and grams per cubic centimetre (g/cm^3) for solids. (For tips on converting from one unit to another, refer to Appendix C, **Table 4**.)

density the mass per unit volume of a substance; it is a scalar quantity with SI units of kg/m^3

(a) solid

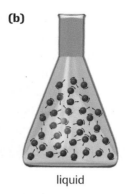
(b) liquid

(c) gas

Figure 5

The kinetic molecular theory of matter, also called the particle theory, states that all matter is made of molecules and that these molecules are constantly in motion. The motion increases as the temperature of the matter increases.

(a) Molecules in a solid have a very high level of attraction for one another, so they are not free to move around. However, they do vibrate about fixed positions. At a given temperature, a solid has a fixed shape and volume.

(b) Molecules in a liquid have a high level of attraction for one another, but they are relatively free to move around. They move to take the shape of their container, colliding with one another and with their surroundings.

(c) Molecules in a gas have a low level of attraction for one another; they move around quickly and easily, colliding with one another and with their surroundings.

Based on the definition of density, the equation for density using symbols is

$$D = \frac{m}{V}$$

where D is the density,
      m is the mass, and
      V is the volume.

LEARNING TIP

Mass–Volume Graphs

A graph of mass (on the vertical axis) versus volume of several samples of the same pure substance is a straight line.

The slope of that line is $\frac{\Delta x}{\Delta y}$ or $\frac{\Delta m}{\Delta V}$, which is constant and equals the density of the substance. You can identify what the slope of a line on a graph represents by looking at the units of the slope calculation.

▶ **SAMPLE problem 1**

Calculating Density

A quantity of helium gas at 0 °C with a volume of 4.00 m³ has a mass of 0.712 kg. (In this case, the helium gas is at atmospheric pressure.) Determine the density of this sample of helium gas.

Solution

$V = 4.00 \text{ m}^3$

$m = 0.712 \text{ kg}$

$D = ?$

$$D = \frac{m}{V}$$

$$= \frac{0.712 \text{ kg}}{4.00 \text{ m}^3}$$

$$D = 0.178 \frac{\text{kg}}{\text{m}^3}$$

The density of the helium gas sample is 0.178 kg/m³.

Because density is a characteristic property of a substance, any sample of a pure substance has the same density. It does not matter how large or small the sample is, where the sample is taken from, or on what part of the planet it is measured. The density of a sample of an unknown pure substance can also be used to identify the substance. The densities of several common solids, liquids, and gases are listed in **Table 1**. (Note that the temperatures of *all* samples must be stated when comparing densities. It is also important to state the pressure of a gas because the density increases if more particles of the gas are compressed into the same space.)

Table 1 Densities of Some Common Substances

Substance	State	Density (kg/m³ or g/L)	Density (g/cm³ or g/mL)
hydrogen	gas (0 °C)	0.089	8.9×10^{-5}
helium	gas (0 °C)	0.178	1.78×10^{-4}
air	gas (0 °C)	1.29	1.29×10^{-3}
cork	solid	240 (varies)	0.24
ethyl alcohol	liquid	789	0.789
ice	solid (0 °C)	920	0.920
water	liquid (4 °C)	1000	1.00
saltwater	liquid (0 °C)	1030 (varies)	1.03
glycerin	liquid	1260	1.26
aluminum	solid	2700	2.70
iron	solid	7860	7.86
brass	solid	8500	8.50
copper	solid	8950	8.95
mercury	liquid	13 600	13.6
gold	solid	19 300	19.3

Notes: Values listed are at 20 °C unless otherwise stated. Gases are at atmospheric pressure.

Compressibility is the ability of the particles of a substance to be pressed closer together. Consider a plastic bottle completely filled with water. It would be very difficult to squeeze any more water into the bottle without causing it to burst. However, if the bottle has only air in it, it would be possible to squeeze more and more air into it (until the bottle bursts). Gases consist of particles that are spread out, so they are highly compressible. Liquids are only slightly compressible, and solids are even less compressible. The property of compressibility influences the choice between hydraulic and pneumatic systems, studied later in the chapter.

compressibility the ability of the particles of a substance to be pressed closer together

DID YOU KNOW ?

Cargo Airships
Large airships, shaped like giant cigars, can rise in air because of the properties of helium gas. A new design, the SkyCat 200, is nearly 200 m long and can carry a load of up to 200 t, twice as much as a Boeing 747 jet. The airship moves forward at about 80 km/h. Helium, the low-density gas used in party balloons, is safe because it is not flammable.

 www.science.nelson.com

Answers

5. (a) 1.3×10^3 kg/m³
   (b) 7.90×10^2 g/L
   (c) 1.0 g/mL
6. 1.6×10^4 kg/m³
8. (a) 72 cm³
   (b) 6.1×10^2 g; 0.61 kg
9. 1.1×10^3 m³
11. (a) 1.28×10^3 g/L
    (b) 1.28×10^3 kg/m³
    (c) 1.63%

▶ **Practice**

Understanding Concepts

5. For each of the following substances, calculate the density in a convenient unit. Using **Table 1**, page 223, identify the substance.
   (a) A volume of 0.50 m³ of a substance has a mass of 6.3×10^2 kg.
   (b) A volume of 5.00 L of a substance has a mass of 3.95×10^3 g.
   (c) A volume of 75 mL of a substance has a mass of 77 g.

6. A mining engineer has a mineral sample that looks like pure gold. The sample is found to have a mass of 4.0×10^2 kg and a volume of 0.025 m³. Determine the density of the sample, and use it to decide whether the sample is pure gold.

7. Rewrite the equation $D = \dfrac{m}{V}$ to solve for (a) m and (b) V.

8. A rectangular solid brass plate is 8.0 cm long, 6.0 cm wide, and 1.5 cm thick.
   (a) Calculate the volume of the plate in cubic centimetres. (The equations for the volumes of various shapes are given in Appendix A1.)
   (b) Calculate the plate's mass in grams and kilograms.

9. One winter it was estimated that one million kilograms of ice had formed on a small lake. Calculate the volume of this mass of ice to two significant digits.

10. Does inflating a party balloon with air prove that air is compressible? Explain your answer.

Applying Inquiry Skills

11. A student performed an investigation to determine the density and identity of a liquid and obtained these measurements:

    mass of graduated cylinder: 12.4 g
    volume of liquid added: 82.0 mL
    mass of cylinder and liquid: 117.4 g

    (a) Calculate the density of the liquid in grams per litre.
    (b) Convert the density to the preferred SI unit.
    (c) Assuming the liquid is glycerin, determine the percent error of the density calculation. (To review percent error, refer to Appendix A1.)
    (d) What are the main sources of error in making measurements in this type of investigation?

Making Connections

12. Use the kinetic molecular theory of matter to explain how solid wax can turn into liquid wax, and then into a gas.

SUMMARY *Properties of Fluids*

- A fluid can flow and take the shape of its container; liquids and gases are both fluids.

- Hydraulics is the study of the mechanical properties of liquids, and pneumatics is the study of the mechanical properties of gases.

- A hydraulic system operates using a liquid under pressure, and a pneumatic system operates using a gas under pressure.

- Density is a property of matter that is important in the study of fluids; it is found using the equation $D = \dfrac{m}{V}$.

- Another important property of matter is compressibility. Gases are highly compressible, whereas liquids are only slightly compressible.

▶ ***Section 5.1*** *Questions*

Understanding Concepts

1. Water can be observed as ice, water, and water vapour. Which of these states can be called a fluid?

2. Explain the difference between hydraulics and a hydraulic system.

3. Explain why temperature and pressure must be considered when stating the density of a gas. (Use the kinetic molecular theory in your answer.)

4. Copy **Table 2** into your notebook and complete the missing information using the most convenient units. Include the equation needed to find each unknown value.

5. A block of metal is 50.0 cm long, 20.0 cm wide, and 15.0 cm high. Its mass is 128 kg.
   (a) Write the block's dimensions in metres.
   (b) Calculate the volume of the block in cubic metres.
   (c) Calculate the block's density.
   (d) What is the likely identity of the metal?

Applying Inquiry Skills

6. Explain how you can demonstrate that a gas can flow.

7. A student measures the mass of a graduated cylinder when known volumes of a liquid are added. The results of the investigation are shown in **Table 3**.
   (a) Plot a graph of mass (vertical axis) versus volume, and draw the line of best fit.
   (b) What is the mass of the graduated cylinder?
   (c) Determine the slope of the line of best fit to two significant digits. What does the slope represent?

Making Connections

8. List three careers in which determining density is an important part of the job. For each career, state the application of the concept of density.

Table 2   For Question 4

	Mass	Volume	Density	Likely State
(a)	8.8×10^2 g	1.5 L	?	?
(b)	0.86 g	1.5 L	?	?
(c)	2.5×10^3 kg	2.5 m³	?	?
(d)	?	1.20×10^{-3} m³	2.70×10^3 kg/m³	?
(e)	8.50 kg	?	1.36×10^4 kg/m³	?

Table 3   Data Table for Question 7

Volume (mL)	Mass (g)
0	68
20	94
40	120
60	146
80	172
100	198

Figure 1
Snowshoes spread the force over a larger area than boots do.

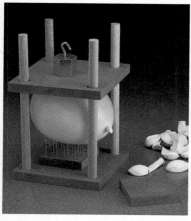

Figure 2
A bed of nails won't burst the balloon.

pressure the magnitude of the force per unit area; the SI unit is the pascal

pascal the SI unit of pressure; symbol Pa; 1.0 Pa = 1.0 N/m²

CAREER CONNECTION

Understanding how and when to apply pressure to muscles and tissue is important to registered massage therapists. RMTs must also learn about human anatomy and nutrition. Clients are varied, for example, people needing rehabilitation services after a car accident or sports injury.

 www.science.nelson.com

A person wearing snowshoes (**Figure 1**) stays on top of the snow, while a person of equal mass wearing boots sinks into the snow. They both exert an equal force on the snow. Why does only one sink? The answer is the difference in pressure: Snowshoes spread the force over a larger surface area, so the pressure on any part of the snow cover is less.

As another example, consider a balloon sandwiched between two boards, each with a single nail pressing against the balloon. The balloon will break because the force of a single nail concentrates the pressure on a small area; this pressure is strong enough that the nail easily pierces the surface. Now consider what will happen if the balloon is sandwiched between a bed of nails, as shown in **Figure 2**. If the nails are spaced appropriately, the force of the nails on the balloon is spread out over the surface area provided by many nails, so it is unlikely that any one nail will break the balloon.

Pressure is the magnitude of the force applied per unit area. In equation form,

$$p = \frac{F}{A}$$

where p is the pressure in newtons per square metre,
F is the magnitude of the force measured in newtons, and
A is the area measured in square metres, perpendicular to the force.

Consider a force of 1.0 N applied over an area of 1.0 m². The pressure in this case is

$$p = \frac{F}{A}$$
$$= \frac{1.0 \text{ N}}{1.0 \text{ m}^2}$$
$$p = 1.0 \, \frac{\text{N}}{\text{m}^2}$$

The unit newtons per square metre (N/m²) is an SI unit. It has been given the name **pascal** (Pa) in honour of French scientist Blaise Pascal (1623–1662), who contributed greatly to our knowledge of fluids, as well as to other fields of learning.

The pressure exerted by a single piece of newspaper resting flat on a table is approximately 1.0 N/m², or 1.0 Pa. Because this is such a small amount of pressure, scientists frequently measure pressure in kilopascals (1.0 kPa = 1.0 × 10³ Pa). (Other units are often used for medical and other purposes.) The pressure a chair exerts upward against you when you are sitting in it is roughly 5 kPa to 10 kPa. ⬧■

▶ **SAMPLE** problem **1**

Calculating Pressure

A flat-bottomed crate is 2.0 m wide, 3.0 m long, and 1.0 m high. The magnitude of its weight (i.e., the magnitude of the force of gravity on the crate) is 1.2×10^4 N.

(a) Calculate the area of the crate in contact with the floor.

(b) Calculate the pressure the crate exerts on the floor in pascals and kilopascals.

(c) How would the pressure change if the box were standing on its end?

Solution

(a) $l = 3.0$ m
$w = 2.0$ m
$A = ?$

$$A = lw$$
$$= 3.0 \text{ m} \times 2.0 \text{ m}$$
$$A = 6.0 \text{ m}^2$$

The area of the crate in contact with the floor is 6.0 m².

(b) $F = 1.2 \times 10^4$ N
$A = 6.0$ m²
$p = ?$

$$p = \frac{F}{A}$$
$$= \frac{1.2 \times 10^4 \text{ N}}{6.0 \text{ m}^2}$$
$$p = 2.0 \times 10^3 \text{ Pa, or 2.0 kPa}$$

The crate exerts a pressure of 2.0×10^3 Pa, or 2.0 kPa, on the floor.

(c) The area in contact with the floor is smaller, so the pressure on the floor is greater. This can be verified using calculations like those in (a) and (b), in which the pressure is found to be 6.0 kPa.

Flooring Damage
Some flooring manufacturers warn that high-heeled shoes with a small surface area can damage wooden floors. The surface area of such a heel is about 50 times smaller than the surface area of a flat heel. Thus, the pressure on the floor is about 50 times greater, enough to dent some hardwood floors.

LEARNING TIP

Mass and Weight
In Sample Problem 1, the magnitude of the force of gravity (or weight) was given. In some problems, the mass is given instead. In those cases, you can apply the equation $F = mg$, where $g = 9.8$ N/kg, to determine the magnitude of the force of gravity on the object.

▶ **TRY THIS** activity

Calculate Your Own Pressure

Design and carry out an experiment to calculate the pressure you exert on the floor when you stand on both feet in the shoes you are currently wearing. (The surface area of your shoes can be a close approximation.) Show your measurements and calculations.

Pressure in Joints
Pressure in bone joints can be high. For example, when you stand on one foot, the pressure in your knee joint can be as high as 1000 kPa. As the pressure on a joint increases, the lubrication of the joint also increases. Healthy human joints are better lubricated than the best bearings!

Answers

3. (a) 1.002×10^5 Pa
   (b) 99 kPa
   (c) 3.0 Pa
4. (a) 8.4×10^2 Pa
   (b) 3.1×10^4 Pa
   (c) 1.2×10^4 Pa
5. (a) 2.7×10^6 Pa; 2.7×10^3 kPa
   (b) 8.8×10^5 Pa; 8.8×10^2 kPa
7. (a) 0.55 N
   (b) 8.0×10^{-3} m²
8. 8.0×10^4 N

gripping side

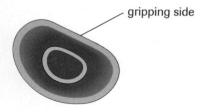

Figure 3
A cross section of a finger

▶ **Practice**

Understanding Concepts

1. Give a reason for each of the following situations, emphasizing the difference between force and pressure:
   (a) An all-terrain vehicle has wide wheels.
   (b) It is much easier to break a walnut shell by hand using two walnuts pressed against each other than to break the shell with your hand alone.

2. Refer back to the *Try This* Activity on page 227. Explain what you observed there.

3. Convert
   (a) 100.2 kPa to pascals
   (b) 9.9×10^4 Pa to kilopascals
   (c) 3.0×10^{-3} kPa to pascals

4. Calculate the pressure in each of the following:
   (a) $F = 4.2 \times 10^2$ N, $A = 0.50$ m²
   (b) $F = 6.2 \times 10^4$ N, $A = 2.0$ m²
   (c) $F = 3.6 \times 10^2$ N, $A = 3.0 \times 10^{-2}$ m²

5. (a) Calculate the pressure, in pascals and kilopascals, applied on the floor by the toe of a ballet dancer's shoe when she balances, briefly, on that toe. Assume that the dancer's mass is 56 kg and the surface area of the tip of her toe is 2.0×10^{-4} m².
   (b) Repeat (a) for a circus elephant standing on two feet. Assume that the elephant's mass is 5.4×10^3 kg and the surface area of each foot is 3.0×10^{-2} m².
   (c) Compare your answers in (a) and (b).

6. Rewrite the equation $p = \dfrac{F}{A}$ to solve for (a) F and (b) A.

7. Calculate the unknown quantities:
   (a) $p = 2.5$ Pa, $A = 0.22$ m², $F = ?$
   (b) $p = 8.0 \times 10^4$ Pa, $F = 6.4 \times 10^2$ N, $A = ?$

8. Assume that the air pressure in a bicycle tire is 4.0×10^2 kPa higher than the air pressure outside the tire and is spread over an area of 0.20 m². Calculate the magnitude of the total force acting on the inside of the tire.

Applying Inquiry Skills

9. Take measurements to determine the pressure a regularly shaped object, such as this textbook, exerts on your desk. (The book should be closed and resting flat on the desk.)

Making Connections

10. If boxing gloves were allowed to be smaller, the sport of boxing would be more dangerous than it is now. Explain why.

11. **Figure 3** shows a cross section of a human finger, with the bone in the middle. The flat side is the gripping side.
    (a) With the index finger and thumb of one hand, tightly squeeze the index finger on your other hand. Which side (the gripping side or the opposite side) hurts more? Why?
    (b) If you are carrying a heavy suitcase by its handle, how does the shape of the finger bones make a difference?

Case Study High-Pressure Water in Industry ▼

Water under pressure is used in many industrial applications, for example, *hydroforming* and *water jet cutting*.

Hydroforming is the process of using water under high pressure to shape metal components. One of the largest fully automated frame manufacturing plants in the world is Magna Structural Systems, located in St. Thomas, Ontario. Magna uses hydroforming, rather than traditional methods such as pressure forming using a hard piece of metal, to manufacture automobile parts.

Currently, the average new car has at least two hydroformed parts. By 2010, the number is expected to rise to at least 10 parts. With more than 16 million cars manufactured each year in North America alone, many millions of hydroformed parts will be required. A typical hydroformed component is shown in **Figure 4(a)**, and the method of applying high-pressure water to form a component is shown in **Figure 4(b)**.

hydroforming the process of using water under high pressure to shape metal components

(a)

(b)

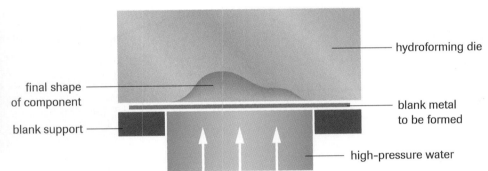

final shape of component

blank support

hydroforming die

blank metal to be formed

high-pressure water

Figure 4
(a) The lower shell of a fuel tank is hydroformed, making it free of seams.
(b) Water under very high pressure forms a metal sheet into the desired shape by pressing the blank metal against the mould, or die.

The water pressure used in a typical hydroforming process can range from about 1.5×10^8 Pa to 7.0×10^8 Pa. (The larger value is close to 7000 times normal atmospheric pressure.) Some advantages of hydroforming are the following:

- It requires fewer operations than traditional methods, such as welding, to make parts into complex shapes.
- It increases the overall strength and stiffness of the component.
- It reduces the amount of material required, thus reducing the weight of the component.

However, hydroforming also has disadvantages:
- The equipment is expensive to set up.
- Hydroforming is slow.
- Operators may get wet.
- Water can be wasted in the process.

Water jet cutting is the process of using water under high pressure and at a high speed to cut manufactured components. This type of cutting has a variety of applications, from trimming the edges of paper as it is being

water jet cutting the process of using water under high pressure and at a high speed to cut manufactured components

manufactured, to cutting holes in pieces of metal up to 30 cm thick, to cutting food products (**Figure 5(a)**). The basic design of a water jet tool used to cut hard materials such as steel, stone, and bullet-proof glass is illustrated in **Figure 5(b)**.

Figure 5

(a) A high-speed water jet can cut food products, leaving sharp, clean cuts.

(b) To cut through hard materials, an abrasive is added to the water jet in the mixing chamber.

(a)

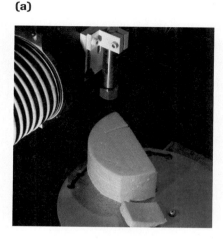

(b)

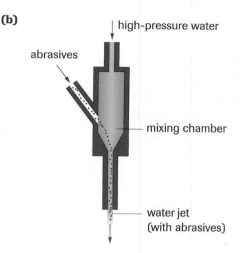

Water jet cutting has a number of advantages:

- It can be started at any location on the component to be cut.
- It is free of dust.
- It does not add stress to material being cut.
- It does not deform the material.
- It is a highly flexible process.
- It can be controlled robotically.
- The machine requires little maintenance compared to mechanical cutters.
- It does not produce high-temperature zones, as laser cutters do.

The process also has disadvantages:

- The cut may be slightly tapered, especially if the component is thick.
- The process is very noisy.
- Cleanup is a big job.
- Cutting waste can be large.
- As with hydroforming, water can be wasted.

When quantitatively analyzing water jet cutting, it is important to know the quantity of water needed per unit time interval. This can be done by first determining the **volume flow rate** (or volume per second) of the water during the process. The equation is

volume flow rate the volume per second of a fluid flow; symbol q_v

$$q_v = \frac{V}{\Delta t}$$

where q_v is the volume flow rate,
V is the volume, and
Δt is the time interval.

If V is measured in cubic metres and Δt in seconds, then q_v is measured in cubic metres per second (m³/s). It can also be measured in other units; litres per second (L/s) and litres per minute (L/min) are commonly used. (Either mass or volume can be used to represent flow rate. The applications in this textbook use volume flow rate.)

▶ **SAMPLE** problem **2**

Calculating Volume

The volume flow rate needed to cut a metal component is 3.3 L/min, and the time interval required is 18 s. Calculate the total volume of water used during the process.

Solution

q_v = 3.3 L/min = 0.055 L/s
Δt = 18 s
V = ?

$$q_v = \frac{V}{\Delta t}$$

$$V = q_v\, \Delta t$$

$$= \left(0.055\ \frac{L}{s}\right)(18\ s)$$

$$V = 0.99\ L$$

The total volume of water needed is 0.99 L.

The speed of the water jet in a cutting process can also be found. We can derive an equation for speed from the equation for volume flow rate: The total volume in a pipe of length l is Al, where A is the cross-sectional area of the pipe. Thus,

$$q_v = \frac{V}{\Delta t}$$

$$= \frac{Al}{\Delta t}$$

$$= A\left(\frac{l}{\Delta t}\right)$$

$$\frac{l}{\Delta t} = \frac{distance}{\Delta t} = v$$

$$q_v = Av$$

$$v = \frac{q_v}{A}$$

Notice that, in addition to deriving an equation for speed, we also have derived a second equation for the volume flow rate ($q_v = Av$), given the cross-sectional area of the pipe and the speed of the water in the pipe.

Calculating the Speed of the Water

Calculate the speed of water in a pipe of radius 0.21 mm if the volume flow rate of the water is 6.0×10^{-6} m³/s.

Solution

$q_v = 6.0 \times 10^{-6}$ m³/s
$r = 0.21$ mm $= 2.1 \times 10^{-4}$ m

$$A = \pi r^2$$
$$= \pi(2.1 \times 10^{-4} \text{ m})^2$$
$$A = 1.385 \times 10^{-7} \text{ m}^2$$

$v = ?$

$$v = \frac{q_v}{A}$$

$$= \frac{6.0 \times 10^{-6} \frac{\text{m}^3}{\text{s}} \text{ m}}{1.385 \times 10^{-7} \text{ m}^2}$$

$$v = 43 \text{ m/s}$$

The speed of the water is 43 m/s.

▶ **Practice**

Understanding Concepts

Answers

13. 1.1×10^8 Pa, 1.1×10^5 kPa

14. (a) 4.9×10^8 Pa; 4.9×10^5 kPa
    (b) 3.9×10^{-5} m³/s
    (c) 6.3×10^{-4} m³

12. (a) State two advantages of hydroforming and two of water jet cutting.
(b) Repeat (a) for disadvantages.

13. During hydroforming, a force of magnitude 1.8×10^6 N is applied to a piece of steel 12 cm × 14 cm. Calculate the pressure, in pascals and kilopascals, on the piece of steel.

14. In a water jet cutting process, a stream of water with the speed 1.6×10^2 m/s drills a hole of radius 0.28 mm with a force of magnitude 1.2×10^2 N. The process lasts for 16 s. Calculate
(a) the pressure, in pascals and kilopascals, of the water on the metal
(b) the volume flow rate of the water
(c) the total volume of water used in the process, in cubic metres

Applying Inquiry Skills

15. Explain how you could use water jet cutting to carve a shape of your choice into a block of ice the size of this textbook.

Making Connections

16. Some limited-production vehicles, such as the Corvette and the Hummer H2, have more hydroformed components than most vehicles.
(a) Research one of these vehicles, and describe what you discover about hydroforming some of the components, such as the frame rail.
(b) Explain why hydroforming is advantageous for limited-production vehicles.

17. The Magna Structural Systems plant is fully automated—the hydroforming is done by computer-controlled robots. Describe the impact of this type of manufacturing on jobs and job safety.

18. Does hydroforming vehicle parts affect the safety and fuel efficiency of the vehicle? Explain your answer.

Atmospheric Pressure

Our atmosphere, which comprises air molecules and other particles, stretches upward in layers. Each layer presses down on the one below. The result is a pressure called **atmospheric pressure**, which is greatest near Earth's surface because of the weight of all the layers above. The pressure is reduced at higher altitudes.

Normally we do not notice atmospheric pressure because the pressure inside our bodies balances the pressure outside. However, our ears are sensitive to changes in atmospheric pressure. You have likely experienced a "pop" in your ears when your altitude above ground level changed rapidly. This can happen when riding in an elevator in a tall building, in a car on a mountain highway, or during takeoff or landing in an airplane. The pop is the result of the change in pressure on either side of the eardrum. Swallowing helps the pressure on either side return to normal.

Atmospheric pressure, like any other pressure, can be measured in pascals and kilopascals. Standard atmospheric pressure, used by scientists for comparison purposes, is 101.3 kPa, or 101 kPa to three significant digits. This is the average atmospheric pressure at sea level (commonly called *one atmosphere*). **Table 1** lists the average atmospheric pressures at elevations above sea level up to 12 500 m; this is the average altitude of the top of the *troposphere*, the atmospheric layer closest to Earth's surface. **Figure 6** shows how pressure depends on altitude up to a much higher altitude of 50 km.

atmospheric pressure the pressure caused by air molecules and other particles above Earth's surface; symbol p_{atm}

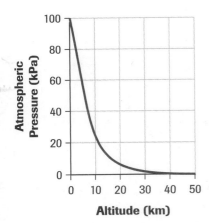

Figure 6
At ground level, the average atmospheric pressure is approximately 100 kPa. The pressure decreases at higher altitudes. The line on the graph represents the average values, which change at each altitude as weather patterns move.

Table 1 Average Pressure at Various Elevations

Altitude (m)	Pressure (kPa)
0	101.3
1500	85.0
3000	70.0
5500	50.0
9000	30.0
12 500	20.0

DID YOU KNOW ?

Atmospheric Pressure Records
The lowest atmospheric pressure recorded at Earth's surface is 87.7 kPa. This was recorded on the island of Guam in the South Pacific, October 12, 1979, during Typhoon Tip. The highest pressure on record is 108.38 kPa, measured in Siberia, December 31, 1968.

Figure 7
Can the cardboard "lid" remain in place?

Analyzing the changes in atmospheric pressure at a specific locality is useful for weather forecasting. A constant pressure indicates little, if any, change in the weather. Generally, decreasing atmospheric pressure signals the approach of a storm; increasing pressure means the approach of fair weather.

▶ **TRY THIS** activity

Demonstrate Atmospheric Pressure

Submerge a flask in water, and tip it so that all the air leaves and the flask completely fills with water. Cover the flask's mouth with a piece of cardboard. While holding the cardboard tightly in place, remove the flask from the water. Let go of the cardboard and hold the flask inverted above the container (**Figure 7**).

Besides weather forecasting, there are numerous applications of atmospheric pressure. The drinking straw is an example. Drawing air out of a straw briefly reduces the pressure inside the straw. The atmospheric pressure on the surface of the liquid in the container is then greater than the pressure inside the straw. It forces the liquid to rise up the straw (**Figure 8(a)**). A suction cup works in a similar way. When moistened and pressed on a smooth, flat surface, air originally inside the cup is forced out past the cup's rim. The air pressure between the cup and the surface is thereby reduced. The atmospheric pressure outside the cup is greater than the pressure inside the cup, forcing the cup against the surface (**Figure 8(b)**). Suction cups are used in industry, for example, to lift plate glass, and in homes, for example, to support towel holders. Another device that operates on the same principle is the eyedropper, used to administer medicine.

DID YOU KNOW ?

The Straw Patent
The drinking straw was patented in 1888. It is an example of an invention that seems so simple that some inventors might not even try to obtain a patent.

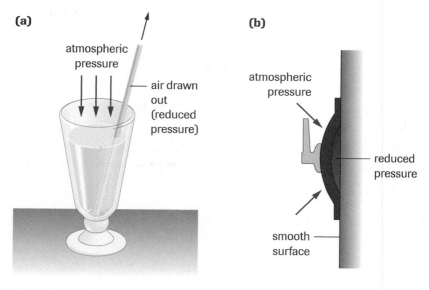

(a)
atmospheric pressure

air drawn out (reduced pressure)

(b)
atmospheric pressure

reduced pressure

smooth surface

Figure 8
(a) A drinking straw   **(b)** A suction cup

▶ **Practice**

Understanding Concepts

19. In Toronto, Ontario's CN Tower, the elevator takes 60 s to rise to the main observation deck, 342 m above ground. If a person's eardrums curve during the ascent, will they curve inward or outward? Explain your answer.

20. A weather forecaster reports that in 6.0 h the atmospheric pressure has changed from 100.7 kPa to 100.1 kPa. What prediction can be made about the weather?

21. Suppose that at sea level a student is able to draw water up a straw to a height of 110 cm. If the same student tried this experiment with the same straw at the top of a mountain, would the water rise to a level higher or lower than 110 cm? Explain your answer.

Applying Inquiry Skills

22. Plot a graph of the data in **Table 1**, page 233, with atmospheric pressure on the vertical axis. From the graph, determine the approximate
(a) atmospheric pressure at 4000 m
(b) altitude at which the pressure is 45 kPa

Answers

22. (a) 62 kPa
    (b) 6400 m

Making Connections

23. The pressure outside a jet aircraft in flight is 25.0 kPa.
(a) What is the elevation of the craft? (Refer to **Table 1**, page 233, or to the graph you drew in question 22.)
(b) What is done to the inside of the aircraft to ensure the comfort of the passengers and crew?

SUMMARY **Pressure**

- Pressure is the force per unit area, measured in pascals (Pa) in the SI. The equation is $p = \dfrac{F}{A}$.

- In hydroforming, water under high pressure blasts metal to form components for cars and for other technologies. In water jet cutting, water under high pressure and at a high speed cuts manufactured components.

- The volume flow rate is the volume of a fluid that flows in a given time interval. There are two equations to determine the volume flow rate: $q_v = \dfrac{V}{\Delta t}$ and $q_v = Av$.

- Atmospheric pressure is the pressure caused by air molecules and other particles above Earth's surface.

- An understanding of atmospheric pressure is important in explaining the operation of devices such as suction cups and in weather forecasting.

Understanding Concepts

1. Explain each of the following statements, emphasizing the difference between force and pressure:
   (a) A paper cut occurs easily.
   (b) The lynx in **Figure 9** has a natural adaptation.

Figure 9
A lynx

2. (a) Express the pascal (Pa) in terms of the newton and metre.
   (b) Explain why the kilopascal is used more frequently than the pascal.

3. Copy **Table 2** into your notebook, and complete the missing information in convenient units. Include the equation needed to find each unknown value.

Table 2   For Question 3

	Force	Area	Pressure
(a)	1.5×10^2 N	6.0×10^{-2} m²	?
(b)	?	6.5 m²	8.8 kPa
(c)	43 N	?	68 Pa

4. Each of the four tires of a car (mass 1.4×10^3 kg) makes contact with the road over a rectangular area 0.15 m × 0.24 m.
   (a) Calculate the surface area of each tire in contact with the road.
   (b) Calculate the magnitude of the force of the car on the road.
   (c) Calculate the pressure, in pascals and kilopascals, that the tires exert on the road. (Assume that the weight of the car is distributed equally over all four tires.)

5. A rectangular roasting pan is 0.45 m long and 0.33 m wide. With a roast resting in the pan, the pressure of the pan on the kitchen countertop is 2.1×10^2 Pa. Calculate the following:
   (a) the area of the pan in contact with the countertop
   (b) the magnitude of the normal force between the pan and the countertop
   (c) the total mass of the roast and the pan

6. The magnitude of the force applied to the desk by the flat bottom of a desk lamp is 44 N; the pressure applied is 1.4 kPa. Calculate the area of the flat bottom of the desk lamp.

7. Explain why atmospheric pressure is greater at Earth's surface than at a high altitude. (Use the kinetic molecular theory in your answer.)

8. The pressure of the water in a hydroforming process is 5.6×10^5 kPa. The sheet of metal being formed is 25 cm × 12 cm. Calculate the magnitude of the force on the sheet.

9. In a water jet cutting process, a stream of water leaves a nozzle of diameter 0.26 mm with the speed 1.4×10^2 m/s. The process lasts for 22 s. Calculate
   (a) the volume flow rate of the water
   (b) the total volume of water used in the process

10. Calculate the volume flow rate of blood in litres per minute for each of the following:
    (a) While resting, a person's heart pumps 12 L of blood in 2.5 min.
    (b) While exercising vigorously, a person's heart pumps 85 L of blood in 3.4 min.

11. Calculate the volume flow rate of water in a hose of radius 2.5 cm if the speed of the water in the hose is 86 m/s.

12. Describe how you would get the maximum amount of liquid into a turkey baster or an eyedropper (**Figure 10**).

Figure 10
Using an eyedropper

Applying Inquiry Skills

13. Determine the pressure between the floor and a chair you are sitting in when your feet are raised off the floor. Use appropriate measurements of the surface area of the bottom of your chair legs. Show your calculations of the pressure.

Making Connections

14. A person wants to glue two flat strips of wood together. Clamps will be needed to hold the strips tightly together while the glue dries. How can this person avoid dents from the clamps in the strips of board?

5.3 Investigation

Inquiry Skills

○ Questioning	● Conducting	● Evaluating
● Predicting	● Recording	● Communicating
○ Planning	● Analyzing	○ Synthesizing

Pressure in Liquids

If an egg and an egg carton were simultaneously lowered deep into the ocean, the carton would become crushed before the egg. The egg's shape allows it to withstand high pressure. For that reason, a rounded shape is applied in the design of deep-sea devices, such as submarines, submersibles, and oxygen tanks for divers (**Figure 1**). In this investigation, you will study how the pressure beneath the surface of a liquid depends on various factors, including the depth at which the pressure is measured.

Figure 1
The design of an underwater submersible ensures it can withstand the high pressure deep below the surface of an ocean.

The way in which you measure pressure in this investigation depends on the equipment available. One way to measure the pressure in various directions beneath the surface of a liquid is to place a pressure-sensitive diaphragm over the mouth of a thistle tube. The tube can then be connected to a computer probe (a *differential pressure transducer*) or a type of pressure meter called a *manometer*.

Save the data you collect in this investigation so you can analyze it further in section 5.4, where you will compare the theoretical values of the pressures with the measured values.

Question

How does the pressure beneath the surface of a liquid depend on (i) the orientation of the object under pressure, (ii) the depth of the liquid, (iii) the size of the container, and (iv) the density of the liquid?

Prediction

(a) Predict answers to all four parts of the Question.

Materials

For each student:
safety goggles
graph paper

For each group or for a class demonstration:
fish tank or similar container at least 30 cm in depth
manometer, support stands, and clamps (**Figure 2**)
  or gauge pressure probe connected to a computer
  interface
thistle tube with a disposable rubber diaphragm
elastic band to secure the diaphragm
three 1000-mL graduated cylinders
metric ruler
a liquid with a density lower than water
  (e.g., methanol)
a liquid with a density higher than water
  (e.g., glycerin)
cleaning materials (e.g., paper towels)

Procedure

Note: The instructions are written for a manometer to determine the pressure. If a pressure probe is used instead, use it wherever the instructions call for the manometer.

1. Fill the tank with water to a depth of 30 cm, and set up the manometer as shown in **Figure 2**. Ensure that the manometer reads zero when the diaphragm is in the air, just above the water. (If the reading is not zero, remove the diaphragm and reattach it.)

2. Insert the diaphragm into the water to a depth of 10 cm beneath the surface; position it so that it faces horizontally. Record the reading on the manometer. Now orient the diaphragm in other directions, including vertically upward and downward (maintain the same depth). Record all depth and manometer readings.

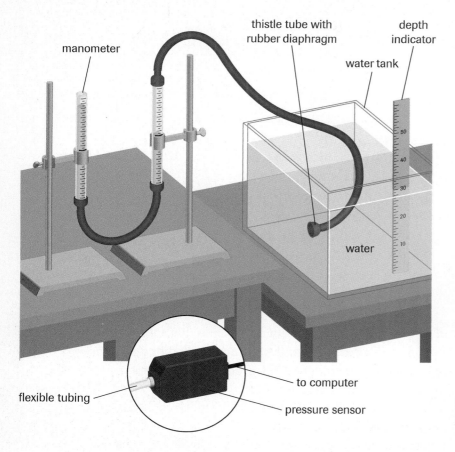

manometer

thistle tube with
rubber diaphragm

depth
indicator

water tank

water

to computer

flexible tubing

pressure sensor

Figure 2
For steps 1 to 3 of Investigation 5.3

3. Aim the diaphragm downward, and lower it to a series of depths: 5 cm, 10 cm, 15 cm, and so on, up to 30 cm. Record the pressures.

4. Add water to a graduated cylinder until its depth is the same as the depth of the water in the tank, and repeat step 3. Dry the thistle tube and diaphragm before going on to the next step.

5. Put on your goggles. Repeat step 3 using the graduated cylinder that contains methanol. Use paper towels to clean and dry the thistle tube and diaphragm before going on to the next step.

6. Repeat step 3 using the graduated cylinder that contains glycerin. Discard the diaphragm and elastic band, and use paper towels to clean and dry the tubing.

Analysis

(b) How does the pressure at a specific depth beneath the surface of the water depend on the orientation of the diaphragm? Give evidence to support your answer.

(c) How does the pressure in a liquid depend on the size of the container? Give evidence to support your answer.

(d) On a single graph, plot the pressure (vertical axis) versus depth beneath the surface for each of the three liquids tested. Draw the line of best fit for each liquid. State how the pressure beneath the surface depends on the depth of the liquid.

(e) Based on your data, does the pressure beneath the surface increase, decrease, or stay the same when the density of the liquid is greater? (Densities are listed in **Table 1**, section 5.1, page 223.) Give evidence to support your answer.

Evaluation

(f) Comment on the accuracy of your predictions.

(g) Describe any problems you had with the measurements in this investigation. What did you do to minimize those problems?

Airplane pilots understand the importance of knowing the altitude of the airplane. This is especially important in conditions of poor visibility, such as when flying through clouds. A type of pressure meter called an **altimeter** uses atmospheric pressure to keep track of an airplane's altitude above sea level (**Figure 1**). The altimeter is just one of many applications of measuring atmospheric pressure.

You learned that atmospheric pressure is due to layers of air molecules and other particles in the atmosphere (section 5.2), and that the pressure beneath the surface of a liquid increases as the depth increases (Investigation 5.3). At any particular position in a static fluid, the height of a column of the fluid that produces a specific pressure is called the **static pressure head**. In this section, you will learn how the static pressure head and various pressures are measured and applied in a variety of situations.

altimeter an instrument that uses atmospheric pressure to determine the altitude of an airplane above sea level

static pressure head the height of the fluid in a column above a position with a specific pressure

barometer an instrument that measures atmospheric pressure

aneroid barometer an instrument that measures atmospheric pressure without the use of a liquid

Measuring Atmospheric Pressure

The instrument that measures atmospheric pressure is called a **barometer**. It was invented in Italy around 1640 by Evangelista Torricelli (**Figure 2**). Torricelli used mercury in his barometer. Although mercury barometers are still used today, they are not as convenient and safe as the **aneroid barometer**. (The word *aneroid* means "without liquid.") An aneroid barometer is an enclosed container with thin metal walls that are sensitive to pressure changes (**Figure 3**). A needle attached to the container indicates the atmospheric pressure. An aneroid barometer can be used both as a weather gauge and as an altimeter.

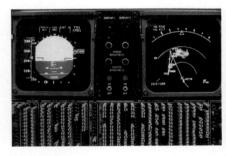

Figure 1
A typical altimeter

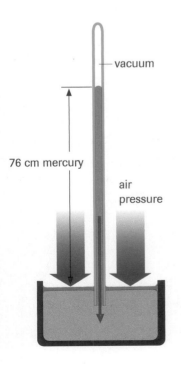

Figure 2
Evangelista Torricelli (1608–1647), a student of Galileo Galilei, made the first barometer by filling a very long glass tube with water so that no air could get into it. The tube, closed at the top and open at the bottom, was placed in a water-filled container. The atmospheric pressure was great enough to hold about 10.3 m of water in the tube—the height of a three-storey building, not very practical for a science laboratory. So Torricelli tried the same experiment using liquid mercury, which has a density about 13.6 times that of water. He found that the tube needed to be only about 76 cm (0.76 m) long.

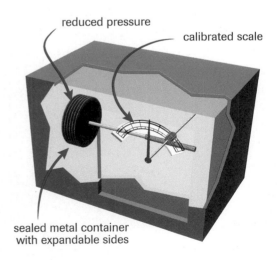

Figure 3
An aneroid barometer. As air pressure drops, the container expands, moving the bar attached to the needle.

The atmospheric pressure at a specific altitude limits the height to which water can be pumped from a well using a suction pump. At an atmospheric pressure of 101 kPa, the maximum height is about 10.3 m. Thus, even with a well-made pump, a water well can be no deeper than 10.3 m. This type of well, called a *shallow well*, was common before fuel-powered and submersible electric pumps became available.

Another effect of atmospheric pressure is that the liquid in connected vessels remains at a constant level (**Figure 4**). The observation that "water seeks its own level" has been applied for centuries by people designing structures that must be horizontally level.

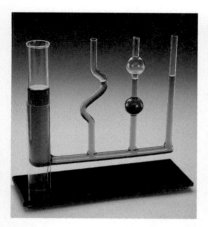

Figure 4
Despite the different shapes and diameters of the connected containers, the liquid is at the same level in all the containers.

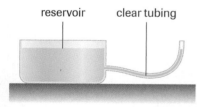

reservoir   clear tubing

Figure 5
This "water level" has a reservoir that remains stationary while the flexible tubing is manipulated.

▶ TRY THIS activity   *Design a Level Ice Rink*

Imagine you are designing an outdoor ice rink the size of your physics classroom, with boards that are 25 cm high surrounding the rink. You apply the principle that water in connected vessels reaches a common level. You can model this task in the classroom, as described below.

(a) Describe how you would use the device shown in **Figure 5** with very long, flexible tubing so that you can draw a line around the inside walls of your classroom at a constant height above the floor.

(b) Describe how you can ensure that, using your method, water will not spill onto the floor.

(c) With your teacher's permission, test your method. (Don't leave any permanent marks on the wall!) Describe any problems you encounter.

▶ Practice

Understanding Concepts

1. State an advantage of each of the following:
   (a) a mercury barometer compared with a water barometer
   (b) an aneroid barometer compared with a mercury barometer

2. The best suction pump in the world can pull liquid mercury in an open tube no higher than about 76 cm. Explain why.

Applying Inquiry Skills

3. Before pouring concrete for the base of a storage shed, you want to make sure that the base is level. Describe how you could ensure a level base by applying the fact that water in connected vessels reaches a common level.

Making Connections

4. Some people think that the only way to get liquid flowing in a siphon is to create a partial vacuum in the tube by drawing on one end of it. But that is not the only way. Suggest another way. Explain how your solution works.

Measuring Gauge Pressure

Pressure readings other than atmospheric pressure are often required when working with gases and liquids. Instruments that measure these values do not measure the true, or **absolute pressure**; rather, they measure the **gauge pressure**, the difference between the absolute pressure and the atmospheric pressure. Thus,

gauge pressure = absolute pressure − atmospheric pressure, or

absolute pressure = gauge pressure + atmospheric pressure

Using symbols, these equations are

$$p_g = p_{abs} - p_{atm} \qquad \text{or}$$
$$p_{abs} = p_g + p_{atm}$$

For example, if air is added to a bicycle tire until the gauge pressure is 405 kPa, and the atmospheric pressure is 101 kPa, then the absolute pressure in the tire is 506 kPa.

One type of instrument used to measure gauge pressure is a **manometer**. (This was one of the instruments suggested for measuring pressure in Investigation 5.3.) A simple U-shaped manometer, which consists of two pieces of glass tubing (each about 30 cm long) connected by a rubber tube, is shown in **Figure 6(a)**. Water is added to a depth of about 15 cm from the bottom of the instrument. If the pressure on one side of the manometer increases, as in **Figure 6(b)**, the water level on that side falls, while the water level on the other side rises.

absolute pressure the true pressure, or the sum of the atmospheric and gauge pressures; symbol p_{abs}

gauge pressure the difference between the absolute pressure and the atmospheric pressure; symbol p_g

manometer an instrument used to measure gauge pressure

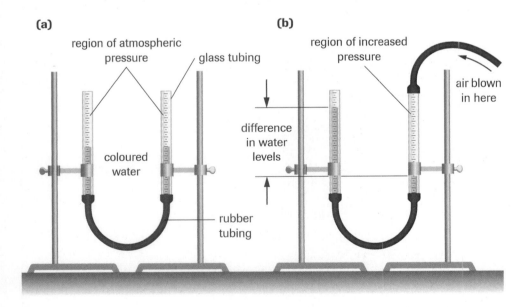

Figure 6
(a) A manometer
(b) Applying additional pressure on one side causes the liquid to rise on the other.

The tubes can be calibrated using a reference manometer, or the gauge pressure can be found using the following derivation: The pressure at the bottom of a column of liquid of height h is given by $p = \dfrac{F}{A}$. From Chapter 1, the magnitude of the force of gravity on the liquid is $F = mg$. Thus,

$$p = \frac{mg}{A}$$

However, from $D = \dfrac{m}{V}$, $m = DV$. Therefore, the pressure is

$$p = \frac{DVg}{A}$$

Finally, we substitute $V = Ah$:

$$p_g = Dhg$$

where p_g is the gauge pressure in pascals,
$\qquad$ D is the density of the liquid in kilograms per cubic metre,
$\qquad$ h is the height of the column in metres, and
$\qquad$ g is the gravitational force constant (9.8 N/kg).

In the case of the U-shaped manometer, h is the *difference* in height between the two levels of liquid. This height is the static pressure head of the liquid relative to the lower level.

▶ *SAMPLE* problem *1*

Pressure Difference

An air pump is connected to one tube of a manometer containing water ($D = 1.0 \times 10^3$ kg/m³). Pressure is increased until the static pressure head is 15 cm. What is the gauge pressure?

Solution
$D = 1.0 \times 10^3$ kg/m³
$h = 15$ cm $= 0.15$ m
$g = 9.8$ N/kg
$p_g = ?$

$$p_g = Dhg$$

$$= \left(1.0 \times 10^3 \; \frac{kg}{m^3 m^2}\right)(0.15 \, m)\left(9.8 \; \frac{N}{kg}\right)$$

$$= 1.5 \times 10^3 \; \frac{N}{m^2}$$

$$p_g = 1.5 \times 10^3 \text{ Pa, or 1.5 kPa}$$

The gauge pressure is 1.5×10^3 Pa, or 1.5 kPa.

▶ **TRY THIS** activity   ***Test Your Grip***

A grip tester is a specially designed manometer (**Figure 7**). If one is available, use it to test the strength of your grip.

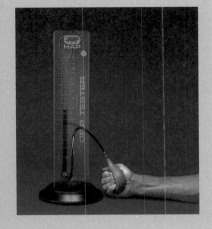

Figure 7
A grip tester

Calculating the gauge pressure for known static pressure heads in liquids is important when designing structures such as dams. **Figure 8** shows the cross section of a typical dam. To understand the shape of the dam, consider the factors that affect the gauge pressure beneath the surface: the density of the liquid (D), the height or depth of the liquid (h), and the gravitational force constant (g). For water at a particular location, both D and g are constant. So the gauge pressure of the water behind a dam depends on the depth, h. Thus, the deeper the water, the greater the gauge pressure, and the thicker the dam must be to withstand that pressure.

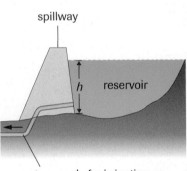

Figure 8
The dam must be built such that it is thickest at the base, where the water pressure is greatest. The static pressure head, h, is the height of the water in the reservoir behind the dam.

▶ **SAMPLE** problem **2**

Absolute Pressure

Calculate the absolute pressure in the water behind a dam with a static pressure head of 18.0 m. The atmospheric pressure at the dam's location is 1.00×10^5 Pa.

Solution

$p_{atm} = 1.00 \times 10^5$ Pa
$D = 1.00 \times 10^3$ kg/m³
$h = 18.0$ m
$g = 9.80$ N/kg
$p_{abs} = ?$

$$p_{abs} = p_{atm} + p_g$$
$$= p_{atm} + Dhg$$

$$= 1.00 \times 10^5 \text{ Pa} + \left(1.00 \times 10^3 \frac{\text{kg}}{\text{m}^3\text{m}^2}\right)(18.0 \text{ m})\left(9.80 \frac{\text{N}}{\text{kg}}\right)$$

$$= 1.00 \times 10^5 \text{ Pa} + 1.76 \times 10^5 \text{ Pa}$$
$$p_{abs} = 2.76 \times 10^5 \text{ Pa, or 276 kPa}$$

The absolute pressure is 2.76×10^5 Pa, or 276 kPa.

Answers

5. 306 kPa
6. 361 kPa
7. (a) −66.6 kPa
   (b) 34 kPa
8. 51.3 kPa
10. 1.21×10^5 Pa; 121 kPa

(a)

(b)

(c)

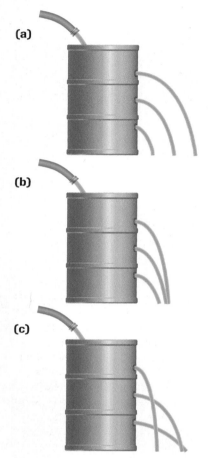

Figure 9
For question 11

DID YOU KNOW?

Early Blood Pressure Experiment
The first recorded experiment to measure blood pressure was made by Stephen Hales in Britain in 1733. He connected a 3.0-m vertical glass tube to a horse's artery using the trachea of a goose as a connecting tube. (A trachea is the tube made of tissue that carries air to the lungs.) After puncturing the artery with a sharpened goose quill, he observed that the blood rose 2.4 m above the horse's heart.

 www.science.nelson.com

▶ *Practice*

Understanding Concepts

5. The gauge pressure of a car tire is 205 kPa, and the atmospheric pressure is 101 kPa. Calculate the absolute pressure of the tire.

6. If the atmospheric pressure is 102 kPa, and the absolute pressure of a bicycle tire inflated by an air pump is 463 kPa, what is the gauge pressure of the tire?

7. A vacuum pump removes air from one side of a manometer that uses mercury ($D = 1.36 \times 10^4$ kg/m³). The other side of the manometer is open to the atmosphere, which exerts pressure on the mercury. The resulting static pressure head of the mercury is 0.500 m.
   (a) Calculate the gauge pressure in the manometer, in kilopascals.
   (b) If the atmospheric pressure is 101 kPa, what is the absolute pressure on the vacuum pump side of the manometer?

8. At a location where the atmospheric pressure is 101.3 kPa, determine the pressure caused by a static pressure head of 5500 m of air. (Hint: According to **Table 1**, page 233 in section 5.2, the atmospheric pressure at 5500 m is 50.0 kPa. In this case, you cannot apply the equation $p_g = Dgh$ because the air density is not constant over such a large change in elevation.)

9. Two freshwater lakes are of equal depth; each has a dam. One lake is 1.0 km long and the other is 2.0 km long. How does the pressure at the base of one dam compare to the pressure at the base of the other dam? Explain your answer.

10. Calculate the absolute pressure, in pascals and kilopascals, at a depth of 2.00 m in a swimming pool if the atmospheric pressure is 1.01×10^5 Pa.

Applying Inquiry Skills

11. A can has three holes, but as water drains out of the holes, a continuous source keeps filling it from the top.
    (a) Which of the diagrams in **Figure 9** best illustrates how water would flow from the holes? Explain your choice.
    (b) Test your answer to (a) by poking three equal-sized holes in the side of a plastic juice container and performing an appropriate experiment. Work outside or use a sink to catch the water. If your answer was wrong, explain why, and give a correct answer.

Making Connections

12. A pressure sensor connected to a data recorder can be dragged across the bottom of a lake or river to measure depth. The data obtained can be used to draw a cross-sectional profile of the lake or river. Describe the physics that allows the sensor to measure depth.

Measuring Blood Pressure

The human heart pumps blood under pressure to all parts of the body. The amount of pressure exerted by the heart is easy to measure and is an important indicator of a person's general health.

One way to measure blood pressure is to wrap a gauge called a *sphygmomanometer* around the upper arm at the same level as the heart

(**Figure 10**). Two measurements are made: first the maximum blood pressure, called the *systolic pressure,* then the pressure when the heart is resting, called the *diastolic pressure.*

A pump is used to increase the pressure in the rubber cuff so that it rises above the systolic level. Then the rubber cuff pressure is gradually reduced, as the health care practitioner listens through a stethoscope to the flow of blood in the main artery of the arm. When the systolic pressure is reached, the blood begins spurting through the artery, causing a tapping sound that can be detected through the stethoscope. When the pressure has decreased to the diastolic level, the sound disappears.

Normal human blood pressures, which are gauge pressures, range from about 16 kPa (systolic) down to about 11 kPa (diastolic). They are usually written as the ratio 16/11, which is read "16 over 11." A person with high blood pressure, for example, above 19/12, requires medical attention. In practice, blood pressure is measured in units of millimetres of mercury (mm Hg). In these units, the ratio of normal blood pressure is 120/80.

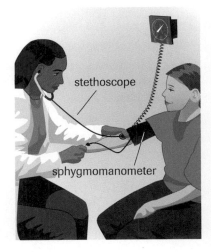

Figure 10
One way to measure blood pressure is to use a sphygmomanometer.

 Practice

Understanding Concepts

13. Explain why it is important to test blood pressure at the same level as the heart.

14. Determine the range of absolute pressures (systolic and diastolic) in a normal human heart at a location where the absolute pressure is 101 kPa.

15. Assume that the density of the air outside the Calgary Tower is 1.28 kg/m³. Calculate the atmospheric pressure difference between the base of the tower and the observation terrace, 191 m higher. (In this case, you can assume that the air density is constant.)

Making Connections

16. Research how blood pressure is monitored in the operating room. Explain how this method differs from the arm cuff method.

Answers

14. 117 kPa to 112 kPa

15. 2.40 kPa

DID YOU *KNOW* **?**

Blood Pressure in a Giraffe
Since the pressure of liquids increases with depth, why doesn't a giraffe faint when it raises its head or bleed when it lowers it? The giraffe has a complex system of valves and blood vessels in its brain, as well as a large heart, which enable it to control the differences in pressure (**Figure 11**).

SUMMARY *Measuring Pressure*

- The static pressure head is the height of the fluid in a column above a position with a specific pressure.
- Atmospheric pressure can be measured using a variety of instruments, including a barometer, which uses a liquid, and an aneroid barometer, which does not.
- Absolute pressure is the sum of the atmospheric and gauge pressures ($p_{abs} = p_g + p_{atm}$).
- Gauge pressure caused by the static pressure head of a liquid can be found by applying the equation $p_g = Dgh$.

Figure 11
Because it has such a long neck, the giraffe needs a specially adapted circulatory system.

 www.science.nelson.com

Understanding Concepts

1. Is an altimeter designed like a manometer, a barometer, or an aneroid barometer? Explain your answer.

2. Is the atmospheric pressure on your index finger greater than, less than, or equal to the atmospheric pressure on your entire hand? Explain your answer.

3. The hose that connects containers A and B in **Figure 12** is filled with liquid.
   (a) At the instant shown, in what direction is the liquid flowing? Why does it flow?
   (b) At what stage will the liquid stop flowing? Why?

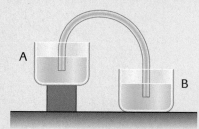

Figure 12
For question 3

4. Copy **Table 1** into your notebook, and complete the missing information in the most convenient units. Include the equation needed to find each unknown value.

Table 1   For Question 4

	p_{abs} (kPa)	p_{atm} (kPa)	p_g (kPa)
(a)	?	101	112
(b)	457	?	355
(c)	204	101	?

5. (a) Use an equation to prove that an atmospheric pressure of 101 kPa can support a static pressure head of water of 10.3 m.
   (b) Determine the static pressure head that the same atmospheric pressure could support with a liquid of density 1.30×10^3 kg/m³.

6. Calculate the gauge difference, in kilopascals, in a mercury manometer if the static pressure head of the mercury is 12.5 cm. (The density of mercury is 1.36×10^4 kg/m³.)

7. The maximum depth recorded for a deep-sea dive is 308 m. Assuming the density of seawater is constant at 1.03×10^3 kg/m³ and the atmospheric pressure is 101 kPa, calculate
   (a) the gauge pressure at this depth, in pascals and kilopascals

   (b) the absolute pressure at this depth, in kilopascals

Applying Inquiry Skills

8. If you measured the gauge pressures in Investigation 5.3, refer to your data. Choose three different trials in which you determined the pressure. For each trial, determine the theoretical gauge pressure using the equation $p_g = Dgh$ and the quantities
   $D_{water} = 1.00 \times 10^3$ kg/m³
   $D_{glycerin} = 1.26 \times 10^3$ kg/m³
   $D_{methanol} = 7.9 \times 10^2$ kg/m³
   Assuming that the calculated gauge pressure is the accepted value, determine the percent error of each measured gauge pressure. (Percent error is discussed in Appendix A1.) Account for any discrepancies.

9. In an experiment to determine the underwater profile across a creek, a scientist in a small boat moving at a constant speed pulls a pressure probe that follows the bottom of the creek. The resulting graph of the gauge pressure versus time is shown in **Figure 13**.
   (a) Calculate the maximum depth of the creek.
   (b) Draw a cross-sectional profile of the creek to show its shape.

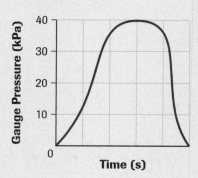

Figure 13
For question 9

Making Connections

10. The dive described in question 7 took just 12 min for the descent. (The diver was hooked to a weighted sled during the descent.) However, the ascent took more than 9.5 h. Research the dangers that divers face if they ascend too quickly. Describe the dangers in terms of pressure changes.

A car hoist in an automotive service centre is a hydraulic system (**Figure 1**). A hydraulic system makes use of an important property of liquids: They cannot be (easily) compressed. The relationship between pressure and enclosed fluids was formulated by French scientist Blaise Pascal. His formulation is called *Pascal's principle*.

Pascal's Principle
Pressure applied to an enclosed liquid is transmitted equally to every part of the liquid and to the walls of the container.

Pascal applied his principle in the design of a device called the **hydraulic press**. **Figure 2** illustrates how such a press works. A small downward force applied to the small movable piston can produce a large upward force on the large movable piston. According to Pascal's principle, the pressure (p_S) on the small piston equals the pressure (p_L) on the large piston, or $p_S = p_L$. Thus, since $p = \dfrac{F}{A}$, we can write

$$\frac{F_S}{A_S} = \frac{F_L}{A_L}$$

where "S" means small and "L" means large.

Figure 1
A car hoist applies Pascal's principle. (The hoist's hydraulic system is controlled by a second system, which is pneumatic. It is not considered in discussing Pascal's principle.)

hydraulic press a device in which a small force on a small piston is transmitted through an enclosed liquid and applies a large force on a large piston

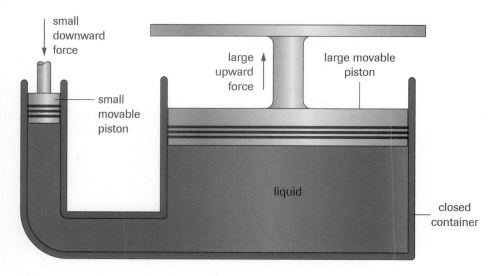

Figure 2
According to Pascal's principle, the small downward force results in a large upward force.

The principle of the hydraulic press is applied in devices as simple as a medical syringe and as complex as a car's braking system. In each case, the small movable piston must move a greater distance than the large piston in order to move the same volume of liquid in each cylinder.

▶ **SAMPLE problem 1**

Pressure in a Hydraulic Press

A car of mass 1.4×10^3 kg is hoisted on the large cylinder of a hydraulic press. The area of the large piston is 0.22 m², and the area of the small piston is 0.013 m².

(a) Calculate the magnitude of the force of the small piston needed to raise the car (at a constant, slow speed) on the large piston.

(b) Calculate the pressure, in pascals and kilopascals, in this hydraulic press.

Solution

(a) The magnitude of the force on the large piston is the magnitude of the car's weight ($F_L = mg$).

$m = 1.4 \times 10^3$ kg

$g = 9.8$ N/kg

$A_L = 0.22$ m²

$A_S = 0.013$ m²

$F_S = ?$

$$\frac{F_S}{A_S} = \frac{F_L}{A_L}$$

$$F_S = \frac{F_L A_S}{A_L}$$

$$= \frac{(mg)A_S}{A_L}$$

$$= \frac{(1.4 \times 10^3 \, \cancel{kg})\left(9.8 \, \frac{N}{\cancel{kg}}\right)(0.013 \, \cancel{m^2})}{0.22 \, \cancel{m^2}}$$

$$F_S = 8.1 \times 10^2 \text{ N}$$

The force on the small piston is 8.1×10^2 N.

(b) Because the pressure in the liquid is constant, the data for either piston can be used to solve for the pressure.

$F_S = 8.1 \times 10^2$ N

$A_S = 0.013$ m²

$p_S = ?$

$$p_S = \frac{F_S}{A_S}$$

$$= \frac{8.1 \times 10^2 \text{ N}}{0.013 \text{ m}^2}$$

$$p_S = 6.2 \times 10^4 \text{ Pa, or } 62 \text{ kPa}$$

The pressure in the hydraulic press is 6.2×10^4 Pa, or 62 kPa.

▶ *TRY THIS* activity   *Test Pascal's Principle*

Figure 3 shows a device with two pistons, one large and one small, that can be used to test Pascal's principle for a liquid. Using a ruler, determine the surface area of each piston. (Recall that the area of a circle is given by πr^2, where r is the radius.) Use force sensors pushing on the large and small pistons to determine the forces involved. Using this apparatus or one similar to it, take measurements and make calculations to test Pascal's principle, verifying that $p_S = p_L$.

Figure 3
A two-cylinder device for testing Pascal's principle

▶ **Practice**

Understanding Concepts

1. Explain why Pascal's principle does not apply to gases.

2. Rewrite $\dfrac{F_S}{A_S} = \dfrac{F_L}{A_L}$ to solve for (a) F_S, (b) A_S, (c) F_L, and (d) A_L.

3. Copy **Table 1** into your notebook. Complete the information in the most convenient units. Include the equation needed to find each unknown value.

Table 1   For Question 3

	F_S (N)	A_S (m²)	F_L (N)	A_L (m²)
(a)	?	0.25	1.8×10^3	1.5
(b)	65	?	5.2×10^2	0.64
(c)	5.4×10^2	0.15	?	1.2
(d)	1.4×10^2	2.0×10^{-2}	7.7×10^3	?

4. In a mining operation, a hydraulic jack can exert a maximum force of magnitude 2.2×10^3 N on the small piston. The surface area of the small piston is 0.10 m², and the surface area of the large piston is 2.0 m². Calculate
   (a) the magnitude of the maximum force that can be used to lift a load
   (b) the mass of that load

Making Connections

5. Many diesel trucks are equipped with compression release engine braking systems, also known as "Jake brakes." Research these brakes and describe their advantages.

  www.science.nelson.com

Answers

3. (a) 3.0×10^2 N
   (b) 8.0×10^{-2} m²
   (c) 4.3×10^3 N
   (d) 1.1 m²

4. (a) 4.4×10^4 N
   (b) 4.5×10^3 kg

Since the invention of cars about 100 years ago, the technology of braking systems has changed many times. The world's first assembly-line production car, the Model T Ford (**Figure 4**), used bands surrounding the transmission to brake the car. The bands were controlled by the rightmost floor pedal. This technology was abandoned because it was not as effective as a brake system linked to the wheels.

The Model A Ford (**Figure 5**), first available in 1928, had mechanical brakes on its four wheels. These brakes were the forerunners of *drum brakes,* which are attached to the wheels and were the standard design in the mid-twentieth century. The drum brakes shown in **Figure 6** are hydraulic brakes that apply Pascal's law. When the brake pedal is pushed, the pressure on the brake fluid in the hydraulic system is increased. This pressure is applied to the brake shoes, which exert a pressure on the brake linings. The linings rub against the brake drum, slowing the wheels. This rubbing, or friction, causes a large build-up of thermal energy, which results in a reduced braking force. When the brake pedal is released, the springs pull the shoes and linings away from the drum.

Figure 4
Between 1909 and 1927, more than 15 million Model T Fords were built.

Figure 5
From 1928 to 1931, around 4 million Model A Fords were built.

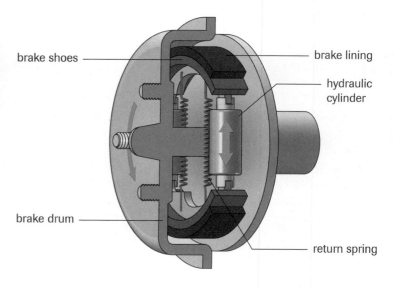

brake shoes — brake lining — hydraulic cylinder — brake drum — return spring

Figure 6
Drum brakes are used on the rear wheels of many cars.

Disk brakes are found on most modern cars, especially on the front wheels, where greater braking force is needed. Like drum brakes, disk brakes apply Pascal's principle in a hydraulic system. **Figure 7** shows the overall design of a hydraulic disk brake system. When the brake pedal is pushed, the pressure on the brake fluid in the master cylinder is increased. This pressure is transmitted through the connectors to apply pressure on the fluid in eight other cylinders, two on each wheel. The increased pressure on the brake pads exerts a frictional resistance on the wheel disks, slowing down the car. Friction generates thermal energy, but that energy is transferred away quickly by air flowing over the disk or through holes in the disk. Modern brake systems can be computer-controlled, giving the added safety feature of antilock braking.

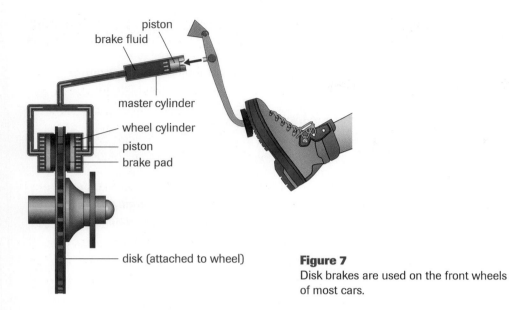

piston
brake fluid

master cylinder

wheel cylinder
piston
brake pad

disk (attached to wheel)

Figure 7
Disk brakes are used on the front wheels
of most cars.

▶ *Practice*

Understanding Concepts

6. Describe the advantages of today's car braking systems compared to the
   braking system used on the Model T Ford.

7. In a hydraulic braking system, a force of magnitude 25 N is exerted on the
   piston in the master cylinder. This piston has a surface area of 5.0 cm².
   Calculate the magnitude of the force exerted on each brake piston having an
   area of 95 cm².

Answer

7. 4.8×10^2 N

Making Connections

8. Research the mechanics of today's power-assisted brakes and antilock
   brakes.
   (a) How do these features improve the performance of the braking system?
   (b) How do these features improve driver safety?

 www.science.nelson.com

SUMMARY *Pascal's Principle*

- Pascal's principle states that pressure applied to an enclosed liquid is
  transmitted equally to every part of the liquid and to the walls of the
  container.

- Pascal's principle can be written as an equation: $\dfrac{F_S}{A_S} = \dfrac{F_L}{A_L}$.

- Pascal's principle is applied in hydraulic systems, for example, a car's
  braking system.

Understanding Concepts

1. **Figure 8** shows a horizontal plastic tube with a straight stopper plugging the left end and an angled stopper plugging the right end.
   (a) If you slowly push in the stopper at the left end, what will happen if the tube is filled with air? What will happen if the tube is filled with water? Explain why there is a difference.
   (b) Relate this situation to Pascal's principle.

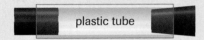

plastic tube

Figure 8
For question 1

2. The principle of the hydraulic press is used to raise and lower a dentist's chair. Assume that the pressure everywhere in the liquid is 15.0 kPa. The small piston has an area of 8.00×10^{-3} m^2, and the large piston has an area of 7.00×10^{-2} m^2. Calculate the magnitude of the force on each piston.

3. In a hydraulic jack, a force of magnitude 1.5×10^2 N is applied to a piston with an area of 6.6 cm^2. Calculate the force exerted on the larger piston, whose area is 33 cm^2.

4. In a hydraulic press, a small input force can be used to obtain a large output force. Is nature giving us something for nothing? Explain your answer.

Applying Inquiry Skills

5. An experiment to test Pascal's principle is carried out with a two-cylinder system. The radius of the small piston is 2.0 cm, and the radius of the large piston is 4.0 cm. The force sensor at the small piston registers 12 N when the force sensor at the large piston registers 48 N. Does this experiment verify Pascal's principle? Show your calculations.

Making Connections

6. What impact has changing brake technology had on driving on Canadian roads? In your answer, include consideration of speed limits and driving safety.

7. Hydraulic jacks are used in heavy equipment in mining, industry, and manufacturing. Research one application of the hydraulic jack. Explain how the jack applies Pascal's principle; include a diagram in your explanation.

5.6 Investigation

Two- and Three-Cylinder Fluid Systems

Syringes and similar devices can be used to study the variables in an enclosed fluid system. **Figure 1(a)** shows a set of cylinders, pistons, valves, and hoses, which control forces and pressures in an enclosed system. **Figure 1(b)** shows a typical experimental syringe with a movable piston in a cylinder. In this investigation, you will observe the action of a hydraulic and a pneumatic system.

In Part A, you will learn how to control the pressure in experimental syringes and determine the properties of a pneumatic system with air as the fluid. In Part B, you will use the experimental syringes to determine the properties of a hydraulic system with water as the fluid. In Part C, you will explore how valves can be used to control a pneumatic system.

(a)   **(b)**

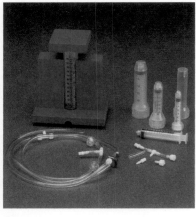

Figure 1
(a) An example of a kit used to study fluid systems
(b) A syringe designed for experiments

Questions

How do hydraulic and pneumatic systems compare? How can valves be used to control fluid flow in a pneumatic system?

Predictions

(a) Predict how the action of a pneumatic system with a long hose connection compares to the action of one with a short hose connection.

(b) Predict how the action of a hydraulic system with a long hose connection compares to the action of one with a short hose connection.

(c) In how many locations can an on–off valve be placed in the fluid system in **Figure 2**? Would the control effects be the same at each location?

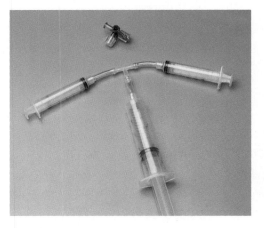

Figure 2
The on–off valve can be inserted at various positions.

Materials

For each student:
safety goggles
apron

For each group of three or four students:
2 large syringes (e.g., 20 mL)
2 small syringes (e.g., 5 mL)
3 short lengths of tubing (e.g., 3 cm long, 6 mm in diameter)
one long length of tubing (e.g., 40 cm long, 6 mm in diameter)
straight connector
T connector
2-way (on–off) valve
metric ruler
2 force sensors
beaker with water
basin to catch water spills
sponge and paper towels

 Wear safety goggles when using syringes. Always aim pressurized syringes away from yourself and others.

Procedure

Part A

1. Put on your safety goggles. Connect the two large syringes using a short length of tubing (**Figure 3**). Try to pull one piston out. If it does not move, determine what must be done before connecting the tubing.

Figure 3
Connecting the large syringes

2. As you depress one piston, what happens to the other piston? Record your observation.

3. Try holding one piston steady while moving the other piston. Describe what you observe.

4. Take measurements and make calculations needed to determine the pressure on each piston when you apply a force to each one in turn. (Hint: Use the metric ruler and two force probes.)

5. Replace the short tubing with the long tubing and repeat steps 2 to 4.

6. Place two short lengths of tubing connected to the straight connector between the two large syringes. Repeat steps 2 to 4.

7. Join one large syringe to the two small syringes using the T connector and three short lengths of tubing. Depress the piston in the large syringe. Record your observations.

Part B

8. Put on your apron. With the same apparatus used in steps 1 to 3, fill the cylinders with water so they contain no air bubbles. (Work over a basin. Fully depress one piston and remove the other piston completely. Pour water from the beaker into the cylinder until the cylinder is completely full, then insert the piston.) Repeat steps 2 to 4.

9. Repeat step 6 using water.

10. Repeat step 7 using water.

11. Clean and dry all the components.

Part C

12. Design a way to control air with the 2-way valve placed at one position (**Figure 2**). Draw a diagram of the system. Move the pistons one at a time, and record what you observe.

13. Place the valve in a different location, and repeat step 12. Test all possible positions.

Analysis

(d) Summarize the changes you observed in the pneumatic system when you changed its components in Part A.

(e) Summarize the changes you observed in the hydraulic system when you changed its components in Part B.

(f) How does a hydraulic system compare with a pneumatic system in its operation?

(g) Was Pascal's principle verified in both pneumatic and hydraulic systems? Give evidence to support your answer.

(h) Describe how a 2-way valve can be used as a control device in a pneumatic system. (A diagram will help your description.)

Evaluation

(i) Evaluate your predictions.

(j) Is an air system or a water system more difficult to experiment with? Explain your answer.

Synthesis

(k) When using a syringe to inject liquid vaccine, the health care practitioner taps the side of the syringe and depresses the piston slightly before injecting the patient. Explain why.

The hydraulic press and its applications in section 5.5 and the two- and three-cylinder systems in Investigation 5.6 are examples of a **fluid system**, an arrangement of components used to transmit and control forces in a fluid. The initial source of energy for the system can be an electric motor or other device that drives a pump (for liquids) or a compressor (for gases). The pump or compressor transforms mechanical forces into fluid forces. The final output of the system is the **actuator**, a device that transforms fluid forces back into mechanical forces.

You have seen how pressure is distributed in the liquid in a hydraulic press. But how is the pressure controlled? To learn about the overall operation of a fluid system, we will look at the complete cycle of the car hoist. This hydraulic system also has pneumatic components, which control the pressure in the hydraulic components.

Figure 1 shows the components of a car hoist system and describes the function of each component. The components link together to form a circuit in which the gas (air) and the liquid (oil) can flow. An important advantage of this system is that only a small electric motor is needed to pump air into the storage tank. (A huge motor would be needed to pump air to raise the car directly.) Whenever needed, stored air can be used to operate the hoist. Starting at the electric motor, you can follow the operation of the system:

- The motor runs the air compressor.

- The compressor forces the air into the air storage tank.

- The air flows through a transmission line to the on–off valve, and then through a filter to the 3-way valve.

- The air exerts a pressure on the oil in the first cylinder.

- According to Pascal's principle, the pressure in the oil is transmitted to the hydraulic cylinder to raise the car.

Symbols are used to draw circuit diagrams of fluid systems. **Figure 2** shows the circuit diagram of the car hoist as the load is being raised. Once again, if you start at the electric motor, you can follow the flow of the air and the oil in the system.

Study **Figure 1** and **Figure 2** to determine what must be done to lower the car on the hoist. If the 3-way valve is pushed in, the pressurized air in the first cylinder will escape through the air outlet, allowing the oil to flow from the hydraulic cylinder back into the first cylinder. (Notice that a pump is not needed to push in the hydraulic cylinder because the weight of the car does that.) This operation is shown in **Figure 3(a),** and the corresponding circuit diagram is shown in **Figure 3(b).**

fluid system an arrangement of components used to transmit and control forces in a fluid

actuator a device that transforms fluid forces into mechanical forces

DID YOU *KNOW* ❓

Actuators in Driver Simulators
To simulate driving conditions, some advanced simulator vehicles use actuators connected to the axles instead of wheels. These hydraulic legs provide bumps and shakes, and even a realistic feel when turning.

 GO www.science.nelson.com

LEARNING *TIP*

Valve Symbols and Uses
Valves are important as control devices in fluid systems. Understanding valve symbols, such as those in **Figures 1** and **2**, is easier if you study the variety of valves described in Appendix C, **Table 5**. As you study the diagrams, look for patterns to help you remember the uses of the valves.

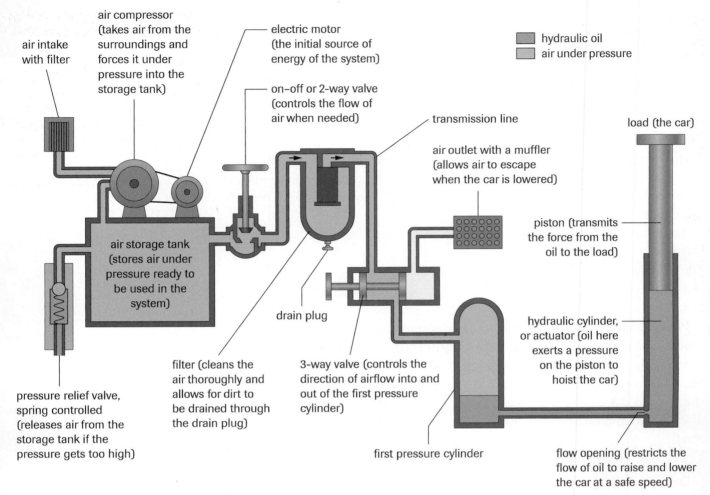

hydraulic oil
air under pressure

air intake with filter

air compressor (takes air from the surroundings and forces it under pressure into the storage tank)

electric motor (the initial source of energy of the system)

on–off or 2-way valve (controls the flow of air when needed)

transmission line

air outlet with a muffler (allows air to escape when the car is lowered)

load (the car)

piston (transmits the force from the oil to the load)

air storage tank (stores air under pressure ready to be used in the system)

drain plug

hydraulic cylinder, or actuator (oil here exerts a pressure on the piston to hoist the car)

pressure relief valve, spring controlled (releases air from the storage tank if the pressure gets too high)

filter (cleans the air thoroughly and allows for dirt to be drained through the drain plug)

3-way valve (controls the direction of airflow into and out of the first pressure cylinder)

first pressure cylinder

flow opening (restricts the flow of oil to raise and lower the car at a safe speed)

Figure 1
A car hoist fluid system with the load being raised

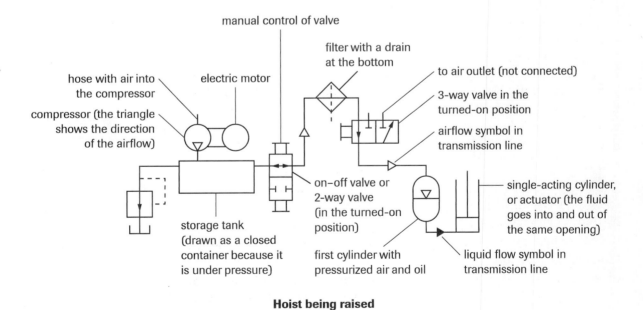

manual control of valve

filter with a drain at the bottom

to air outlet (not connected)

hose with air into the compressor

electric motor

3-way valve in the turned-on position

compressor (the triangle shows the direction of the airflow)

airflow symbol in transmission line

on–off valve or 2-way valve (in the turned-on position)

single-acting cylinder, or actuator (the fluid goes into and out of the same opening)

storage tank (drawn as a closed container because it is under pressure)

first cylinder with pressurized air and oil

liquid flow symbol in transmission line

Hoist being raised

Figure 2
The circuit diagram of the car hoist system with the car being raised

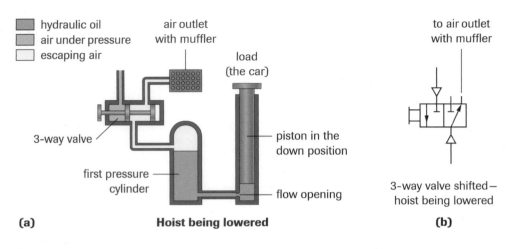

- hydraulic oil
- air under pressure
- escaping air

air outlet with muffler

load (the car)

3-way valve

first pressure cylinder

piston in the down position

flow opening

(a)

Hoist being lowered

to air outlet with muffler

3-way valve shifted— hoist being lowered

(b)

Figure 3
(a) The car hoist system with the car being lowered
(b) Circuit diagram symbol of the 3-way valve

Fluid systems have distinct advantages:

- The pipes and hoses used for transmission lines provide the flexibility to transfer the fluid to a variety of locations.

- A large variety of valves can be used, which allows for greater versatility of control.

> **Practice**

Understanding Concepts

1. The car hoist in **Figures 1** and **2**, page 256, is called an "air-over-oil" system. Explain why.

2. In a car hoist system, state the function of the
   (a) compressor
   (b) filter
   (c) 3-way valve

3. Look at the valve just before the first pressure cylinder in **Figures 1**, **2**, and **3**. Its full name is a "3-way, 2-position valve with a manual control." (For detailed information about valves, refer to Appendix C, **Table 5**.)
   (a) How many openings does the valve have?
   (b) What positions can the valve have?
   (c) What symbol indicates that the control is manual?
   (d) Is the valve's full name justified? Explain your answer.

4. (a) What symbol is used to show the direction of flow of a fluid?
   (b) How can you tell whether the fluid is a gas or a liquid?

5. As the car on the hoist is being lowered, does the air in the system get recycled back to the storage tank for use in the next cycle? Explain your answer.

6. In the car hoist, the pressure on the oil in the first cylinder is caused by air pressure. For other applications, the pressure on the liquid in this type of cylinder can also be provided by a spring or by gravity using a weight. Referring to Appendix C, **Table 5**, draw the circuit diagram symbol for spring- and weight-controlled pressure devices.

robot an automated device, often computer-controlled, that performs a task

Robotic Applications

A **robot** is an automated device, often computer-controlled, that performs a task. Robots can be operated by hydraulic or pneumatic systems, which are driven by electric motors. The many tasks performed by robots can be classified into two general types. One type is repetitive tasks, such as loading and unloading parts, capping bottles, and welding car components on assembly lines (**Figure 4**). The other is handling hazardous materials, such as toxic chemicals or nuclear waste.

(a)

(b)

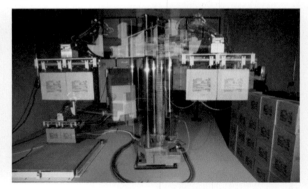

Figure 4
(a) Automobile production with robots
(b) Robot arms loading and unloading boxes

Robots have both advantages and disadvantages:

- In repetitive jobs, robot productivity is high, even though robots don't necessarily work faster than humans.

- Robots designed to perform a specific task may be efficient for that task, but they aren't flexible if the task changes.

- Once programmed, robots are very accurate.

- Hydraulically operated robots do not produce sparks as an electric robot might, so they can be used in situations where sparks are dangerous, such as in spray painting.

- Hydraulically operated robots can lift very heavy loads.

- Hydraulically operated robots need a warm-up period and require careful maintenance.

- Pneumatically operated robots are easy to design; they are also flexible because air can be easily pumped anywhere in a manufacturing plant.

- Pneumatically operated robots are easier to maintain than hydraulically operated robots. (For example, air leaks are easier to repair than oil leaks.)

- Pneumatically operated robots may have a slow reaction time, as you learned in Investigation 5.6. This spongy action is due to the compressibility of gases.

- The design, use, and maintenance of robotic systems require well-trained designers and service workers.

One example of a hydraulic robotic system is one that bends pieces of metal into a V shape (**Figure 5**). In a manufacturing plant, the pieces of metal are brought to the robot by an assembly line. The system is likely hydraulic and uses a double-acting cylinder, that is, a cylinder with two openings. At the instant shown in **Figure 5**, the piston is being pushed toward the metal component being shaped; this is the *power stroke*. As soon as this stroke is complete, the *refill stroke* begins as the liquid flows in the opposite direction. A complete cycle of the robotic press involves both power and refill strokes. The cycle is repeated over and over on the assembly line to create many uniform parts.

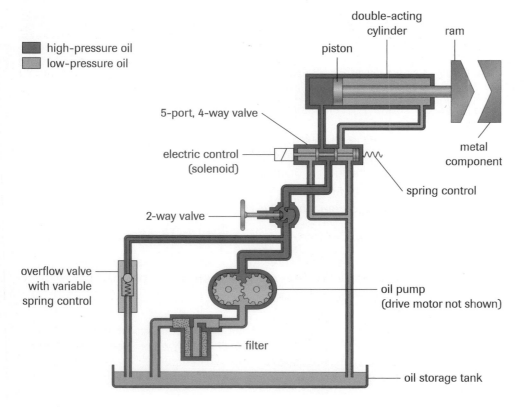

Figure 5
A robotic press with a double-acting cylinder as an actuator

Circuit Diagrams

Using appropriate symbols, draw the circuit diagram that corresponds to the robotic press shown in **Figure 5**.

Solution

The symbols for most of the components are the same as the symbols used in the circuit diagrams in **Figures 2** and **3**. The symbol for the double-acting cylinder is a simple extension of the single-acting cylinder. The required circuit diagram is shown in **Figure 6**.

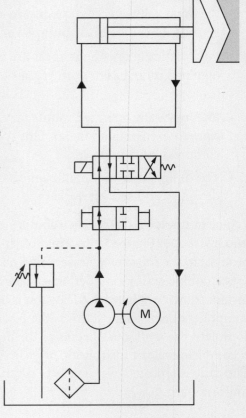

Figure 6
The circuit diagram for Sample Problem 1

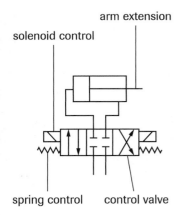

Figure 7
A circuit diagram of the arm extension of a robot

solenoid control

arm extension

spring control   control valve

▶ *Practice*

Understanding Concepts

9. Name two advantages and two disadvantages of using robots for industrial purposes.

10. **Figure 7** shows a circuit diagram of a small part of a robot, in this case, the arm extension.
    (a) What types of controls are used to operate the valves?
    (b) How many positions and how many ports does the valve have?
    (c) Is the cylinder single- or double-acting? How do you know?

11. Using appropriate symbols, draw the circuit diagram for the robotic press (**Figures 5** and **6**) during the refill cycle.

Applying Inquiry Skills

12. Refer to your diagrams and observations in Part C of Investigation 5.6, page 254. Draw circuit diagrams of two different setups you tested.

Making Connections

13. Robots are used in space and underwater exploration. Research Canada's involvement in one of these uses. Identify and analyze some of the social and economic consequences of the uses you researched.

 www.science.nelson.com

| SUMMARY | *Fluid Systems* |

- A fluid system transmits forces from a pump or compressor through control components to an actuator, which performs a task.

- The easy, safe transmission of fluids gives fluid systems great flexibility.

- Various types of valves can be used to give a fluid system great versatility.

- Standard symbols are used to draw circuit diagrams of fluid systems.

- Robotics applies the principles of fluid systems, with the possible addition of computer controls.

▶ **Section 5.7 Questions**

Understanding Concepts

1. Draw circuit symbols for a reservoir and a receiver. Why is one open and the other closed?

2. What are the main uses of valves in a fluid system?

3. Name and draw the symbol for the type of device that controls
   (a) a hand-held hose sprinkler
   (b) the hand pump used to inflate a sphygmomanometer

4. Name a device that has a valve with both manual control and spring control.

5. **Figure 8** shows the fluid system of an airplane landing gear in the down position and two circuit diagrams related to the landing gear.
   (a) What is the energy source that drives the oil pump?

(b) What must the pilot do to raise the landing gear?
(c) Which circuit diagram shows the landing gear in the up position? Explain your answer.

Making Connections

6. Would you recommend a hydraulically operated or pneumatically operated robot for handling hazardous materials? Why?

7. If all manufacturing tasks were performed by a robot, would there be any jobs for humans in manufacturing? Explain your answer.

8. (a) Identify two applications of hydraulic or pneumatic systems that affect your life.
   (b) How has your quality of life been affected by those systems?

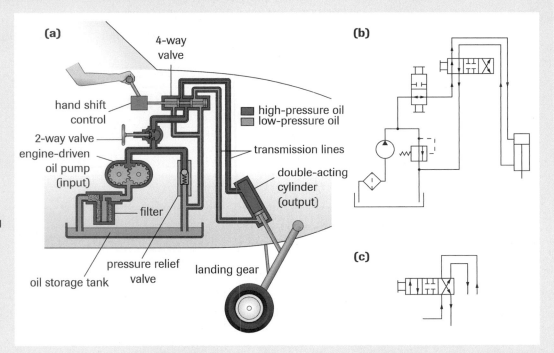

Figure 8
(a) The airplane landing gear in the down position
(b) Circuit diagram for the landing gear in position 1
(c) Circuit diagram for the landing gear in position 2

Until now, you have solved fluid system problems involving three or four variables. A more complete analysis combines several variables for a single system. To do this, you will quantitatively analyze the system depicted in **Figure 1**, which is a cylinder with a piston that can exert a pressure to perform a task. This type of analysis can be applied to a variety of systems in which the variables are known or can be measured. **Table 1** summarizes the variables used in the chapter, as well as some useful equations and conversions.

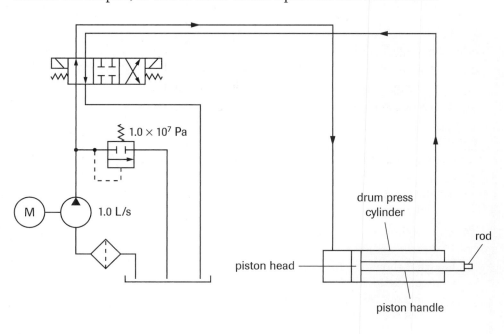

Figure 1
A cylinder press

Table 1   Variables, Equations, and Conversions

Variable	Symbol	SI Unit	Other Possible Units
volume	V	m³	cm³; mL; L
volume flow rate	q_v	m³/s	cm³/s; mL/s; L/min
speed	v	m/s	cm/s
time interval	Δt	s	min; h
pressure	p	Pa	kPa
force	F	N	kN
area	A	m²	cm²
work	W	J	kJ; MJ
displacement	Δd	m	cm; mm
power	P	W	kW; MW

Equations and Conversions

V_{cylinder} = the product of the cylinder's cross-sectional area (πr^2) and length or height (h):

$V_{\text{cylinder}} = Ah = (\pi r^2)h$

1 cm³ = 1 mL and 1 L = 1000 mL

1 m² = 10⁴ cm²

▶ **SAMPLE** problem **1**

Calculating the Time Interval of a Complete Cycle

The cylinder press in **Figure 1** has a piston that moves back and forth regularly. The rod extended from the piston performs a repetitive task on an assembly line as it pushes components to the next stage of assembly. The known quantities related to its operation are

q_v = 1.0 L/s (provided by the pump)

p = 1.0 × 10⁷ Pa

r_{rod} = 2.0 cm

r_{handle} = 3.0 cm

r_{head} = 8.0 cm

h = length of piston stroke (each direction) = 34 cm

Calculate the following:

(a) the volume of liquid moved into the cylinder during the power stroke

(b) the volume of liquid moved into the cylinder during the refill stroke

(c) the total time of a complete cycle of the power and refill strokes

Solution

(a) During the power stroke, the liquid flows into and fills the cylinder to the left of the piston.

h = 34 cm

r_{head} = 8.0 cm

V_{power} = ?

$$V = \pi r^2 h$$
$$= \pi (8.0 \text{ cm})^2 (34 \text{ cm})$$
$$= 6834 \text{ cm}^3$$
$$= 6834 \text{ mL (conversion from \textbf{Table 1})}$$
$$V_{power} = 6.8 \text{ L}$$

The volume during the power stroke is 6.8 L.

(b) During the refill stroke, the liquid fills the cylinder to the right of the piston, which has a lower volume due to the volume taken up by the handle itself:

h = 34 cm

r_{handle} = 3.0 cm

V_{refill} = ?

$$V_{handle} = \pi r^2 h$$
$$= \pi (3.0 \text{ cm})^2 (34 \text{ cm})$$
$$= 961 \text{ cm}^3$$
$$V_{handle} = 961 \text{ mL}$$

The volume of the liquid is $V_{refill} = V_{power} - V_{handle}$ = 6834 mL − 961 mL = 5873 mL, or 5.9 L.

(c) We can divide the volume of each stroke by the flow rate to determine the individual times, and then add the times to determine the total time:

$q_v = 1.0$ L/s

$\Delta t = ?$

$$\Delta t = \frac{V_{power}}{q_v} + \frac{V_{refill}}{q_v}$$

$$= \frac{6.8 \cancel{L}}{1.0 \frac{\cancel{L}}{s}} + \frac{5.9 \cancel{L}}{1.0 \frac{\cancel{L}}{s}}$$

$$= 6.8 \text{ s} + 5.9 \text{ s}$$

$$\Delta t = 12.7 \text{ s}$$

The total time for a complete cycle is 12.7 s.

▶ **SAMPLE** problem **2**

Calculating Force and Work

Using the data in Sample Problem 1, calculate

(a) the magnitude of the force applied by the rod to the component during the power stroke

(b) the work done by the liquid in the cylinder during the power stroke

Solution

(a) The force applied to the rod's left end comes directly from the force of the liquid on the piston during the power stroke:

$p = 1.0 \times 10^7$ Pa

$r_{head} = 8.0$ cm $= 0.080$ m

$F = ?$

$$p = \frac{F}{A}$$

$$F = pA$$

$$= p(\pi r^2)$$

$$= (1.0 \times 10^7 \text{ Pa})(\pi)(0.080 \text{ m})^2$$

$$F = 2.0 \times 10^5 \text{ N}$$

The magnitude of the force applied by the rod is 2.0×10^5 N.

(b) $F = 2.0 \times 10^5$ N

$\Delta d = 34$ cm $= 0.34$ m

$W = ?$

$$W = F\Delta d$$

$$= (2.0 \times 10^5 \text{ N})(0.34 \text{ m})$$

$$W = 6.8 \times 10^4 \text{ J}$$

The work done by the liquid is 6.8×10^4 J.

> ▶ **SAMPLE** problem **3**
> ### Calculating the Power

Using the data in Sample Problems 1 and 2, calculate the power during the power stroke.

Solution
$W = 6.8 \times 10^4$ J
$\Delta t = 6.8$ s
$P = ?$

$$P = \frac{W}{\Delta t}$$

$$= \frac{6.8 \times 10^4 \text{ J}}{6.8 \text{ s}}$$

$$P = 1.0 \times 10^4 \text{ W}$$

The power of the system during the power stroke is 1.0×10^4 W.

To gain experience in designing and testing your own pneumatic system, you can perform Activity 5.9.

> ▶ **Practice**

Understanding Concepts

1. A cylinder press has the same design and features as the one in **Figure 1**, page 262, except that the length of the piston stroke is 75 cm and the volume flow rate provided by the pump is 2.0 L/s. Calculate
   (a) the volume of liquid, in litres, moved into the cylinder during the power stroke
   (b) the time interval of the power stroke
   (c) the total time for one cycle of operation

2. For the cylinder press in **Figure 1**, page xxx, calculate
   (a) the magnitude of the force applied to the rod side of the piston during the refill stroke
   (b) the work done by the fluid in moving the piston during the refill stroke
   (c) the power during the refill stroke

Answers

1. (a) 15 L
   (b) 7.5 s
   (c) 14 s

2. (a) 1.7×10^5 N
   (b) 5.9×10^4 J
   (c) 1.0×10^4 W

SUMMARY   *Work and Power in Fluid Systems*

- Fluid systems can be analyzed quantitatively to determine such variables as time for each cycle, work done, and power generated. Time is calculated using the equation $\Delta t = \dfrac{V}{q_{\mathrm{v}}}$.

Understanding Concepts

1. **Figure 2** is a robotic cylinder press with a regular back-and-forth motion. Using the quantities in the figure, calculate
   (a) the volume of liquid moved into the cylinder during the power stroke
   (b) the volume of liquid moved into the cylinder during the refill stroke
   (c) the total time of a complete cycle of the power and refill strokes
   (d) the magnitude of the force applied by the rod to the component during the power stroke
   (e) the work done by the liquid in the cylinder during the power stroke
   (f) the power generated during the power stroke

2. A student-made pneumatic muscle, initially 28.5 cm long, is suspended vertically from a support stand, as shown in **Figure 2** in Activity 5.9, page 268. A 0.20-kg mass is suspended from the muscle.

A pump takes 1.2 s to fill the muscle to a pressure of 2.2×10^2 kPa, after which the muscle is 23.8 cm long. Calculate
   (a) the work done by the muscle in raising the mass
   (b) the power generated by the muscle in raising the mass

Applying Inquiry Skills

3. (a) Describe how you would design a pneumatic muscle that opens jar lids. Draw a diagram of the design.
   (b) How would you test your design to determine how well it can accomplish its task?
   (c) Who would find such a tool useful?

Making Connections

4. Identify some limitations of the pneumatic muscle designs described in Activity 5.9 and in question 2 above. Do industrial robots have similar limitations? Explain your answer.

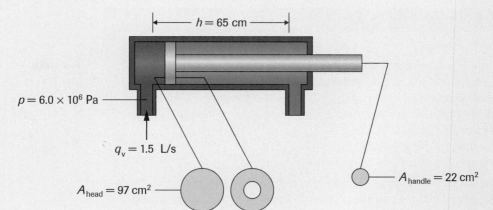

$h = 65$ cm

$p = 6.0 \times 10^6$ Pa

$q_v = 1.5$ L/s

$A_{head} = 97$ cm^2

$A_{handle} = 22$ cm^2

Figure 2
For question 1

5.9 Activity

Design a Pneumatic Muscle

A sphygmomanometer, used to measure blood pressure, can be called a *pneumatic muscle,* or an *air muscle.* Air pumped into the cuff causes it to tighten around your arm. A similar design can be used for other tasks, such as opening lids on jars.

Pneumatic muscles are useful for a few reasons: Their mass is low, yet they are strong. They are also pliable, so they can be designed for a variety of tasks. In this activity, you will build and test a pneumatic muscle, then modify and redesign it to perform another task.

Materials

For each student:
safety goggles

For each group of three or four students:

Part A

soft rubber tubing (about 15 cm long when stretched)
strong plastic mesh or netting, preferably cylindrical
  in shape
fishing line or strong string (if the mesh is rectangular
  rather than cylindrical)
scissors
plastic tubing
pump (e.g., a bicycle tire pump or an air mattress
  pump)
2 nylon cable ties
metric ruler or metre stick
pressure gauge

Part B

all of the above, plus
electronic balance
support stand and clamps
100-g mass
stopwatch

Part C

all of the above, plus
receiver (such as a 0.5-L plastic pop or water bottle)
one 3-way valve

Check the condition of the rubber and plastic tubing, and do not use if deteriorated.

Wear goggles when working with pressurized devices.

Procedure

Part A

1. Put on your safety goggles. Stretch the soft rubber tubing to determine the length of plastic mesh to be cut. If the mesh needs to be tied to form a cylinder, use fishing line or strong string.

2. Insert one end of the plastic tubing a short way into the rubber tubing. Insert the soft rubber tubing into the cut length of mesh. Secure the two ends of the muscle with nylon cable ties; form a loop at each end. Make sure that one end is closed completely and that the end with the plastic tubing is open (**Figure 1**).

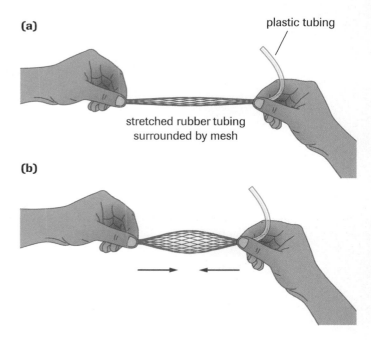

Figure 1
(a) Soft rubber tubing is stretched and then surrounded by plastic mesh. The ends of the mesh are formed into loops to link the ends of the muscle to external parts or to each other.
(b) When air is introduced into the muscle, the middle of the muscle bulges outward, so the muscle contracts, pulling on the ends with a large tension force.

3. Measure the length of the muscle (excluding the ties). Connect the pump to the plastic tubing and an air pressure gauge. Pump air into the muscle until the gauge pressure is no more than 300 kPa. Measure the new length of the muscle. (If the length did not decrease, you must start again. Make sure that the soft rubber tubing is stretched more *before* securing the plastic mesh with the cable ties.) Record your observations and measurements, and release the air in the muscle.

Part B

4. Using the electronic balance, record the mass of the muscle built in steps 1 and 2.

5. Suspend the muscle from a clamp on a support stand, and attach a 100-g mass to the lower end of the muscle (**Figure 2**). Design a way to measure the time the fluid system (the pump, muscle, etc.) takes to raise the mass as the gauge pressure in the muscle is increased from zero to 300 kPa (or another safe value). Also measure the distance the mass is raised. Describe your method and record your measurements.

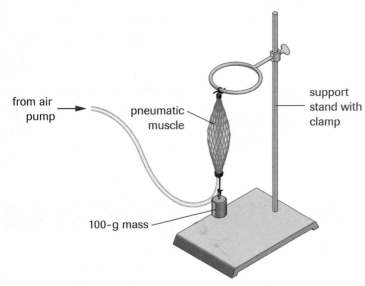

Figure 2
Setting up the muscle to perform the task of lifting a known mass

Part C

6. Brainstorm ideas on how you can modify your design in Parts A and B to improve the muscle's lifting action and to build a more complete fluid system. The system should include a receiver (a plastic pop or water bottle), at least one control valve, any adaptors needed, and connecting tubing. Draw a diagram of your system, and have your teacher approve your design.

7. Devise a test to determine the power of the redesigned muscle as it performs a task. Perform the test, and record the measurements and calculations.

Analysis

(a) What is a simple test to judge whether a pneumatic muscle will work when activated?

(b) Determine the ratio of the muscle mass to the load mass for the mass raised in step 5. Compare this ratio to the ratio found by other groups.

(c) Calculate the work done in raising the mass in step 5.

(d) Calculate the power generated in raising the mass in step 5.

(e) Draw a circuit diagram of the pneumatic system you designed in Part C. Use the proper circuit symbols.

(f) Summarize the results of the tests you used to determine the work done and power generated by the pneumatic muscle.

Evaluation

(g) Evaluate the process your group used in building, testing, and modifying the fluid system used in this activity.

(h) If you redid this activity, how would you change the design and testing?

Synthesis

(i) How could your pneumatic muscle be used in a robotic device?

Key Understandings and Skills

5.1 Properties of Fluids

- **Fluid statics** is the study of fluids at rest.
- The study of the mechanical properties of fluids includes **hydraulics** (for liquids) and **pneumatics** (for gases).
- The properties **density** and **compressibility** of fluids determine the selection of **hydraulic** and **pneumatic systems**.

5.2 Pressure

- **Pressure** is the force per unit area; its SI unit is the **pascal** (Pa).
- **Hydroforming** uses water under high pressure to form metal components.
- **Water jet cutting** uses high-speed water under high pressure to cut materials; **volume flow rate** can be used to analyze flow variables in this process.
- The average **atmospheric pressure** at sea level is 101.3 kPa.

5.3 Investigation: Pressure in Liquids

- The factors that affect the pressure beneath the surface of a liquid can be determined experimentally.

5.4 Measuring Pressure

- Instruments that measure pressure include the **barometer**, **aneroid barometer**, **altimeter**, and **manometer**.
- The **absolute pressure** is the sum of the **gauge pressure** and the atmospheric pressure.

- The **static pressure head** is the height of a column of a fluid that produces a specific pressure at any particular position in a static fluid.

5.5 Pascal's Principle

- The pressure applied to an enclosed (incompressible) liquid is transmitted everywhere in the liquid. One application of this principle is the **hydraulic press**.

5.6 Investigation: Two- and Three-Cylinder Fluid Systems

- The properties of hydraulic and pneumatic systems as well as their control with valves can be explored experimentally.

5.7 Fluid Systems

- Hydraulic and pneumatic systems transmit forces from a source through control components to an **actuator**, which uses fluid forces to perform a task.
- Fluid systems, including **robots**, are represented by circuit diagrams.

5.8 Work and Power in Fluid Systems

- Typical variables that can be analyzed in fluid systems are pressure, force, volume flow rate, time interval of a cycle, work, and power.

5.9 Activity: Design a Pneumatic Muscle

- A pneumatic (or air) muscle can be designed, built, tested, evaluated, and modified to perform a specific task.

Key Terms

5.1	**5.2**	**5.4**	**5.5**
fluid	pressure	altimeter	Pascal's principle
hydraulics	pascal	static pressure head	hydraulic press
pneumatics	hydroforming	barometer	
fluid statics	water jet cutting	aneroid barometer	**5.7**
hydraulic system	volume flow rate	absolute pressure	fluid system
pneumatic system	atmospheric pressure	gauge pressure	actuator
density		manometer	robot
compressibility			

Key Equations

5.1

- $D = \dfrac{m}{V}$

- $q_v = \dfrac{V}{\Delta t}$

- $q_v = Av$

5.4

- $p_{abs} = p_g + p_{atm}$
- $p_g = Dgh$

5.5

- $\dfrac{F_S}{A_S} = \dfrac{F_L}{A_L}$

5.8

- $\Delta t = \dfrac{V}{q_v}$

5.2

- $p = \dfrac{F}{A}$

Problems You Can Solve

5.1

- Define and identify examples of a fluid, a hydraulic system, and a pneumatic system.
- Given any two of density, mass, and volume, determine the third quantity.

5.2

- Given any two of pressure, force, and area, determine the third quantity.
- Given any two of volume flow rate, volume, and time interval, determine the third quantity.
- Given any two of volume flow rate, area, and speed, determine the third quantity.
- Explain the cause of atmospheric pressure, providing evidence for your explanation.
- Describe some of the ways in which atmospheric pressure affects our daily lives.

5.3

- State how the pressure in a static liquid depends on the density of the liquid and the depth beneath the surface.

5.4

- Describe and compare instruments that measure pressure.
- Distinguish between gauge pressure and absolute pressure.

5.5

- State Pascal's principle, and describe how it is applied in common hydraulic devices.
- Given any three of the four areas and forces in a hydraulic press, calculate the fourth quantity.

5.6

- Describe the properties of two- and three-cylinder pneumatic and hydraulic systems.

5.7

- Identify, state the function of, and recognize the circuit symbols of the main components of hydraulic and pneumatic systems.
- Draw a circuit diagram of a simple hydraulic or pneumatic system.

5.8

- Given the volume flow rate, the pressure in the system, and the dimensions of a liquid-operated cylinder press, calculate the time of a complete cycle of the piston, the work done by the liquid on the piston, and the power generated by the liquid.

5.9

- Describe how to test a pneumatic muscle to determine the work it does and the power it generates in performing a lifting task.

▶ *MAKE* a summary

Most of the concepts, equations, and applications
studied in this chapter can be summarized in
diagrams, like the one in **Figure 1**. Sketch the
following, and add as many key understandings, skills,
terms, equations, and applications as you can on each
sketch:

- a mountain and a valley (to show atmospheric
  pressure, measuring atmospheric pressure, applying
  atmospheric pressure, etc.)

- a dam and a reservoir (to show density, the static
  pressure head, gauge pressure, etc.)

- a car (to show gauge pressure in tires, applications
  of Pascal's principle, the use of pneumatic tools,
  etc.)

- a car hoist (to show a fluid system, calculations
  involving Pascal's principle, etc.)

- a cylinder press (to show quantitative analysis of
  fluid systems, etc.)

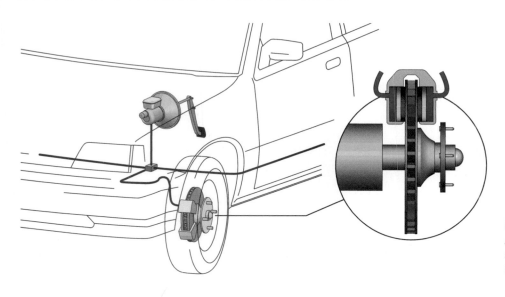

Figure 1
A drawing of a car can be used to
summarize some of the concepts in
Chapter 5.

Write the numbers 1 to 9 in your notebook. Indicate beside each number whether the corresponding statement is true (T) or false (F). If it is false, write a corrected version.

1. Liquids and gases are both highly compressible because their molecules flow.

2. Pressure is a vector quantity measured in the SI unit pascal (Pa).

3. Pouring water into a moulded piece of metal or plastic and letting it freeze is an example of hydroforming.

4. In a swimming pool, the pressure at a depth of 1.0 m is the same in all directions.

5. The static pressure head is measured in the SI unit metres.

6. The maximum height that water can be raised by atmospheric pressure is 10.3 m.

7. In general, hydraulic systems act faster than pneumatic systems.

8. Pascal's principle can be applied to liquids but not to gases.

9. An actuator transforms fluid forces into mechanical forces.

Write the numbers 10 to 15 in your notebook. Beside each number, write the letter corresponding to the best choice.

10. The order of substances from maximum to minimum compressibility is
    (a) air, water, steel   (c) water, air, steel
    (b) steel, water, air   (d) air, steel, water

11. A helium-filled balloon rises in the atmosphere. The density of the helium
    (a) remains the same because the mass and volume remain the same
    (b) decreases because the mass decreases and the volume remains the same
    (c) decreases because the volume increases and the mass remains the same
    (d) remains the same because the volume increases at the same time as the force of gravity on the mass decreases

12. When you shift from standing on two feet to standing on one foot,
    (a) both the normal force and the pressure are half the original values
    (b) both the normal force and the pressure are twice the original values
    (c) the normal force is the same but the pressure is twice as much
    (d) the normal force is twice as much but the pressure is half as much

13. If the atmospheric pressure is 100 kPa, and the gauge pressure in a tire is 300 kPa, then the absolute pressure in the tire is
    (a) 100 kPa   (c) 300 kPa
    (b) 200 kPa   (d) 400 kPa

14. For the hydraulic press in **Figure 1**,
    (a) $p_A = p_B$; $|F_1| = |F_2|$
    (b) $p_A = p_B$; $|F_1| < |F_2|$
    (c) $p_A < p_B$; $|F_1| < |F_2|$
    (d) $p_A > p_B$; $|F_1| > |F_2|$

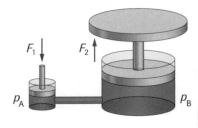

F_1 ↓ F_2 ↑

p_A p_B

Figure 1

15. The fluid circuit symbols in **Figure 2** represent, respectively,
    (a) storage tank, double-acting cylinder, filter, and compressor
    (b) pressure relief valve, single-acting cylinder, filter, and cylinder press
    (c) pressure relief valve, single-acting cylinder, filter, and compressor
    (d) storage tank, single-acting cylinder, reservoir, and pump

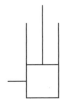

Figure 2

Understanding Concepts

1. For each of the following, calculate the density, and state whether the substance is likely a gas or a liquid:
   (a) $m = 1.8$ kg; $V = 1.4$ m^3
   (b) $m = 1.2$ kg; $V = 1.1 \times 10^{-3}$ m^3

2. Explain why a sharp hypodermic needle hurts less than a dull one.

3. When a full bottle of wine is corked, for safety reasons, a small air space should be left between the liquid and the cork. Explain why.

4. Explain the cause of atmospheric pressure.

5. Name the type of device shown in **Figure 1**, and describe the cause of the flow of water.

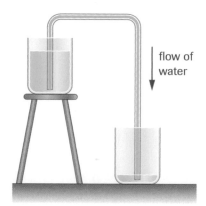

flow of water

Figure 1

6. Astronauts aboard a space vehicle can use suction-cup shoes to "walk" around the interior walls of the pressurized vehicle. Would those shoes stick to the exterior walls of the vehicle? Explain your answer.

7. When you lie on the beach, air pressure produces a force over your body equivalent to the force of gravity on a 5000-kg truck. However, you have no trouble getting up from the sand. Why not?

8. Does a brick apply the same pressure to a desk if the brick is on its end, its side, or its base?

9. Two laboratory syringes are linked, first with a 20-cm tube and then with a 40-cm tube.
   (a) When used as a pneumatic system, how do the reaction times compare?
   (b) When used as a hydraulic system, how do the reaction times compare?
   (c) Why is there a difference between (a) and (b)?

10. Assume that a student's weight has a magnitude of 7.8×10^2 N.
    (a) The area of the sole of one shoe is 0.020 m^2. What pressure does the student apply to the floor when standing on one foot?
    (b) If the student puts on a snowshoe whose area is 0.21 m^2, how much pressure is applied to the snow?
    (c) Explain the advantage of using snowshoes to walk on snow.

11. In a water jet cutting process, water in a hose of radius 0.35 mm has a volume flow rate of 4.0×10^{-6} m^3/s. Calculate
    (a) the volume of water used in 24 s
    (b) the speed of the water in the hose

12. An absolute pressure of 3.0×10^5 Pa inside a car tire is exerted over a surface area of 1.2 m^2.
    (a) Calculate the magnitude of the force caused by the absolute pressure on the inside of the tire.
    (b) If the atmospheric pressure is 1.0×10^5 Pa, calculate the magnitude of the net force on the inside of the tire.

13. A *bathyscaphe* is an underwater vessel used for researching the ocean floor. The small glass porthole at the bottom of a bathyscaphe must withstand extremely high pressures. Calculate the gauge pressure at the depth of 11.6 km, reached by one such vessel in 1960. (Assume that the density of the saltwater in this case is constant at 1.03×10^3 kg/m^3.) Express your answer in pascals and megapascals.

14. Describe how Pascal's law is applied to the operation of a hydraulic press.

15. In a hydraulic press, if the pressure in the small cylinder is increased by 25 kPa, what happens to the pressure in the large cylinder? How do you know?

16. In a hydraulic press, the area of the small piston is 0.12 m^2 and the area of the large piston is 0.89 m^2. The magnitude of the force on the small piston is 65 N.
    (a) Calculate the pressure in this press.
    (b) Calculate the magnitude of the force on the large piston.

(c) If a platform were attached to the large piston so that you could stand on it, would it support your weight? Show your calculations.

17. Name the component of a fluid system that
    (a) stores the liquid at atmospheric pressure
    (b) transmits the fluid
    (c) transforms fluid forces into mechanical forces

18. **Figure 2** shows a circuit diagram in which a 4-way valve controls a fluid system.

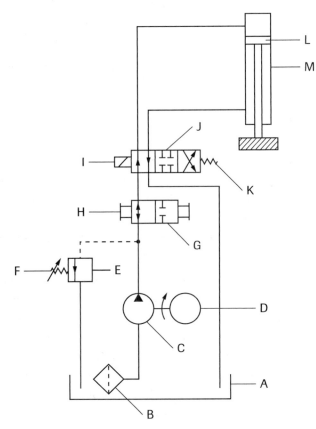

Figure 2

(a) Write the letters A to M in your notebook, and beside each letter name the component shown in the diagram.
(b) Does the system use a liquid or a gas? How can you tell?
(c) At the instant shown, $d_{piston} = 19.4$ cm is the fluid exerting an upward or a downward force on the piston? Explain how you know. $p = 8.5 \times 10^6$ Pa

19. A robotic cylinder press is designed with a piston that presses against a mould during the power stroke and stays there for 17 s while molten plastic is poured into the mould and allowed to cool. Then the piston moves back during the refill stroke. **Figure 3** shows the variables involving the regular back-and-forth motion. Calculate the following quantities:
    (a) volume of liquid moved into the cylinder during the power stroke
    (b) volume of liquid moved into the cylinder during the refill stroke
    (c) total time of a complete cycle (including the power stroke, the cooling period, and the refill stroke)
    (d) magnitude of the force applied by the rod to the component during the power stroke
    (e) work done by the liquid in the cylinder during the power stroke
    (f) power during the power stroke

Applying Inquiry Skills

20. (a) Estimate the total volume (in cubic metres) of the air in your classroom.
    (b) Estimate the total mass (in kilograms) of air in the room.
    (c) Use the volume estimate in (b) and the density of air (1.29 kg/m^3) to calculate the mass of the air. How good was your estimate in (a)?

21. Drink the contents of an individual cardboard juice container using the straw, then continue drawing on the straw. Explain what happens using physics principles from this chapter. Remove your mouth from the straw, and again explain what happens. (This procedure can be repeated until the apparatus wears out.)

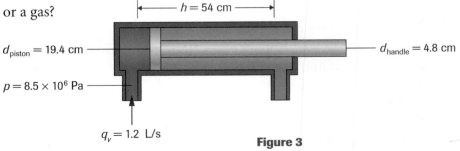

$h = 54$ cm

$d_{handle} = 4.8$ cm

$q_v = 1.2$ L/s

Figure 3

22. A *Cartesian diver* (**Figure 4(a)**) can be used to demonstrate the effects on a diver when the pressure applied to the surface of the water changes. (**Figure 4(b)** is a student-designed version. The eyedropper, when partially filled with water and capped, barely floats in a larger plastic graduated cylinder.) Obtain a Cartesian diver and learn how to control it. Explain the physics principles involved.

(a)   **(b)**

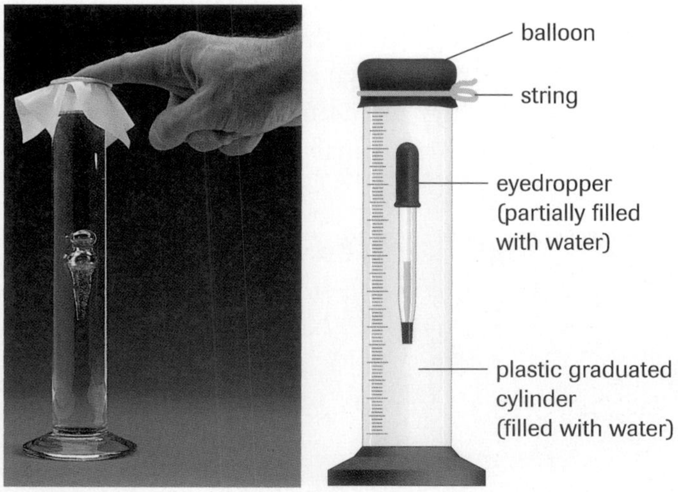

balloon

string

eyedropper
(partially filled
with water)

plastic graduated
cylinder
(filled with water)

Figure 4
(a) A commercially available Cartesian diver
(b) A student-made Cartesian diver

23. You are asked to design a pneumatic muscle that is stronger than the one described in Activity 5.9 (page 267). You decide to test two models, the first using two muscles side by side and the second using one large muscle.
    (a) Sketch the setup you would need to test the strength of each model.
    (b) What safety precautions should be followed during the test?
    (c) What design could you use to change a lifting muscle to a clamping muscle?

Making Connections

24. A student who has not studied physics designs a snorkel to use while exploring depths of up to 3.0 m below the surface of water. The student soon discovers that it is impossible to breathe through the snorkel tube at any depth over 1.0 m. Explain why the snorkel design does not work. (A diagram will help.)

25. A pressure cooker is an airtight pot used to cook food quickly using steam (water vapour) under pressure. The radius of the cover of one pressure cooker is 0.21 m. The steam's pressure is 303 kPa.
    (a) Calculate the absolute pressure in the pressure cooker.
    (b) Calculate the force of the steam on the lid.
    (c) Use the kinetic molecular theory to explain why the food in a pressure cooker cooks faster than it would in a regular pot of boiling water.
    (d) Extend your answer in (c) to explain why hard-boiling an egg takes longer at a high altitude than at sea level.

26. Identify one important way in which hydraulics or pneumatics is applied in
    (a) building construction
    (b) mining
    (c) road construction
    (d) stores that sell large boxed items

27. How would the principles of hydraulics and/or pneumatics be applied in the following careers?
    (a) scuba instructor
    (b) meteorological instrument repair technician
    (c) chef at a large resort

28. *Vacuum bagging* is a method of attaching layers of components on a surface using the pressure of the atmosphere. By creating a partial vacuum on one side of a component, the atmosphere exerts a pressure on the other side. This process is especially useful in applying layers on curved surfaces, such as on surfboards, snowboards, and sailboards, where clamps would not work well. Research vacuum bagging in magazines or on the Internet.
    (a) In a sketch, illustrate the layers of materials in a vacuum bagging process.
    (b) What is a typical pressure used in the process? What devices can be used to achieve such a pressure?
    (c) Describe difficulties in the process that you would have to overcome in vacuum bagging your own sailboard or other device.
    (d) Describe what else you discovered.

 www.science.nelson.com

6

Fluid Dynamics

In this chapter, you will be able to

- define and describe, with examples, the concepts related to laminar flow and turbulent flow of fluids
- experimentally verify that the pressure exerted by a fluid changes when its speed changes
- state Bernoulli's principle and explain some of its applications in such fields as technology, transportation, sports, and medicine

Getting Started

Understanding fluid motion has many practical applications, such as improving the streamlining of cars, trucks, motorcycles, and other vehicles.

In this chapter, you will learn many things about this important branch of science, for example, how researchers test the design and properties of vehicles and other structures by placing them in large wind tunnels, like the one in **Figure 1**. Computer simulations are also used to analyze fluid motion around these objects.

Figure 1
Huge fans blow air through this wind tunnel. The wind forms patterns as it moves past objects placed in its path. Observing these wind patterns helps researchers analyze the properties and designs of vehicles and other objects.

You will learn how the study of fluid motion helps us analyze fluids such as water and natural gas moving through a pipe, and objects such as a baseball moving through air or a submarine moving through water. You will learn why fluid flow is sometimes turbulent and sometimes smooth. As well, you will have the opportunity to discover what happens to the pressure in a flowing fluid when the speed of the fluid changes. The principles you explore in this chapter have practical applications, not only in the design of vehicles, but also in sports activities such as baseball, golf, and skiing, in explaining airplane flight, and a variety of processes in nature.

1. (a) What does the expression "as slow as molasses in January" mean? Include the word *viscosity* in your answer.
   (b) Name two liquids found in the kitchen that have a high viscosity and two that have a low viscosity.
   (c) What happens to the viscosity of honey as its temperature increases?

2. A coach's whistle has a little bead inside, but an inexpensive toy whistle does not.
   (a) How do the sounds of these two whistles compare?
   (b) In which of the two whistles is the motion of the air more turbulent? Does the turbulence help or hinder the sounds produced?

3. The train in **Figure 2** is highly streamlined. What features make it so?

4. How does the speed of the flowing water in a river depend on the width of the river? Explain your answer.

Figure 2
One of Japan's high-speed "bullet trains"

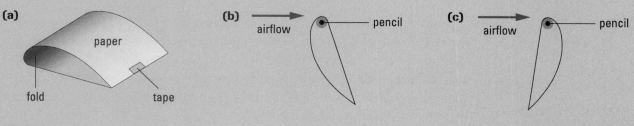

▶ *TRY THIS* activity   *Testing Wing Shapes*

Obtain a piece of paper about 10 cm × 20 cm, and fold it into the shape of an airplane wing, as shown in **Figure 3(a)**. Tape the ends together. Hold the middle of the wing with a pencil and blow across the wing, as shown in **Figure 3(b)**. Repeat the procedure with the wing facing the opposite way, as in **Figure 3(c)**.

(a) Which position of the wing gives the greater "lift"?

(b) Why must an airplane reach a high speed before it can take off?

(a)
paper
fold
tape

(b)
airflow
pencil

(c)
airflow
pencil

Figure 3

A log ride is smooth as long as the water flows evenly. However, the flow and the ride can become rough where there are dips and sharp curves (**Figure 1**). The factors that affect the smooth or rough flow of a fluid are part of the study of fluids in motion, called **fluid dynamics**.

As a fluid flows, the forces of attraction between the molecules cause internal friction, or resistance to the flow. The fluid's **viscosity** is a measure of this resistance to flow. A fluid with a high viscosity, such as liquid honey, has a large amount of internal resistance and does not flow readily. A fluid with a low viscosity, such as water, has low internal resistance and flows easily. Viscosity depends not only on the nature of the fluid, but also on the fluid's temperature. As the temperature of a liquid increases, the viscosity generally decreases because the particles of the liquid have more energy and flow more easily. As the temperature of a gas increases, however, the viscosity generally increases because the particles of the gas collide with each other more often, making it more difficult for them to flow in one direction.

Figure 1
The design of a log ride determines its smoothness.

fluid dynamics the study of the factors that affect fluids in motion

viscosity the property of a fluid that determines its resistance to flow; a high viscosity means a high resistance to flow

> ▶ **TRY THIS** activity **Viscosity**

Observe the effect of temperature changes on the viscosity of various products from your home (e.g., vegetable oil, honey, clear shampoo, syrup) when placed in stoppered test tubes provided by your teacher (**Figure 2**). Each tube should also contain a marble. Obtain a test tube that has been placed in a cold water bath, invert it, and measure the time it takes for the marble to travel through the liquid. Compare your results with the time it takes for the marble to travel through the liquid in a test tube that has been placed in a hot water bath.

 Wear gloves when handling the liquids. Exercise care when using hot water.

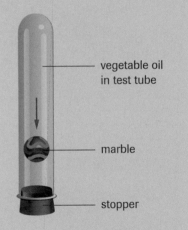

vegetable oil in test tube

marble

stopper

Figure 2
After you invert the test tube, the marble moves slowly downward through the oil.

As a fluid flows, the fluid particles interact with their surroundings and experience external friction. For example, as water flows through a pipe, the water molecules closest to the walls of the pipe experience a frictional resistance that reduces their speed to nearly zero. Measurements show that the water speed varies from the wall of the pipe (minimum speed) to the centre of the pipe (maximum speed). If the speed of a fluid is slow and the adjacent regions flow smoothly over one another, the flow is called a **laminar flow**

laminar flow fluid flow in which adjacent regions of fluid flow smoothly over one another

(a)  **(b)**

Figure 3
The length of each vector represents the magnitude of the fluid velocity at that point. The longer the arrow, the greater the velocity.
(a) Water in a pipe
(b) Air around a cone

(**Figure 3(a)**). The flow of a fluid such as air passing around a smooth object is also called laminar flow (**Figure 3(b)**).

Laminar flow is usually difficult to achieve because, as the fluid flows through or past an object, the flow becomes irregular, resulting in whirls called *eddies* (**Figure 4**). Eddies are common in **turbulent flow**, a fluid flow with a disturbance that resists the fluid's motion (e.g., water encountering boulders in a river). A fluid undergoing turbulent flow loses kinetic energy as some of the energy is transformed into thermal energy and sound energy. The likelihood of turbulence increases as the velocity of the fluid relative to its surroundings increases.

(a) **(b)** **(c)**

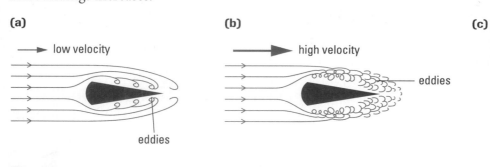 

Figure 4
(a) Low turbulence at a low velocity
(b) Higher turbulence at a higher velocity
(c) You can see examples of turbulent flow in this view of Jupiter's atmosphere.

Turbulence in fluids moving in tubes or pipes can be reduced in various ways. For example, in a sewage system, small amounts of liquid plastic can be injected into the system. The plastic particles are slippery and mix with the sewage particles, reducing the liquid's viscosity and preventing the sewage from sticking to the sewer pipe and walls. The plastic particles make it easier for the pumps to transfer the sewage. A similar method can be used to reduce the turbulence of water ejected from fire hoses, allowing the water-jet to stream farther. This is especially advantageous for fighting fires in tall buildings. Liquid plastic can also be added to the bloodstreams of people who have problems with blood flow. This treatment helps reduce turbulence in the blood, which reduces the chance that the blood will stop flowing.

Studies of large structures, such as bridges and submarines, can be conducted using models in a wind tunnel or a water tank. These studies can then be used to create computer simulations for further analysis. Large quantities of data on flow patterns can be stored and analyzed. Then

DID YOU KNOW ?

The Benefits of Turbulence
In some cases, air turbulence is beneficial, for example, in the coach's whistle mentioned on page 277. When you blow the whistle, the bead gets knocked around and increases the turbulence, which improves the quality of the sound and increases its loudness. Air entering the mouthpiece of a wind instrument also needs some turbulence in order to vibrate easily.

turbulent flow fluid flow in which a disturbance resists the fluid's motion; it results when fluids cannot move smoothly around or through objects

CAREER CONNECTION

Practical nurses have a background in physiology, anatomy, and computer applications. Employment opportunities exist in a variety of areas, for example, hospitals, retirement homes, doctors' offices, and in industry.

GO www.science.nelson.com

Figure 5
A model of a bridge is tested in the National Research Council's Aerodynamics Laboratory in Ottawa.

Figure 6
Toronto City Hall

modifications to the model can be made and tested before construction begins. An example of a model is shown in **Figure 5**.

Turbulence is often created by high-rise buildings in urban areas. When high-speed winds encounter tall buildings, the buildings direct the fast-moving air from near the roof downward to street level. At street level, gusts of wind can have devastating effects on pedestrians. To help overcome this problem, scientists build models of proposed structures and their surroundings and test the models in a wind tunnel. After analyzing observed problems, they may make alterations to the lower part of the building or add wind barriers (trees and shrubs).

For example, when the new city hall in Toronto, Ontario, was designed, wind tunnel testing of the design revealed that a circular podium between the two towers was necessary to reduce the wind gusts at street level (**Figure 6**). Several years after the structure was built, the city decided to add a jogging path around the podium, probably without realizing how gusty the winds would be there. Unfortunately, on one occasion, a gust of wind picked up a portion of the track and injured a family on the track.

▶ **Practice**

Understanding Concepts

1. For each of the following liquids, state whether the viscosity is high or low:
   (a) skim milk
   (b) liquid honey
   (c) whipping cream
   (d) methyl alcohol (antifreeze)

2. Compare the speeds of the top and bottom of the bulge where the viscous fluid in **Figure 7** leaves the beaker. Relate this pattern to laminar flow.

3. Give an example of how turbulence is reduced in each of the following:
   (a) medicine
   (b) firefighting
   (c) construction of high-rise buildings

Applying Inquiry Skills

4. Describe how you would use common household items to safely test different shapes or structures to determine the turbulence around them in windy conditions.

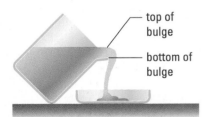

top of bulge

bottom of bulge

Figure 7
For question 2

Making Connections

5. If you were installing small wind turbines at Toronto's City Hall (shown in **Figure 6**, page 280), where would you place them? Explain your answer.

6. Explain why pilots wait a fixed period of time before taking off after another plane has taken off.

7. Shelters for livestock are built to control snowdrifts. Research the design of these shelters.
   (a) Draw a diagram of the shelter design. On your diagram, show the predominant wind direction and the snowdrift patterns likely to develop.
   (b) What must the livestock owner be aware of when using this type of shelter?

 www.science.nelson.com

SUMMARY *Laminar and Turbulent Flow*

- Fluid dynamics is the study of the factors affecting fluid motion.

- Laminar fluid flow is smooth; it results when both the internal resistance (viscosity) and external resistance (caused by contact with the surroundings) are low.

- Turbulent flow is a disturbance in fluid flow that causes irregularities.

- Wind tunnel or water tank tests on models of structures, as well as analysis using computer simulations, help improve the factors that affect fluid flow, thus reducing unwanted turbulence.

▶ *Section 6.1 Questions*

Understanding Concepts

1. Name four liquids of differing viscosity, and arrange them in a list from lowest to highest viscosity. (Do not include liquids mentioned so far in this chapter.)

2. Is fluid flow more likely to be laminar in a pipe with a smooth interior or a corroded interior? Explain your answer.

3. Is the viscosity of syrup higher at 5 °C or at 55 °C? Use the particle theory of matter to explain why.

Applying Inquiry Skills

4. Shortly after a snowstorm with high winds, you notice snowdrifts that display a variety of shapes.
   (a) Where would you look for examples of eddies in the snow?
   (b) Where would you look for examples of laminar flow of the blowing snow?

(c) If possible, create a photographic portfolio of examples of turbulent and laminar flow in snow, sand, or dust.

Making Connections

5. Explain why parachute instructors and hot-air balloon pilots always check the wind conditions before a flight.

6. Why are compressor stations required at regular intervals along the cross-Canada natural gas pipeline? (These stations are located approximately 200 km apart; they do not add any gas to the pipeline.)

7. Motor oils are made with different viscosities (e.g., SAE 20 and SAE 50) and sometimes with a range of viscosities (e.g., SAE 10W40).
   (a) Which oils are used in the summer? in the winter?
   (b) What is the advantage of 10W40 oil?

Figure 1
Speed skating is an example of a sport that requires high-tech designs to maximize the speed.

drag the forces that act against an object's motion through a fluid

streamlining the process of reducing the turbulence experienced by an object moving rapidly relative to a fluid

---

LEARNING TIP

Try a computer simulation to help visualize drag.

 www.science.nelson.com

At speeds that often exceed 50 km/h, speed skaters can encounter air resistance and turbulence. One way to reduce this resistance and turbulence is to crouch; another is to wear low-friction clothing (**Figure 1**).

The forces that act against an object's motion through a fluid are called **drag**. Drag is a form of frictional resistance. The importance of drag forces can be observed in sports activities and the transportation industry, as well as in nature. The main technique used to reduce drag is *streamlining*. **Streamlining** is the process of reducing turbulence by altering the design, which includes shape and surface features, of an object that moves rapidly relative to a fluid. Streamlined flow is the same as laminar flow.

▶ **TRY THIS** activity | **The Effects of Altering Shapes**

To study an example of the effect of streamlining, you can use a tea light in a sand tray and a piece of paper about 15 cm × 20 cm.

(a) Predict what will happen to the flame of the tea light when air is blown toward the paper, as illustrated in **Figure 2**.

(b) Keeping the paper a safe distance from the flame, verify your prediction and explain what occurs.

 Place the tea light in a sand tray before lighting it. Keep the paper at least 20 cm from the candle.

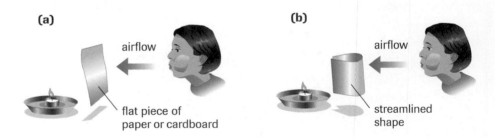

(a) airflow — flat piece of paper or cardboard

(b) airflow — streamlined shape

Figure 2
Testing streamlining

Fish, birds, and many other animals that move quickly in water or air provide excellent examples of streamlining. In fact, scientists study animal streamlining closely and try to apply their findings to technology. The transportation industry in particular devotes much research to trying to improve streamlining to reduce drag on cars, trucks, motorcycles, trains, boats, submarines, airplanes, spacecraft, and other vehicles. Although streamlining often enhances the appearance of a vehicle, its more important functions are to improve safety in windy or turbulent conditions and reduce fuel consumption.

Streamlining is an experimental science that relies heavily on large wind tunnels, water tanks, and computer simulations for its research. **Figure 3(a)** shows a series of fans that propel air in a wind tunnel used to research the streamlining of automobiles. **Figure 3(b)** shows how a fan directs air along a tunnel, around two corners, then through a smaller tunnel. As the air moves into the smaller tunnel it accelerates, reaching speeds up to 100 km/h, and flows past the test automobile or model. It then returns to the fan to be recirculated. Researchers view the action from behind an adjacent glass wall (shown in the diagram) and analyze the turbulence around the model. Pressure-sensitive beams, electronic sensors, drops of coloured water, small flags, and plumes of smoke are among the devices used to detect turbulence.

(a)

(b)

corner vanes (direct the air around corners)

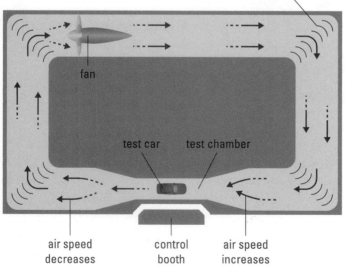

fan

test car   test chamber

air speed
decreases

control
booth

air speed
increases

Figure 3

(a) Fans in a wind tunnel. Testing large objects rather than small models requires more airflow, which is provided by a series of large fans.

(b) A typical wind tunnel arrangement used to analyze the streamlining of automobiles

DID YOU KNOW ?

Unique Wind Tunnels
Imagine an enclosed tunnel built for cyclists who simply sit on their bikes and let the wind in the tunnel carry them along for several kilometres. This type of tunnel is proposed for linking cities in Holland, a country where 80% of the population own bikes. Fans above the tunnel would provide a tailwind for the cyclists. Of course, two tunnels would be needed to allow two-way traffic. If you were on a bike in such a tunnel, what posture would give you maximum speed?

 www.science.nelson.com

Figure 5
(a) Sharkskin, showing the grooves. This patch of skin is magnified to about 3000×.
(b) A thin plastic coating with three grooves per millimetre reduces the drag of a metal surface passing through water.

The best streamlining features discovered through research are applied to the design of cars, like the one illustrated in **Figure 4**. Similar features are found on other vehicles.

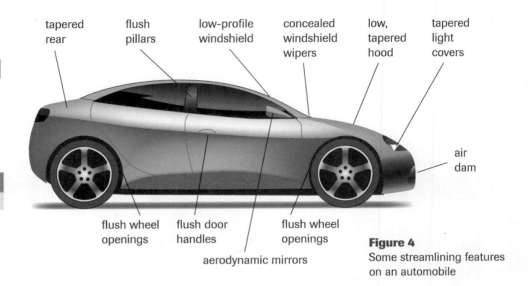

Figure 4
Some streamlining features on an automobile

It has long been assumed that perfectly smooth surfaces and hidden joints are the best means of reducing drag. However, nature has provided a clue that this is not necessarily so. Sharks are obviously well adapted to moving through water with reduced drag. A microscopic view of the skin of some species of fast-moving sharks reveals that the skin has tiny grooves parallel to the flow of water (**Figure 5(a)**). This feature reduces the tendency of the water to stick to the skin, thus reducing drag. Based on this finding, the surfaces of submarines are now covered with a thin, plastic coating with fine grooves (**Figure 5(b)**). This coating reduces drag and increases the submarine's maximum speed. Competitive swimmers wear suits with similar technology; the suits reduce drag even though they are not smooth.

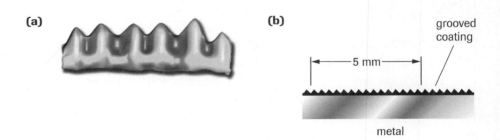

Researchers have found another way to streamline submarines: To reduce the tendency of water particles to stick to a submarine's hull, compressed air is forced out from a thin layer between the hull and its porous outer skin. Millions of air bubbles then pass along the submarine, preventing sticking and thus reducing the drag. Some of the discoveries applied to submarines can also be used for boats, ships, and aircraft.

> **Practice**

Understanding Concepts

1. Imagine you are driving a motorcycle travelling around 60 km/h. You extend your left arm to signal a turn, and you feel the drag caused by air resistance.
   (a) Sketch your hand in the position in which it feels maximum drag.
   (b) Sketch your hand in the position in which it experiences maximum streamlining.

2. **Figure 6** shows the test of a scale model of a truck with its trailer attached.
   (a) Describe the design features that cause turbulent flow.
   (b) What improvements would provide better streamlining?

Applying Inquiry Skills

3. Students are asked to design and test a way to prevent snowdrifts from piling up in an airport runway. Two designs are shown in **Figure 7**.
   (a) From which direction in the diagrams do the predominant winds come?
   (b) How would you test the designs to see how effective they are?
   (c) Based on your observations in the *Try This* Activity titled The Effects of Altering Shapes, page 282, which design do you think would work better? Explain your answer.
   (d) Sketch another design that you would submit for testing.

Making Connections

4. (a) Name three animals (not named so far) that are streamlined.
   (b) Name a human-made technology (not described so far) that is shaped like each animal you named in (a).

5. Sketch a conifer tree and a deciduous tree.
   (a) In the winter, which tree is more streamlined?
   (b) Use your sketches to show why a winter storm with heavy wet snow and high winds would be devastating to a deciduous tree if it kept its leaves all winter.

Figure 6
This model was tested in the National Research Council's wind tunnel in Ottawa. In this test, smoke plumes show airflow patterns.

(a)

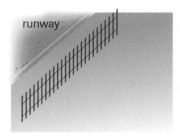

(b)

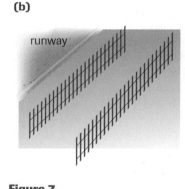

Figure 7
(a) Single fence design
(b) Double fence design

Case Study Drag Coefficients

To communicate the results of their research in a simple fashion, scientists determine a number, called the *drag coefficient* (symbol C_d), for each vehicle tested. To understand the range of values of this coefficient, consider the following two extremes: For a highly streamlined airplane wing, $C_d = 0.050$. For an open parachute, which is designed for maximum drag, $C_d = 1.35$. Most other C_d values lie between these extremes.

During the 1930s, when cars were not streamlined and gasoline was less expensive, the average C_d for cars was about 0.70. Today, the average value has dropped to less than 0.40, although some test models have C_d values reported to be as low as 0.15.

(a)

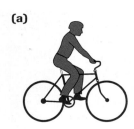

(b)

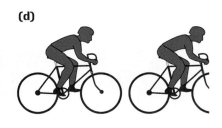

(c)

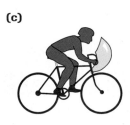

(d)

(e)

Figure 8
(a) Commuter bicycle, upright position, $C_d = 1.1$
(b) Aerodynamic components, crouched position, $C_d = 0.83$
(c) Partial fairing, crouched position, $C_d = 0.70$
(d) Closely following another bicycle (drafting), $C_d = 0.50$
(e) Vector single (three wheels), complete fairing, $C_d = 0.10$

Consider the difference made by the way the rider sits on a bicycle (**Figure 8**). When the rider is sitting in an upright position, the drag coefficient of the bicycle and rider can be as high as 1.1. However, when the rider is in the crouched, racing position on a streamlined racing bicycle, the C_d is reduced to about 0.83. The C_d can be further reduced by following another cyclist closely. The lower air resistance in this case allows a C_d as low as 0.50. (Refer to **Figure 4(a)** in section 6.1, page 279, to see how fluid flow behind an object doubles back. This produces a zone where a closely following object experiences less air resistance.) Some cycling enthusiasts use a lightweight shielding, called *fairing*, which creates a streamlined envelope around cycle and cyclist and reduces the C_d to an amazing 0.10.

▶ *Practice*

Understanding Concepts

6. (a) Draw a diagram of an egg so that its drag coefficient is lowest when it is moving to the right.
   (b) At what location in a ski jump run would a competitive ski jumper use the "egg position?" Why?

7. In which sports, besides speed skating and ski jumping, do the athletes try to reduce the drag coefficient to a minimum?

8. **Figure 9** shows a cyclist in a wind tunnel.
   (a) What is revealed by the smoke streamer seen above the cyclist?
   (b) Describe features in the photograph that reduce drag.
   (c) Estimate the cyclist's drag coefficient in this case using the values given in **Figure 8**.

Figure 9
How can you tell that this cyclist is in a wind tunnel?

Making Connections

9. Small cars and motorcycles can get better gasoline consumption by following a transport truck relatively closely.
   (a) Explain why the gasoline consumption improves.
   (b) Explain the dangers of this practice.

10. Using the Internet or another resource, research drag for Formula 1 racing cars.
    (a) How is the drag coefficient calculated?
    (b) Why do the drag coefficients of these cars tend to be so much higher than the coefficients of passenger cars?

GO   www.science.nelson.com

SUMMARY   *Streamlining*

- Drag on an object moving in a fluid results from frictional resistance and turbulence acting on the object.

- Streamlining reduces drag by allowing the fluid to flow smoothly around an object.

- Applications of streamlining are found in many areas, for example, in nature and in the fields of transportation and sports.

DID YOU KNOW ?

Every Bit Counts
Researchers do everything they can to reduce drag, even if the gain is as little as 1% or 2%. For example, when shaved heads are compared with heads covered with low-friction material, they find that the material provides slightly less drag. In comparing wheel designs, researchers find that spoke shape has a measurable effect. For front wheels, oval spokes are slightly better than round and flat spokes in most situations. However, flat spokes provide the least drag for straight-line racing (such as in triathlon racing). For the rear wheels, the disk design provides the least drag.

▶ *Section 6.2 Questions*

Understanding Concepts

1. Discuss the types of features used on each of the following vehicles to reduce drag:
   (a) sport motorcycle
   (b) large passenger airplane
   (c) locomotive
   (d) bobsled
   (e) racing car

2. Describe three ways in which drag can be reduced in bicycle racing.

Applying Inquiry Skills

3. Small flags, smoke plumes, and other methods can be used to study an object in a wind tunnel.
   (a) How could you experimentally observe the flow of air into or out of the air vents in your classroom?
   (b) How might you benefit from knowing about the airflow in a room?

Making Connections

4. The solar collectors on the *International Space Station* are rectangular in shape and are about the size of a football field. Explain why they do not need to be streamlined.

5. Describe three instances where drag is intentionally increased. (You can use **Figure 10** as your first example.)

Figure 10
For question 5

The Effects of Fast Fluid Flow

When a fluid flows at a high speed, interesting effects happen in and near the fluid. You saw an example of these effects in the Chapter 6 introductory activity, page 277. In that case, you could not see the fluid (the air), but you could easily see the effect as the paper wing in one case moved higher as the air blew faster.

In this investigation, you will use flowing air to discover the effects of an increase in the speed of the air.

Inquiry Skills

○ Questioning	● Conducting	● Evaluating
● Predicting	● Recording	● Communicating
● Planning	● Analyzing	● Synthesizing

Question

What happens to the pressure in a fluid when the fluid's speed increases?

Prediction

(a) Predict an answer to the Question, giving reasons.

Figure 1
Kinetic pop cans

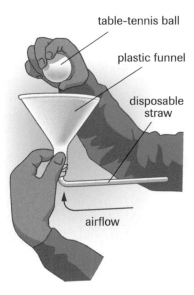

Figure 2
Ball control

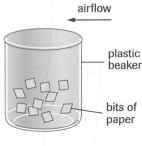

Figure 3
Beaker cleaner

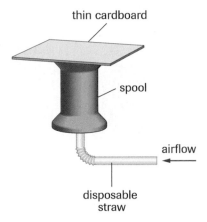

Figure 4
Blowing the cover

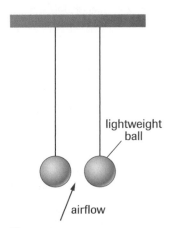

Figure 5
Attraction or repulsion?

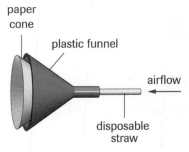

Figure 6
The air filter

(b) Look carefully at **Figures 1** to **9**. For each setup, write what you predict you will see when the air moves as shown.

Experimental Design

In Part A, you will participate in and/or observe several activities set up in the classroom in which air moves around or through an object. Depending on the equipment available, some of the activities may be class demonstrations. In Part B, you will apply what you learned in Part A to solve a challenge. In Part C, you can design your own demonstration to show the effects of increased speed of a fluid and share it with the class.

Materials

For each student:

safety goggles

a disposable straw, preferably a flexible one (to be used by one student only and then discarded at the end of the investigation)

Part A

clean-up materials (sponges and paper towels)

for all other apparatus and materials, refer to the labels in the diagrams

Part B

small coin

saucer

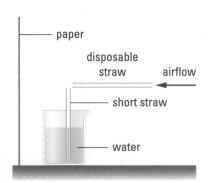

Figure 7
Water art

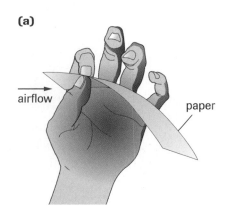

Figure 8
(a) Airflow below the paper
(b) Airflow above the paper

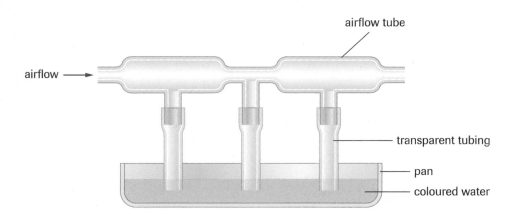

Figure 9
A flowmeter

Part C

materials of your own choice to design a safe
 demonstration

 Wear safety goggles. Do not share straws.

Procedure

Part A

1. Put on your safety goggles. In a small group, go
    to the first lab station assigned by your teacher.
    Observe what happens when the air moves
    according to the airflow shown in the diagram.
    Record your observations. Clean up the station.

2. When signalled by your teacher, move to the next
    station and repeat step 1. Continue until you
    have visited all the stations.

3. Answer questions (c), (d), (f), (g), and (h) at the
    right before proceeding to Part B.

Part B

4. Solve the following challenge: Place a small coin
    flat on the lab bench or a countertop. Place a
    saucer about 5 cm beyond the coin, as shown in
    Figure 10. Without touching the coin, move the
    coin onto the saucer. (Hint: Apply what you
    learned in Part A. You may need to hold the
    saucer steady.)

Figure 10
Manipulating money

Part C

5. Design a safe demonstration to show what
    happens when air (or water) moves quickly.
    Create a catchy title for the demonstration.
    Be prepared to share your demonstration
    with your class.

Analysis

(c) In Part A, what happened to the pressure of the
     air when the air moved quickly? Give two
     examples to justify your answer.

(d) Did any observations in Part A not appear to fit
     a pattern? Explain your answer.

(e) For Part B, draw a diagram showing how the
     coin got onto the saucer.

Evaluation

(f) How accurate were your predictions in (a)
     and (b)?

(g) Which observations did you find most difficult
     to understand or explain?

Synthesis

(h) Choose one of the trials in Part A, and apply the
     particle theory of matter to explain what you
     observed.

The speed of a moving fluid has an effect on the pressure exerted by the fluid. Consider water flowing under pressure through the pipe illustrated in **Figure 1**. As the water flows from the wide section to the narrow section, its speed increases. This effect is also seen in rivers that flow slowly through widely spaced banks but speed up when passing through a narrow gorge. The effect can be verified experimentally, as you discovered in Investigation 6.3.

The water flow in **Figure 1** accelerates as the water molecules travel from region A into region B. The acceleration is caused by an unbalanced force, but what is the source of the acceleration? The answer lies in the pressure difference between the two regions. The pressure (or force per unit area) must be greater in region A than in region B in order to accelerate the molecules as they pass into B.

These concepts were analyzed in detail by the Swiss scientist Daniel Bernoulli (1700–1782). His conclusions became known as *Bernoulli's principle.*

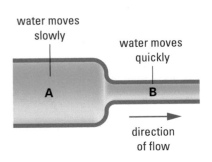

Figure 1
The speed of water flowing under pressure in a pipe depends on the pipe's diameter.

Bernoulli's Principle
Where the speed of a fluid is low, the pressure is high. Where the speed of a fluid is high, the pressure is low.

A device used to demonstrate pressure differences in a fluid at various speeds is shown in **Figure 2**. The same conclusions can be drawn from viewing this apparatus. By comparing the liquid level in the three vertical columns, we can compare the different speeds of the liquid in the different parts of the horizontal pipe. (In the diagram, the base of each vertical column is located at the same level so that the effect of gravity need not be considered.)

There are many applications of Bernoulli's principle in technology, transportation, sports, and other fields. As you read about the examples that follow, think about how they relate to the activities in Investigation 6.3.

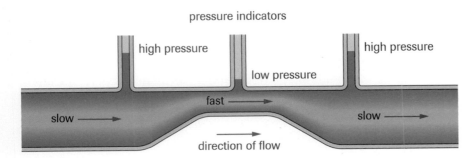

Figure 2
The pressure of the water depends on its speed.

Figure 3
A paint sprayer

Consider a paint sprayer (**Figure 3**). Air from a pump moves rapidly across the top end of a tube, reducing the pressure in the tube. Atmospheric pressure forces the paint up the tube to be mixed with the flowing air, creating a spray.

The upward force, or *lift*, on an airplane wing can be explained by applying Newton's third law of motion and Bernoulli's principle. An airplane wing (**Figure 4**) is flatter on the bottom and more curved on the top. As the wing moves forward, air deflects off the bottom of the wing. The wing thus exerts a downward action force on the air, and the air exerts an upward reaction force on the wing. Furthermore, the speed of the air near the deflection is less than the speed of the streamlined flow of air over the wing. Thus, the pressure above the wing is low, and the pressure below the wing is high. The pressure difference adds to the lift on the wing.

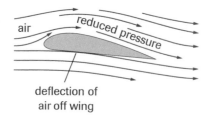

Figure 4
An airplane wing

DID YOU KNOW

The Physics of Flight
There is more to airplane flight than lift, as explained by Newton's third law of motion and Bernoulli's principle. Control components such as flaps are also needed. These components are examples of hydraulic systems; they aid in takeoff, landing, changing altitude, and steering the plane.

GO www.science.nelson.com

▶ **TRY THIS** activity   ***Paper Airplanes***

You can learn a lot about lift and control in airplane flight by designing and testing paper airplanes. With your teacher's permission, create and safely test a paper airplane.

(a) Compare your design and flight distance with those of other students.

(b) Compare your flight distance with the maximum distance in public, indoor competitions, which can be greater than 40 m!

🖐 **Perform the tests in an empty space, aiming the planes away from observers.**

Figure 5
Jet engine fan

Turbine blades in jet engines, windmills, fans, and so on, apply Bernoulli's principle (**Figure 5**). The blade shape resembles an airplane wing. As the fan rotates, the air above the more curved part of each blade moves faster than the air above the flatter part. Thus, the pressure is higher on the flatter side of the blade, and air is forced to move through the fan in the direction from higher pressure to lower pressure.

Some skis used in ski-jumping competitions have flexible rubber extensions at their rear. As the skier glides through the air, the air above the skis travels faster than the air below. The result is a higher pressure below the skis than above, which creates an upward lift on the skis (**Figure 6**).

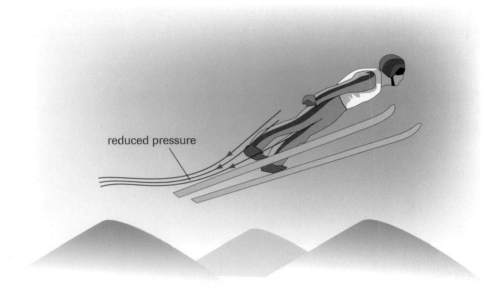

reduced pressure

Figure 6
Ski-jumping

Throwing a curveball is also an application of Bernoulli's principle. In **Figure 7(a)**, a baseball is thrown in the direction shown. Relative to the ball, the air is moving backward. When the ball is thrown with a clockwise spin, as viewed from above, the air near the ball's surface on the left is dragged along with the ball in the opposite direction to the main airflow (**Figure 7(b)**). To the left of the moving ball, the speed of the air is slow, so the pressure is high. The ball is forced to curve to the right, following the path shown in **Figure 7(c)**. Most of the ball's curve takes place in the last few metres of its path. This last-minute curve tricks the batter, which is the object of throwing a curveball.

LEARNING TIP

Curveball Simulation
You can pitch a curveball online and see the airflow around the baseball. You can choose from a few locations, including Mars!

GO   www.science.nelson.com

(a)

direction of throw

air   air

(b)

(c)

original direction   curved path
of throw   of ball

air   air

Figure 7
(a) A ball thrown without spin is not deflected.
(b) Air is dragged around the surface of a baseball thrown with a spin. In this case, the spin is clockwise when viewed from above the ball.
(c) Because the speed around a spinning ball is not equal on both sides, the pressure is not equal. The ball is deflected in the direction of the lower pressure.

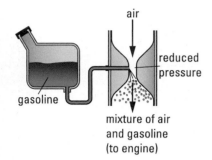

Figure 8
A carburetor used in a small engine

A carburetor is a device that applies Bernoulli's principle to control the air–fuel mixture fed to small engines, like those used in snow blowers and lawnmowers. (It serves the same function as fuel injectors in cars.) A carburetor has a barrel in which airflow controls the amount of gasoline sent to the engine. **Figure 8** shows air flowing by the gasoline intake. The fast-moving air has reduced pressure, so the gasoline, which is under atmospheric pressure, is forced into the carburetor. There it mixes with the air and goes to the engine.

▶ **TRY THIS** activity

Measuring the Speed of Water

(a) Design and perform an experiment to measure the linear speed of water as it leaves a horizontal hose of known diameter, as shown in **Figure 9(a)**. (Hint: Time the collection of a measured volume of water to determine the volume flow rate, q_v, then divide that value by the area of the nozzle. This will give you the speed. Recall from section 5.2, page 231, that $v = \dfrac{q_v}{A}$.

(b) Extend the experiment to find the speed of the water jet when the original fluid pressure is constant but the area of the end of the nozzle is halved, as shown in **Figure 9(b)**.

(c) Relate what you observed in this activity to Bernoulli's principle.

 Perform this activity outdoors where spills will not cause damage.

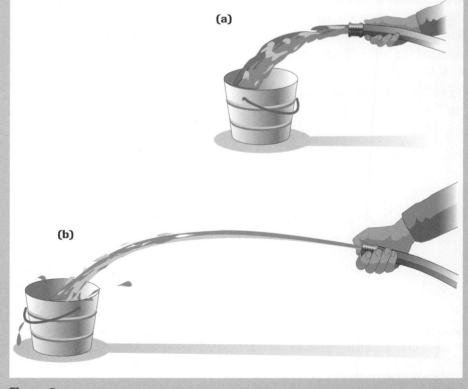

Figure 9
(a) Water flow with a low pressure   (b) Water flow with a high pressure

▶ **Practice**

Understanding Concepts

1. Explain the following statements in terms of Bernoulli's principle:
   (a) A tarp, which covers a dumpster with no top, bulges outward as the dumpster is towed along a highway.
   (b) A fire in a fireplace burns better when the wind is blowing outside.

2. Apply Bernoulli's principle to explain what you observed in Investigation 6.3, page 288, **Figures 1**, **3**, **4**, **5**, and **6**.

3. Animals that live underground, such as prairie dogs and gophers, require air circulation in their burrows. To provide enough circulation, these creatures make one burrow entrance higher than the other, as shown in **Figure 10**. Explain how this design helps increase air circulation.

Figure 10

4. Devices described in the text that apply Bernoulli's principle to turbine blades are used in air. List devices with turbine blades that are used in water.

Applying Inquiry Skills

5. (a) In a storm with extremely high winds, what might happen to windows if all windows and doors are closed tightly? Explain your answer.
   (b) How would you test your answer to (a) in a laboratory situation?

Figure 11

Making Connections

6. (a) A baseball (viewed from above) is thrown as indicated in **Figure 11**. If the ball is spinning counterclockwise, determine the approximate direction of the ball's path. Use diagrams in your explanation.
   (b) Research the spit ball. What is it, and why was it banned from baseball?

7. Research the meanings of the golfing terms "slice" and "hook." What causes slices and hooks, and what should a golfer do to prevent them?

8. During winter, airplane components are carefully checked for ice buildup before takeoff. If necessary, de-icing procedures are used (**Figure 12**). Explain the dangers of ice buildup on aircraft.

Figure 12
De-icing an airplane

Extension

9. (a) If you performed the *Try This* Activity titled Measuring the Speed of Water, page 294, relate your calculations in (a) and (b) from the activity to the *equation of continuity*, illustrated in **Figure 13**. This equation states that the volume flow rate remains constant or continuous for a particular flow. Thus,

$$q_1 = q_2 \qquad \text{or} \qquad A_1 v_1 = A_2 v_2$$

where A_1 is the cross-sectional area of the first pipe or opening,
v_1 is the speed of the water in the first pipe or through the first opening,
A_2 is the cross-sectional area of the second pipe or opening, and
v_2 is the speed of the water in the second pipe or through the second opening.
(equation applies to incompressible fluids with constant density only)

   (b) Water in a garden hose with a cross-sectional area of 3.4 cm² is travelling at 4.5 m/s. Determine the speed of the water coming from a nozzle with a cross-sectional area of 1.2 cm² attached to the hose. (Apply the equation of continuity.)

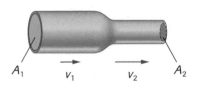

Figure 13
Where the area is large, the speed is slow; where the area is small, the speed is fast.

Answer

9. (b) 13 m/s

- Bernoulli's principle states that where the speed of a fluid is low, the pressure is high, and where the speed of a fluid is high, the pressure is low. This principle is applied in throwing a curveball and in many other situations.

▶ *Section 6.4 Questions*

Understanding Concepts

1. It is unwise to stand close to a fast-moving subway train. Explain why in terms of Bernoulli's principle.

2. **Figure 14** shows a device called a Venturi flowmeter, used to measure the speed of gas flowing through a tube. Explain how its design relates to Bernoulli's principle.

3. Imagine you are able to test your throwing skills in a large laboratory with all its air removed. You are given a table-tennis ball and a Frisbee to throw. You find that the table-tennis ball goes much farther than it does in a room with air, but the Frisbee flies comparatively poorly. Explain these observations.

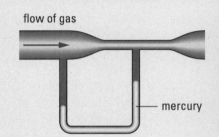

flow of gas

mercury

Figure 14
A Venturi flowmeter

Figure 15

Applying Inquiry Skills

4. Explain how you would use a manometer as a meter to measure the flow rate of moving air. (The manometer was described in section 5.3, page 241.)

Making Connections

5. A spoiler on the rear of a car (**Figure 15**) has a cross-sectional shape that resembles an airplane wing.
   (a) Does the spoiler shown result in an upward or downward force on the rear of the moving car? Explain your answer.
   (b) To help keep a car stable when travelling around curves, which way should the spoiler be installed? Explain your answer.

   (c) Some spoilers are designed just for appearance (i.e., they provide none of the advantages that spoilers on racing cars provide). Describe the disadvantage of having this type of spoiler.

6. Windsurfing boards are designed for maximum speed and stability in a variety of wind conditions. However, the problems created by high winds are different from problems created by low winds. Research windsurfing boards, and find out how the vented nose on the board reduces the tendency of the board to lift off and "tail walk" in high winds.

 www.science.nelson.com

Key Understandings and Skills

6.1 Laminar and Turbulent Flow

- Fluid flow can be **laminar** or **turbulent**. Wind tunnels, water tanks, and computer simulations help researchers analyze laminar and turbulent flows.

6.2 Streamlining

- **Streamlining** is used to reduce the **drag** and turbulence that objects experience while moving in a fluid.

6.3 Investigation: The Effects of Fast Fluid Flow

- Many examples illustrate how the pressure exerted by a fluid changes when the speed of the fluid changes.

6.4 Bernoulli's Principle

- Where the speed of a fluid is low, the pressure is high. Where the speed of a fluid is high, the pressure is low. This principle has many applications in technology, transportation, sports, and medicine.

Key Terms

6.1
fluid dynamics
viscosity
laminar flow
turbulent flow

6.2
drag
streamlining

6.4
Bernoulli's principle

Problems You Can Solve

6.1

- Contrast and compare laminar flow and turbulent flow of fluids.
- Identify factors that affect fluid flow and describe ways of achieving smooth fluid flow.

6.2

- Identify the main ways in which objects moving in a fluid can be streamlined.

6.3

- Describe how you could safely demonstrate that the pressure exerted by a fluid decreases as the speed of the fluid increases.

6.4

- State Bernoulli's principle, and give an example of its application in airplane flight, automobile transportation, sports activities, medicine, and technology.

▶ MAKE a summary

Think of an object whose design could be analyzed in a wind tunnel or water tank. (One example is shown in **Figure 1**; however, you might prefer something related to sports, medicine, technology, underwater exploration, and so on.) On a piece of notepaper, draw a design of your object that would experience a high level of drag.

Create a second design of the same object, but this time make it highly streamlined. On each diagram, show as many features related to fluids in motion as you can. Include key understandings, key words, and various ways of observing fluid flow in a wind tunnel or water tank.

(a)

(b)

Figure 1
Possible truck designs

Write the numbers 1 to 8 in your notebook. Indicate beside each number whether the corresponding statement is true (T) or false (F). If it is false, write a corrected version.

1. Fluid dynamics is the study of the factors that affect fluids in motion.

2. A liquid with low viscosity has a high amount of internal friction.

3. Viscosity is a type of friction that fluids experience.

4. In laminar flow in an enclosed pipe, the fluid's speed is slowest near the pipe's interior surface and fastest in the middle of the pipe.

5. In laminar flow in a pipe, the fluid's pressure is greatest in the middle and least near the pipe's interior surface.

6. A dimpled golf ball can travel farther than a smooth ball because it encounters less turbulence.

7. Streamlining is a method of reducing drag.

8. When a fluid travels faster, the pressure it exerts increases.

Write the numbers 9 to 15 in your notebook. Beside each number, write the letter corresponding to the best choice.

9. At room temperature, the viscosities of whipping cream, milk, and maple syrup, from lowest to highest, are
   (a) milk, whipping cream, maple syrup
   (b) maple syrup, milk, whipping cream
   (c) whipping cream, milk, maple syrup
   (d) maple syrup, whipping cream, milk

10. Streamlining is a
   (a) force of fluid friction
   (b) way of increasing turbulence
   (c) way of decreasing turbulence
   (d) none of the above

11. Air rushing past your ear will
   (a) cause your eardrum to move outward
   (b) cause your eardrum to move inward
   (c) have no effect on your eardrum because your eardrum does not move
   (d) have no effect on your eardrum because your eardrum is neither a gas nor a liquid

12. Refer to **Figure 1**. If v represents the speed of the fluid,
   (a) $v_A > v_B > v_C$
   (b) $v_B > v_A > v_C$
   (c) $v_C > v_B > v_A$
   (d) $v_C > v_A > v_B$

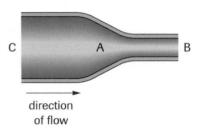

direction
of flow

Figure 1

13. In **Figure 1**, if P represents the pressure in the fluid,
   (a) $P_A > P_B > P_C$
   (b) $P_B > P_A > P_C$
   (c) $P_C > P_B > P_A$
   (d) $P_C > P_A > P_B$

14. If you want a baseball to curve to the right, you should cause it to spin
   (a) horizontally counterclockwise as viewed from above
   (b) horizontally clockwise as viewed from above
   (c) vertically, with the top spinning away from you
   (d) vertically, with the top spinning toward you

15. Using P for pressure and v for speed, in **Figure 2**
   (a) $P_A > P_B$ and $v_A > v_B$
   (b) $P_A < P_B$ and $v_A > v_B$
   (c) $P_A < P_B$ and $v_A < v_B$
   (d) $P_A > P_B$ and $v_A < v_B$

airflow

Figure 2

An interactive version of the quiz is available online.

 www.science.nelson.com

Understanding Concepts

1. Explain why the viscosity of a liquid changes when its temperature changes. In your explanation, apply the particle theory of matter.

2. A small steel ball is falling in vegetable oil at a certain speed. Will the speed be greater if the oil is at 20 °C or at 60 °C? Explain your answer.

3. (a) Which shape in **Figure 1** would result in the greatest laminar flow of the air? Explain why.
   (b) Assume that each shape represents the top view of a flagpole and the straight line represents a flag. Which shape of flagpole would yield the greatest flutter of the flag? Explain your answer.

4. Describe the main ways that athletes achieve streamlining in individual, high-speed sports. Identify the sports.

5. **Figure 2** shows a wind tunnel for testing aircraft components in atmospheric icing conditions. Why is this wind tunnel so much smaller than the wind tunnels used to test cars?

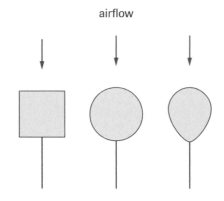

Figure 1

6. **Figure 3** illustrates two types of fish that have different streamlining features. One type can accelerate quickly to get away from predators but is a poor distance swimmer. The other type accelerates more slowly but can migrate long distances. Which fish is which? Explain why.

7. In **Figure 4**, the water and the vertical straw are in a vacuum chamber (i.e., a chamber from which the air can be removed).
   (a) Before the air is removed, explain how the water can spray the paper and why.
   (b) Will the water spray work when the air has been removed from the chamber? Explain your answer.

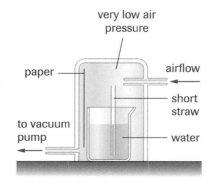

Figure 4

Figure 2
A low-temperature wind tunnel at the National Research Council in Ottawa

(a) skipjack tuna

(b) sea raven

Figure 3

8. You are travelling in a small car on a two-lane highway. A large transport truck approaches you in the opposite lane. Apply physics principles to explain why you should leave a wide space between the truck and your vehicle.

9. Given the choice between taking off into the wind (a headwind) or with the wind (a tailwind), which choice should the pilot of an aircraft make? Explain why.

10. (a) When you blow out candles on a birthday cake, do you do so with a wide open mouth or a narrow one? Explain why.
    (b) Can the situation in (a) be considered an example of Bernoulli's principle? Explain your answer.

11. You throw a baseball eastward. The ball has a fast clockwise spin when viewed from above. In which direction does the ball tend to swerve? Explain your reasoning and include a diagram.

12. A golfer gives a golf ball a powerful backspin; the bottom of the ball spins toward the direction of the travel of the ball. Explain, with a diagram, how this affects the flight of the ball.

13. (a) Why are pumps required at regular intervals along oil pipelines? (These pumps do *not* add any oil to the pipeline.)
    (b) Describe how Venturi flowmeters could be used to determine the distances required between the pumps.

Applying Inquiry Skills

14. Explain how you could use water flowing from a faucet to illustrate laminar and turbulent flows.

15. A simple fan can be used in the classroom for airflow experiments.
    (a) Does it matter if the fan is in front of or behind the object being tested? Give a reason for your answer.
    (b) How would you test your answer in (a)?

16. Design a toy based on the principles presented in this chapter. Draw a diagram of your design. With your teacher's permission, build and test the toy in a safe manner.

17. A physics teacher places several small Styrofoam™ balls into a large plastic beaker.
    (a) Using this setup, how can the teacher safely demonstrate Bernoulli's principle?
    (b) Explain how this demonstration works, and include a diagram.
    (c) How can the principle shown in this demonstration be applied to spraying liquid fertilizer on a rose garden?

Making Connections

18. If you look at a refrigerator's air vent cover, you notice that dust accumulates there easily.
    (a) Why does dust gather at a location where air is moving more rapidly than its surroundings? (Use Bernoulli's principle in your answer.)
    (b) Explain why it is wise to keep vent covers clean.

19. **Figure 5** shows two plumbing designs for a kitchen drain.
    (a) What is the purpose of the curved portion of the U-shaped drain?

(a)

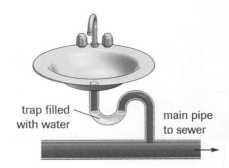

trap filled with water

main pipe to sewer

(b)

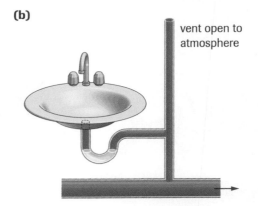

vent open to atmosphere

Figure 5

(b) Which design would you recommend for a plumbing system? Explain your answer. (In your explanation, use Bernoulli's principle.)

20. **Figure 6** shows the design of a typical Bunsen burner.
   (a) Explain why the flame does not burn cleanly if the air vents at the base of the burner are closed. In your answer, apply Bernoulli's principle.
   (b) Design a candleholder, and include a drawing. Relate the design of a Bunsen burner to your design.

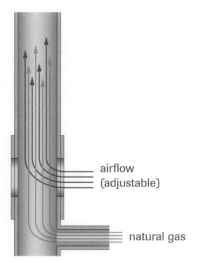

airflow (adjustable)

natural gas

Figure 6

21. **Figure 7** illustrates a portion of an artery partially clogged with plaque.
   (a) As the blood flows from A to B, what happens to its speed? its pressure?
   (b) If the pressure at B becomes very low, the external pressure (at C and D) causes the artery to close. At that point, what happens to the speed and pressure at B?
   (c) Explain how a doctor can detect plaque in arteries by listening for flutter through a stethoscope.

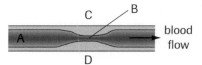

C   B

A

blood flow

D

Figure 7

PERFORMANCE TASK

Your completed task will be assessed according to the following criteria:

Process

- Draw up detailed plans and safety considerations for the design, tests, and modifications of the system or model.
- Choose appropriate research tools, such as books, magazines, and the Internet.
- Choose appropriate materials to construct the system or model.
- Carry out the construction, tests, and modifications of the system or model appropriately and safely.
- Analyze the process (as described in the Analysis).
- Evaluate the task (as described in the Evaluation).

Product

- Demonstrate an understanding of the relevant physics principles, laws, and equations.
- Submit a report containing the design plans for the system or model, as well as test results and any related calculations.
- Use terms, symbols, equations, and SI metric units correctly.
- Demonstrate that the final product works.

Hydraulics, Pneumatics, and Streamlining

As you learned in this unit, hydraulic systems use a liquid under pressure to provide power, and pneumatic systems use a gas under pressure. Because water and air are plentiful and generally safe to use, they are appropriate choices for a student-designed hydraulic or pneumatic system. Examining a number of examples of such systems will help you prepare for Option 1 of this task, which is to design, construct, evaluate, and analyze a hydraulic or pneumatic system that accomplishes a specific task.

In Option 2, the properties of fluids in motion can be studied in a water tank, a controlled airflow from a fan, or a wind tunnel, if one is available. You can study scale models of structures, as illustrated in **Figure 1.**

(a)

(b)

object —

plastic tube —
at an angle

Figure 1
(a) Using air as the fluid
(b) Using water as the fluid

In Option 3, you can propose your own independent performance task and have your teacher approve it. To review how to evaluate and assess research tasks, you can refer to the Unit 1 and Unit 2 Performance Tasks (pages 114 and 206, respectively).

Before choosing an option, you should decide what you want to accomplish. For example, you might want to create a system that shows the operation of a vehicle braking system or a construction crane. Alternatively, you might decide to invent a new device, such as a water fountain. You might want to test your own design for the shape of a windscreen on a motorcycle. Or you might think of an independent project in which you research applications of hydraulics, pneumatics, or streamlining and build a model or create a position paper related to the research. Once you have decided what problem to work on, you will find it easier to brainstorm ideas for a solution.

If you build a device or model, use materials that are readily available, inexpensive, and safe (bottles and cylinders should be plastic, not glass).

The Task

Option 1   A Hydraulic or Pneumatic System

Your task is to design, construct, evaluate, and analyze a hydraulic or pneumatic system that accomplishes a specific task. Here you apply the principles presented in Chapter 5.

- In a group, decide what task you want your system to accomplish and whether you will use a hydraulic or a pneumatic system to perform it.

- With your teacher, decide on the safety precautions required, as well as the criteria that will be used to evaluate the system. For example, the criteria might include originality of design, wise use of materials, usefulness of the system, appearance, and the process used to complete the task. The maximum force or pressure achieved by the system is another optional criterion.

 Design your system, and have your teacher approve the design.

- Construct and test the system, and modify it until you are satisfied with its performance.

- If appropriate, analyze the forces and pressures involved in operating the system.

Option 2   A Model System

Your task is to design, test, and analyze two models and to determine how their properties of streamlining compare when tested in a water tank or in a controlled airflow. You are expected to design one model that is intentionally poorly streamlined and another that features the best streamlining you can design. This option applies the principles from Chapter 6.

- In a group, decide what type of object(s) you want to test in the water tank or controlled airflow.

- With your teacher, decide on the safety precautions required as well as the criteria that will be used to evaluate the models and the tests you intend to perform on them. The criteria might include originality of design, wise use of materials, usefulness of the device modelled, appearance, and the process used to complete the task.

 Design the models, and have your teacher approve your designs.

- Build and test the models, and then modify them until you are satisfied with the results.

- Analyze the features of the models you tested. Give detailed evidence of drag and streamlining using diagrams to illustrate.

Option 3   Independent Study

Your task is to propose an independent study task involving at least one important application of topics in Unit 3. Your proposal should include the following:

- a description of what you want to research

- a list of resources you intend to use

- a set of appropriate Analysis questions

- a set of appropriate Evaluation questions

- a list of Assessment criteria, including the Process and Product assessment

Have your teacher approve your choice before you begin the research.

Analysis*

(a) What physics principles apply to the design and use of your system or model?

(b) How can you judge whether your system or model was successful?

(c) What can the system or model be used for?

(d) What careers are related to the manufacture and use of your system or model?

(e) What safety precautions did you follow in building and testing your system or model?

(f) After testing your system or model, how did you modify it to improve it?

(g) How could the process you used in this task be applied in business or industry?

(h) List problems you encountered while building the system or model, and explain how you solved them.

Evaluation*

(i) How does your system or model compare with the designs of other groups? (Some criteria to consider are originality, usefulness, appearance, and the wise use of materials.)

(j) Evaluate the tools you used in constructing the system or model.

(k) If you repeated this task, how would you modify the process to obtain a better final product?

* The Analysis and Evaluation given are for Options 1 and 2 only. Option 3 requires an independent set of questions and criteria.

1. Write the letters (a) to (f) in your notebook. Beside each letter, write the word that best completes the blank(s).
   (a) The science and technology of the mechanical properties of gases is called ___?___.
   (b) ___?___ is the mass per unit volume of a substance.
   (c) The newton per square metre is given the name ___?___.
   (d) Two equations used to determine volume flow rate are ___?___ and ___?___.
   (e) Laminar flow and ___?___ flow are the same.
   (f) As the speed of a fluid increases, the pressure ___?___.

2. Write the letters (a) to (h) in your notebook. Beside each letter, write the word or phrase that corresponds to each of the following:
   (a) a general term for any gas or liquid
   (b) the study of fluids at rest
   (c) the study of fluids in motion
   (d) the substance whose density is 1.00×10^3 kg/m^3
   (e) a standard of 101.3 kPa
   (f) a barometer that does not require a liquid
   (g) the property of a fluid that determines its resistance to flow
   (h) the forces that act against an object's motion through a fluid

3. Write the letters (a) to (e) in your notebook. Beside each letter, write the letter from A to J that corresponds to each of the following terms:
   (a) Bernoulli's principle
   (b) static pressure head
   (c) gauge pressure
   (d) Pascal's principle
   (e) compressibility

   A the sum of the absolute and atmospheric pressures
   B the ability of the particles of a substance to be compressed
   C applied in the operation of a hydraulic press
   D applied in the operation of a bicycle tire pump
   E the principle of action–reaction pressures
   F the pressure exerted by a given height of fluid in a column
   G applied in the operation of a Venturi meter
   H the difference between absolute and atmospheric pressures
   I the mass per unit volume of a substance
   J the height of a column of fluid that gives a particular pressure

Write the numbers 4 to 17 in your notebook. Indicate beside each number whether the corresponding statement is true (T) or false (F). If it is false, write a corrected version.

4. When comparing gases, the densities should be stated for a known temperature and pressure.

5. A density of 1.2 g/L is the same as a density of 1.2×10^3 kg/m^3.

6. One pascal is equivalent to one newton per square metre.

7. Pneumatic pressure systems with short connecting hoses react more quickly than those with long connecting hoses.

8. A pressure of "two atmospheres" is equivalent to 202 kPa.

9. An aneroid barometer uses liquid in a tube with a closed top to measure atmospheric pressure.

10. The pressure beneath the surface of a liquid depends only on h (the depth) and g (the gravitational force constant).

11. A cold-water tap is an example of a two-way valve.

12. The gauge pressure of a tire is less than the absolute pressure in the tire.

13. Pascal's principle is applied in the design of airplane wings.

14. Bernoulli's principle is applied in the use of the hydraulic press.

15. Streamlining increases turbulence.

16. When water flows from a large pipe into a smaller pipe, the speed of the water decreases.

17. An airplane wing is shaped so that the speed of the air above it is greater than the speed of the air beneath it.

Write the numbers 18 to 27 in your notebook. Beside each number, write the letter corresponding to the best choice.

18. A gas such as air is
    (a) compressible because its molecules are far apart and have a high attraction for each other
    (b) compressible because its molecules are far apart and have a low attraction for each other
    (c) not compressible because its molecules are far apart and have a high attraction for each other
    (d) not compressible because its molecules are far apart and have a low attraction for each other

19. A pressure of 2 kPa is equivalent to
    (a) a pressure of 2000 Pa
    (b) a force per unit area of 2000 N/m^2
    (c) a force per unit area of 2 kN/m^2
    (d) all of the above

20. Of the following absolute pressures, the one most likely found in a vacuum cleaner hose when the vacuum is in operation is
    (a) 51 kPa   (c) 202 kPa
    (b) 101 kPa   (d) 404 kPa

21. The slope of the line on the graph in **Figure 1** is
    (a) a pressure of 100 Pa
    (b) a force of magnitude 100 N
    (c) a force of magnitude 10 N
    (d) a pressure of 10 Pa

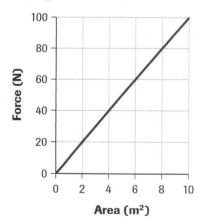

Figure 1

22. Volume flow rate can be measured in which of the following symbols:
    (a) kg/m^3 (c) m^3/kg
    (b) m^3/s   (d) N/m^2

23. Static pressure head is
    (a) a pressure measured in pascals
    (b) a force measured in newtons
    (c) an area measured in square metres
    (d) a height measured in metres

24. At sea level, the height of mercury in a mercury barometer is 76 cm. At the top of a mountain, the height is
    (a) greater than 76 cm because there is less air pushing down on the manometer, so the mercury can rise higher
    (b) less than 76 cm because there is less air pushing down on the mercury, so it cannot rise higher
    (c) equal to 76 cm because the density of mercury is the same at any altitude
    (d) greater than 76 cm because the force of gravity is lower at the top of the mountain

25. The valve in **Figure 2** has
    (a) two positions and two ports
    (b) two positions and three ports
    (c) two positions and four ports
    (d) three positions and three ports

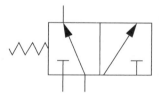

Figure 2

26. A liquid with a high viscosity
    (a) has high internal friction and flows slowly
    (b) flows quickly and easily
    (c) flows more quickly as its temperature decreases
    (d) is highly compressible

27. The lift on an airplane wing is caused by a combination of
    (a) a reaction force of the air and a reduced pressure on the bottom of the wing
    (b) a reaction force of the air on the bottom of the wing and a reduced pressure on the top
    (c) a reaction force of the air and a reduced pressure on the top of the wing
    (d) a reaction force of the air on the top of the wing and a reduced pressure on the bottom

Understanding Concepts

1. Explain the difference between fluid dynamics and fluid statics.

2. The propane in a propane tank for a barbecue is a liquid, yet the propane that burns in the barbecue is a gas. How can the propane be both types of fluid? (Include the kinetic molecular theory of matter as part of your answer.)

3. (a) Calculate the density, in the most convenient unit, of each of the following samples:
   sample A: $m_A = 3.9$ g; $V_A = 3.0$ L
   sample B: $m_B = 7.9$ kg; $V_B = 6.0$ m^3
   (b) Are A and B gases or liquids? Explain your answer.
   (c) Could samples A and B be the same substance? How do you know?

4. Explain the difference between pressure and force. (Consider the meaning, type of quantity, and SI units.)

5. A homeowner wants to hammer the edge of a doorframe so the door will close more easily. It is wise to hold a wooden block against the doorframe before hitting it with a hammer. Apply physics principles to explain why.

6. The apparatus in **Figure 1** can be used to demonstrate the action of human lungs. A thin rubber diaphragm is stretched over the bottom of the jar.

Figure 1

(a) What stage of a breath cycle is shown in the diagram?

(b) What must you do to show the opposite stage? Describe what you would observe.
(c) Explain how this apparatus demonstrates breathing.

7. Could you drink from a straw on the Moon, where there is no atmosphere? Explain your answer.

8. A piece of Styrofoam can withstand 27 kPa of pressure without being crushed. A box of weight 6.0×10^3 N is to be placed on the Styrofoam.
   (a) Calculate the minimum area that the bottom of the box must have to prevent the Styrofoam from being crushed.
   (b) If the box is placed on its side (area = 0.10 m^2), will the Styrofoam be crushed? Explain your answer.

9. In a hydroforming process, the volume flow rate in a pipe of radius 3.6 cm is 0.24 m^3/s. Each blast of water during the process lasts for 15 s. Calculate
   (a) the total volume of water used for one blast
   (b) the speed of the water in the pipe

10. An aerosol can contains gas on top of a liquid. Explain how the gas causes the liquid to come out of the spray nozzle.

11. State the effect of each of the following on pressure:
    (a) decreasing the depth beneath the surface of a liquid
    (b) increasing the density of a liquid
    (c) going from a large lake to a swimming pool at the same depth

12. (a) If you dive from a given depth to double that depth in a swimming pool, what happens to the pressure on your ears?
    (b) If you dive the same depth in seawater as in the swimming pool, how will the pressure in your ears compare?

13. An eye specialist measures the gauge pressure in a patient's eyes and finds that one eye has a pressure of 2.1 kPa and the other has an abnormal pressure of 6.5 kPa. If the atmospheric pressure is 101.1 kPa, what is the absolute pressure in each eye?

14. Explain why a barometer that uses water needs to be 13.6 times higher than a barometer that uses mercury.

15. (a) Calculate the magnitude of the force caused by an atmospheric pressure of 101 kPa on a window that is 0.50 m square.
    (b) Explain why the window doesn't break.

16. Hold one hand high above your head and the other hand at your side. Now look at the veins on the backs of your hands. What do you observe? Explain why this happens.

17. There is a folk story in which a Dutch boy became a hero when he held back the North Sea by putting his finger into a hole in a dike. Is a person strong enough to hold back a sea? Explain your answer. (Assume the hole in the dike was about a metre beneath the surface of the water.)

18. In **Figure 2**, a person blowing into tube A would cause water to rise up tube B.
    (a) If the static pressure head of the water in B is 0.31 m above the water in the flask, what is the difference in pressure between the two water levels?
    (b) If the atmospheric pressure is 102 kPa, what is the absolute pressure in the flask?

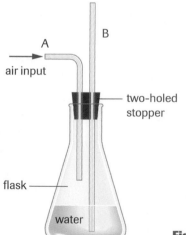

A
air input

B

two-holed stopper

flask

water

Figure 2

19. Explain the design of the dam in **Figure 3**.

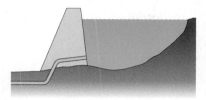

Figure 3

20. A diver may experience a popping sensation in the ears when diving to a depth greater than about 1 m. Explain why this happens.

21. Assume that 52% of a person's body weight is above the fifth lumbar vertebra. For a 64-kg person with a cross-sectional area of that vertebra of 3.0×10^{-3} m², determine the pressure in kilopascals at that location.

22. (a) State Pascal's principle.
    (b) How is this principle applied in the operation of a car braking system?
    (c) How is this application superior to the braking design used in the earliest cars manufactured?

23. Give advantages of using robots rather than humans to explore the water and mineral content of the surface of Mars.

24. **Figure 4** shows two designs of a cylinder press intended for different applications. Both are operated in a liquid power system with the same size pump, the same pressure in the liquid, and the same volume of liquid moved with each stroke.
    (a) What is the difference between the two designs?
    (b) Which design would be better for an application that requires a long time interval between power strokes?

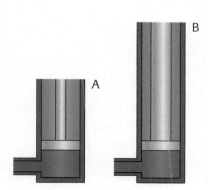

B

A

Figure 4

25. (a) Choose the appropriate symbols from **Figure 5** to draw a circuit diagram of a hydraulic power system that moves a piston back and forth in a cylinder.
   (b) How would your design change for a pneumatic rather than a hydraulic system?
   (c) What is the function of the relief valve?

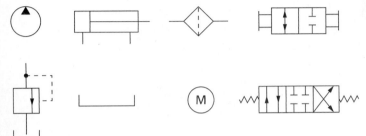

Figure 5

26. A robotic cylinder press moves back and forth in a regular cycle of power stroke and refill stroke. You are given the following quantities related to the system:
   - the volume flow rate
   - the pressure in the system
   - the cross-sectional area of the head of the piston
   - the cross-sectional area of the handle of the piston
   - the length of the piston stroke in each direction
   (a) Write the symbol used for each of the given quantities.
   (b) Write the equation you would use to determine
      - the volume of liquid that flows into the cylinder during the power stroke
      - the time interval for the power stroke
      - the magnitude of the force applied to the piston by the liquid during the power stroke
      - the work done by the liquid in the cylinder during the power stroke

27. (a) Describe the features that reduce drag on the luge toboggan and rider in **Figure 6**.
   (b) How would the drag coefficient of the toboggan and rider change if the temperature rose above 0 °C?

Figure 6

28. In the sport of sprint cycling, some athletes use outerwear with fine ribbing on the shoulders and back.
   (a) What is the purpose of the ribbing?
   (b) Name an animal that has a skin with a similar design.

29. You serve a volleyball eastward. The ball has a fast clockwise spin when viewed from above. In which direction does the ball tend to curve? Show your reasoning.

Applying Inquiry Skills

30. You have been challenged to move the water in a plastic cup into a sink using only your fingers and a straw. You cannot touch the cup or sink.
   (a) How can you do it?
   (b) Explain why your procedure should work.

31. In a sketch, illustrate how you would test Pascal's principle using a two-cylinder system and any other apparatus you need.

32. You are asked to design a pneumatic muscle that picks up a pencil the way an elephant's trunk picks up a log.
   (a) Which gas would you use in this pneumatic system? What are its advantages?
   (b) Would you use one large muscle or several small components linked together? Draw a sketch to illustrate your answer.

33. Describe how you could compare the viscosities of various liquids by applying the concept of volume flow rate. Include safety considerations in your answer.

Making Connections

34. The shape of our front teeth is different from the shape of our molars. Use physics principles to explain the difference.

35. If you cut your finger, would the bleeding stop more readily if you placed your finger high above your head or low toward the floor? Explain your answer.

36. A homeowner, cleaning up after a meal, places leftover wine into a flat-bottom glass bottle and pounds a cork into the neck. The radius of the neck is 1.1 cm and the radius of the bottom part of the bottle is 3.6 cm. With no air space, the bottle acts like a hydraulic press. The base can withstand a force of 700 N before breaking.
    (a) If the magnitude of the force on the cork is 75 N, what is the pressure in the bottle?
    (b) Will the base of the bottle break? Show your calculations.
    (c) The most common base is illustrated in **Figure 7**. Explain this shape using physics principles.

Figure 7

37. Your dentist and dental hygienist regularly use hydraulic and/or pneumatic systems and apply Bernoulli's principle.
    (a) List some examples of hydraulic and/or pneumatic systems found in a dental office.
    (b) The device that the dentist uses to remove excess liquids from your mouth is an actuator. Based on the actuator illustrated in **Figure 8**, describe how this device applies Bernoulli's principle.

38. How can the design of the actuator in **Figure 8** be altered to allow a shop vacuum cleaner to vacuum water in a flooded basement? Draw your design.

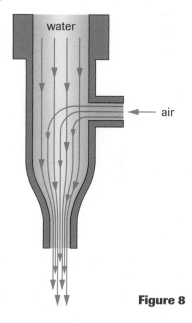

water

air

Figure 8

39. A car's braking system is a hydraulic power system. If it were a pneumatic power system, how might pushing on the brake pedal feel? Explain why.

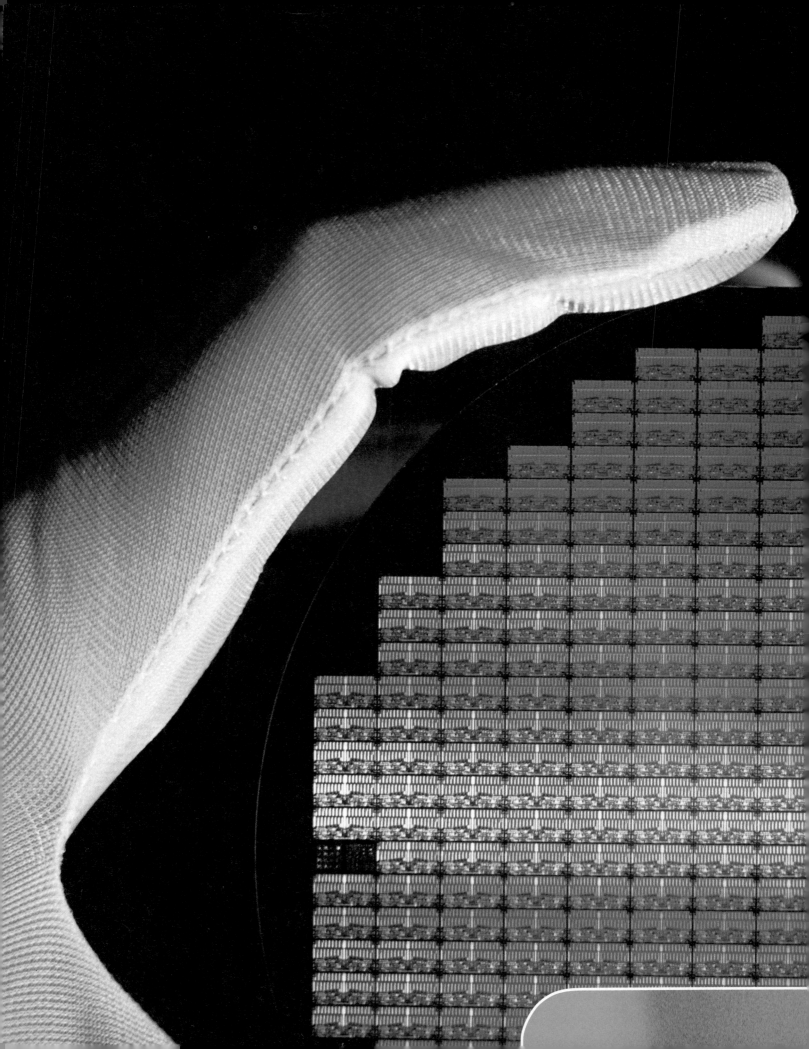

Electricity and Electronics

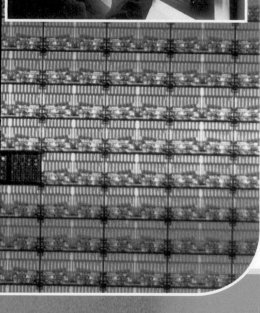

Electricity is an integral part of our lives. Electrical energy is transformed into light energy to light our streets and buildings. It is transformed into thermal energy to cook food and operate incubators. It is converted into chemical energy to electroplate metal components of cars with layers of copper, nickel, and chromium. Electrical energy is also turned into mechanical energy to operate blenders, drills, lathes, and countless other devices.

Electrical energy is commonly used in the electrical circuits of electronic devices, especially computers. Computers are the brain of car, airplane, and robotic control systems. They help us design cars and buildings, operate robots in industry, run instant banking systems, control pacemakers, compose music, operate cell phones, and run household appliances. What you learn about electricity and electronics will be useful as you consider a variety of careers.

In this unit, you will study the components that make up electrical and electronic circuits. You will construct and analyze electrical circuits. You will also explore the development and application of electrical technologies.

This unit is divided into two chapters. Chapter 7 deals with electricity, and Chapter 8 deals with electronics. What you study in these two chapters will lead to the Unit Performance Task, in which you can design an electronic device to perform a specific task.

▶ Overall Expectations

In this unit, you will be able to

- understand common applications of electrical and electronic circuits, and the function and arrangement of the components used in those circuits
- construct, analyze, and troubleshoot electrical circuits by using schematic diagrams and appropriate tools and measuring equipment
- investigate the development and application of electrical technologies
- identify and describe science- and technology-based careers related to electricity and electronics

Knowledge and Understanding

1. (a) **Figure 1** is a model of a copper atom. Copy the diagram into your notebook, and label as many components as you can.
   (b) Which particle produces the electric current in wire conductors?
   (c) Is the particle named in (b) positively charged, negatively charged, or neutral?

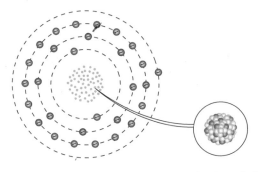

Figure 1

2. Give examples to show you understand the difference between static and current electricity.

3. In the equation $V = IR$,
   (a) What do the symbols V, I, and R stand for?
   (b) What are the units used to measure these quantities?
   (c) Draw an electrical circuit diagram in which a DC battery is used to operate a light bulb. Include the instruments used to measure the voltage across the bulb and the current through it.

4. What are some examples of voltages used in and around a home? For each voltage, name an appliance or device that uses that voltage.

5. In an electrical circuit, state the function of
   (a) the power source
   (b) a switch
   (c) a fuse

6. What is the difference between an analog scale and a digital scale? Draw a diagram to illustrate each. (You can draw thermometers, voltmeters, or other instruments.)

7. **Figure 2** shows two electrical circuits.
   (a) How many paths for electric current does each circuit have?
   (b) State what the symbols numbered 1 to 7 represent.

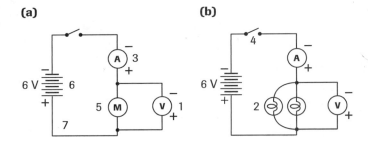

Figure 2

(c) Which circuit forms the basis of the type of circuit used in homes? What is a major advantage of this type of circuit?

Inquiry and Communication

8. Name the instrument used in the laboratory to measure
   (a) electric current
   (b) electric potential difference

9. You are given a 9-V cell, two switches, two light bulbs, and connecting wires.
   (a) Draw a diagram showing how to connect these components so that each bulb is controlled separately by a switch.
   (b) Repeat (a) using symbols in a circuit diagram.

10. Describe the difference between an open circuit and a short circuit. (Diagrams may be used.)

Math Skills

11. Determine the percent efficiency of a 60-W light bulb that uses 11 kJ of electrical energy to produce 2.1 kJ of light energy.

12. Determine the unknown x in each of the following, and express your answer as a proper fraction:
   (a) $\dfrac{1}{x} = \dfrac{1}{2} + \dfrac{1}{8}$ (b) $\dfrac{1}{x} = \dfrac{1}{2} + \dfrac{1}{4} + \dfrac{1}{8}$

Technical Skills and Safety

13. **Figure 3** shows an analog ammeter that has three scales and three negative terminals. State the current when the wire is connected to the
   (a) 0–50-mA scale
   (b) 0–500-mA scale
   (c) 0–5-A scale

14. Why is it unwise to connect the positive and negative terminals of a battery together?

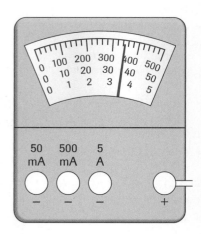

Figure 3

Making Connections

15. If a fuse in an electrical circuit burns out, what steps would you take to troubleshoot the situation?

16. Match the following activities to either electricity or electronics:
   (a) repairing hydro lines after an ice storm
   (b) troubleshooting a computer chip board
   (c) inspecting the work done by electricians who wire a new house
   (d) training technicians who test cell phones
   (e) the situation shown in **Figure 4**

Figure 4

Current Electricity

Getting Started

- define and describe the concepts and units related to electrical systems
- describe the function of basic electrical circuit components
- compare direct and alternating current
- analyze and describe the operation of electrical devices that control other systems
- analyze, in quantitative terms, circuit problems involving electric current, potential difference, and resistance
- measure electric current, potential difference, and resistance
- safely construct electrical circuits
- draw schematic diagrams of electrical circuits
- analyze electrical circuits using Ohm's law and Kirchhoff's rules
- design, construct, and evaluate an electrical circuit to perform a function
- analyze electrical circuits to identify faults, and suggest corrections
- describe applications of electrical circuits and identify energy transformations
- identify proper safety procedures to be followed when working with electrical circuits and potential electrical hazards

Electrical energy is very useful, but only when used wisely and safely. Every year, fires started by electrical problems result in loss of life and property (**Figure 1**). In addition to learning about the concepts and relationships related to current electricity, you will learn how to troubleshoot electrical problems.

In this chapter, you will compare direct and alternating current, and analyze electric current, potential difference, resistance, power, and energy. You will also discover how electrical components can be joined to create circuits that accomplish specific functions. You will perform experiments with laboratory equipment or use computer simulations to analyze electrical circuits. Then you will apply what you learn by designing and constructing a basic electrical circuit. The principles you explore in this chapter are applied in the design, troubleshooting, and safe use of many devices you use every day.

Figure 1
This fire, like so many started by the unsafe use of electrical energy, could have been prevented.

⚡ *REFLECT* on your learning

1. Which of the following devices use direct current and which use alternating current: a hair dryer; a portable CD player; a car headlight; a refrigerator; the lights in your classroom; the light in a camera flash.

2. A circuit breaker is labelled 15 A.
   (a) What does the "A" represent?
   (b) What quantity does it measure?

3. If you already know the quantity of electrical energy used, what other quantity do you need to calculate the quantity of electrical power? (Assume the power is constant.)

4. What safety features are illustrated in the electrical outlet in **Figure 2**?

5. Describe the energy transformations and write the energy-transformation equations for using (a) a kettle and (b) a flashlight.

6. Which is more dangerous to a person's safety, a high electric voltage or a high electric current? Explain your answer.

Figure 2
An electrical outlet

▶ *TRY THIS* activity   *Electrical Safety*

Your teacher will either give you printed diagrams or set up lab stations with electrical components or devices. For each diagram or setup,

(a) identify any hazards and explain how to avoid them

(b) describe the purpose of any tools (Photographs of some tools are shown in **Figure 3**.)

(c) describe the purpose of any safety apparatus

Be prepared to discuss the hazards and the use of tools and safety apparatus in class.

Figure 3

Figure 1
When exposed to solar energy, the solar cells in the arrays generate an electric current. This current powers the station and charges the batteries, which are used when the *ISS* is not in sunlight.

current electricity the flow of electric charges

electrical circuit an arrangement of components that transform electrical energy into some other form of energy in an electrical device

open circuit an electrical circuit that is not complete, so there is no current

source an energy-transformation device that transforms one form of energy into electrical energy; also called an energy source

electrical conductor a substance through which electrons can easily move

load a device that transforms electrical energy into another form of energy

control a switch for starting and stopping the current

Every time you use a computer, listen to the radio, or use a microwave oven, you are using **current electricity**, which is the flow of electric charges. The energy that is transformed into the electrical energy in current electricity can come from a variety of sources (Chapter 3). For example, on the *International Space Station*, solar arrays receive radiant energy from the Sun and transform that energy into electrical energy (**Figure 1**).

Recall from earlier studies that all substances are composed of atoms, a model of which is shown in **Figure 2**. The electron has a negative charge, the proton has a positive charge, and the neutron has a neutral charge. When wires conduct electric current, the electrons already in the wires get pushed through them. ("Electric current" is used informally here to mean the flow of electric charges. A more formal definition will be introduced in section 7.2.)

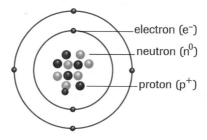

electron (e⁻)
neutron (n⁰)
proton (p⁺)

Figure 2
A simple model of a carbon atom

Electric current becomes useful when it is controlled in an *electrical circuit*. In an **electrical circuit**, an arrangement of components transforms electrical energy into another form of energy in an electrical device. A basic electrical circuit, like the one in **Figure 3**, consists of three main parts: the *source* (or *energy source*), *conductors,* and *load*, and sometimes an optional part, a *control*.

The switch in **Figure 3** is open, resulting in an **open circuit**. This means that the electric current does not flow. As soon as the switch is closed, the circuit

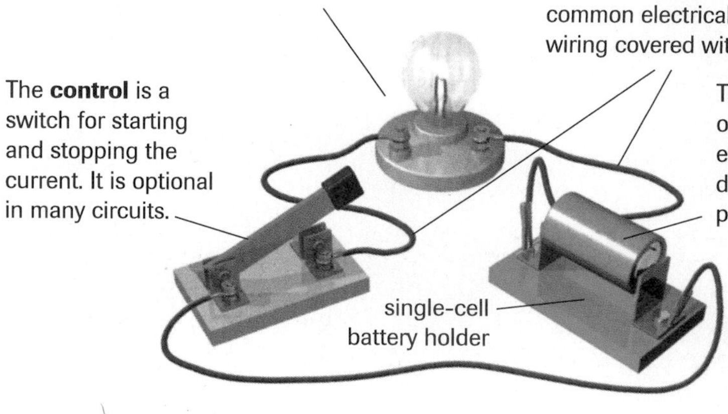

The **load** is a device that transforms electrical energy into another form of energy, such as kinetic energy, thermal energy, light energy, or sound energy.

An **electrical conductor** transmits the electrical energy, so the conductor must be made of a material that allows electrons to flow freely in a path. A common electrical conductor is copper wiring covered with electrical insulation.

The **control** is a switch for starting and stopping the current. It is optional in many circuits.

The **source** transforms one form of energy into electrical energy (e.g., a dry cell, a battery, and a power supply).

single-cell battery holder

Figure 3
A simple electrical circuit

becomes a complete path, or **closed circuit**, which allows the electrons to be pushed along. If the conductor from one side of the cell is joined directly to the other side of the cell so that the load is bypassed, a **short circuit** results. A short circuit can cause the cell to overheat and burn out. It can also cause the conductor to get dangerously hot because of the large current generated.

As you learned in Chapter 3, energy transformations can be summarized using energy-transformation equations. When the circuit in **Figure 3** is closed, chemical potential energy in the cell (the source) is transformed into electrical energy, which in turn is transformed into useful light energy and thermal energy in the load. The energy-transformation equation is

$$E_{chemical} \rightarrow E_{electrical} \rightarrow E_{light} + E_{thermal}$$

> ▶ **TRY THIS** activity

Analyzing Simple Electrical Circuits

Obtain a battery-operated flashlight and take it apart.

(a) Draw a diagram of the electrical components of the flashlight.

(b) How does the circuit in the flashlight compare with the circuit shown in **Figure 3**?

In Chapter 5, you drew circuit diagrams for hydraulic and pneumatic systems. However, the symbols used to draw electrical circuits are different from those symbols. **Figure 4** illustrates the simple circuit from **Figure 3** in standard symbols. Refer to Appendix C, **Table 6**, for some of the many symbols used to draw electrical circuits.

> ▶ **Practice**

Understanding Concepts

1. Name a source that generates electrical energy from
   (a) chemical potential energy
   (b) gravitational potential energy
   (c) radiant energy

2. Name five electrical loads that you have experienced in the past few days.

3. Write the energy-transformation equation for each of the following:
   (a) operating a solar-powered calculator
   (b) heating water in an electric kettle
   (c) listening to a portable CD player

4. Draw the electrical circuit symbol for
   (a) a variable energy source
   (b) a variable resistor

Applying Inquiry Skills

5. Some novelty stores sell clocks that operate when connected to food, such as potatoes or fruit (**Figure 5(a)**). When two electrodes are inserted into the food, it acts like an electrolyte in a chemical cell. (An electrolyte is a substance that acts as a conductor in a water solution.) With your teacher's permission, create your own chemical cell as shown in **Figure 5(b)**; use a

closed circuit an electrical circuit forming a complete path for the current

short circuit an error in which a conductor is connected across a source or other component, bypassing the load; sometimes just called a "short"

LEARNING TIP

Comparing Circuits
An electrical circuit can be compared to the circuit of an enclosed fluid system, described in Chapter 5. The electrical source corresponds to the pump or compressor; the conductors correspond to the transmission lines; the switches correspond to the valves; and the load corresponds to the actuator.

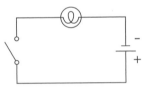

Figure 4
The circuit diagram of the electrical circuit in **Figure 3**

small light bulb to test the electrical output. Determine whether the cell can light up a small light bulb or an LED.

(a) Describe what you discover about your chemical cell.

(b) List the disadvantages of this source of energy.

(c) List the advantages of this source of energy.

Figure 5

(a) The potatoes in this clock transform chemical potential energy into electrical energy.

(b) Electrodes made of different metals, inserted into a piece of fruit, can generate electrical energy.

(a)

(b)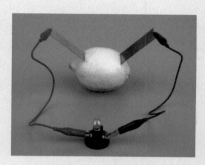

SUMMARY   *Electrical Circuits*

- In current electricity, an energy source pushes electrons in a circuit through conductors and control devices; at the load, electrical energy is transformed into another form of energy.

- Electrical circuits are drawn using symbols.

- Energy-transformation equations summarize the transformations that occur when energy is transformed into electrical energy, which in turn is transformed into useful forms of energy in an electrical load.

► Section 7.1 Questions

Understanding Concepts

1. (a) Which electric charges flow in the conductors of electrical circuits?
   (b) What charge do these particles have?

2. Write the energy-transformation equation for each of the following:
   (a) A battery operates an electric motor on a toy car.
   (b) Two metal electrodes are inserted into a lemon and connected to a light bulb, lighting it up.
   (c) An electric fan cools a room on a hot day.

Applying Inquiry Skills

3. (a) Using appropriate circuit symbols, draw an open circuit diagram of an electrical circuit in which a cell is used to operate an electric light bulb. The current is controlled by a switch.
   (b) Repeat (a) as a closed circuit.

Making Connections

4. Another means of transforming mechanical energy into electrical energy is *piezoelectricity*. In this method, a mechanical force applied to layers of crystals forces charges to opposite sides. Piezoelectric technology is used in crystal microphones, some radiation detectors, force plates for testing the techniques of athletes, force sensors, and some cigarette lighters. Research piezoelectricity, and briefly describe one application.

 www.science.nelson.com

5. Primary cells and secondary (rechargeable) cells are examples of chemical cells. Investigate the historical development and use of chemical cells.
   (a) Describe what you discover about voltaic cells.
   (b) Describe how modern cells are safer than the first chemical cells developed.
   (c) How do the costs and availability of non-rechargeable and rechargeable cells and batteries compare?
   (d) What are the environmental impacts of using present-day cells and batteries?

Electric current can be compared to the volume flow rate of water (**Figure 1**). Recall from Chapter 5 that the volume flow rate is expressed in units of volume per second, for example, litres per second (L/s). **Electric current** (symbol I) is a measure of the number of electric charges that pass by a particular point in a circuit each second. The SI unit of measurement of electric current is the **ampere**, or **amp** (symbol A). This unit is named in honour of French physicist André Marie Ampère (1775–1836).

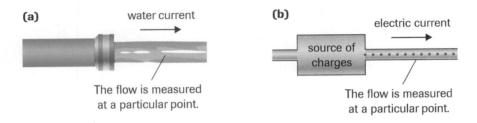

(a) water current

The flow is measured at a particular point.

(b) electric current

source of charges

The flow is measured at a particular point.

Figure 1
(a) Volume flow rate of water is a measure of the volume of water flowing past a given point each second.
(b) Electric current is a measure of the number of electric charges flowing past a given point each second.

electric current a measure of the number of electric charges that pass by a particular point in a circuit each second; current = $\dfrac{\text{charge}}{\text{time}}$

ampere (amp) the SI unit of measurement of electric current; symbol A

direct current (DC) an electric current in a single direction

alternating current (AC) an electric current that reverses direction periodically

Some visualize electric current as an actual continuous flow of charges that start at the energy source, travel through wires to the load, and then return to the source. Electric current does not happen that way. Atoms with electrons in their outer regions exist everywhere in the circuit. The electrons throughout the wires are ready to move as soon as the switch is turned on, just like water is ready to flow before a tap is turned on. To produce a current, all the electrons need is a push (the electric force) from the source, which occurs as soon as the circuit is closed. As a result of the push, the electrons that are in the load give up energy.

If electric charges move in a path without ever reversing direction, the current is called **direct current**, or **DC** (**Figure 2**). Battery-powered devices, such as cell phones, personal CD players, palm pilots, and electrical systems in cars, use DC. Most of the laboratory activities in this chapter also use DC.

If the charges in a circuit are forced to reverse direction periodically, the current is called **alternating current**, or **AC** (**Figure 3**). This is the type of current generated at electrical generating stations. The generators force the charges in the wire conductors back and forth. The electric charges apply forces on neighbouring charges throughout the circuit. Most of our industrial plants and household appliances use AC.

LEARNING *TIP*

DC Cells and Batteries
A chemical cell transforms chemical potential energy into DC. Two or more cells connected together form a battery of cells, called a battery. Cells and batteries can be classified as primary or secondary. A primary cell cannot be reactivated, but a secondary cell or battery can be.

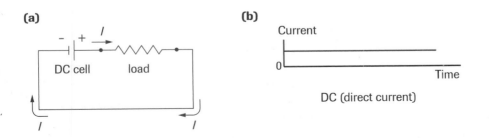

(a)

DC cell   load

(b)

Current

0

Time

DC (direct current)

Figure 2
(a) The circuit diagram of a simple DC electrical circuit
(b) The current–time graph of DC

Figure 3
(a) The circuit diagram of a simple AC electrical circuit
(b) The current–time graph of AC

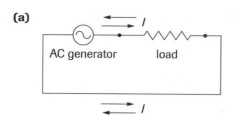

(a)

AC generator   load

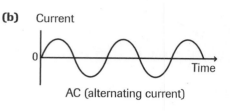

(b) Current

0

Time

AC (alternating current)

AC is used in our electrical systems because, with our present technology, it is easier to transmit than DC. Electrical energy produced by generators can be transmitted long distances by large networks of power lines and transformers.

Small portable devices use very low currents. For example, a battery-powered calculator may operate with a current of 0.002 A (2 mA). Home appliances use higher currents. For example, a colour TV may operate at 4 A. Appliances that produce a lot of heat require even higher currents. **Table 1** lists typical currents of some common electrical loads.

DID YOU KNOW ?

AC Frequencies
AC generators in North America force the electric charges to reverse direction at a frequency of 60 Hz. Some European countries use a frequency of 50 Hz. In the early days of electricity in North America, AC was produced at 25 Hz. When the system switched to 60 Hz, all devices with electric motors had to be changed, at great cost and inconvenience.

  www.science.nelson.com

Table 1   Examples of Electric Currents of Common Loads

Electrical Device	Electric Current (A)	Type of Current
battery-powered wristwatch	1.3×10^{-4}	DC
battery-powered calculator	2×10^{-3}	DC
electric clock	0.16	AC
flashlight	0.4	DC
light bulb (60 W)	0.5	AC
colour TV	4.0	AC
electric drill	4.5	AC
stove element	6.8	AC
vacuum cleaner	8.5	AC
oven element	11	AC
hair dryer	12	AC
toaster	13	AC
water heater	27	AC
car starter motor	500	pulsating DC

▶ **Practice**

Understanding Concepts

1. State whether AC or DC is used to operate a
   (a) home TV
   (b) portable computer game
   (c) car TV
   (d) fluorescent light bulb

Applying Inquiry Skills

2. Symbols should communicate ideas in a quick, simple way. Does the symbol for AC generators communicate effectively? (Refer to **Figure 3(a)** or Appendix C, **Table 6**.) Explain your answer.

Measuring Electric Current

When the direction of DC was first defined, scientists did not know that it was the electrons that pushed each other along the wires. By convention, they agreed that electric current was the flow of positive charges. The resulting current is still called **conventional current**, with the symbol I (**Figure 4(a)**). Conventional current starts from the positive terminal of the source, moves through the circuit, then returns to the negative terminal of the source. The confusing part is that in metal conductors it is the electrons, the negative charges, that flow. This current is called **electron flow**, with the symbol e^-. Thus, electron flow is opposite in direction to conventional current. As shown in **Figure 4(b)**, electron flow starts from the negative terminal of the source, and electrons everywhere in the circuit get pushed along, entering the positive terminal of the source.

(a)

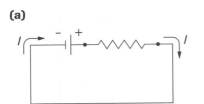

(b)

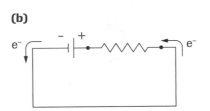

The instrument used to measure electric current is called an **ammeter**. It must be connected directly in the path of the moving charges. This type of connection is called a **series connection**. **Figure 5** shows the correct way to connect it. In the circuit shown, current leaves the positive terminal of the source; farther along the circuit, the current enters the positive terminal of the ammeter. If the ammeter is analog, its needle swings to the right on the scale. (If the ammeter is connected incorrectly, the needle moves backward, and the meter could be damaged.) If the ammeter is digital, it gives a positive reading. At the load (a light bulb in this case), the electrical energy is transformed into light energy and thermal energy. At the source, charges gain energy and once again push on the charges in the circuit.

LEARNING TIP

Current in Liquids, Gases, and Plasmas
Conventional current and electron flow apply to the flow of charges in solids. In liquids, gases, and plasmas (ionized gases), both positive and negatives charges can flow. For example, in the plasma in a fluorescent tube, positive charges and negative charges move in opposite directions; the total current is the sum of the individual currents. This is an important reason why we still use the concept of conventional current today, even though we know that in metal conductors it is the negative charges that flow.

conventional current the flow of positive charges in a circuit; symbol I

electron flow the flow of negative charges in a circuit; symbol e^-

Figure 4
(a) The direction of conventional current, I
(b) The direction of electron flow, e^-

ammeter the instrument used to measure electric current

series connection an electrical connection in which the current in a circuit moves in one path

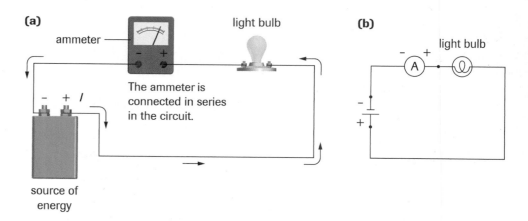

(a)

ammeter

light bulb

The ammeter is connected in series in the circuit.

_ + I

source of energy

(b)

light bulb

Figure 5
(a) The ammeter is very sensitive and must be connected as shown.
(b) The corresponding circuit diagram

▶ **TRY THIS** *activity*

Analog and Digital Ammeters

Analog ammeters have a single scale or, in some cases, multiple scales. Each scale corresponds to a particular range of currents. Digital ammeters and digital multimeters have a selector knob that indicates a select range of currents. To learn or review how to interpret analog scales and how to use digital ammeters and multimeters, refer to Appendix A1. Obtain both an analog and a digital ammeter or multimeter. Draw sketches of the meters, and indicate how you would connect them in a simple electrical circuit to measure the current.

▶ **Practice**

Understanding Concepts

4. Compare and contrast conventional current and electron flow.

5. In a series connection, what is the maximum number of paths available for the movement of charges?

6. Convert the following measurements as indicated, assuming two significant digits:
    (a) 650 mA = ? A   (c) 4500 mA = ? A
    (b) 85 mA = ? A   (d) 0.095 A = ? mA

Applying Inquiry Skills

7. Describe, in words, how the positive and negative terminals of an ammeter are connected in a circuit relative to the positive and negative terminals of the source.

Making Connections

8. (a) Copy the circuit diagram in **Figure 5(b)** into your notebook, and show how the conductors could be connected to cause a short circuit across the light bulb.
    (b) Explain why a short circuit would damage the ammeter.

Answers

6. (a) 0.65 A
    (b) 0.085 A
    (c) 4.5 A
    (d) 95 mA

SUMMARY   *Electric Current*

- Electric current, measured in amps, is a measure of the number of electric charges that pass by a particular point in a circuit each second; $current = \dfrac{charge}{time}$.

- Direct current (DC) is used in most portable or battery-operated devices. Alternating current (AC) is used in most household and industrial applications.

- Conventional current (I) leaves the positive terminal of a source and moves through the circuit. Electron flow (e^-) moves in the opposite direction.

- Electric current is measured with an ammeter connected in series at some point in the circuit.

▶ *Section 7.2* *Questions*

Understanding Concepts

1. (a) Compare the motion of electric charges in AC and DC.
   (b) Name three devices not yet mentioned that use DC and three that use AC.

2. Convert the following measurements as indicated, assuming three significant digits:
   (a) 2.51 mA = ? A
   (b) 0.995 A = ? mA
   (c) 885 mA = ? A

Applying Inquiry Skills

3. Using proper circuit symbols, draw an electrical circuit showing a DC cell, a switch, a light bulb, and the instrument used to measure the electric current through the bulb. Show all positive and negative signs, and indicate the directions of the conventional current and electron flow.

4. Do you get what you pay for when you buy a cell or a battery? Describe the quantities you must know or measure in an investigation to compare the value of a variety of D cells or a similar product.

Making Connections

5. For each of the following currents, name one electrical device that could have that current. Justify your answers.
   (a) 0.4 A (DC)
   (b) 7.0 A (AC)
   (c) 25 A (AC)
   (d) 5.0 mA (DC)

To help visualize how energy is distributed in an electrical circuit, we can compare an electrical circuit to an isolated fluid system. In the water system in **Figure 1(a)**, the circulating water causes the water wheel to rotate. The pump exerts a force on the water, and it does work to raise the water to the higher level. At that level, the water has gravitational potential energy. The water then falls toward the wheel, and the gravitational potential energy is transformed into the kinetic energy of the rotating wheel. As long as the water circuit operates, the wheel continues to rotate. Because the system is isolated, the water flow rate (e.g., 5 L/min) is the same at every point in the circuit.

(a)

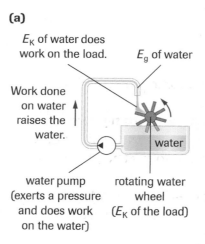

E_K of water does work on the load.

E_g of water

Work done on water raises the water.

water

water pump (exerts a pressure and does work on the water)

rotating water wheel (E_K of the load)

(b)

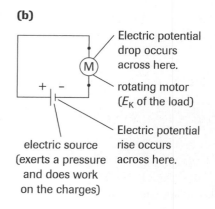

M

Electric potential drop occurs across here.

rotating motor (E_K of the load)

+ −

electric source (exerts a pressure and does work on the charges)

Electric potential rise occurs across here.

Figure 1
(a) In this water flow system, a pump does work on the water, giving it gravitational potential energy.
(b) Comparing an electrical circuit to the water flow system.

electric potential rise a measure of the amount of energy per charge given by the source; symbol ΔV

electric potential drop a measure of the amount of energy per charge given to a load; symbol ΔV

electric potential difference either a potential rise or a potential drop; symbol ΔV

volt the SI unit of electric potential difference; symbol V

voltmeter the instrument used to measure electric potential difference

parallel connection an electrical circuit in which the current follows two or more paths

In the electrical circuit in **Figure 1(b)**, the source, which corresponds to the water pump, does work on the electric charges, giving them electrical energy. The charges push on adjacent charges throughout the circuit. **Electric potential rise** is a measure of the amount of energy per charge given by the source. At the load, electrical energy is transformed into mainly kinetic energy. The load, in this case the motor, corresponds to the water wheel. **Electric potential drop** is a measure of the amount of energy per charge given to a load. An **electric potential difference** is either a potential rise or a potential drop.

Both potential rise and potential drop are represented by the symbol ΔV because there is a change in potential from one side of a source or load to the other side. The SI unit of electric potential difference is the **volt** (symbol V). This unit is named after the Italian scientist Alessandro Volta (1745–1827), hence the term *voltage* for electric potential difference. (Similarly, electric current is sometimes called *amperage*.)

The instrument used to measure electric potential difference is called a **voltmeter**. To measure potential rise, the voltmeter must be connected across the source. To measure potential drop, the voltmeter must be connected across the load. This type of connection, called a **parallel connection**, is

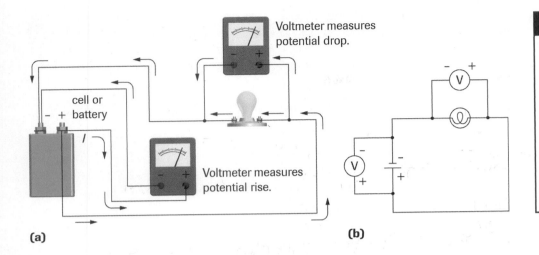

(a) Voltmeter measures potential drop.

cell or battery

Voltmeter measures potential rise.

(a)

(b)

Figure 2
(a) Measuring electric potential rise and potential drop
(b) The corresponding circuit diagram

illustrated in **Figure 2**. Compare this parallel connection to the series connection for ammeters in section 7.2, **Figure 5**, page 322.

Notice in **Figure 2** that the electric current leaves the positive terminal of the source and enters the positive terminal of each voltmeter. This will cause a positive reading in a digital voltmeter; in an analog voltmeter, it will cause the needle to move to the right.

DC cells and batteries have low potential rises. AC circuits in North American households have two potential rises: either 120 V or 240 V. Examples of these voltages are listed in **Table 1**.

Table 1 Examples of Voltages of Common Loads

Electrical Device	Electric Voltage (V)	Type of Circuit
signal from human brain	5.0×10^{-5}	DC
neuron in nervous system	0.075	DC
flashlight	3.0	DC
camera with flash	6.0	DC
smoke alarm	9.0	DC
boat battery	12	DC
some car batteries	43	DC
DVD player	120	AC
electric oven	240	AC
car's spark plugs	25 000	DC
local transformer station	44 000	AC
high-voltage transmission lines	500 000	AC
lightning	1×10^9	DC

Understanding Concepts

1. Describe, in words, how the positive and negative terminals of a voltmeter are connected in a circuit relative to the terminals of the source.

2. Convert the following measurements as indicated, assuming two significant digits:
   (a) 500 mV = ? V
   (b) 1.5 kV = ? V
   (c) 64 µV = ? V
   (d) 0.15 V = ? mV

Applying Inquiry Skills

3. Obtain an analog voltmeter and a digital voltmeter. Draw sketches of the meters and compare them to analog and digital ammeters. (To review how to read analog scales, refer to Appendix A1.)

Making Connections

4. (a) What potential rise is provided by cells that are labelled AAA, AA, A, C, and D (**Figure 3**)?
   (b) In what devices would you find these cells?

5. A motorboat's lights have been left on, draining its 12-V battery to the point that the engine won't start (**Figure 4**). However, a voltmeter reading indicates that the battery is still 12 V. Explain how this is possible. (Hint: Include the concept of "cranking current" in your answer.)

Figure 3

Figure 4

Answers

2. (a) 0.50 V
   (b) 1.5×10^3 V
   (c) 6.4×10^{-5} V
   (d) 1.5×10^2 mV

SUMMARY **Electric Potential Difference**

- Electric potential difference, or voltage, is a measure of the amount of energy per charge given by a source (potential rise) or given to a load (potential drop).

- Electric potential difference is measured in volts with a voltmeter connected in parallel across the device.

▶ *Section 7.3 Questions*

Understanding Concepts

1. To compare electric current and potential difference, set up and complete a table with the following titles: Definition; Symbol; SI Unit and Symbol; Instrument Used to Measure Quantity; Symbol of Instrument; Other Facts.

2. Convert the following measurements as indicated, assuming three significant digits:
   (a) 125 mV = ? V
   (b) 126 000 mV = ? V = ? kV

3. You are the director of a troupe of actors planning a one-act play for a grade 6 science class studying electricity. Your troupe must act out what happens to the electric charges in a DC electrical circuit in which the load is a portable CD player. You want the students to understand the energy transformations in the circuit. How would you design the play?

Applying Inquiry Skills

4. Using proper circuit symbols (Appendix C, **Table 6**), draw an electrical circuit showing a DC cell, a switch, a light bulb, and the electrical instruments used to measure electric current and potential differences. Show all positive and negative signs, and indicate the direction of conventional current.

Making Connections

5. For each of the following voltages, name one electrical device that could have the voltage. Justify each answer.
   (a) 3.0 V (DC)
   (b) 120 V (AC)
   (c) 240 V (AC)
   (d) 9.0 V (DC)

6. Battery testers and indicators are used for DC sources, such as 9-V batteries and car batteries (**Figure 5**). Do these testing devices indicate the electric current, voltage, or both? Explain your answer.

Figure 5
A standard car battery

electric resistance a measure of how much an electrical component opposes the flow of electric charges; symbol *R*

You have seen that electric current and potential difference can be compared with quantities that express water flow. But it is important to relate electric current and potential difference to each other. A quantity that does this is *resistance*. **Electric resistance** (symbol *R*) is a measure of how much an electrical component opposes the flow of electric charges. The greater the resistance, the greater the amount of energy each charge gives up as it passes through the component.

An incandescent light bulb is a good example of electric resistance. **Figure 1** shows the inside of a bulb. The thicker outer wires have low resistance and gain little energy from the electric charges. The thin, coiled wire strung across the top has a high resistance. It gains a large amount of energy from the charges, becomes hot, and emits light energy. If the coiled wire were replaced with straight copper wiring, the bulb would not become hot.

A material with a low resistance is a good electrical conductor (discussed in section 7.1); it readily allows the transfer of electric charges. A material that prevents the transfer of electric charges is an **electrical insulator**. **Table 1** lists common electrical conductors and insulators.

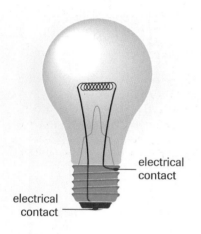

electrical contact

electrical contact

Figure 1
The thin coiled wire is used because of its high resistance.

electrical insulator a material that prevents the transfer of electric charges

The SI unit of measurement of electric resistance is the **ohm** (symbol Ω, the Greek letter omega). This unit is named in honour of German physicist Georg Simon Ohm (1787−1854). Resistance can be measured using an ohmmeter or the resistance setting on a multimeter. Resistance values of typical loads are given in **Table 2**. Conducting wires have a very low resistance. Appliances used to generate heat and light require a higher electric current, so they tend to have a relatively low resistance. Devices used to control electrical and electronic devices, such as wireless phones and calculators, require a very low current, so they tend to have a high resistance.

ohm the SI unit of electric resistance; symbol Ω

Table 1   Electrical Conductors and Insulators*

Conductors	Insulators
aluminum	air (dry)
copper	glass
gold	paper
iron	rubber
nickel	silk
silver	wool

*(in alphabetical order)

Table 2   Typical Resistances of Electrical Loads

Load	Resistance (Ω)
75-m extension cord (16-gauge copper)	1.0
toaster oven	8.6
food dehydrator	26
coffee grinder	100
light bulb (100 W)	144
person in bathtub	500
calculator	1500
dry person	600 000

Resistors are devices that have a known resistance. They are often used in electrical and electronic devices and in science laboratories. Two common materials used to make resistors are wire and granulated carbon. **Figure 2** illustrates three types of resistors: wire-wound resistance coils of constant value, colour-coded carbon resistors of constant value, and a variable wire-wound resistor called a *rheostat*.

resistor a device that has a known resistance

(a)

(b)

(c)

Figure 2
(a) Resistance coils
(b) The resistance in colour-coded resistors can be determined by interpreting the colour rings.
(c) Variable resistance (or rheostat)

▶ **SAMPLE** problem **1**

Colour-Coded Resistors

Using the data in **Table 3**, determine the resistance value and tolerance for the colour-coded resistor in **Figure 3**.

Solution
From left to right, coloured rings 1 and 2 are the significant digits; ring 3 is the multiplier or divider; and ring 4 is the tolerance. For the resistor in **Figure 3**,

ring 1 is brown = 1   ring 3 is orange = 1000
ring 2 is red = 2   ring 4 is silver = ± 10%

The resistance, including tolerance, is 12 × 1000 Ω ± 10%, or 1.2 × 10^4 Ω ± 10%.

Table 3   Colour-Coded Resistors

Ring Colour	Digits (Rings 1 and 2)	Multiplier or Divider (Ring 3)	Tolerance (Ring 4)
black	0	10^0 or 1	–
brown	1	10^1 or 10	–
red	2	10^2 or 100	–
orange	3	10^3 or 1000 (1 k)	–
yellow	4	10^4 or 10 k	–
green	5	10^5 or 100 k	–
blue	6	10^6 or 1000 k (1 M)	–
violet	7	10^7 or 10 M	–
grey	8	10^8 or 100 M	–
white	9	10^9 or 1000 M (1 G)	–
gold	–	10^{-1} or 1/10	± 5%
silver	–	10^{-2} or 1/100	± 10%
no colour	–	–	± 20%

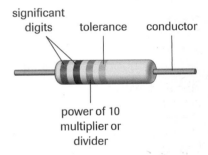

significant digits   tolerance   conductor

power of 10 multiplier or divider

Figure 3
A colour-coded resistor

Understanding Concepts

1. Using examples, describe the difference between
   (a) electrical conductors and insulators
   (b) a fixed resistance and a rheostat

2. Look at the pattern of load resistances in **Table 2**, page 328. Would a 60-W light bulb have a lower or higher electric resistance than a 100-W bulb? Explain your answer.

Applying Inquiry Skills

3. Three colour-coded resistors are used in an experiment. State each resistance and tolerance if the colours are
   (a) red; red; black; silver
   (b) blue; orange; black; gold
   (c) brown; green; brown

4. An ohmmeter can be used to measure the resistance of the human body, which varies with the amount of moisture on the skin. This is the principle behind the *polygraph,* or lie detector. A stressful situation, such as telling lies, causes one to perspire. Perspiration lowers the electric resistance of the skin. With your teacher's permission, you can test your own resistance using an ohmmeter or a multimeter. First test the resistance with dry skin and then with moist skin.

Answers

3. (a) $22\ \Omega \pm 10\%$
   (b) $63\ \Omega \pm 5\%$
   (c) $150\ \Omega \pm 20\%$

Ohm's Law

Electric resistance can be measured directly using an ohmmeter (or the resistance setting on a multimeter). But more commonly, it is calculated using the relationship between electric resistance, current, and potential difference developed by Ohm. This relationship is called *Ohm's law.*

Ohm's Law
For many devices, the ratio of the electric potential difference across a resistor to the current through it is constant if the temperature remains constant. The constant value is the resistance.

Because resistance is the ratio of the potential drop to the current, Ohm's law can be written in equation form:

$$\text{resistance} = \frac{\text{potential difference}}{\text{current}}, \quad \text{or} \quad R = \frac{\Delta V}{I},$$

where ΔV is measured in volts (V), I in amps (A), and R in ohms (Ω).

A resistor that obeys Ohm's law is called *ohmic.* A resistor that does not obey Ohm's law, because its resistance changes as the temperature changes, is called *nonohmic.* In experiments, a graph of electric potential drop (vertical axis) versus current for an ohmic resistor is a straight line (**Figure 4**). The slope of the straight line, $\dfrac{\Delta V}{\Delta I}$, equals the resistance of the resistor. The voltage–current graph of a nonohmic resistor does not remain straight. You can use this fact to test resistors in Investigation 7.5.

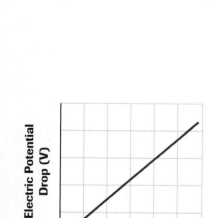

Figure 4
The graph of electric potential drop versus current for a resistor that obeys Ohm's law

▶ **SAMPLE** problem **2**

Calculating Resistance

A heating coil on an electric stove uses 25 A of current from a 240-V circuit. Calculate the coil's resistance.

Solution

$\Delta V = 240$ V
$I = 25$ A
$R = ?$

$$R = \frac{\Delta V}{I}$$

$$= \frac{2.4 \times 10^2 \text{ V}}{25 \text{ A}}$$

$$R = 9.6 \ \Omega$$

The coil's resistance is 9.6 Ω.

▶ **Practice**

Understanding Concepts

5. Calculate the resistance in each of the following cases:
   (a) $\Delta V = 3.0$ V; $I = 0.40$ A
   (b) $\Delta V = 120$ V; $I = 8.0$ A
   (c) $\Delta V = 240$ V; $I = 32$ A
   (d) $\Delta V = 9.0$ V; $I = 5.0$ mA

6. Rearrange the equation $R = \frac{\Delta V}{I}$ to solve for (a) ΔV and (b) I.

7. The current through a calculator is 4.0×10^{-3} A. If the resistance of the calculator is 1.5×10^3 Ω, what is the potential drop across it?

8. The potential drop across a 26-Ω food dehydrator is 120 V. Calculate the current through the dehydrator.

Answers

5. (a) 7.5 Ω
   (b) 15 Ω
   (c) 7.5 Ω
   (d) 1.8×10^3 Ω

7. 6.0 V

8. 4.6 A

SUMMARY *Electric Resistance and Ohm's Law*

- Electric resistance is a measure of how much an electrical component opposes the flow of charges.
- A good electrical insulator, which prevents the flow of electric charges, is a poor electrical conductor.
- Ohm's law states that if the temperature of a resistance remains constant, the ratio of the potential drop across it to the current through it is constant and equals the resistance. In equation form, $R = \frac{\Delta V}{I}$.

Understanding Concepts

1. Explain why copper is a good choice for electrical wiring.

2. Copy **Table 4** into your notebook, and complete it. Include the equation used to solve for each unknown.

Table 4   For Question 2

	R (Ω)	ΔV (V)	I (A)
(a)	?	1.5	0.25
(b)	?	2.4×10^2	21
(c)	2.86	?	3.15
(d)	15	6.0	?

3. An electric toaster uses 9.6 A of current from a 120-V household circuit. Calculate the toaster's resistance.

4. Calculate the current through a 95-Ω coffee grinder used in a 120-V household circuit (**Figure 5**).

Figure 5

5. The current in a 24-Ω curling iron is 5.0 A. Calculate the potential difference of the circuit.

Applying Inquiry Skills

6. Describe how you would test a resistor experimentally to determine whether it obeys Ohm's law.

7. (a) You are experimenting with an electrical circuit, and you smell something burning. What should you do?
   (b) What might cause the burning smell?

8. (a) The colours on a colour-coded resistor are brown, green, black, silver. State the resistance and tolerance of the resistor.
   (b) State the colour code of a resistor whose resistance is 470 Ω $\pm$ 5%.

Making Connections

9. (a) Using the data in **Table 2**, page 328, determine the current, in milliamps, through the body when exposed to a 120-V household circuit, both when the body is dry and when the body is in a bathtub.
   (b) Use your answer in (a) to discuss safety concerns and precautions.

7.5 Investigation

Testing Resistors

Electrical circuits in the lab can be controlled to test Ohm's law to determine, for any resistor, whether $R = \dfrac{\Delta V}{I}$. Computer-simulated circuits can also be used to accomplish this objective. Whether you use actual or simulated circuits, you can use your graphing skills to analyze the data to check Ohm's law.

Question

(a) Make up a question for this investigation.

Prediction

(b) If a resistor in a circuit obeys Ohm's law, predict the shape of a graph of the electric potential drop (vertical axis) versus the electric current.

(c) Repeat (b) for a resistor that does *not* obey Ohm's law.

Materials

For each group of three or four students:
3 fixed resistors of different values (e.g., 100 Ω, 50 Ω, and 25 Ω)
low-voltage DC light source (such as the 12-V DC light bulb in some ray boxes)
3 dry cells or variable power supply
voltmeter
ammeter
switch
connecting wires

For each student:
graph paper

Procedure

1. Set up a table of data to record the labelled resistance, the potential rise, and the current for each of the three resistors and the light bulb tested in this investigation.

2. Set up the circuit as shown in **Figure 1**, using the first resistor, R_1, and a low potential rise. Leave the switch open until you have checked the circuit.

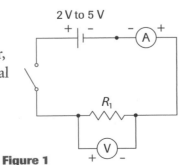

2 V to 5 V

R_1

Figure 1

3. Close the switch momentarily, and measure the current through the resistor and the potential drop across it. Record your observations.

4. Repeat steps 2 and 3 using the same resistor but a higher potential rise, and finally a still higher potential rise.

5. Repeat steps 2 to 4 using resistors R_2 and R_3.

6. Repeat steps 2 to 4 using the low-voltage light bulb and the DC source that operates it.

 When unplugging the power supply, grasp the plug and pull it from the wall receptacle. Do not pull on the cord.

Do not leave the current on for longer than needed. Wires may become overheated.

Do not exceed the voltage prescribed by your teacher.

Do not connect the terminals without a load; a short circuit will result.

Analysis

(d) On a single graph of potential difference (vertical axis) versus current, plot the results found using resistors 1, 2, and 3. Draw and label three separate straight lines of best fit, one for each resistor.

(e) Calculate the slope of each line of best fit.

(f) Compare the slopes to the values of the labelled (known) resistances of the resistors.

(g) On a separate graph, plot the potential difference–current data for the light bulb tested. Analyze the graph.

(h) How can graphing be used to determine whether a resistor is ohmic or nonohmic?

Evaluation

(i) How accurate were your predictions in (a), (b), and (c)?

(j) Did your results verify or refute Ohm's law? Explain your answer.

Figure 1 shows two ways of connecting a household circuit. In the first circuit, there is only one path for the electric current, so all four appliances are either off or on at the same time. In the second circuit, the number of paths is the same as the number of appliances, so each appliance has its own switch. The second circuit is an example of the usual wiring in a household. (When you do Investigation 7.7, you will discover more than one reason why this is so.)

(a)

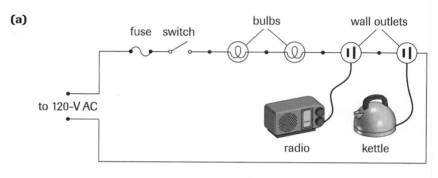

(b)

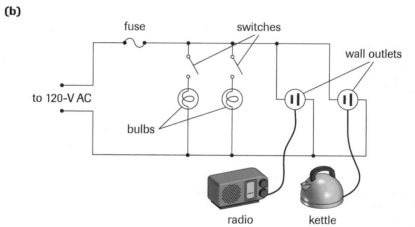

Figure 1

(a) A single path for all electric current

(b) Individual paths for each appliance

Figure 1(a) is an example of a *series circuit,* in which all the current follows the same path. **Figure 1(b)** is an example of a *parallel circuit,* in which the current has two or more paths to follow. There can also be both series and parallel connections in the same circuit. **Figure 2** shows a series–parallel circuit with a DC source.

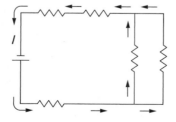

Figure 2

DC circuit with resistors connected in both series and parallel

Cells in Series and in Parallel

You can compare the difference in potential rise in two cells connected in series and two cells connected in parallel. You need two identical cells, a voltmeter, a switch, a load (e.g., a 22-Ω resistor), and connecting wires.

(a) Predict the total potential rise of two cells connected first in series and then in parallel.

- Set up the circuit shown in **Figure 3(a)** to test the potential rise of the cells, one at a time. Record the values.
- Set up the circuit with the cells in series, as shown in **Figure 3(b)**; quickly close the switch, then open it, and record the total potential rise.
- Set up the circuit with the cells in parallel, as shown in **Figure 3(c)**; quickly close the switch, then open it, and record the total potential rise.

(b) Communicate what you discovered using an appropriate method (e.g., a written description, diagrams, or equations).

 Do not connect the terminals without a load; a short circuit will result.

Do not leave the switch closed after you have taken a reading.

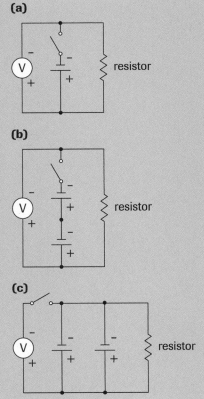

(a)

(b)

(c)

Figure 3

Kirchhoff's Rules

German physicist Gustav Kirchhoff (1824−1887) first analyzed series and parallel circuits and formulated the rules that are now called *Kirchhoff's current rule (KCR)* and *Kirchhoff's voltage rule (KVR)*. Kirchhoff's rules can be applied to solve a variety of electrical circuit problems.

Kirchhoff's Current Rule (KCR)
At any junction point in an electrical circuit, the total current into the junction equals the total current out of the junction (**Figure 4(a)**).

Kirchhoff's Voltage Rule (KVR)
In any complete path in an electrical circuit, the sum of the potential rises equals the sum of the potential drops (**Figure 4(b)**).

(a)
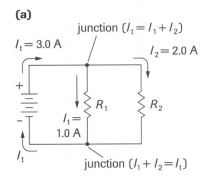

junction ($I_t = I_1 + I_2$)
$I_t = 3.0$ A
$I_2 = 2.0$ A
R_1 R_2
$I_1 = 1.0$ A
I_t
junction ($I_1 + I_2 = I_t$)

(b)
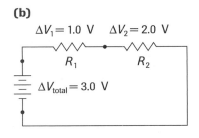

$\Delta V_1 = 1.0$ V $\Delta V_2 = 2.0$ V
R_1 R_2
$\Delta V_{total} = 3.0$ V

Figure 4
(a) Illustrating KCR. $I_{total} = I_t = I_1 + I_2$, where the symbol for current using a single subscript, such as I_1, means the current in a particular branch of the circuit, such as through R_1.
(b) Illustrating KVR. $\Delta V_{total} = \Delta V_1 + \Delta V_2$, where the symbol ΔV_1 means the electric potential drop across component 1, in this case R_1.

► **SAMPLE** problem **1**

Applying Kirchhoff's Current Rule

Calculate the electric current I_2 for the circuit in **Figure 5**.

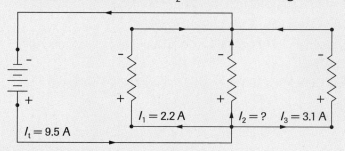

Figure 5

Solution

$I_{total} = I_t = 9.5$ A

$I_1 = 2.2$ A

$I_3 = 3.1$ A

$I_2 = ?$

Applying KCR to the circuit,

$$I_t = I_1 + I_2 + I_3$$
$$I_2 = I_t - I_1 - I_3$$
$$= 9.5 \text{ A} - 2.2 \text{ A} - 3.1 \text{ A}$$
$$I_2 = 4.2 \text{ A}$$

The electric current I_2 is 4.2 A.

► **SAMPLE** problem **2**

Applying Kirchhoff's Voltage Rule

Calculate the electric potential drop ΔV_2 for the circuit in **Figure 6**.

Solution

$\Delta V_{total} = V_t = 36$ V

$\Delta V_1 = 12$ V

$\Delta V_3 = 13$ V

$\Delta V_2 = ?$

Applying KVR to the circuit,

$$\Delta V_t = \Delta V_1 + \Delta V_2 + \Delta V_3$$
$$\Delta V_2 = \Delta V_t - \Delta V_1 - \Delta V_3$$
$$= 36 \text{ V} - 12 \text{ V} - 13 \text{ V}$$
$$\Delta V_2 = 11 \text{ V}$$

The electric potential drop ΔV_2 is 11 V.

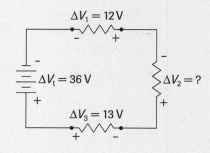

Figure 6

Understanding Concepts

1. Calculate the unknown electric currents for the circuit in **Figure 7**. Show your steps and reasoning.

2. Calculate the unknown electric potential differences in the circuits in **Figure 8**. Show your steps and reasoning.

Answers

1. $I_3 = 2.3$ A; $I_1 = 3.4$ A
2. (a) 6.6 V
   (b) 2.8 V
3. (b) 6.0 V

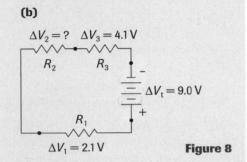

$I_1 = ?$ $I_2 = 1.1$ A $I_3 = ?$

$V_t = 6.0$ V

$I_t = 3.4$ A   **Figure 7**

(a)

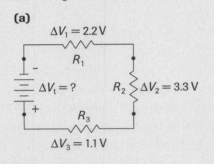

$\Delta V_1 = 2.2$ V
R_1
$\Delta V_t = ?$ $R_2 \gtrless \Delta V_2 = 3.3$ V
R_3
$\Delta V_3 = 1.1$ V

(b)

$\Delta V_2 = ?$ $\Delta V_3 = 4.1$ V
R_2 R_3
$\Delta V_t = 9.0$ V
R_1
$\Delta V_1 = 2.1$ V   **Figure 8**

Applying Inquiry Skills

3. Four 1.5-V cells are connected in series.
   (a) Using the appropriate symbols, draw a circuit diagram showing how you can measure the total potential rise given by the cells. Include all positive and negative signs.
   (b) Calculate the total potential rise.

equivalent resistance a single resistance that can replace all the resistances in an electrical circuit while maintaining the same current when connected to the same source; symbol R_{total} or R_t

Resistors in Series

When analyzing any electrical circuit with two or more resistors, it is convenient to determine the circuit's **equivalent resistance**, R_{total} or R_t, which is a single resistance that can replace all the resistances yet produce the same current when connected to the same source. Consider **Figure 9(a)**, which is a circuit with three resistors connected in series. Using ΔV to represent potential rise or potential drop, and applying KVR to the circuit,

$$\Delta V_t = \Delta V_1 + \Delta V_2 + \Delta V_3$$

Applying Ohm's law in the form $\Delta V = IR$ to each potential difference,

$$I_t R_t = I_1 R_1 + I_2 R_2 + I_3 R_3$$

Applying KCR, $I_t = I_1 = I_2 = I_3$, to the circuit, the currents cancel out, leaving only the resistances:

$$R_t = R_1 + R_2 + R_3$$

For any number n of resistors connected in series, the equivalent resistance is

$$R_t = R_1 + R_2 + ... + R_n$$

where R_t is the equivalent or total resistance, as illustrated in **Figure 9(b)**.

(a)

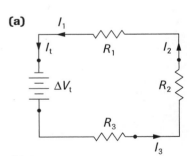

I_1
I_t R_1 I_2
ΔV_t R_2
R_3
I_3

(b)

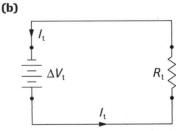

I_t
ΔV_t R_t
I_t

Figure 9

Equivalent Resistance in a Series Circuit

Calculate the equivalent resistance in a series circuit containing the following resistances in series: 26 Ω, 11 Ω, and 18 Ω.

Solution

$R_1 = 26\ \Omega$
$R_2 = 11\ \Omega$
$R_3 = 18\ \Omega$
$R_t = ?$

$$R_t = R_1 + R_2 + R_3$$
$$= 26\ \Omega + 11\ \Omega + 18\ \Omega$$
$$R_t = 55\ \Omega$$

The equivalent resistance is 55 Ω.

Notice in Sample Problem 3 that the equivalent or total resistance (55 Ω) is greater than any of the resistances connected in series. This occurs because there is only one path for the charges, so the total resistance increases when there are more resistances in the path.

A popular application of resistors connected in series is the *personal digital assistant* (**Figure 10(a)**). As the user writes on the pressure-sensitive pad with a stylus, writing appears where the stylus has made contact. The pad's two layers, top (T) and bottom (B), are electrical conductors. These layers remain separated except where the stylus presses down at some position, P. At that instant, the current is from T to B, and the equivalent resistance encountered by the current is the sum of the resistance in the top layer, R_T, and the resistance in the bottom layer, R_B (**Figure 10(b)**). At some new position, P′, the new equivalent resistance is the sum of the series resistances, $R'_T + R'_B$ (**Figure 10(c)**). Software interprets each set of resistances (R_T, R_B, R'_T, R'_B, etc.) and stores the information in the assistant's built-in computer.

(a)

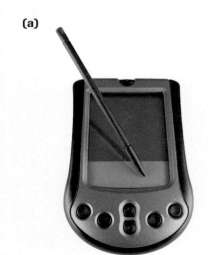

(b)

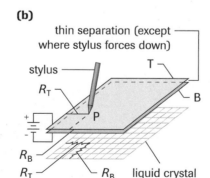

(c)

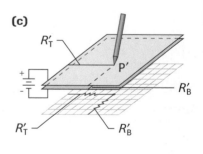

Figure 10
(a) A personal digital assistant
(b) The series resistances when the stylus presses at point P
(c) The series resistances when the stylus presses at point P′

▶ **Practice**

Understanding Concepts

4. Calculate the equivalent resistance in each of the following:
   (a) A 22-Ω, a 25-Ω, and a 47-Ω resistor are connected in series.
   (b) Two 28-Ω light bulbs and two 21-Ω heaters are connected in series.

5. Calculate the unknown resistance in each of the following:
   (a) A 15-Ω, a 27-Ω, and an unknown resistor are connected in series, yielding an equivalent resistance of 54 Ω.
   (b) Two identical bulbs of unknown resistance are connected in series with a 21-Ω heater and a 22-Ω heater, producing an equivalent resistance of 99 Ω.

Answers

4. (a) 94 Ω
   (b) 98 Ω

5. (a) 12 Ω
   (b) 28 Ω each

6. (a) 68 Ω

Making Connections

6. Two strings of Christmas tree lights are connected in series. The first string has eight 4.0-Ω bulbs in series, and the second has twelve 3.0-Ω bulbs in series.
   (a) Calculate the equivalent resistance of the two strings of bulbs.
   (b) If one bulb burns out, what happens to the remaining bulbs on the string? Is this an advantage or a disadvantage? Explain your answer.

Resistors in Parallel

When resistors are connected in parallel, the equivalent resistance can be found by using a different equation from the one used for series resistors. Consider **Figure 11(a)**.

(a)

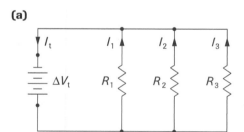

(b)

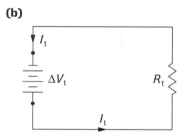

Figure 11

Applying KCR to the circuit,

$I_t = I_1 + I_2 + I_3$

Applying Ohm's law in the form $I = \dfrac{\Delta V}{R}$ to each term,

$$\frac{\Delta V_t}{R_t} = \frac{\Delta V_1}{R_1} + \frac{\Delta V_2}{R_2} + \frac{\Delta V_3}{R_3}$$

According to KVR, the potential differences are all equal in a parallel circuit, so they cancel out, leaving only the resistances:

$$\frac{1}{R_t} = \frac{1}{R_1} + \frac{1}{R_2} + \frac{1}{R_3}$$

For any number n of resistors connected in parallel, the equivalent resistance is found by using the equation

$$\frac{1}{R_t} = \frac{1}{R_1} + \frac{1}{R_2} + \cdots + \frac{1}{R_n}$$

where R_t is the equivalent or total resistance, as illustrated in **Figure 11(b)**.

▶ **SAMPLE** problem **4**

Equivalent Resistance in a Parallel Circuit

Calculate the equivalent resistance of a circuit with the following resistors in parallel: 5 Ω, 10 Ω, and 30 Ω.

Solution
$R_1 = 5\ \Omega$
$R_2 = 10\ \Omega$
$R_3 = 30\ \Omega$
$R_t = ?$

$$\frac{1}{R_t} = \frac{1}{R_1} + \frac{1}{R_2} + \frac{1}{R_3}$$

$$= \frac{1}{5\ \Omega} + \frac{1}{10\ \Omega} + \frac{1}{30\ \Omega}$$

$$= \frac{6}{30\ \Omega} + \frac{3}{30\ \Omega} + \frac{1}{30\ \Omega}$$

$$\frac{1}{R_t} = \frac{10}{30\ \Omega}$$

$$R_t = \frac{30\ \Omega}{10}$$

$$R_t = 3\ \Omega$$

The equivalent resistance is 3 Ω.

Note that the equivalent or total resistance (3 Ω) in Sample Problem 4 is less than any of the resistances connected in parallel. This is because the electric charges have more than one path to take, so resistance to their flow decreases.

▶ **Practice**

Understanding Concepts

Answers

7. (a) 4 Ω
   (b) 10 Ω
   (c) 10 Ω

7. Calculate the equivalent resistance when the following resistors are connected in parallel:
   (a) 8 Ω, 8 Ω
   (b) 30 Ω, 30 Ω, 30 Ω
   (c) 15 Ω, 60 Ω, 60 Ω

Applying Inquiry Skills

8. Obtain an ohmmeter and two or three resistors of unknown resistances. Measure each resistance using the ohmmeter. Use the data to determine the equivalent resistance if the resistors were connected in parallel. Then connect them in parallel and use the ohmmeter to check your calculations.

Analyzing Series–Parallel Circuits

You have learned several principles of electrical circuits: Ohm's law, KCR, KVR, equivalent resistances, and the properties of series and parallel circuits. These principles can be combined to analyze combined series–parallel circuits in a step-by-step process.

> ▶ **SAMPLE** problem **5**
>
> ### Unknowns in a Combined Circuit

For the circuit shown in **Figure 12(a)**, determine both the current through and the potential drop across each resistor.

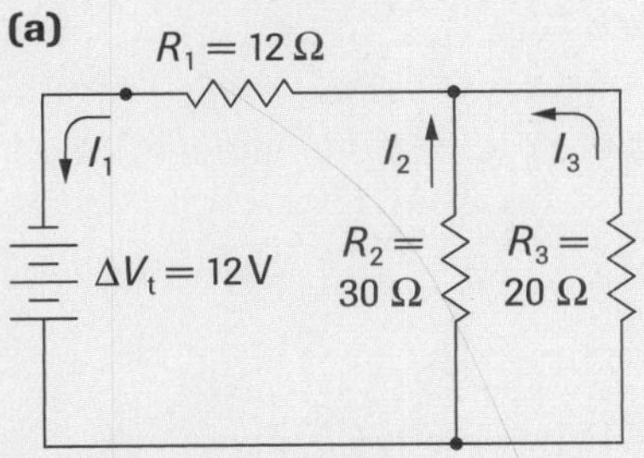

(a)

Figure 12

Solution

The first step is to find the equivalent resistance of the circuit. In this case, the equivalent resistance (R_p) of the two parallel resistors is found, and then added to the series resistor to obtain the equivalent resistance of the entire circuit.

$R_2 = 30\ \Omega$
$R_3 = 20\ \Omega$
$R_p = ?$

$$\frac{1}{R_p} = \frac{1}{R_2} + \frac{1}{R_3}$$

$$= \frac{1}{30\ \Omega} + \frac{1}{20\ \Omega}$$

$$= \frac{2}{60\ \Omega} + \frac{3}{60\ \Omega}$$

$$\frac{1}{R_p} = \frac{5}{60\ \Omega}$$

$$R_p = \frac{60\ \Omega}{5}$$

$$R_p = 12\ \Omega$$

The equivalent resistance is 12 Ω, as illustrated in **Figure 12(b)**.

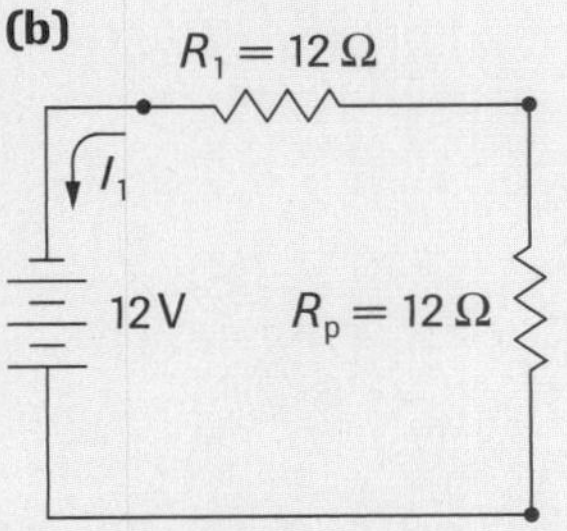

(b)

Figure 12

The equivalent resistance, R_t, of the entire circuit is then

$$R_t = R_1 + R_p$$
$$= 12\ \Omega + 12\ \Omega$$
$$R_t = 24\ \Omega$$

The equivalent resistance is 24 Ω, as illustrated in **Figure 12(c)**. This value is used to find the total current, I_t.

$$I_t = \frac{\Delta V_t}{R_t}$$
$$= \frac{12\ \text{V}}{24\ \Omega}$$
$$I_t = 0.50\ \text{A}$$

(c)

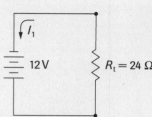

Figure 12

The total current, which passes through R_1, is used to find the potential drop, ΔV_1, across R_1.

$$\Delta V_1 = I_1 R_1$$
$$= (0.50\ \text{A})(12\ \Omega)$$
$$\Delta V_1 = 6.0\ \text{V}$$

Now the potential drop across R_2 equals that across R_3; it can be found by applying KVR to the circuit in **Figure 12(b)**.

$$\Delta V_t = \Delta V_1 + \Delta V_p$$
$$\Delta V_p = \Delta V_t - \Delta V_1$$
$$= 12\ \text{V} - 6.0\ \text{V}$$
$$\Delta V_p = 6.0\ \text{V}$$

The last step is to apply Ohm's law to find the currents through the two parallel resistors.

$$I_2 = \frac{\Delta V_2}{R_2} \qquad\qquad I_3 = \frac{\Delta V_3}{R_3}$$
$$= \frac{6.0\ \text{V}}{30\ \Omega} \qquad\qquad = \frac{6.0\ \text{V}}{20\ \Omega}$$
$$I_2 = 0.20\ \text{A} \qquad\qquad I_3 = 0.30\ \text{A}$$

As a final check, notice that the total current (0.50 A) equals the sum of the currents in the parallel resistors (0.30 A + 0.20 A).

▶ *Practice*

Understanding Concepts

Answer

9. $\Delta V_2 = \Delta V_3 = 6.0\ \text{V}; \Delta V_1 = 3.0\ \text{V};$
   $I_1 = 0.50\ \text{A}; I_2 = 0.30\ \text{A};$
   $I_3 = 0.20\ \text{A}$

9. For the circuit shown in **Figure 13**, find the current through and the potential drop across each resistor.

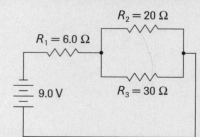

Figure 13

SUMMARY **Series and Parallel Circuits**

- Kirchhoff's current rule (KCR) states that the total current into a junction in an electrical circuit equals the total current out of the junction.

- Kirchhoff's voltage rule (KVR) states that in any complete path in an electrical circuit, the sum of the potential rises equals the sum of the potential drops.

- In a series circuit, the equivalent resistance is $R_t = R_1 + R_2 + ... + R_n$.

- In a parallel circuit, the equivalent resistance is $\dfrac{1}{R_t} = \dfrac{1}{R_1} + \dfrac{1}{R_2} + ... + \dfrac{1}{R_n}$.

- Applying equations for Ohm's law, KCR, KVR, and equivalent resistances helps to solve for unknowns in complex electrical circuits.

▶ **Section 7.6 Questions**

Understanding Concepts

1. Calculate the total potential rise across three 6.0-V batteries that are connected (a) in series and (b) in parallel.

2. Calculate the total current in the circuit in **Figure 14**.

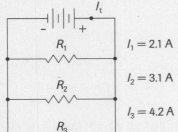

$I_1 = 2.1\ A$

$I_2 = 3.1\ A$

$I_3 = 4.2\ A$

Figure 14

3. Set up a table to compare series and parallel circuits using the following headings: Type of Circuit; Definition; Example; Current; Voltage; Resistance.

4. Calculate the potential rise of the source in **Figure 15**.

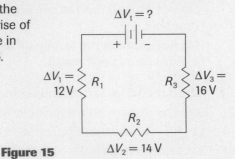

Figure 15

5. (a) Describe how the equivalent resistance of a circuit compares to the individual resistances when the resistances are connected (i) in series and (ii) in parallel.
   (b) In each case, explain why.

6. Calculate the equivalent resistance when 100 Ω, 200 Ω, and 600 Ω are connected (a) in series and (b) in parallel.

7. Draw a circuit diagram with a DC source, a switch, an electric motor, a green light, and a red light. When the motor is off, the green light should be on. When the motor is on, the red light should be on (warning).

8. For the circuit shown in **Figure 16**, find both the current through and the potential drop across each resistor.

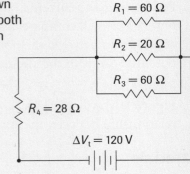

Figure 16

Applying Inquiry Skills

9. The first and third rings of a colour-coded resistor are brown, but you can't decide whether the second ring is red or brown. Describe an experiment you could perform to determine the colour of the ring. (You cannot use an ohmmeter.) Include the equations you need to solve the problem.

Making Connections

10. To boost a discharged car battery, you can connect a battery from a second car. Should the batteries be connected in series or in parallel? Explain your answer, and draw a sketch of the connections.

Resistors in Series and Parallel

In section 7.2, you learned that ammeters must be connected in series, and in section 7.3 that voltmeters must be connected in parallel. Resistors can be connected either in series or in parallel. In this investigation, you will study the properties of each type of load connection.

For this investigation, the following symbols for measuring electric current are introduced: I_a means the current at position "a," I_b means the current at position "b," and so on.

This investigation has two parts. In Part A, you will explore the properties of a series circuit. In Part B, you will explore the properties of a parallel circuit.

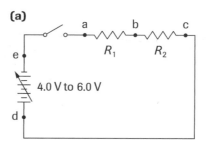

(a)

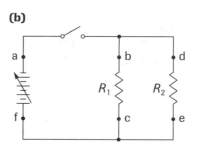

(b)

Figure 1
(a) A series circuit
(b) A parallel circuit

Questions

What are the properties of electrical circuits with the resistors connected in series?

What are the properties of electrical circuits with the resistors connected in parallel?

Prediction

(a) Predict how the currents I_a, I_b, and I_c in **Figure 1(a)** will compare, and predict the relationship between the potential rises and the potential drops in the circuit.

(b) Predict how ΔV_{source}, ΔV_1, and ΔV_2 in **Figure 1(b)** will compare, and predict the relationship between the currents I_a, I_b, and I_d.

Materials

For each group of three or four students:
2 fixed resistors of different sizes
variable power supply (or an alternative)
voltmeter
ammeter
switch
connecting wires

Procedure

Part A

1. Read steps 2 to 5, and draw a circuit diagram for each step. Show the instruments required, as well as all positive and negative signs. Record the resistance values of the resistors on your

diagrams. Have your teacher approve your diagrams before you set up the circuits.

 When unplugging the power supply, grasp the plug and pull it from the wall receptacle. Do not pull on the cord.

Do not leave the current on for longer than needed. Wires may become overheated.

Do not exceed the voltage prescribed by your teacher.

Do not connect the terminals without a load; a short circuit will result.

2. Set up the series circuit shown in **Figure 1(a)** with the instruments needed to measure I_a and ΔV_{source}. Adjust the supply potential rise to a moderate value, about 6.0 V or less, as advised by your teacher. Close the switch briefly, determine I_a in amps, and record the meter readings on your diagram. Open the switch.

3. With ΔV_{source} constant, move the ammeter to measure I_b. Record the data on your diagram.

4. Repeat step 3 to measure I_c.

5. With ΔV_{source} constant, move the voltmeter and use it to measure the potential drops ΔV_1 and ΔV_2. (If you leave the ammeter in the circuit, you can use it to ensure that the current remains constant.) Record the data on your diagram.

Part B

6. Read steps 7 to 10, and draw a circuit diagram for each step. Show the instruments required, all positive and negative signs, and the values of the resistors. Have your teacher approve your diagrams before you set up the circuits.

7. Set up the parallel circuit shown in **Figure 1(b)** with the instruments needed to measure I_a and ΔV_{source}. Adjust the supply potential rise to a moderate value, again about 6.0 V or less, as advised by your teacher. Determine I_a in amps, and record the meter readings on your diagram.

8. With ΔV_{source} constant, move the ammeter to measure I_b. Record the data on your diagram. (Note: If the current is the same as at I_a, your circuit is connected incorrectly. Have your teacher check it.)

9. Repeat step 8 to measure I_d.

10. With ΔV_{source} constant, move the voltmeter and use it to measure the potential drops ΔV_1 and ΔV_2. (If you leave the ammeter in the circuit, you can use it to ensure that the current remains constant.) Record the data on your diagram.

Analysis

(c) In a series circuit, how do the currents I_a, I_b, and I_c compare? Does this verify or refute KCR?

(d) In a series circuit, how does the potential rise ΔV_{source} compare with the sum of the potential drops $(\Delta V_1 + \Delta V_2)$? Does this verify or refute KVR?

(e) In a parallel circuit, how does the total current I_a compare with the sum of the currents through the individual resistors $(I_b + I_d)$? Does this verify or refute KCR?

(f) In a parallel circuit, how do all the potential differences compare? Does this verify or refute KVR?

(g) Apply Ohm's law to calculate the total resistance of the series circuit, $R_{total} = \dfrac{\Delta V_{source}}{I_{total}}$. How does this value compare with the sum $R_1 + R_2$? How does it compare with the individual resistors in the circuit?

(h) Apply Ohm's law to calculate the total resistance of the parallel circuit. How does this value compare with the individual resistors in the circuit?

Evaluation

(i) Comment on the accuracy of your predictions.

(j) Describe the major sources of error in this investigation. What can be done to minimize those sources?

Synthesis

(k) Describe the advantages of a parallel circuit over a series circuit for household use.

Electricity is hazardous; it must be treated with caution. Each year, many lives are lost and much property is damaged because of the careless use of electricity.

The human body is sensitive to all electric currents, even small ones. As you can see in **Table 1**, being subjected to a current as small as 9×10^{-3} A (AC) causes shock. Twice that amount causes muscle paralysis. A current of about 0.1 A (AC) causes *fibrillation*, a rapid, uncontrollable twitching of the heart that prevents circulation and, without medical intervention, results in death.

Table 1 Human Reactions to Electric Currents

Current (A, AC)	Current (A, DC)	Reaction
4×10^{-4}	1×10^{-3}	slight sensation
9×10^{-3}	0.05	shock
0.02	0.07	muscles paralyzed
0.1	0.5	fibrillation and death

Note: Values are approximate.

However, the body's nervous system operates on electrical impulses, so electric current can also save lives. A device called a *defibrillator* sends a relatively large current through the body for a fraction of a second; this current restores a fibrillating heart to its normal heartbeat (**Figure 1(a)**). Electric current can also be used to promote healing and to relieve pain. Pacemakers stimulate the heart electrically; they help people with certain heart problems lead normal lives (**Figure 1(b)**). Physicians use electrical nerve stimulation to treat certain types of pain. A 9-V battery supplies a weak electric current across the patient's skin into the nerve cells beneath the skin's surface. The current stimulates the body's natural ability to fight pain.

(a)

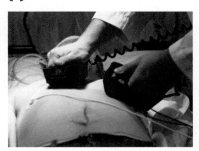

(b)

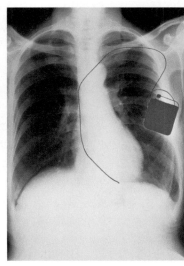

Figure 1
(a) Defibrillators have saved countless lives.
(b) An implanted pacemaker stimulates the muscle cells that control the heart's pumping rate.

Moisture greatly affects the amount of current that passes through a human body. A person who is dry has a resistance on the order of 10^6 Ω, so the current resulting from contact with a 120-V circuit is about 0.12 mA ($I = \Delta V/R$), large enough to cause an unpleasant sensation but not death. However, a person with wet skin has a resistance of only about 1500 Ω, and a person sitting in bathwater has a resistance as low as 500 Ω. Under these circumstances, the current from a 120-V circuit is large enough to be fatal. Thus, electrical devices should not be used near water or moisture. For instance, electric radios, hair dryers, and shavers with cords should never be used or placed near a water-filled bathtub.

Electric currents can be even more dangerous within the human body. For example, in hospitals, some heart patients are connected to heart monitors. Even a tiny electric current of 3×10^{-5} A (30 μA) applied directly to the heart can be fatal, so medical practitioners must exercise extreme caution when handling heart monitors.

As you can see, safety is very important when using electrical circuits. A common electrical problem in everyday life is an **overloaded circuit**, sometimes called an "octopus circuit." When too many appliances are connected to a single circuit, the current increases and the circuit overheats. When the current exceeds the safe limit, the fuse should burn out or the circuit breaker should trip. If this happens, the fault should be corrected, and the circuit breaker should be reset or the burned-out fuse should be replaced with a fuse of the correct size. Consider, for example, the household circuit illustrated in **Figure 2**, in which a 15-A fuse or circuit breaker protects the 120-V circuit. The kettle uses 11.6 A of current and the fruit juicer uses 0.5 A, for a total of 12.1 A. Now if the toaster, which requires 8.3 A, is activated, the current exceeds the allowed 15 A, so the fuse will burn out or the circuit breaker will trip.

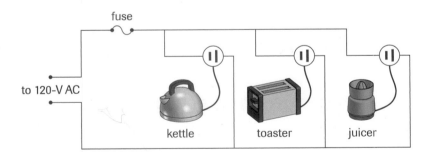

Figure 2

As you saw earlier, another common cause of overheating is a short circuit. A frayed electrical cord or a faulty electrical appliance allows the current to *arc* from one conductor to another with little or no resistance. As a result, the current increases rapidly, and the circuit overheats. If this happens, the fuse should blow or the circuit breaker should trip. The problem should be corrected as soon as it arises.

Extension cords can also overheat, which may result in a fire. Often the less expensive cords can carry no more than 7.0 A of current safely. Thus, only low-power appliances can be connected to these extension cords safely. If the cord has a label attached, it should indicate the safe limit.

DID YOU KNOW ?

Fuses
Modern household circuits are protected with circuit breakers rather than fuses. However, individual appliances often have fuse protection. For example, a glass-top electric stove with four burners and a convection oven has two 30-A fuses for the oven elements, four 20-A fuses for the burners, and two 15-A fuses for the light and clock circuits.

overloaded circuit an electrical circuit in which the current exceeds the circuit's safe limit

CAREER CONNECTION

An interest in the heart can lead to different careers, for example, as a cardiovascular technician as an important part of a medical diagnostic team, or helping animals as a veterinary technician. Both paths lead to rewarding and highly trained professions.

GO   www.science.nelson.com

LEARNING TIP

Household Wiring
Household circuits are wired with conductors with different colours of insulation. The red and black wires are "hot," with an electric potential difference across them of 240 V. The white wires are neutral, and they are grounded to metal water pipes or other grounding bars. The white wires can be connected in a circuit with a red wire or a black one to produce a 120-V circuit. Green wires are added as extra protection of outlets to the ground.

> **Practice**

Understanding Concepts

1. Calculate the current received directly from contact with a 120-V (AC) circuit in each of the following, and state a likely outcome. Refer to **Table 1**, page 346.
   (a) A dry person has a resistance of $6.0 \times 10^5 \ \Omega$.
   (b) A person with wet hands has a resistance of $1.5 \times 10^3 \ \Omega$.
   (c) A person in bathwater has a resistance of $5.0 \times 10^2 \ \Omega$.

2. Compare the fatal external current for a human (**Table 1**, page 346) with the fatal internal current described later in the section. Explain why there is a difference.

3. A 15-A fuse is used to protect a 120-V household lighting circuit. How many 100-W bulbs, each of which uses 0.83 A of current, can be connected to the circuit to cause the circuit breaker to trip?

Making Connections

4. Explain why it is dangerous and extremely foolish to replace a burned-out fuse in a car with a rolled-up piece of foil from a wrapper.

> **TRY THIS activity**

Modelling a Circuit Breaker

The compound bar shown in **Figure 3(a)** is made of two metals that expand at different rates when heated. This control device, which is also called a bimetallic strip, is used in electric irons, coffee makers, toasters, kettles, and frying pans. Most electric heaters and hair dryers have built-in bimetallic switches to interrupt the current if the devices overheat. **Figure 3(b)** shows the design of a control device that operates with a bimetallic strip.

(a) Obtain a compound bar, and observe what happens to it when you place it in a hot water bath. Observe what happens when you place it in a cold water bath. Sketch your observations.

(b) In what way is the device in **Figure 3(b)** a control device?

(c) Describe how you would use the compound bar as a control device (either as a circuit breaker or as an on–off switch). Draw a diagram of the circuit.

(a)

(b)

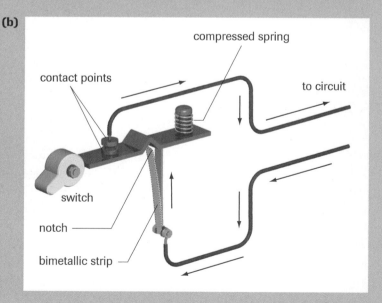

Figure 3
(a) A compound bar
(b) Using a bimetallic strip

▶ **EXPLORE** *an issue*

Electrical Safety Standards

When you shop for electrical products, do you consider safety features? Do you look for the special label plate with CSA or UL symbols, shown in **Figure 4**, that identifies safety-approved appliances? Would you buy a product that does not meet minimal safety standards? These are questions to think about as you explore the issue of electrical safety standards.

Electrical products that are tested and approved have either a CSA label or a UL label. However, not all electrical products sold in Canada have one of these labels. Canadian laws require that electrical products sold in Canada be certified to the relevant CSA standards. Products sold without safety approval can be problematic, as insurance companies might not pay for fire damage caused by faulty, unsafe electrical products. The CSA and UL marks assure the consumer that the products meet minimum safety standards. In this way, consumers are responsible for their own safety.

Understanding the Issue

(a) Why is it important for consumers to consider safety when purchasing electrical products?

(b) Read the label plates on at least 10 electrical devices in your home or on the boxes of electrical devices at a store. Make a table summarizing what you discover about the safety standards of the products.

(c) Look up CSA and UL on the Internet. What do these symbols stand for? Describe what else you discover about CSA or UL.

 www.science.nelson.com

Take a Stand

Based on your research in (b) and (c) above, decide whether you would recommend that only safety-approved devices be sold in Canada, and defend your position. Create a position statement (e.g., a poster board, an essay, a letter, a Web page, a video, or an audio presentation) to summarize your research and stand. Your analysis should include answers to the following questions:

(a) What are the benefits of compulsory safety standards?

Figure 4
The CSA and UL labels are placed on electrical devices and other products that have been tested and approved. Other safety labels are listed in Appendix B1.

(b) What are the drawbacks of compulsory safety standards?

(c) What are the economic, social, and environmental impacts of high safety standards for electrical products?

(d) What careers are related to the implementation of safety standards?

Evaluation

(e) Were the CSA or UL labels easy to find on all the electrical products you checked?

(f) Evaluate the usefulness of the Web sites you used as references.

SUMMARY *Electrical Safety*

- Electrical safety is important because of the dangers to the human body as well as the potential for fire.
- Circuit breakers and fuses protect electrical circuits from overheating when the current becomes too high.

Section 7.8 Questions

Understanding Concepts

1. Explain the electrical safety precautions you would recommend to a person operating an electric lawn mower (**Figure 5**).

Figure 5

2. What is the purpose of a fuse or circuit breaker?

3. Assume that a 12-Ω kettle usually used in a 120-V circuit is plugged in with a 7.0-A extension cord. Would a 15-A circuit breaker provide adequate protection? Explain your answer.

Applying Inquiry Skills

4. A 12-V battery, an ammeter, a 5.0-A fuse, and several 10-Ω lamps are used in an experiment to find the effect of connecting loads in parallel.
   (a) Draw a diagram of the circuit used in the experiment.
   (b) Determine the total resistance and current when the number of lamps connected in parallel is 1, 2, 3, 4, 5, and 6.
   (c) What is the maximum number of lamps that can be connected before the fuse becomes overloaded and burns out?
   (d) Write at least one conclusion for the experiment.

Making Connections

5. Identify three electrical hazards in and around the home, and describe how to avoid them.

6. A Taser gun (**Figure 6**) is a weapon used by police to apprehend criminals, as an alternative to traditional guns. It has hooks that can deliver an electric current that disrupts the way muscles work, causing a person to fall down. Research the Taser gun.

Figure 6
A police Taser gun

   (a) What power and voltage does the gun deliver?
   (b) Describe the advantages of the Taser gun.
   (c) Is the Taser gun an example of a hydraulic or pneumatic system?

Every major appliance sold in Canada has an "Energy Star" or an "EnerGuide" label. This label states the amount of electrical energy consumed per month or per year by the appliance in normal use. The Energy Star label is reserved for appliances with the highest efficiency in their class. When buying appliances, consumers can refer to the labels and make choices that conserve energy (**Figure 1**).

(a)

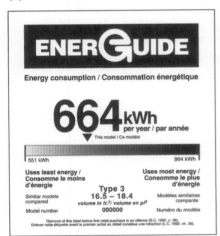

(b)

(c)

Figure 1
A typical EnerGuide label **(a)**. The EnerGuide label on a top-loading washing machine like this one **(b)** may state that the appliance uses 418 kW·h per year, whereas a front-loading washing machine may use 282 kW·h per year **(c)**.

In section 4.1, you learned that power is the rate of transforming energy. The equation that corresponds to this definition is

$$P = \frac{\Delta E}{\Delta t}$$

where P is power measured in joules per second (J/s), or watts (W),
ΔE is energy transformed in joules (J), and
Δt is time interval in seconds (s).

This equation applies to a variety of systems, including mechanical, fluid, and electrical systems. However, another equation can be used to find power in electrical systems:

$$P = \Delta VI$$

where P is electrical power in watts (W),
ΔV is electric potential difference in volts (V), and
I is electric current in amps (A).

One watt is a small amount of power, so electrical power is often stated in kilowatts (1 kW = 10^3 W) or megawatts (1 MW = 10^6 W).

The power ratings of several appliances and some electrical generating stations are listed in **Table 1**. For appliances, these ratings apply to the electrical power needed by the appliances, not the power output.

Table 1　Typical Power Ratings of Electrical Appliances and Generating Stations

Appliance or Generating Station	Electrical Power Rating	Useful Output
stereo	30 W	sound
VCR	40 W	sound, TV signal
TV	180 W	sound, TV signal
computer	200 W	calculations
refrigerator	200 W	heat removal
vacuum cleaner	500 W	air removal
microwave oven	750 W	radiation
toaster	1000 W, or 1 kW	radiant energy
iron	1.2 kW	heat
clothes dryer	5 kW	heat and motion
electric stove	6 kW–10 kW	heat
Niagara Falls generating station	1900 MW	electric current
Pickering nuclear generating stations A and B	4120 MW	electric current
Three Gorges generating station in China (**Figure 2**)	18 200 MW	electric current

Figure 2
The Three Gorges are located along China's Yangtze River. A huge hydroelectric dam is under construction there. The reservoir behind the dam will be up to 175 m deep and 660 km long (about one-third the size of Lake Ontario). Although the dam will provide large quantities of much-needed electrical energy, it is controversial for cultural and environmental reasons: It will bury 21 cities and more than 1000 archaeological sites.

> ▶ **SAMPLE** problem **1**

Calculating Electrical Power

Calculate the power rating for a small colour TV connected to a 120-V circuit and drawing 1.5 A of current.

Solution

$\Delta V = 120 \text{ V} = 1.2 \times 10^2 \text{ V}$

$I = 1.5 \text{ A}$

$P = ?$

$$P = \Delta V I$$
$$= (1.2 \times 10^2 \text{ V})(1.5 \text{ A})$$
$$P = 1.8 \times 10^2 \text{ W}$$

The power rating is 1.8×10^2 W.

> ▶ **Practice**

Understanding Concepts

1. Calculate the power rating for each of the following:
   (a) A 120-V electric sander uses 3.0 A of current.
   (b) An electric can opener operates at 2.2 A in a 1.2×10^2-V circuit.
   (c) A portable radio, using four 1.5-V cells in series, uses a current of 0.60 A.

2. Rearrange the equation $P = \Delta V I$ to solve for (a) ΔV and (b) I.

3. Calculate the potential drop across a 0.90-W calculator that uses a current of 0.10 A.

4. Calculate the current used by a 1.3-kW kettle in a 120-V household circuit.

Answers

1. (a) 3.6×10^2 W
   (b) 2.6×10^2 W
   (c) 3.6 W

3. 9.0 V

4. 11 A

The Cost of Electrical Energy

The cost of any form of energy is an important consideration. To learn how power companies charge customers for electrical energy, we can rearrange the defining equation for power to solve for energy transformed:

energy transformed = power × time interval, or

$$\Delta E = P \Delta t$$

Thus, if we know the power rating (in watts) of an appliance and the time interval in which the appliance is used (in seconds), we can find the energy consumed (in joules).

After the energy consumed has been calculated, it can be used in the following cost equation to find the cost of the electrical energy:

cost = rate × energy consumed, or

cost = rate × ΔE

The rate must be expressed in an appropriate unit, such as ¢/MJ or ¢/(kW·h).

> ▶ **SAMPLE** problem **2**
>
> ### Calculating Cost (Megajoules)
>
> A 1.2-kW hair dryer is used for 5.0 min. Assume the cost rate is 3.6 ¢/MJ.
> Calculate
>
> (a) the energy consumed in joules and megajoules
> (b) the cost of the energy
>
> **Solution**
>
> (a) $P = 1.2 \text{ kW} = 1.2 \times 10^3 \text{ W}$
> $\Delta t = 5.0 \text{ min} = 3.0 \times 10^2 \text{ s}$
> $\Delta E = ?$
>
> $$\Delta E = P\Delta t$$
> $$= (1.2 \times 10^3 \text{ W})(3.0 \times 10^2 \text{ s})$$
> $$\Delta E = 3.6 \times 10^5 \text{ J, or } 0.36 \text{ MJ}$$
>
> The energy consumed is 3.6×10^5 J, or 0.36 MJ.
>
> (b) rate = 3.6 ¢/MJ
> $\Delta E = 0.36$ MJ
> cost = ?
>
> $$\text{cost} = \text{rate} \times \Delta E$$
> $$= \left(3.6 \frac{¢}{\text{MJ}}\right)(0.36 \text{ MJ})$$
> $$\text{cost} = 1.3¢$$
>
> The cost of the energy is 1.3¢.

In many regions, power companies calculate energy consumed in units other than megajoules. Thus, in the equation $\Delta E = P\Delta t$, if power is measured in kilowatts (kW) and time interval in hours (h), energy is stated in kilowatt hours (kW·h).

> ▶ **SAMPLE** problem **3**
>
> ### Calculating Cost (Kilowatt Hours)
>
> A TV rated at 220 W is turned on for 4.5 h. Calculate
> (a) the energy consumed (in kilowatt hours)
> (b) the cost of the energy consumed using a typical rate of 8.9 ¢/(kW·h)
>
> **Solution**
>
> (a) $P = 220 \text{ W} = 0.22 \text{ kW}$
> $\Delta t = 4.5 \text{ h}$
> $\Delta E = ?$
>
> $$\Delta E = P\Delta t$$
> $$= (0.22 \text{ kW})(4.5 \text{ h})$$
> $$\Delta E = 0.99 \text{ kW·h}$$
>
> The energy consumed is 0.99 kW·h.

(b) rate = 8.9 ¢/(kW·h)
    ΔE = 0.99 kW·h
    cost = ?

$$\text{cost} = \text{rate} \times \Delta E$$

$$= \left(8.9 \ \frac{¢}{\text{kW·h}}\right)(0.99 \ \text{kW·h})$$

    cost = 8.8¢

The cost of the energy consumed is 8.8¢.

▶ **TRY THIS** activity

Appliance Energy Consumption

Choose several electrical appliances used in your home. Determine the power rating of each appliance from **Table 1**, page 352, from the appliance's instruction manual or from the label plate on the appliance. Estimate as accurately as possible the amount of time that each appliance would be in use over one year. Then calculate the electrical energy consumed and the cost of that energy for one year. (Use the total cost rate for the production, transmission, and administration of the energy charged by your local electric power company.) Set up a spreadsheet or other appropriate means to summarize your report.

▶ **Practice**

Understanding Concepts

5. For each of the following, calculate the energy consumed (in megajoules) and the cost of the energy. Assume a rate of 2.5 ¢/MJ.
   (a) A 75-W stereo is operated for 16 h.
   (b) An air conditioner rated at 6.5×10^2 W is operated for 8.2 h.

6. For each of the following, calculate the energy consumed (in kilowatt hours) and the cost of the energy. Assume a rate of 9.2 ¢/(kW·h).
   (a) A 0.32-kW drill is used for 0.50 h.
   (b) A 2.5-kW oven is operated for 2.4 h.

Answers

5. (a) 4.3 MJ; 11¢
   (b) 19 MJ; 48¢

6. (a) 0.16 kW·h; 1.5¢
   (b) 6.0 kW·h; 55¢

Applying Inquiry Skills

7. Design a questionnaire that asks people to list the five devices in their homes that consume the most electrical energy in a one-year period, starting with the one that consumes the most. Give the questionnaire to 10 people who are not in your physics class. Tabulate the results, and draw a conclusion as to whether people know which devices consume the most energy.

Making Connections

8. Search the Internet to find out more about appliances that display the Energy Star symbol.
   (a) State the standards that appliances in various categories must meet in order to display the Energy Star label.
   (b) Why do the requirements differ among categories?

(c) Is a low standby power important in any of the categories? How do the Canadian requirements compare with the discussion of standby power in section 4.4, page 194? Explain your answer.

 www.science.nelson.com

9. Some electric power companies decrease their energy charges when the amount of energy consumed by an industrial plant drops to an agreed minimum.
   (a) Why do you think this price scheme exists?
   (b) Is this a wise method of charging for electrical energy? Explain your answer.

SUMMARY *Electrical Power and Energy*

- Electrical power, or the rate of using electrical energy, can be found using the equation $P = \Delta VI$. Its SI unit is the watt.

- The cost of electrical energy can be found using the equation cost = rate $\times \Delta E$, where the rate is given in cents per megajoule (¢/MJ) if the energy consumed is given in megajoules, or in cents per kilowatt hour (¢/(kW·h)) if the energy is given in kilowatt hours.

▶ *Section 7.9* *Questions*

Understanding Concepts

1. Copy **Table 2** into your notebook, and complete it. Include the equation used to find each unknown quantity.

Table 2   For Question 1

	P (W)	ΔV (V)	I (A)	ΔE (J)	Δt (s)
(a)	?	12	0.15	?	25
(b)	18	6.0	?	?	15
(c)	55	?	0.46	9.0 × 10³ J	?

2. Calculate the power rating in each of the following:
   (a) An electronic toy, using a 9.0-V battery, uses 0.20 A of current.
   (b) A 240-V water heater uses 21 A of current.
   (c) A computer printer operates at 0.92 A and 120 V.

3. Calculate the potential drop across a 45-W DVD player that uses a current of 2.5 A.

4. Determine the current used by an electric clock that uses 2.4 W of power in a 120-V household circuit.

5. A 3.8-kW oven operates on a 240-V household circuit.
   (a) Calculate the current total the oven uses.
   (b) Determine the energy consumed (in megajoules) by the oven in 75 min.
   (c) At a rate of 2.5 ¢/MJ, how much does this energy cost?

6. Many light bulbs are designed to burn out after 3000 h of operation. Assume that a 100-W bulb is left on for 3000 h (about 4 months) and that the cost rate of the electrical energy is 9.8 ¢/(kW·h).
   (a) Find the energy consumed (in kilowatt hours).
   (b) Calculate the cost of the energy over the lifetime of the bulb.

Applying Inquiry Skills

7. Electric meters that monitor household energy may be digital or analog (dial).
   (a) If an analog meter or a model of one (**Figure 3(a)**) is available, practise reading its scale.

(b) The first meter reading in **Figure 3(b)** is 47 840 kW·h. Interpret and record the second reading, taken two months later. Assuming a rate of 9.6 ¢/(kW·h), calculate the cost of the electrical energy consumed in the two months.

(a)

(b)

May 1

× 10

July 1

× 10

Figure 3
(a) A model of the dials on an electric meter. The dial on the left rotates counterclockwise, the next one rotates clockwise, the next counterclockwise, and the last clockwise.
(b) The number is read from left to right and then multiplied by 10 to obtain the value in kilowatt hours (kW·h). The energy consumed is the difference between the two readings.

Making Connections

8. A 15-A circuit breaker is used in a 120-V household circuit.
   (a) What is the maximum power available in this circuit?

(b) What is the maximum number of 60-W light bulbs that can operate at the same time on this circuit?

9. Electric power distribution companies pay the market rate for electrical energy and, if possible, pass the charges on to the customers.
   (a) What conditions tend to make the rate rise dramatically?
   (b) In parts of England, the cost of electrical energy used during the day is nearly three times that for nighttime use. Owners of electric water heaters often choose to heat their water overnight because the process requires a large amount of electrical energy. How could this principle be applied in your area to reduce electrical energy use during peak demand periods?
   (c) Research a new technology called "remote interval metering," featured in some suburbs. This system keeps track of the electrical energy used in residential areas during different parts of the day. Do you recommend that this technology be available to all consumers? Explain why or why not.

 www.science.nelson.com

10. (a) Substitute the Ohm's law equation into the equation for electrical power to express electric power in terms of electric current and resistance.
    (b) Use the equation in (a) to determine the power wasted through heat along a low-voltage electrical transmission line with a current of 2.0×10^2 A and a resistance of 0.50 Ω.
    (c) Repeat (b) for a high-voltage transmission line with a current of 2.0 A and the same resistance.
    (d) Are transmission lines more efficient in transmitting electrical energy at a high or low voltage? Does this agree with what you found in section 4.4, page 194?

Troubleshooting Electrical Faults

You can learn a lot by troubleshooting appliances and other devices that have electrical faults. You can bring appliances from home or use those provided by your teacher, as long as they use only low-voltage DC. The appliances can also be simulated on the computer; this method allows you to troubleshoot both AC and DC devices safely. Whether you use real or simulated devices in this investigation, you should follow the process of troubleshooting as outlined in the procedure steps.

Question

(a) Read all the instructions, and then make up one or two questions you can answer for this investigation.

Prediction

(b) Offer a prediction to answer your question(s).

Materials

For each group of two or three students:

at least one low-voltage DC appliance or electrical setup that has a fault (or a troubleshooting computer simulation)

multimeter (**Figure 1**)

tools needed to dismantle components of the appliance (e.g., screwdrivers, pliers, etc.)

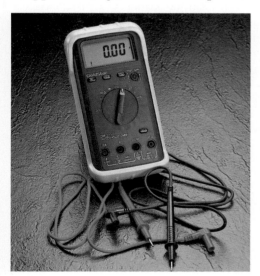

Figure 1
A multimeter can be used to measure current, potential difference, and resistance. Resistance is most important in this investigation.

Inquiry Skills

● Questioning	● Conducting	● Evaluating
● Predicting	● Recording	● Communicating
● Planning	● Analyzing	● Synthesizing

 Do not experiment with any devices that provide these warnings: "Do not remove back," and "No serviceable parts inside," or any similar warnings. These devices probably contain high-voltage components (capacitors) that remain dangerous even after the plug is removed.

Procedure

1. *Researching Background Information*

   • Check the instruction manual (or software data) for information that may help analyze the device.

   • Ask the operator or user to explain any observed difficulties with the device.

 Use only low-voltage DC devices.
   Do not plug any device into a wall outlet.
   For any battery-operated device, be sure that the battery is removed.
   Do not shake the device you are troubleshooting or pull hard on connecting wires or other components. These actions can cause additional problems.

2. *Investigating the Symptoms of the Problem*

   • Start by checking the most obvious source of a problem, such as a burned-out bulb or dead battery.

   • Look at the electrical input and output connections, which are the locations where a fault is most likely to develop. (For example, if a flashlight is not working, the bulb connection may be at fault.)

3. *Identifying Possible Causes of the Problem*

   • Start at the input and work slowly and methodically through the circuit(s), looking for possible causes of the problem. Always look for simple problems first. (For example, the screw holding the wire to a switch may be loose, causing an open circuit.) If you cannot find any, start looking for more complex problems.

- Make a systematic list of the possible causes of the problem.

4. *Testing to Verify the Causes of the Problem*
   - Starting with the most probable cause, test the causes you listed one at a time until you find the fault. Use a meter, such as an ohmmeter or a multimeter on the resistance setting, to test for a short circuit (zero resistance) or an open circuit (infinite resistance) in the input cord or other suspected area.
   - Record the test results in your list of possible causes.

5. *Suggesting a Solution to the Problem*
   - Give a recommendation to solve the problem.

Analysis

(c) List safety precautions you recommend when using tools to troubleshoot electrical devices.

(d) What general safety concerns should a person be aware of when troubleshooting electrical devices?

(e) You plug a lamp into a wall outlet and turn on the light, but the light does not go on. List the steps you would take to troubleshoot the problem. (Assume the first three steps don't work on the first try.)

(f) Give reasons why troubleshooting is a useful life skill, even if you do not intend to repair your own electrical appliances.

Evaluation

(g) Are you satisfied with your question(s) and prediction(s)? Explain your answer.

(h) If you were to start this investigation again, what would you do to improve it?

Synthesis

(i) In real-life troubleshooting, a repair or maintenance technician carries the process further than was done in this investigation. Describe what the technician would do.

CAREER CONNECTION

Electricians—both commercial and residential—can earn good hourly rates. Not only do electricians test and troubleshoot electrical systems, but they also maintain, design, and replace them. Some examples of where electricians are needed are for renovations, in construction, and working with heat, lighting, and alarm systems.

GO www.science.nelson.com

Key Understandings and Skills

7.1 Electrical Circuits

- **Current electricity** is the flow of electric charges; in an **electrical circuit** the charges flow from a **source** of energy through an **electrical conductor** to a **load**.
- A switch is a **control** that allows an **open circuit** to become a **closed circuit**.
- Circuit symbols are used to draw electrical circuit diagrams.
- Energy-transformation equations can be used to summarize the operation of an electrical circuit.

7.2 Electric Current

- **Electric current** (I), with the SI unit **ampere** or **amp** (A), is a measure of the number of charges per second passing a point in a circuit.
- In **direct current (DC)**, the electric charges do not reverse direction; in **alternating current (AC)**, the charges reverse direction periodically.
- **Conventional current** is the flow of positive charges; it is opposite in direction to **electron flow**.
- In a circuit, electric current is measured with an **ammeter** connected in **series**.

7.3 Electric Potential Difference

- **Electric potential difference** or **voltage** (ΔV) is a measure of the energy per charge given by a source (**potential rise**) or given to a load (**potential drop**); its SI unit is the **volt** (V).
- Voltage is measured with a **voltmeter** connected in **parallel** across a source or a load.

7.4 Electric Resistance and Ohm's Law

- **Electric resistance** (R) is a measure of how much an electrical component opposes the flow of charge; its SI unit is the **ohm** (Ω).
- **Ohm's law** states that the ratio of the potential drop across a resistor to the current through it is constant and equals the resistance.

7.5 Investigation: Testing Resistors

- A graph of experimental data can be used to determine whether a resistor obeys Ohm's law.

7.6 Series and Parallel Circuits

- In a **series circuit**, the current moves along one path; in a **parallel circuit**, the current moves along two or more paths.
- **Kirchhoff's current rule** (**KCR**) states that the total current into a junction in an electrical circuit equals the total current out of the junction.
- **Kirchhoff's voltage rule** (**KVR**) states that in any complete path in an electrical circuit, the sum of the potential rises equals the sum of the potential drops.
- The **equivalent resistance** (R_t) of resistances can be found using the appropriate equation for resistors connected in series or in parallel.

7.7 Investigation: Resistors in Series and Parallel

- The properties of series and parallel circuits can be investigated experimentally.

7.8 Electrical Safety

- Electrical hazards must be identified and corrected because of the dangers to the human body as well as fire hazards.
- An **overloaded circuit** is one with too many loads connected in parallel. This hazard, which causes overheating, is reduced by using a circuit breaker or a fuse.

7.9 Electrical Power and Energy

- Electrical power is the rate of transforming electrical energy; it can be found using two different equations.
- The cost of electrical energy is the product of the rate and the electrical energy consumed.

7.10 Investigation: Troubleshooting Electrical Faults

- Troubleshooting faults in electrical circuits requires researching background material, investigating symptoms, identifying possible causes, testing the possible causes, and suggesting a solution.

Key Terms

7.1
current electricity
electrical circuit
open circuit
source
electrical conductor
load
control
closed circuit
short circuit

7.2
electric current
ampere (amp)
direct current (DC)
alternating current (AC)
conventional current
electron flow
ammeter
series connection

7.3
electric potential rise
electric potential drop

electric potential
  difference
volt
voltmeter
parallel connection

7.4
electric resistance
electrical insulator
ohm
resistor
Ohm's law

7.6
Kirchhoff's current rule
  (KCR)
Kirchhoff's voltage rule
  (KVR)
equivalent resistance

7.8
overloaded circuit

Key Equations

7.4

- $R = \dfrac{\Delta V}{I}$

7.6

- $R_t = R_1 + R_2 + \ldots + R_n$
  (resistors in series)

- $\dfrac{1}{R_t} = \dfrac{1}{R_1} + \dfrac{1}{R_2} + \ldots + \dfrac{1}{R_n}$
  (resistors in parallel)

7.9

- $P = \Delta VI$
- $\Delta E = P\Delta t$
- $\text{cost} = \text{rate} \times \Delta E$

Problems You Can Solve

7.1

- Describe energy transformations involving the generation or use of electrical energy, and write the corresponding energy-transformation equations.
- State the main components of an electrical circuit, and draw their symbols as used in schematic circuit diagrams.

7.2

- Recognize typical currents of electrical devices that use either DC or AC.
- Use correct symbols to draw circuit diagrams that show how to measure electric current, indicate both conventional current and electron flow on the diagrams, and show all positive and negative terminals of the instruments.

7.3

- Recognize typical voltages of the loads of common electrical devices that operate on either DC or AC.
- Use proper circuit symbols to draw circuit diagrams to show how to measure electric potential difference, and show all positive and negative terminals of the instruments.

7.4

- Recognize typical resistances of electrical loads.
- Interpret the rings on a colour-coded resistor to determine the resistance.
- Given any two of resistance, potential difference, and current, determine the third quantity.

7.5

- Experimentally determine the effect on the current through a fixed resistor when the voltage across the resistor is changed.
- Use graphing of the resulting experimental data to determine the resistance of the resistor and interpret the voltage–current graph to determine whether the resistor is ohmic or nonohmic.

7.6

- Given the voltage of individual cells, determine the total voltage when the cells are connected in series or parallel.
- Apply KCR, KVR, Ohm's law, and the equivalent resistance equations to determine unknown currents, voltages, and resistances in electrical circuits.

7.7

- Construct electrical circuits with a source, a switch, resistors, a voltmeter, an ammeter, and connecting wires to experimentally analyze circuits with resistors in series and parallel.

7.8

- Describe situations that pose dangers when using electrical circuits and devices.
- Describe ways to avoid electrical hazards.
- Describe how you can use a compound bar to design a control device in an electrical circuit.

7.9

- Given any two of electric power, energy, and time interval, determine the third quantity.
- Given any two of electric power, voltage, and current, determine the third quantity.
- Determine the cost of the electrical energy needed to operate a device for a certain time interval, given the power of the device and the cost rate of the energy.

7.10

- Describe the main steps you would take to troubleshoot an electrical device that does not function properly.

▸ *MAKE a summary*

Figure 1 shows the start of a concept map summarizing Chapter 7. Copy the diagram onto a blank page. Add details to the concept map to indicate what you have learned in the chapter. Include as many key understandings, skills, terms, and equations from the chapter as possible. Also show how the concepts link to each other.

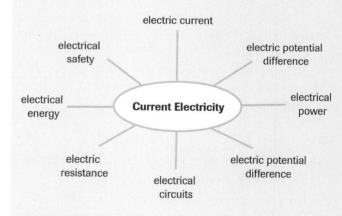

Figure 1
A concept map is a useful tool to organize your ideas.

Write the numbers 1 to 10 in your notebook. Indicate beside each number whether the corresponding statement is true (T) or false (F). If it is false, write a corrected version.

1. The particles that travel in the conducting wires in an electrical circuit are positively charged protons.

2. If a coin is connected across the terminals of a nearly dead 9-V battery, a short circuit results.

3. To measure electric current, an ammeter must be connected in series in the circuit.

4. If two 1.5-V DC cells are connected in series, the total electric potential rise is 3.0 V.

5. Digital voltmeters are connected in series with a resistor, but analog voltmeters are connected in parallel across the resistor.

6. A good electrical insulator is a poor electrical conductor.

7. In a series circuit with three resistors, the current has three paths to follow.

8. The equivalent resistance of two 4.0-Ω resistors connected in series is 2.0 Ω.

9. Electrical energy can be measured in joules, megajoules, or kilowatt hours.

10. A typical rate for the cost of electrical energy is between $6 and $12 per kilowatt hour.

Write the numbers 11 to 17 in your notebook. Beside each number, write the letter corresponding to the best choice.

11. An electrical circuit that operates correctly is a(n)
    (a) short circuit
    (b) closed circuit
    (c) overloaded circuit
    (d) open circuit

12. The term "alternating current" means that the
    (a) electric current, measured in amperes, increases and decreases regularly
    (b) electric potential difference, measured in volts, increases and decreases regularly
    (c) electric current can travel in more than one path
    (d) electric current moves back and forth regularly in the circuit

13. The wall receptacle where you plug in an electric desk lamp has these properties:
    (a) 120 V, 60 Hz AC
    (b) 240 V, 120 Hz AC
    (c) 9.0 V DC
    (d) 120 V DC

14. As long as the switch in **Figure 1** is open,
    (a) $A_1 = A_2 > A_3$
    (b) $A_1 > A_2 > A_3$
    (c) $A_1 > A_2 = A_3$
    (d) $A_1 = A_2 = A_3$

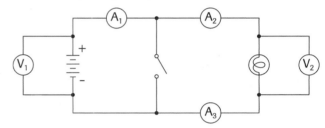

Figure 1

15. As long as the switch in **Figure 1** is open,
    (a) The circuit is a closed circuit and $V_1 = V_2$.
    (b) The circuit is a closed circuit and $V_1 > V_2$.
    (c) The circuit is an open circuit and $V_1 < V_2$.
    (d) The circuit is an open circuit and $V_1 = V_2$.

16. When the switch in **Figure 1** is closed, the light bulb
    (a) goes out, and A_1 becomes dangerously high
    (b) goes out, and A_1 drops to zero
    (c) becomes brighter, and $V_2 > V_1$
    (d) becomes brighter, and $V_2 = V_1$

17. The equivalent resistance of four 4.0-Ω resistors connected in parallel is
    (a) 1.0 Ω
    (b) 4.0 Ω
    (c) 8.0 Ω
    (d) 16 Ω

Understanding Concepts

1. Explain why copper is a good choice for electric wiring.

2. State the function of the source and the switch in an electrical circuit.

3. Write the energy-transformation equation for each device shown in **Figure 1**.

(a)   **(b)**

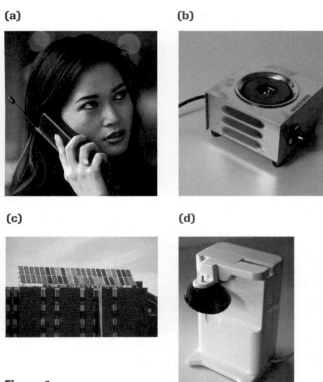

(c)   **(d)**

Figure 1

4. Give an example of an electrical device with the following energy transformations:
   (a) electrical energy → kinetic energy
   (b) electrical energy → chemical potential energy → electrical energy → sound energy

5. An electric can opener used in a 120-V circuit has a resistance of 110 Ω. Calculate the can opener's current. (Assume two significant digits.)

6. An electric razor has a resistance of 18 Ω and a current of 0.28 A. Calculate the potential drop across the razor.

7. A handheld video game, operating on 9.0 V, has a current of 7.5 mA. Determine the game's
   (a) resistance
   (b) power rating
   (c) energy consumed in 15 min

8. Explain why voltmeters must be connected in parallel while ammeters must be connected in series in a circuit.

9. Copy **Figure 2** into your notebook, and add the following:
   (a) ammeters to find I_b, I_c, and I_e
   (b) voltmeters to find the potential rise across the source and the potential drop across the two resistors in parallel
   (c) a fuse at "a"
   (d) all positive and negative terminals
   (e) arrows indicating the conventional current and the electron flow

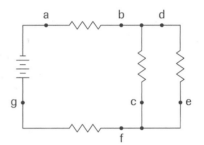

Figure 2

10. In **Figure 3**, $R_1 = 25\ \Omega$. The potential drop across R_1 is 12 V; across R_2, it is 24 V. Determine
    (a) the ntotal potential rise of the source
    (b) the current through the resistors
    (c) the resistance of R_2

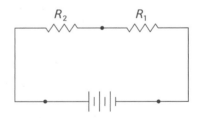

Figure 3

11. In **Figure 4**, find
    (a) the current through R_1
    (b) the resistances of R_1 and R_2
    (c) the total resistance of the circuit

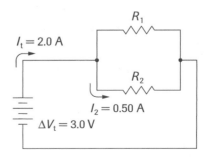

Figure 4

12. Find the current through and the potential drop across each resistor in the circuit in **Figure 5**. Assume two significant digits.

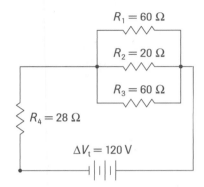

Figure 5

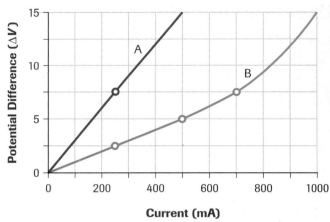

Figure 6

13. Calculate the total potential rise when three 9.0-V batteries are connected (a) in series and (b) in parallel.

14. List the advantages of connecting appliances in parallel in a household circuit.

15. Explain the purpose of a circuit breaker or fuse.

16. Calculate the current in an electric radio that uses 3.6 W of power in a 120-V circuit.

17. Calculate the electric potential in a 35-W stereo CD player that has a current of 2.5 A.

18. A 2.2-kW stove burner is used for 24 min. The cost of the electrical energy is 3.5 ¢/MJ. Calculate
    (a) the energy consumed, in megajoules
    (b) the cost of the energy

19. A homeowner leaves eight 200-W floodlights on for 12 h each night for 150 nights. At a cost of 8.8 ¢/ (kW·h), find
    (a) the energy consumed, in kilowatt hours
    (b) the cost of the energy consumed

Applying Inquiry Skills

20. An experiment is performed to find the relationship between current and potential for two resistors, A and B. The data collected are plotted in the graph in **Figure 6**.
    (a) Find the resistance of each resistor for a current that does not exceed 500 mA.
    (b) What happens to B when the current exceeds 500 mA?

(c) Describe two major sources of error in this type of experiment.

21. In a lab practical test, your teacher hands you three resistors that are labelled, but one label is incorrect. You must design a single circuit with no more than four instruments to determine which resistor is incorrectly labelled. Draw a circuit diagram showing how you would solve the problem, and describe your solution.

22. Explain how you would design an electrically powered hand dryer for use in a public washroom.

Making Connections

23. Most Canadian homes use hot-water tanks that hold between 150 L and 270 L of water. The water, which is heated by electric current or natural gas, is kept heated, ready for use. An alternative method of heating water, commonly used in Europe, heats the water only as it is being used. This "tankless heating" uses natural gas.
    (a) Which of the two technologies operates in standby mode?
    (b) Which energy source, electric current or natural gas, is more efficient at heating water quickly? How can you tell?
    (c) Describe the main advantages and disadvantages of using tankless water heating technology. (Research will help.)

 www.science.nelson.com

chapter

Electronics

In this chapter, you will be able to

- define and describe the concepts and units related to electronic systems
- describe the function of basic electronic circuit components, including semiconductor devices
- draw schematic diagrams of electronic circuits and construct real or simulated electronic circuits
- analyze and describe the operation of electronic devices that control other systems
- distinguish between, and explain the functions of, analog and digital systems
- describe examples of miniaturized circuits that serve a particular function in electronic circuits
- investigate the use and historical development of an electronic device

Getting Started

We are surrounded by applications of electronics. Appliances, communication equipment, and tools that use electronic components affect our lives in many ways. Perhaps the computer is the electronic innovation that has changed our lives most (**Figure 1**). Computers are used in everything from kitchen appliances to watches, and even in the measurement of athletic performance. For example, cyclists can monitor their heart rate, average speed, and other quantities by wearing a transmitter and receiver. The receiver (in the form of a wrist watch) consists of a computer that contains a battery, a circuit board, a memory chip, a visual output (the display), and an audio output (the beeper). This chapter explores the functions of these and other electronic components, and describes how they are applied to many useful devices.

The chapter builds on the concepts presented in Chapter 7. What you learned about electric current, potential difference, and resistors will be applied here. Your skill in drawing schematic diagrams of electrical circuits will be extended to include basic electronic circuits. You will also investigate the properties of common electronic components.

REFLECT on your learning

1. In your notebook, create a table with the headings shown in **Table 1**, followed by six rows. In the first column, list the following abbreviations, then complete as much of the table as you can: LED; DVD; RAM; CPU; ROM; HDTV. As you go through this chapter, update your table.

Table 1 Typical Abbreviations in Electronics

Abbreviation	Meaning	Uses of the Device

2. Use each of the following terms in a sentence that demonstrates its meaning in the context of electronics:
   (a) detector
   (b) digital
   (c) photosensitive
   (d) data reception
   (e) amplify
   (f) miniaturization

3. A "semiconductor" is an example of an electronic component. How do the conductivity properties of a semiconductor compare with those of a conductor and an insulator?

4. When a camera flash attachment is turned on, there is a buzzing sound. It rises in pitch for a few seconds until the flash is ready.
   (a) Describe what is happening just after the switch is turned on.
   (b) Is it wise or unwise to leave the flash on at all times? Explain why.

5. (a) List ways in which computers have affected your life.
   (b) How do you think computers will affect your life in the future?

Figure 1
Assembling components of
personal computers

Charging and Discharging

Your teacher will perform a demonstration using the
following materials: hard plastic base; one large flat metal
disk and one small disk, each with an insulating handle;
wool or fur; electroscope. Read each step, and predict
what will happen. Observe the demonstration, and then
record your observations.

- Rub the base vigorously with the wool or fur, leaving it
  with a negative charge. Hold the small conductor by
  the handle so that it is just above the base. Note
  whether the conductor is charged by moving it toward
  the electroscope from a distance of several
  centimetres.

- Repeat the previous step, but this time hold the
  conductor close to the base and then touch the edge
  of the conductor with a finger (i.e., ground the
  conductor). Test the charge using the electroscope.

- Charge the large conductor using the grounding
  method described above. Test the charge on the
  conductor by using the electroscope.

- With the charge still on the large conductor, move the
  conductor away from the base and try to discharge it.
  Test any remaining charge on the conductor using the
  electroscope.

(a) Which materials used in this activity are electrical
conductors and which are insulators?

(b) Describe an effective method for charging a
conductor.

(c) How can a conductor be discharged?

(d) Does the maximum size of the charge on a
conductor depend on the size of the conductor?
Explain your answer.

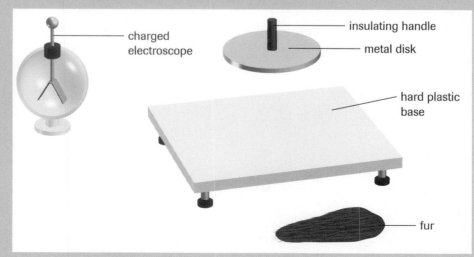

charged
electroscope

insulating handle

metal disk

hard plastic
base

fur

Figure 2
Charging a conductor and testing
the charge

electronics the study of the controlled conduction of electrons, or current, through gases, solids, and vacuums

Electronics is the study of the controlled conduction of electrons, or current, through gases, solids, and vacuums. In this chapter, we concentrate on current conducted through solids, called *solid-state electronics*. This has become an important and exciting area of physics. In particular, the use of *semiconductors* has grown rapidly; they are now the basic building blocks of many essential electronic components (**Figure 1**).

semiconductor a solid, crystalline substance that conducts electric current better than an insulator but not as well as a conductor

A **semiconductor** is a solid, crystalline substance that conducts electric current better than an insulator but not as well as a conductor. For example, semiconductors conduct electric current better than paper, wood, and air but not as well as copper wire. A **semiconductor device** is an electronic component made of a semiconductor. The advantage of using a semiconductor is that the current can be controlled in ways not possible in a conductor. This allows for numerous applications, as you will see in this chapter.

semiconductor device an electronic component made of a semiconductor

LEARNING TIP

The Prefix "Semi"
The prefix "semi" provides a clue about the semiconductor's conduction properties. "Semi" means "partial." Many common words and terms have the same prefix, for example, semifinals, semigloss paint, semipermeable membranes, semiprecious stones, semisweet chocolate, and semitones in music.

Figure 1
Aircraft rely on electronic technology. Flight variables are monitored electronically, and aircraft can be flown and landed automatically using electronic instruments.

Many substances can be used as semiconductors. **Table 1** describes some of the more common substances and their applications. Some of the materials are mined, while others are crystals grown in laboratories.

Table 1 Common Semiconductors

Material	Properties	Applications*
silicon	• can be used as a pure crystal • impurities can be added to give special properties	wafers; chips; diodes; transistors; integrated circuits
gallium arsenide	• needs lower voltage than silicon • conducts current faster than silicon • not affected by X rays or gamma rays • more expensive than silicon	light-emitting diodes; infrared light-emitting diodes; laser diodes; infrared detectors; amplifiers; integrated circuits
selenium	• more sensitive than most materials to solar radiation • withstands surges of high voltage better than other materials	photocells; changing AC to DC
germanium	• requires added impurities to give desirable qualities • semiconductor pioneer material • can be damaged by high temperatures	diodes; transistors
metal oxides	• useful when low electric current needed • permit high-speed switching, thus useful in computing • can be damaged by static electricity	some diodes and transistors; miniature integrated circuits

* You will learn more about these applications later in the chapter.

The reason that some materials are better conductors than others relates to electron configuration. The outer electron region of an atom is called the *valence region*, or the *valence shell*. An electron in this region is called a *valence electron*. Silver atoms have one valence electron (**Figure 2(a)**). That electron is not tightly bound to the atom's nucleus, so it is free to move to an adjacent silver atom in an electric current. This makes silver an excellent electrical conductor. Germanium atoms, however, have four valence electrons (**Figure 2(b)**). These electrons are more strongly attracted to the nucleus of the atom and do not easily shift from one atom to the next.

(a)

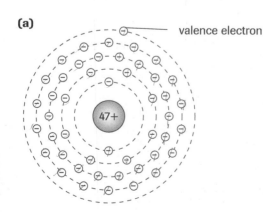

valence electron

(b)

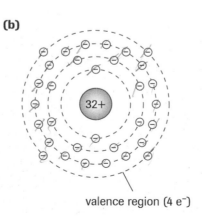

valence region (4 e⁻)

Figure 2
(a) Silver
(b) Germanium

(a)

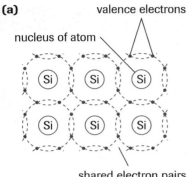

valence electrons

nucleus of atom

shared electron pairs

(b) hole

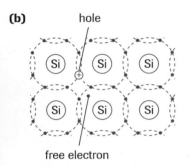

free electron

Figure 3
(a) A simple model of the crystalline structure of silicon
(b) Crystalline silicon with a free electron and a hole

pure semiconductor a semiconductor made of crystalline material with no impurities

free electron an electron that has broken away from an atom in a semiconductor; it has a negative charge

hole the vacancy left by a free electron in a semiconductor; it has a positive charge

impurities substances added to semiconductors to increase their ability to conduct electric current

doping the process of adding impurities to semiconductors

donor atom a doping atom with five valence electrons

n-type semiconductor a crystalline material with free electrons from donor impurities

To understand how semiconductors work, it is helpful to understand the crystal structure of a **pure semiconductor**, that is, one with no impurities, such as pure silicon (**Figure 3(a)**). The four outer, or valence, electrons of each atom are shared with the surrounding atoms. Therefore, the atoms tend to be stable at low temperatures. In other words, the electrons don't move easily, as they would in a copper wire. However, with a high energy input, such as an increase in temperature, the atoms become agitated, and some electrons gain enough energy to break free of the rigid structures.

If an electron breaks free from an atom, it becomes a **free electron**. In a semiconductor, the vacancy that the free electron leaves is called a hole. Think of the hole as a temporary area where an electron is very strongly attracted by the electric force of the nucleus. (In a conductor, the valence electrons are not as strongly attracted to the nucleus.) Since the electron has a negative charge, the hole is defined as having a positive charge. See **Figure 3(b)**.

To increase the number of electrons that become free, tiny amounts of other substances, called **impurities**, are added to the semiconductor. The process of adding these impurities is called **doping**. One type of doping results in increased electron flow, while another type results in increased hole flow. In both cases, the greater the amount of doping, the greater the current in the semiconductor.

Arsenic (As) has five valence electrons (**Figure 4(a)**). Each arsenic atom has one more electron than is needed to bond with silicon, a semiconductor. That extra electron is easily freed from the atom. Thus, when a tiny amount of arsenic is added to a semiconductor to increase conductivity, the arsenic is called a **donor atom**. If a semiconductor has enough donor atoms, the free electrons can migrate in a current (**Figure 4(b)**). Since the majority of the current in the doped material is electron flow, the material is called an **n-type semiconductor**, because of the negative charge on the electrons. **Table 2** lists features of n-type semiconductors.

(a)

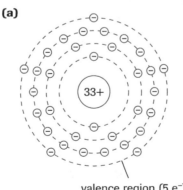

valence region (5 e⁻)

(b)

extra electron can move in a current

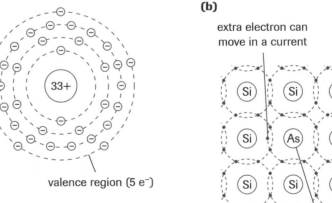

nucleus of donor atom (arsenic)

Figure 4
(a) Arsenic has five valence electrons.
(b) In a silicon crystal doped with arsenic, four of arsenic's valence electrons bond with silicon, freeing arsenic's fifth valence electron. For effective doping, approximately one arsenic atom is needed for every million silicon atoms.

Table 2   Features of *n*-Type Semiconductors

typical doping element	arsenic
symbol	As
number of valence electrons	5
type of atom	donor
majority flow in material	electrons
reason for the name *n*-type	negative charge (e⁻)

LEARNING TIP

Intrinsic and Extrinsic Semiconductors
Some reference sources call a pure semiconductor an *intrinsic* semiconductor. The corresponding name for a doped semiconductor is an *extrinsic* semiconductor.

You can probably predict the type of atom needed to produce a ***p*-type semiconductor**, which has a majority of holes that migrate in the doped material. In this case, the atom has three valence electrons. An example is the element gallium (Ga), shown in **Figure 5(a)**. This type of atom is called an **acceptor atom** because it has holes that accept electrons from adjacent atoms. As the electrons migrate (**Figure 5(b)**), they leave holes (positives) behind. The uses of *n*-type and *p*-type semiconductors are presented later in the chapter.

p-type semiconductor a crystalline material with holes from acceptor impurities

acceptor atom a doping atom with three valence electrons

(a)

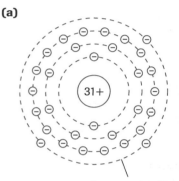

valence region (3 e⁻)

(b)

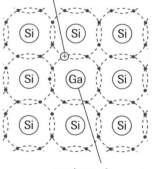

hole accepts a moving electron

nucleus of acceptor atom (gallium)

Figure 5
(a) Gallium has three valence electrons.
(b) In a silicon crystal doped with gallium, three of silicon's four valence electrons bond with gallium, leaving a hole that can accept more electrons.

▶ **Practice**

Understanding Concepts

1. Arrange the following substances in order of ability to conduct current, from high to low: silicon, paper, copper.

2. (a) How many valence electrons are there in an atom of copper and an atom of silicon (**Figure 6**)?
   (b) Explain how the number of valence electrons affects a material's electrical conductivity.

(a)   **(b)**

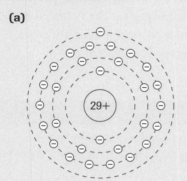 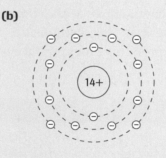

Figure 6
(a) Copper
(b) Silicon

3. What can be done to a pure semiconductor material to free an electron?

4. Describe how the electric current in a semiconductor differs from the electric current in a metal conductor.

5. In a table similar to **Table 2**, summarize the features of *p*-type semiconductors.

6. Three materials are available to manufacture electronic components. Their atoms have these characteristics: X: three valence electrons; Y: four valence

electrons; Z: five valence electrons. Which substance(s) would you choose
(a) as a pure semiconductor? Explain why.
(b) to form an *n*-type semiconductor? Explain why.
(c) to form a *p*-type semiconductor? Explain why.

Applying Inquiry Skills

7. (a) Explain how the electric current in a *p*-type semiconductor depends on the number of acceptor atoms.
   (b) Sketch a graph to illustrate the relationship in (a).

Making Connections

8. According to **Table 1**, page 369, which semiconductors would be good choices for each of the following? Give a reason.
   (a) a graphing calculator
   (b) a solar array on the Hubble Space Telescope
   (c) the control panel lights on a medical imaging machine

SUMMARY *Semiconductors*

- Semiconductors, such as silicon and germanium, have electrical conductivity properties between those of conductors and insulators.

- In a circuit made with a semiconductor, electrons and holes in the material flow in an electric current.

- The electric current in a semiconductor can be increased by increasing the temperature (within limits) and by doping the material with impurities. Donor impurities produce *n*-type semiconductors, and acceptor impurities produce *p*-type semiconductors.

▶ *Section 8.1* **Questions**

Understanding Concepts

1. State the main difference(s) between each of the following pairs:
   (a) a semiconductor and a semiconductor device
   (b) the valence region of a semiconductor and the valence region of a conductor
   (c) the valence region of a pure semiconductor and the valence region of a doped semiconductor
   (d) a pure semiconductor and a doped semiconductor
   (e) a hole and an electron

2. Compare and contrast the direction of hole movement with the direction of conventional current.

3. What property of selenium makes it a good choice for use in photocells?

4. In a diagram, show the bonding of valence electrons in a doped semiconductor containing acceptor atoms.

Making Connections

5. According to **Table 1**, page 369, which of the semiconductors would not be good choices for each of the following items? Give a reason.
   (a) an electronic thermometer used to measure the temperature in a blast furnace
   (b) an electronic weather data transmitter located on a hill that is often struck by lightning
   (c) an electronic audio greeting card to be sold at the lowest possible cost

Semiconductors are important in the design and function of many electronic components, such as *diodes* (**Figure 1**). A **diode** is an electronic device formed by joining two differently doped semiconductors. The word diode is formed from *di* (meaning two) and *ode* (from electrode). In a *p-n diode*, a *p*-type semiconductor is joined to an *n*-type semiconductor, and the interface where they join is called the *junction*. The type of diode described here is a junction diode.

When these materials are joined and placed in a circuit, as shown in **Figure 2(a)**, an electric current can be set up. In this circuit, the negative terminal repels the electrons in the *n*-type material. These electrons migrate across the junction and fall into the holes in the *p*-type material (**Figure 2(b)**). Each electron that migrates across the junction leaves a vacancy (a hole) behind. This results in a hole migration opposite in direction to the electron migration. At the same time, the positive charges migrate across the junction, toward the negative terminal as a current. **Figure 2(c)** shows the circuit diagram for a *p-n* diode. Notice that the direction of the current in the symbol for the diode is the same as the direction of conventional current. The diode symbol is one of several symbols used for electronic components. Refer to **Table 7** in Appendix C for other symbols for electronic components.

The connection of the source to the diode in **Figure 2** is called **forward bias** because the current can be the maximum possible. The positive terminal of the source is connected to the *p* end of the diode, and the negative terminal is connected to the *n* end.

Figure 1
Diodes have two lead wires attached.

diode an electronic device formed by joining two differently doped semiconductors

forward bias the connection of the source to a diode such that positive is connected to *p* and negative to *n* to allow maximum current

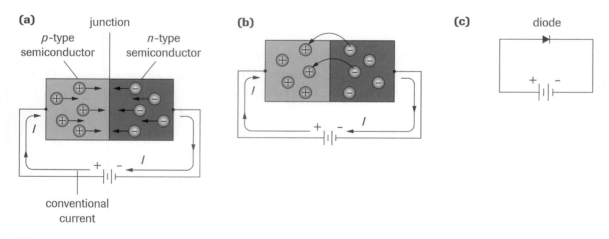

(a) junction
p-type semiconductor   *n*-type semiconductor
conventional current

(b)

(c) diode

Figure 2
(a) The negative terminal repels the electrons, forcing them to migrate through the *n*-type semiconductor and across the junction of the diode. As the electrons migrate in one direction, the holes migrate in the opposite direction.
(b) The electrons temporarily fill the holes as migration continues.
(c) The circuit diagram of the setup

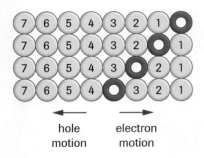

Figure 3
Candy can help us visualize the fact that electron migration to the right results in hole migration to the left. Each row in the diagram occurs a short time after the previous row.

hole motion   electron motion

To help visualize how electron migration in one direction in a diode can result in hole migration in the opposite direction, refer to **Figure 3**. The red candy represents the hole in each step as time progresses; the blue-green candy represents the electrons. As the electrons migrate to the right in this case, the red candy, or hole, migrates to the left.

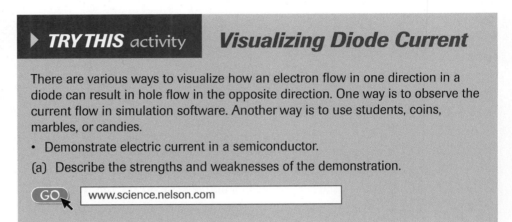

▶ **TRY THIS** activity   *Visualizing Diode Current*

There are various ways to visualize how an electron flow in one direction in a diode can result in hole flow in the opposite direction. One way is to observe the current flow in simulation software. Another way is to use students, coins, marbles, or candies.

• Demonstrate electric current in a semiconductor.

(a) Describe the strengths and weaknesses of the demonstration.

GO   www.science.nelson.com

reverse bias the connection of the source to a diode such that negative is connected to *p* and positive to *n* to prevent current

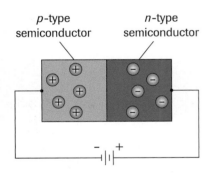

Figure 4
When the terminals of the source are reversed, the current in the diode is zero, or nearly zero. This connection is reverse bias.

If the battery terminals are reversed, as shown in **Figure 4**, the positive terminal of the source attracts the electrons. At the same time, the negative terminal attracts the holes in the *p*-section. This reversal makes it nearly impossible for electrons to migrate across the junction between the materials. This type of connection is called **reverse bias**, and it results in minimum current. You can explore the effects of forward and reverse bias in simple electronic circuits in Activity 8.3.

The graph in **Figure 5** shows that when the diode is connected in forward bias, even a small change in the voltage can result in a large change in current. However, with the diode connected in reverse bias, current is practically zero until the voltage is relatively high. These bias characteristics allow diodes to be used as switches, with current turned on and off up to millions of times each second. This is a major advantage of electronic circuits over electrical circuits with mechanical switches. (Bias characteristics are also applied in the design and use of transistors, presented in section 8.4.)

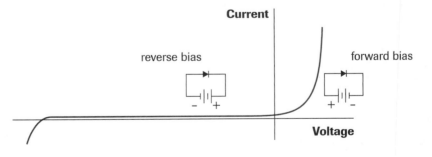

Figure 5
The current–voltage graph of a typical junction diode

When input voltages are not extreme, the current in a diode can flow in one direction only. Thus, the *p-n* diode can be used in **rectifier circuits** to convert AC to DC, as required in many electronic circuits. The easiest rectifier to understand is the *half-wave rectifier circuit,* illustrated in the circuit diagram in **Figure 6(a)**. The input is AC, which has both positive and negative voltages. The output across the resistor is only the positive half of the input, so it is DC. (See the graphs in **Figure 6(b)** and **(c)**.) The output of a *full-wave rectifier circuit* is shown in the graph in **Figure 6(d)**. Notice that the negative input voltages have been flipped around to become positive voltages. In both types of rectifier, one or more extra components are needed to smooth out the ripples in order to give a smooth DC. (You will learn how this is done in section 8.6.)

Rectifier circuits are used when the incoming electrical energy, for example, from a wall outlet, is AC but needs to be changed into DC. Recall from Chapter 7 that our audio systems (CD players, FM radio tuners, etc.), TVs, microwave ovens, computers, and other electronic appliances use DC.

Rectifier circuits are also important in the type of fuel cell technology in which AC must be changed to DC to separate hydrogen from oxygen in water. (The basic design of a fuel cell was described in section 3.7, page 162.) To separate the hydrogen and oxygen molecules in water, DC is applied to chemical cells in a process called *electrolysis.* In these cells, the hydrogen molecules collect at the negative terminals, and the oxygen molecules collect at the positive terminals. The oxygen gas is vented to the atmosphere, but the hydrogen gas is compressed and stored in tanks to be used in cars, buses, electrical appliances, and so on. This process allows electrical energy from the AC grid to be used at off-peak hours to produce clean-burning hydrogen for many uses.

Another useful type of semiconductor diode is a **light-emitting diode**, or **LED**. As energetic electrons migrate across the junction and combine with holes, some of the energy from the electrons is transformed into radiant energy (**Figure 7**). In many cases, the radiant energy is red light, which we often see in the visible displays on clocks, CD players, VCRs, car instrument panels, and so on. In other cases, the radiant energy is invisible infrared radiation, IR, so the diode is called an **infrared light-emitting diode**, or **ILED**. This type of diode is used in remote controls for TVs, VCRs, radios, and stereos.

rectifier circuit an electronic component that converts AC to DC

(a)

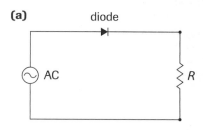

(b)

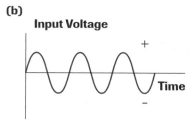

(c)

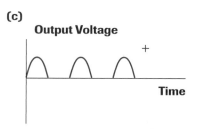

(d)

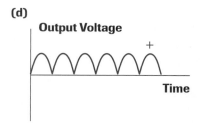

Figure 6
(a) A half-wave rectifier circuit
(b) The AC input voltage with both positive and negative voltages
(c) The output or DC voltage of a half-wave rectifier circuit with only positive voltage
(d) The output or DC voltage of a full-wave rectifier circuit, again with only positive voltage

LEARNING TIP

Graphs Involving Rectifier Circuits
The graphs of the rectifier circuit in **Figure 6** are voltage–time graphs. The shapes of the graphs would be the same if input and output currents were plotted rather than input and output voltages.

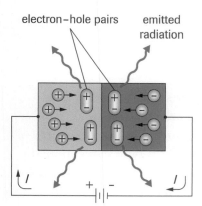

Figure 7
When the source is connected in forward bias to an LED, the current causes the diode to glow.

light-emitting diode (LED) a diode that transforms electrical energy into visible radiant energy

infrared light-emitting diode (ILED) a diode that transforms electrical energy into infrared energy

▶ *TRY THIS* activity

Testing Conductivity Using an LED

In this activity, you will test the conductivity properties of various materials using a circuit containing an LED. You will need an LED, a 220-Ω resistor, a 9-V battery, a circuit board (optional), connecting wires, an iron nail, a plastic millimetre ruler, and two metal wood screws. The resistor is connected in series with the LED so the electric current from the 9-V battery does not damage the LED.

Figure 8 illustrates the positive and negative terminals of one type of diode. The negative terminal lead (also called the *cathode*) is closest to the flat part of the diode's base; it is also shorter than the positive lead. Recall that, in order to be forward biased, the positive terminal (or *anode*) of the LED must be connected to the positive terminal of the battery.

• Construct the circuit illustrated in **Figure 9**. Make sure all components are well connected.

• Place the ends of an iron nail between the screws. Describe what you observe.

• Place the plastic ruler between the screws. Describe what you observe.

• Reposition the LED in the reverse bias direction. Repeat the previous two steps. Describe what you observe.

(a) How could your circuit be used to test electrical switches and fuses?

(b) Your circuit could have been constructed with an incandescent light bulb instead of an LED. On the Internet, research the benefits of using an LED instead of conventional incandescent bulbs in electronic equipment. Summarize your findings in a table.

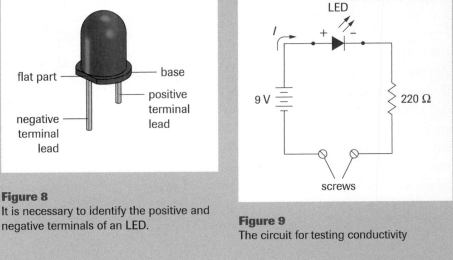

Figure 8
It is necessary to identify the positive and negative terminals of an LED.

Figure 9
The circuit for testing conductivity

GO www.science.nelson.com

photodiode a semiconductor device that converts radiant energy into electrical energy; also called a *photocell,* a *photoelectric cell,* or a *solar cell*

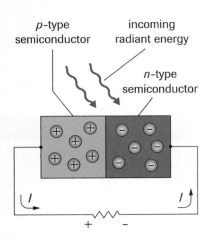

Figure 10
Using a diode as a photocell

A **photodiode**, or *photocell,* is a diode that converts radiant energy into electrical energy. Rather than requiring a source of electric potential difference, this diode generates the potential difference. The radiant energy of the incoming light causes some of the electrons in the *n*-type semiconductor to break loose and migrate across the junction to the *p*-type semiconductor. When connected to a load, an electric current can then flow (**Figure 10**).

> ### Practice

Understanding Concepts

1. Explain, with the help of diagrams, why the electric current in a diode depends on the way the diode is connected to the supply potential. Label the diagrams with the appropriate bias terminology.

2. **Figure 11** shows the DC output of a rectifier circuit in a synthesizer. Draw a diagram of the AC input.

3. Write the energy-transformation equation for the operation of an LED.

4. Refer to **Figure 7**, page 375, which shows the operation of an LED. What happens if the terminals of the source are reversed? Explain why.

Applying Inquiry Skills

5. You are given an LED, a low-voltage battery, and connecting wires. How do you decide which way to connect the LED so that it lights up? Draw a circuit diagram of the setup.

Making Connections

6. (a) Describe the energy transformations that occur in the collection and use of hydrogen in a fuel cell that requires a rectifier circuit. Write the corresponding energy-transformation equation.
   (b) Follow the links from the Nelson Web site to Stuart Energy's Web site. Draw and label a flow diagram of the Stuart Energy fuel cell that requires a rectifier circuit.
   (c) What type of fuel cell does not require a rectifier circuit?

    www.science.nelson.com

7. The wavelengths of infrared radiation are longer than the wavelengths of visible light. Also, long-wavelength waves travel around obstacles and corners more easily than short-wavelength waves. Use these facts to explain whether to use an LED or an ILED for
   (a) a TV remote control
   (b) lights on an instrument panel

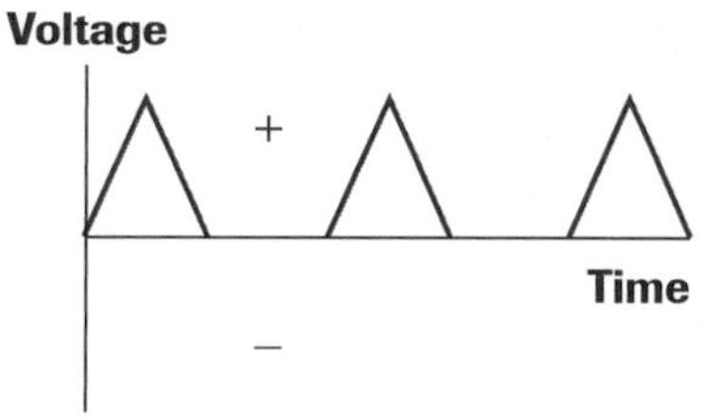

Voltage

Figure 11
This output is a "triangular waveform."

SUMMARY **Diodes**

- A diode, which consists of two semiconductors, allows electric current in a circuit in one direction only.

- Diodes can act as high-speed switching devices because they can change rapidly from forward to reverse bias.

- Diodes can be used to convert AC to DC (in rectifiers), to transform electrical energy into radiant energy (in LEDs and ILEDs), and to transform radiant energy into electrical energy (in photodiodes).

Understanding Concepts

1. What is joined at the junction of a diode?

2. Will the light bulb work in the electronic circuit shown in **Figure 12**? If so, why? If not, what would you change to make it work?

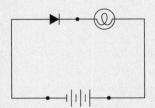

Figure 12

3. **Figure 13** shows the input signal of a rectifier circuit.
   (a) Is the input signal AC or DC?
   (b) What is needed to rectify the input?
   (c) Draw the output signal in a half-wave rectifier.
   (d) Repeat (c) for a full-wave rectifier.

Voltage

Figure 13
This input signal is called a "saw-tooth waveform."

4. Write the energy-transformation equation for the operation of a photodiode.

5. **Figure 14** shows the design of a smoke alarm in which light from an LED can scatter (or reflect) off any smoke particles that enter the chamber.
   (a) Describe the function of the diode in this control circuit.
   (b) Explain why the alarm sounds when smoke enters the chamber.

Applying Inquiry Skills

6. Using appropriate symbols, draw the following electronic circuits (use **Table 7**, Appendix C for reference):
   (a) A *p-n* diode is connected in series with a DC battery, a knife switch, and a resistor.
   (b) A photodiode operates when connected to a small DC source.
   (c) A photodiode is connected to a voltmeter.

7. Describe how you could use students to model the crystalline nature of the semiconductor components and show the flow of holes and free electrons.

Making Connections

8. *Laser diodes* are special LEDs that produce radiant energy with an extremely precise wavelength. Research laser diodes, and describe one application that you discover.

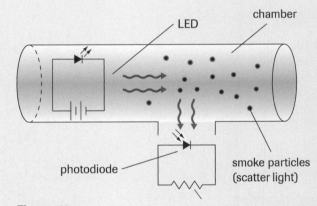

Figure 14
A smoke alarm with an LED

Using Diodes

This activity has two parts. In Part A, you will measure the resistance of a diode. In Part B, you will construct three circuits, each containing two diodes. Then you will determine the effects of the diodes in the circuits.

You must first identify the positive and negative leads on a diode: A black, white, or red band on the diode body can identify the negative lead (**Figure 1**). In some cases, the diode symbol is drawn on the diode body itself. Conventional current flows from the negative terminal; thus current flows from left to right in the diode in **Figure 1**.

Questions

Does the resistance of a diode depend on how it is connected in a circuit?

How can diodes be used to control the current in an electronic circuit?

Predictions

(a) Predict whether the resistance of the diode in reverse bias in **Figure 2** is larger than, smaller than, or the same as the resistance in forward bias.

(b) Predict whether the lamp in **Figure 3** will turn on when the switch is closed.

(c) Predict whether lamp L1, lamp L2, or both lamps L1 and L2 in **Figure 4** will turn on when the switch is closed.

(d) For the lamps in **Figure 5**, predict whether (i) all the lamps will turn on, (ii) none of the lamps will turn on, or (iii) a particular combination of lamps will turn on.

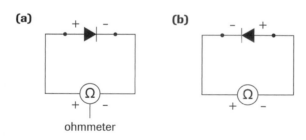

positive terminal (anode) — **coloured band** — **negative terminal (cathode)**

Figure 1
The negative (or cathode) lead is the one closest to the coloured band.

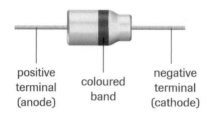

(a) — **(b)**

ohmmeter

Figure 2
(a) Resistance in forward bias
(b) Resistance in reverse bias

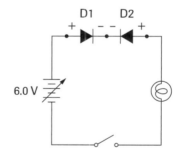

D1   D2

6.0 V

Figure 3

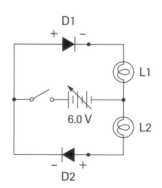

D1
L1
6.0 V
L2
D2

Figure 4

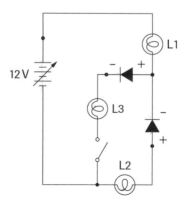

L1
12 V
L3
L2

Figure 5

Materials

For each group of three or four students:
ohmmeter or digital multimeter with resistance
   setting
two diodes
single-pole single-throw switch
variable DC power supply (to 12 V)
three 6-V, 60-mA lamps

 When unplugging the power supply, grasp the plug and pull it from the wall receptacle. Do not unplug the power supply by pulling on the cord.

Do not leave the current on for longer than needed. Wires may become overheated.

Do not exceed the voltage prescribed by your teacher.

Do not connect the terminals without a load; a short circuit will result.

Procedure

Part A

1. Switch the ohmmeter or digital multimeter to the "ohms × 10" range.

2. Choose a diode to be tested. Connect the leads across the diode as shown in **Figure 2(a)**. Measure and record the resistance of the diode.

3. Reverse the leads to the diode as shown in **Figure 2(b)**. Measure and record the resistance of the diode.

Part B

4. Construct the circuit in **Figure 3**. Close the switch and record your observations.

5. Construct the circuit in **Figure 4**. Close the switch and record your observations.

6. Construct the circuit in **Figure 5**. Close the switch and record your observations.

Analysis

(e) How does a diode's resistance in reverse bias compare with its resistance in forward bias? Give experimental evidence to support your answer.

(f) Describe three different ways that a pair of diodes can be used to control the current in an electronic circuit. Give experimental evidence to support each answer.

Evaluation

(g) Evaluate your predictions.

Synthesis

(h) Generally, if the resistance of a diode in reverse bias is less than 100 times the resistance in forward bias, the diode is in poor condition and should be discarded. Determine whether the diode(s) you tested should be discarded. Show your calculation(s).

Like diodes, *transistors* are made of semiconductors and are found in almost all electronic devices (**Figure 1**). A **transistor** is a control device that amplifies a small input signal into a larger output signal. It consists of one type of semiconductor sandwiched between two regions of the opposite type of semiconductor. As shown in **Figure 2**, there are two choices for this type of transistor: *p-n-p* and *n-p-n* transistors.

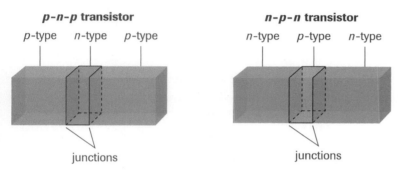

Figure 2
Two types of transistors

Figure 1
Transistors have three lead wires attached.

transistor an electronic control device that amplifies a small input signal into a larger output signal

base the centre section of a transistor; made of one type of semiconductor

emitter the thinner of the outside sections of a transistor; made of the opposite type of semiconductor to the base material

collector the thicker outside section of a transistor; made of the same type of semiconductor as the emitter

The centre section of a transistor is the **base**; it is extremely thin, often about 1 μm (1×10^{-6} m, which is about 1/50 the thickness of a human hair). The outer sections compose the **emitter**, which is the thinner of the two, and the **collector**, which is the thicker one. **Figure 3(a)** shows a *p-n-p* transistor, and **Figure 3(b)** shows the corresponding circuit diagram with an arrow on the emitter indicating the direction of conventional current.

Notice in **Figure 3(a)** that the left side of the transistor resembles a *p-n* diode connected in forward bias. In this part of the circuit, a small change in the source voltage can cause a large change in the current (**Figure 4**). The right side of the transistor resembles an *n-p* diode connected in reverse bias. This means that, once the collector current begins, the voltage output in the

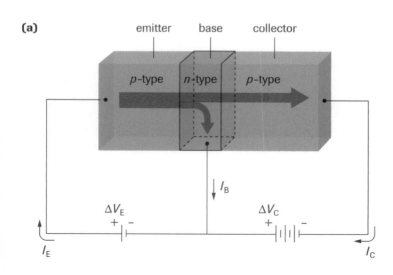

(a)

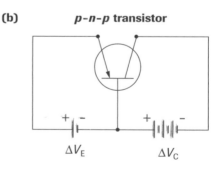

Figure 3
(a) A *p-n-p* transistor in a circuit
(b) The circuit diagram of the transistor circuit

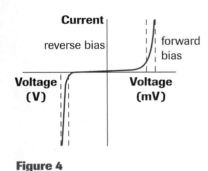

Figure 4

LEARNING TIP

Field-Effect Transistors
Another common type of transistor is the *field-effect transistor,* or *FET.* An FET controls electron and hole migration by varying the electric field, which explains its name. In an FET, the emitter is called the source, the base is called the gate, and the collector is called the drain.

LEARNING TIP

Transistor Amplifiers
The transistor amplifier described here has the input signal at the emitter. The circuit is called a "common-base amplifier" because the base is between (or common to) the emitter and collector. This design produces an output signal with the same positive and negative cycles as the input signal. Two other circuits are possible: the common-emitter circuit and the common-collector circuit. In these cases, the input signal is at the base. These circuits act as current amplifiers; the amplification is the ratio of the collector current to the base current.

collector circuit will be larger than the voltage input in the emitter. Notice that the current directions correspond to conventional current. This makes sense because conventional current is the flow of positive charges, that is, the flow of holes.

Similar reasoning can be applied to an *n-p-n* transistor. In that case, we say electron flow rather than hole flow, and the ΔV terminals are reversed. To practise identifying transistor components, you can perform Activity 8.5.

Sound systems for musicians are a specific use of a transistor amplifier. A small voltage signal can be picked up from an electric guitar, amplified in a transistor circuit, and sent to an output speaker. **Figure 5** shows a *p-n-p* transistor circuit with a variable input signal and the output signal across the resistor to the loudspeaker. Small changes in the voltage input to the transistor in the emitter circuit result in large changes in the voltage output in the collector circuit. The output is a loudspeaker with a resistance, R. The transistor amplifier shown is a voltage amplifier. Amplification occurs because the collector voltage is larger than the emitter voltage.

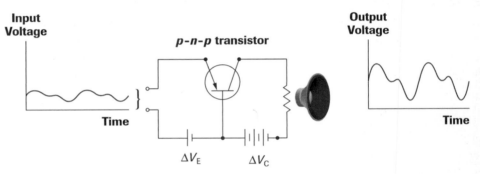

Figure 5
This *p-n-p* transistor amplifies a small input voltage into a large output voltage.

> ▶ **Practice**

Understanding Concepts

1. (a) If the collector in a transistor is made of a *p*-type semiconductor, what type of material is used in the transistor's base? in the emitter?
   (b) Name the transistor in (a).

2. Refer to **Figure 3**, page 381.
   (a) Explain why I_C exceeds I_B in an *n-p-n* transistor.
   (b) How does the magnitude of I_E compare with the magnitudes of I_C, I_B, and the sum of $I_C + I_B$? Show why. (Hint: Recall KCR.)

Making Connections

3. The law of conservation of energy says that when energy changes from one form to another, energy is neither created nor destroyed. Does the fact that the output voltage of a transistor is greater than the input voltage violate this law? Explain your answer. (Hint: What do ΔV_E and ΔV_C have to do with the operation of the type of transistor described in this section?)

Case Study *From Vacuum Tubes to Transistors*

When you turn on a modern radio, you hear sound almost immediately. In addition, your radio does not require frequent replacement of parts to keep it working. However, radios manufactured before electronic components were available needed a warm-up period before they produced sound. They also often required replacement of burned-out parts. These older radios were made with components called *vacuum tubes*, which had to reach a certain temperature before they could operate as designed. The tubes also had to be replaced the same way we replace light bulbs today. Other devices, such as TVs and audio amplifiers, also used vacuum tubes.

Albert Einstein (1879−1955) played a role in the early development of vacuum tubes. In 1905, he first explained the operation of the vacuum tube setup shown in **Figure 6**. In this early type of photocell, incoming radiant energy strikes a metal surface. If the energy is great enough, electrons near the surface break free. These emitted electrons are attracted to the plate on the opposite side of the tube because it is attached to the positive terminal of the source. Thus, electrons move through a vacuum, which is normally a very good electrical insulator. The resulting electric current is measured with the ammeter.

In **Figure 6**, the source of energy to free the electrons is radiant energy. Electrons can also be freed from a metal surface by applying thermal energy. Many vacuum tubes in older radios and TVs used this method. The basic design of this type of vacuum tube is shown in **Figure 7**. In this example, the circuit with the heating element is separate from the circuit with the emitter and collector.

A major disadvantage of vacuum tubes was the heat they generated. The excess heat wasted energy and shortened the lifespan of the tubes. The need for better ways of controlling electronic circuits led to research into more reliable, solid-state devices. An important step was the invention of the transistor by Bell Telephone engineers John Bardeen, Walter H. Brattain, and William B. Shockley in 1948. The transistor sparked the electronics revolution. Since then, electronic components have continued to shrink in size while growing in power, and the revolution continues to this day.

Today's electronic components have many advantages over vacuum tubes:

- They are solid state, so they are much less fragile than glass vacuum tubes.

- Unlike tubes, they do not require heating filaments. As a result, they operate at a lower temperature, and reduced heat means higher efficiencies and longer life expectancy.

- They are less expensive to build than tubes.

- They require less electric potential input, often between 1 V and 25 V, as compared with at least 100 V for many vacuum tubes. This makes them safer to operate.

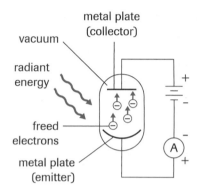

Figure 6
In this early vacuum tube, radiant energy strikes one metal surface, causing electrons to escape. The electrons from this "emitter" are attracted to the positive plate, or "collector." (The emitter and collector are also called the cathode and anode, respectively.)

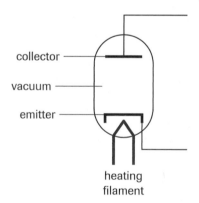

Figure 7
In this vacuum tube design, thermal energy causes the electrons on the surface of the emitter plate to escape and move through the vacuum to the positive collector plate.

Viewing Vacuum Tubes
To investigate how technologies related to TV, radio, and other devices have changed, you can visit Web sites that show many types of vacuum tubes.

 www.science.nelson.com

- They are compact. A single, tiny electronic component may include many transistors and perform the function of several vacuum tubes and related components.

Solid-state electronics have disadvantages as well:

- They can be easily damaged by large changes in temperature.

- They can be easily damaged if applied voltages exceed the suggested value, or if the voltage is reversed.

- Disposing of electronic equipment in the garbage is hazardous for the environment. One of the major risks is that chemicals can leak out into the soil and eventually contaminate water supplies.

- Semiconductors are not used in some high-voltage applications. These include the microwave tube in a microwave oven and the picture tube in TV sets with tube technology.

▶ Practice

Understanding Concepts

4. Explain why electronic devices are more portable than devices that contain vacuum tubes. Give at least two reasons.

5. Which devices are safer to use, those with electronic components or those with vacuum tubes? Explain why.

Making Connections

6. The release of electrons in the early vacuum tube shown in **Figure 6** is called the photoelectric effect. In 1921, Einstein earned the Nobel Prize in physics for explaining this effect. Explain why the name "photoelectric effect" is appropriate.

7. Read the instruction manual accompanying an electronic device, such as a computer. At what temperature range does the device operate best? What do you conclude?

SUMMARY *Transistors*

- A transistor is a device in which one semiconductor separates two other semiconductors. It takes an input signal in the emitter and produces an amplified signal in the collector.

- Semiconductor devices have largely replaced the vacuum tubes that were formerly used in electronic devices.

▶ *Section 8.4 Questions*

Understanding Concepts

1. What is joined at the junctions of a transistor?

2. **Figure 8** shows an electronic circuit diagram.
   (a) Name the electronic device in the circuit. How can you identify the type?
   (b) Copy the diagram into your notebook. Draw and label as many components of the diagram as you can, including the electron flow and hole flow (or current) wherever possible.
   (c) Explain how this circuit amplifies an input signal.

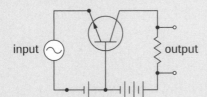

Figure 8

3. Explain why solid-state devices require less power to operate than vacuum-tube devices.

4. List reasons why electronic devices are less expensive to build and operate than vacuum tube devices.

Applying Inquiry Skills

5. Using appropriate symbols, draw an electronic circuit in which a *p-n-p* transistor amplifies the voltage signal from an AC generator. The load is a resistor with a voltmeter across it.

6. (a) Describe how you would investigate the change in cost of the read-only memory (ROM) in computers over the past 20 years. (For example, you can express the cost in dollars per megabyte.)
   (b) Perform the investigation, and summarize what you discover in a table or a graph.

7. The amount of electronic waste ("e-waste") discarded in dump sites is a serious environmental issue. Some efforts are being made to recycle e-waste; for example, refill kits are available for computer printer ink cartridges. Make up a questionnaire to address the environmental impact of the disposal of electronic equipment. Your questions should determine what people understand about the issue, what they currently do to address it, and what changes they would be willing to make.

Making Connections

8. A common element used in doping is arsenic. What is a major disadvantage to the environment of using this substance? (Hint: A dictionary will help you answer this question.)

Identifying Transistor Components

This activity will help you understand transistor components. You can examine transistors in the laboratory or in pictures in catalogues and on the Internet. Some configurations of transistors are illustrated in **Figure 1** for reference. If a label does not indicate whether a transistor is *n-p-n* or *p-n-p*, you can look at the arrow labelled on the emitter to determine the type. If the arrow points from the emitter into the base, the section identified must be *p-n*, so the transistor is a *p-n-p* type. (A memory aid for this orienation is pn = <u>p</u>ointing i<u>n</u>.) If the arrow points from the base to the emitter and out of the transistor, the transistor is an *n-p-n* type. (A memory aid for this orientation is npn = <u>n</u>ot <u>p</u>ointing i<u>n</u>.)

Procedure

1. Number the transistors available as 1, 2, 3, …, so that you can tabulate their properties. Read step 2, and then set up a table to record the data.

2. For each transistor:
   - Determine its type (*n-p-n* or *p-n-p*).
   - Identify which leads are the emitter, base, and collector.
   - Draw and label a diagram of the transistor; show the length and shape of each lead (straight, angled, spread, etc.).
   - Identify any other features labelled on the transistor.

Analysis

(a) Summarize the features of the transistor that enable you to identify its characteristics.

Synthesis

(b) In what college courses or types of careers would experience in analyzing transistors be helpful?

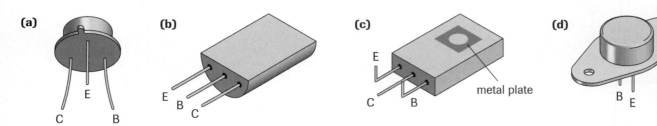

Figure 1
(a) With a small tab: The emitter is closest to the tab.
(b) Semicircular: The emitter is usually on the left.
(c) Flat, with a metal plate for cooling: With the plate facing upward, the emitter is on the left.
(d) In this design, the case acts as the collector.

When you turn on certain electronic devices, such as a camera flash attachment or a small power supply, you hear a high-pitched sound. This sound means that something is being charged (**Figure 1**). That something is a **capacitor**, a device used to store electric charge and thus electrical energy. Once the capacitor is fully charged, it can be used to operate a device, for example, a flashbulb.

The first capacitor, called the *Leyden jar*, was designed in 1746 in Leyden, the Netherlands. A Leyden jar is layered like a sandwich, with a glass or plastic insulator separating two metal conductors (**Figure 2(a)**). When the Leyden jar is grounded on the outside and touched to an electrostatic generator, electrons repelled from the generator charge the inner metal plate. This negative charge repels electrons in the outer metal plate to the ground (**Figure 2(b)**). When the Leyden jar is removed from the generator, it has a negative charge on the inside and a positive charge on the outside. If a conducting wand with an insulating handle is used to discharge the jar, the result is a strong electric shock between the jars and the wand (**Figure 2(c)**).

Figure 1
An electronic camera flash

capacitor a device used to store electric charge and thus electrical energy

(a)

(b)

(c)

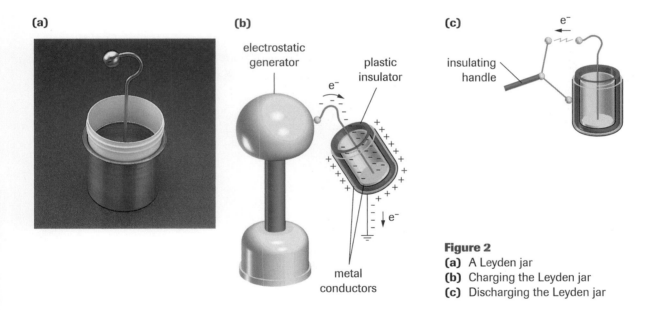

electrostatic generator

plastic insulator

insulating handle

e⁻

metal conductors

Figure 2
(a) A Leyden jar
(b) Charging the Leyden jar
(c) Discharging the Leyden jar

Many of today's capacitors are much smaller, and they consist of two conducting plates separated by an insulating material. The plates are sometimes parallel, as in **Figure 3(a)**, and sometimes cylindrical (formed by rolling up parallel plates), as in **Figure 3(b)**. To determine the properties of a capacitor as it is being charged and discharged, you can perform Activity 8.7.

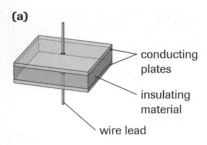

(a)

conducting plates

insulating material

wire lead

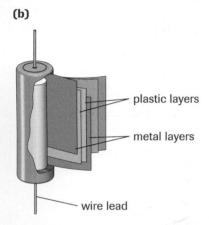

(b)

plastic layers

metal layers

wire lead

Figure 3
(a) A parallel plate capacitor
(b) A cylindrical capacitor

capacitance the ratio of the charge on the capacitor to the voltage across its plates

farad the SI unit of capacitance; symbol F

A capacitor must be connected to a voltage source in order to become charged. The ratio of the charge on the capacitor to the voltage across its plates is its **capacitance**. The SI unit for capacitance is the **farad** (F), named in honour of English physicist Michael Faraday (1791−1867), who made many discoveries related to electricity and magnetism. A capacitor that can store a large charge for each volt of potential difference across its plates has a large capacitance. A capacitance of 1.0 F is very large—a capacitor of this size can be used in backup systems during power failures. Smaller values are often stated in microfarads ($\mu F = 10^{-6}$ F), nanofarads (nF $= 10^{-9}$ F), and picofarads (pF $= 10^{-12}$ F). Picofarad capacitors are often used in electronic components.

One factor that affects capacitance is the surface area of the plates. As the surface area increases, the capacitance increases. (You discovered this in the *Try This* Activity, page 367.) In a *fixed capacitor*, the surface areas of the plates are constant. In a *variable capacitor*, one set of plates can move relative to a stationary set, so the area is adjustable (**Figure 4**). This type of capacitor is used in most radios to tune in to different stations. Adjusting the radio setting causes the movable plates to slide between the stationary plates. This movement changes the capacitance and allows the radio receiver to pick up the specific signals sent out by a radio station.

Another factor that affects capacitance is the distance separating the plates: the smaller the distance, the greater the capacitance. This relationship is applied in the operation of many computer and music keyboards (**Figure 5**). Each key is attached to a piston, which in turn is attached to the movable top plate of a capacitor. When you press the key, the movable top plate of the capacitor moves closer to the fixed bottom plate, increasing the capacitance. This change is detected by an electronic circuit, which tells the computer to store the information.

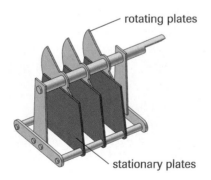

rotating plates

stationary plates

Figure 4
One design of a variable-area capacitor

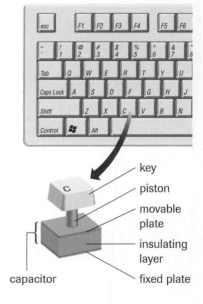

key

piston

movable plate

insulating layer

capacitor

fixed plate

Figure 5
When you press on the letter on this keyboard, the plate separation of the capacitor beneath that letter is reduced.

TRY THIS activity

Properties and Uses of Capacitors

Research capacitors on the Internet, and set up a table to summarize the main properties and uses of several of the capacitors you find. Include a sketch of the shape of each capacitor.

GO www.science.nelson.com

Capacitors are often used to smooth out the ripples of DC currents after they have gone through a rectifier (discussed in section 8.2). In **Figure 6**, a capacitor is connected in parallel across the load of the half-wave rectifier circuit. With each half-wave cycle, the capacitor becomes charged, keeping the voltage nearly constant for the remainder of the cycle. The resulting output is shown in the voltage–time graph.

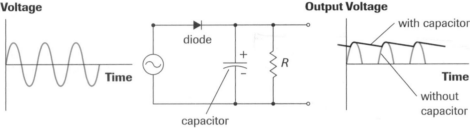

Figure 6
When a capacitor is connected to a half-wave rectifier circuit, the output voltage becomes fairly smooth.

Figure 7 shows how a capacitor can be linked to a photodiode to act as a switch in an alarm. An infrared beam strikes the photodiode continuously, causing the capacitor to remain fully charged. In this state, there is no current, so the resistance device (alarm bell, flashing light, or other warning device) is not activated. When the infrared beam is interrupted, the capacitor begins discharging, setting off the alarm.

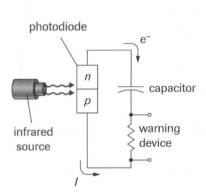

Figure 7
A simple alarm circuit

 Practice

Understanding Concepts

1. Describe what happens to the capacitance of a capacitor when
   (a) the surface area of the plates decreases
   (b) the distance between the plates increases

2. The input of a DC power supply in your physics lab is AC. How can you ensure a nearly constant output of DC voltage?

3. Describe how a capacitor could be used in the design of a force sensor.

4. Convert the following capacitor values as indicated:
   (a) 5.5×10^{-6} F = ? μF
   (b) 3.4×10^{-11} F = ? pF
   (c) 6.1×10^{-10} F = ? nF

Answers

4. (a) 5.5 μF
   (b) 34 pF
   (c) 0.61 nF

Figure 8

Applying Inquiry Skills

5. Describe how you could use paper and aluminum foil to construct a model of a cylindrical capacitor. Draw a diagram of your design. (Do not try to use your model as a capacitor; it could be dangerous!)

Making Connections

6. An electronic flash attachment for a camera stores charge from the camera's battery. When the flash is activated, the capacitor discharges, lighting the bulb.
   (a) Are the discharge times long or short? Explain why.
   (b) High-speed flash photography was used to take the photograph in **Figure 8**. Explain the operation of the capacitor in this case.

SUMMARY *Capacitors*

- A capacitor stores electric charge and thus electrical energy.
- The SI unit of capacitance is the farad (F). Most capacitors have a capacitance in the range of microfarads, nanofarads, or picofarads.
- Capacitance depends on the area and separation of the plates.
- Capacitors are common electronic components used in rectifier circuits, computer printers, computer chips, and many other devices.

> ### *Section 8.6 Questions*

Understanding Concepts

1. Under what conditions does an electric current exist in an electronic circuit with a capacitor?

2. Is the following statement true or false: "As a capacitor in a circuit is being charged, the electric current in the circuit does not actually pass *through* the capacitor." Explain your answer.

3. A light bulb is connected in series with a capacitor, a switch, an ammeter, and a DC battery. (The results of Activity 8.7 will help you understand the concepts in this question.)
   (a) Draw a circuit diagram of this setup.
   (b) When the switch is closed, what happens to the electric current as time passes? Why does the current change?
   (c) Draw a new diagram without the power supply to show the discharging of the capacitor.
   (d) State what happens to the electric current in (c) as time passes.

4. Describe how the following pairs of components can be used in an electronic circuit:
   (a) a capacitor and a resistor
   (b) a capacitor and a photodiode

Applying Inquiry Skills

5. (a) Describe how materials like those used in the *Try This* Activity, page 367, can be adapted to create a model of a flat plate capacitor that holds a charge.
   (b) Draw a diagram of the design, and explain how it works. With your teacher's permission, test your design.

6. How can you use protractors to demonstrate the changing surface area of a variable capacitor? Draw a diagram of your design.

Making Connections

7. Research defibrillators (mentioned in section 7.8, page 346), and identify the role of the capacitor in delivering a controlled electric current to restore a patient's heartbeat. Describe what you discover, giving the capacitor's approximate voltage and discharge times.

Charging and Discharging a Capacitor

Capacitors are important components in electronic circuits. In this activity, you will explore how a capacitor can be charged and discharged and the properties it displays during those processes.

 It is important to distinguish between polarized capacitors and nonpolarized capacitors. A polarized capacitor has distinct positive and negative plates. For safety reasons, these plates *must* be connected properly across a power supply or voltmeter (positive to positive and negative to negative). This precaution does not apply to nonpolarized capacitors.

It is not necessary to understand all the labels on a capacitor to learn from this activity. However, you need to know the following:

- "F" stands for farad, which is the SI unit of the charge per voltage of a capacitor. Values you might find in a laboratory are 1.0 F, 1.0 mF, and 1.0 μF.
- Check the voltage rating of the capacitor. You should never exceed the value labelled.
- The circuit symbol for a capacitor is ─┤├─ .

 - **Do not short-circuit the terminal of a capacitor.**
- **Have your teacher test the charge on the capacitor before you use it.**
- **Do not use an electrolytic capacitor. Without careful control, it can overheat and explode.**

Question

What are the properties of a capacitor as it is being charged and as it is being discharged?

Prediction

(a) Predict what will happen to the brightness of the light bulb in **Figure 1(a)** after the switch is closed. Also draw the graph of $\Delta V_{capacitor}$ (vertical axis) versus time that you think is right for this setup.

(b) Repeat (a) for the circuit in **Figure 1(b)**.

(c) Predict what the graph of $\Delta V_{capacitor}$ versus time will be for the circuit in **Figure 1(c)**.

(a)

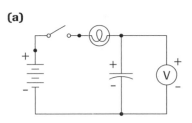

(b)

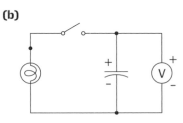

(c)

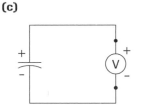

Figure 1
(a) Charging a capacitor with a load in the circuit
(b) Discharging the capacitor with a load in the circuit
(c) Observing the charged capacitor's voltage with no load

Materials

For each group of three or four students:
capacitor
low-voltage power supply or battery
light bulb (either 1.5 V or 3.0 V)
knife switch
voltmeter
connecting wires
stopwatch

For each student:
graph paper

Procedure

1. Obtain a fully *discharged* capacitor from your teacher. Leaving the switch open, set up the circuit in **Figure 1(a)**.

2. Close the switch and observe both the light bulb and the voltmeter. Describe the change in brightness of the bulb, and record the maximum voltmeter reading.

3. Remove the voltmeter and power supply from the circuit in step 1, and open the switch. Remember the safety precautions.

4. With the switch open, set up the circuit in **Figure 1(b)**. Close the switch and observe both the light bulb and the voltmeter. Describe the change in brightness of the bulb, and record the maximum voltmeter reading.

5. Set up the circuit in **Figure 1(c)**. Close the switch, and observe the voltmeter. Record the maximum voltmeter reading, and describe what happens to the voltmeter as time elapses.

Analysis

(d) During charging, how does the maximum value of $\Delta V_{capacitor}$ observed compare with the rated voltage of the capacitor?

(e) Explain your observations in step 2. (Hint: Apply the fact that like charges repel and opposite charges attract.)

(f) Plot $\Delta V_{capacitor}$ versus time graphs to summarize your findings in steps 2, 4, and 5. In each case, choose one of the graphs shown in **Figure 2**, but label the axes appropriately.

(g) Based on what happens to the light bulb as the capacitor is being charged and discharged, sketch current versus time graphs for steps 2 and 4.

(h) Explain why a capacitor should *not* be short-circuited. (Relate your answer to what you discovered in step 5.) In each case, you can choose one of the graphs shown in **Figure 2**, but label the axes appropriately.

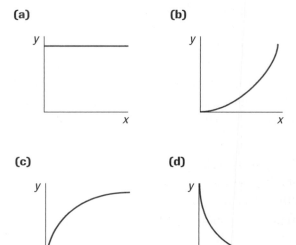

(a)   **(b)**

(c)   **(d)**

Figure 2
Possible graphs of relationships

Evaluation

(i) How accurate were your predictions?

(j) Describe any difficulties you had in taking the measurements in this activity.

Synthesis

(k) The capacitors recommended for use in this investigation are made with safety in mind. They will simply stop working if the input voltage exceeds a safe value. However, standard electrolytic capacitors will explode if used incorrectly. What conditions would cause such an explosion?

If you listen to a song played on an acoustic guitar (i.e., an "unplugged" guitar) and then played by an electronic synthesizer using the guitar mode, you can likely hear the difference. Sounds from an acoustic guitar can be called *analog signals.* Sounds from a musical synthesizer can be called *digital signals* because they are controlled by a digital circuit. As you can see in **Figure 1**, the analog signal is wavy and smooth. The digital signal is "sliced up," so it only becomes smooth if it has several slices. In the digital signal, what we actually hear is a sort of average of the amplitudes of the "slices." In a diagram of a digital signal, this average is drawn as an *envelope*, which is a smooth line that joins the midpoints of the tops of the slices. ⬩▮

Analog Signals

An **analog signal** is a signal that varies continuously with time. In an analog circuit, the input signal, the transmitted signal, and the output signal are all analog. A typical analog circuit is the telephone system illustrated in **Figure 2**.

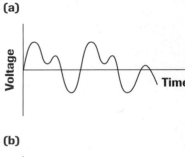
(a)

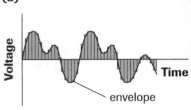
(b)
envelope

Figure 1
(a) A typical analog sound signal
(b) A typical digital sound signal with the envelope showing what we actually hear

analog signal a signal that varies continuously with time

CAREER CONNECTION

Many Ontario community colleges offer Electronics Engineering Technology programs. With this diploma, you can choose from many exciting and interesting careers, for example, designing sound equipment or creating the designs on computer.

GO www.science.nelson.com

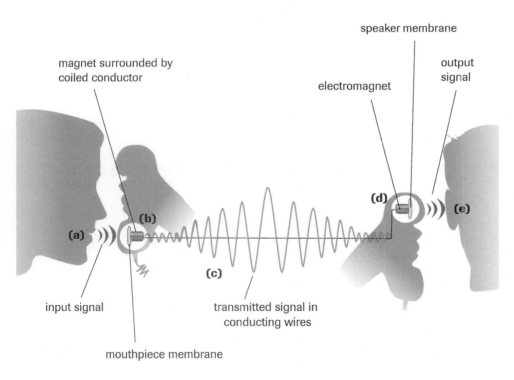

speaker membrane

magnet surrounded by coiled conductor

electromagnet

output signal

input signal

mouthpiece membrane

transmitted signal in conducting wires

Figure 2
(a) As you speak into a microphone, sound waves cause the mouthpiece membrane to vibrate smoothly back and forth with the sound waves. **(b)** The microphone transforms sound energy into electrical energy, forcing a coiled conductor surrounding a magnet to move. As the conducting coil vibrates, a changing electric current is produced in the conductor. **(c)** The varying signal is sent along the conducting wires to the receiver, in this case, the phone's speaker. **(d)** The speaker has an electromagnet with a coil that vibrates according to the changing current signals. **(e)** The vibrating coil causes the speaker membrane to vibrate, thus transforming electrical energy back into sound energy.

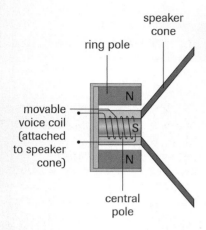

ring pole
speaker cone
movable voice coil (attached to speaker cone)
N
S
N
central pole

Figure 3
A cutaway view of a loudspeaker

(a)

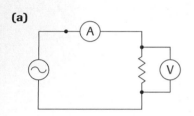

(b)

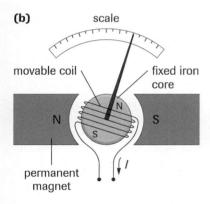

scale
movable coil
fixed iron core
N   S
permanent magnet
I

Figure 4
(a) Analog meters in a simple electrical circuit
(b) Design of a movable-coil galvanometer

▶ **TRY THIS** *activity*   **A Vibrating Speaker**

Obtain a loudspeaker and locate the permanent magnet at the rear; it has a coiled conductor surrounding it (**Figure 3**). Attach a sound signal generator (or other audio source) to the speaker and generate sounds. Observe as the speaker cone vibrates in response to the changing electrical signals in the speaker coil.

Note that the telephone system in **Figure 2** uses magnets. *Electromagnetism* is the interaction of electric and magnetic fields. This interaction is smooth and continuous, so it is commonly used to produce and detect analog signals. Electromagnetism is not covered in this course, but you should be aware of two important related facts:

- Wherever the magnetic field near a conductor changes, electric current is produced in the conductor. (It is the resulting changing current that is transmitted in the telephone wires in a traditional telephone system.)

- A changing current in a coiled conductor produces a magnetic field around the conductor. (It is this changing magnetic field that interacts with the magnet in the receiving end of a telephone system.)

Other analog systems also apply the interaction of electric and magnetic forces. For example, if a variable power supply provides a varying input in the circuit in **Figure 4(a)**, the analog voltmeter and analog ammeter respond smoothly to changes in input. These analog meters are based on the operation of an analog *galvanometer*, shown in **Figure 4(b)**. A changing electric current in the movable coil causes the coil to move the needle that points to the scale.

▶ **Practice**

Understanding Concepts

1. Is the envelope of a digital signal sliced up or smooth? Draw a sketch to illustrate your answer.

2. Explain why an analog signal is called a continuous-time signal.

Applying Inquiry Skills

3. Obtain an analog meter, such as a voltmeter, an ammeter, or a galvanometer. Look closely at the construction. Draw and label a diagram showing the structure, including enough detail to show its basic operation.

Making Connections

4. A magnetic-tape recorder is used for storage and playback of sound. Research the structure and operation of a magnetic-tape system that uses analog signals. Draw a diagram to show how the tiny particles in the magnetic coating of the tape can be influenced by the input signals of an electromagnetic pickup coil.

Digital Signals

A **digital signal** is a signal represented as on and off pulses. The pulses are also referred to as bits or as the digits one and zero. Each one or zero takes a distinct time interval, so a digital signal can also be called a discrete-time signal. ("Discrete" means "distinct.") Compare this with the continuous-time analog signal.

To see how digital signals work, look at the audio system illustrated in **Figure 5**. The microphone receives the analog sound waves from the singer's voice, producing the input signal. An electrical circuit connected to the microphone *samples,* or takes a small amount of, the input signal thousands of times per second and generates a digital signal. This process is called **analog-to-digital conversion**. The digital signals are transmitted through wires in an electrical circuit or by means of electromagnetic waves. Then the process is reversed as the signals are converted back to analog signals. This process is called **digital-to-analog conversion**. In the audio system, the analog signals are the sound waves coming from the loudspeaker.

digital signal a signal represented as on and off pulses, bits, or zeroes and ones

Bits and the Binary System
The word "bit" is formed from the term binary digit. "Binary" refers to the number system to the base 2. Only two symbols are needed for a binary system, 1 and 0. That makes the binary system ideal for "on–off" computer circuits, where 1 means on and 0 means off.

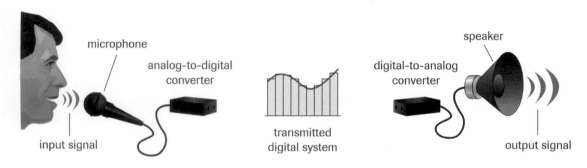

Figure 5
An audio system with both analog and digital signals

microphone, *analog-to-digital converter*, *input signal*, *transmitted digital system*, *digital-to-analog converter*, *speaker*, *output signal*

If both the input and final output signals of the system in **Figure 5** are analog, why use a digital signal between them? You will see why when you consider the *advantages of digital information*:

- Digital signals can be reproduced more accurately, thus reducing unwanted "noise" in the signals.
- Digital signals can be copied countless times, with no reduction in quality.
- Digital signals can be stored easily.
- Digital signals can be processed in order to modify or "clean up" the signals. This process, called *digital signal processing (DSP),* allows receivers of voice, digital, and image communications to download data that is far superior to most analog signals.

The quality of an analog-to-digital conversion depends on two main factors: *sampling resolution* and *sampling rate*. The **sampling resolution**

analog-to-digital conversion changing analog signals to digital signals

digital-to-analog conversion changing digital signals to analog signals

sampling resolution the maximum number of states available for a digital signal; also called resolution

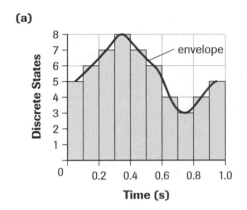
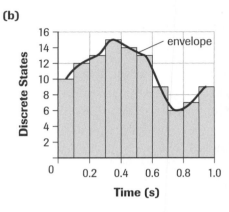

(a)

Figure 6
(a) A sampling resolution of $2^3 = 8$
(b) A sampling resolution of $2^4 = 16$

sampling rate the frequency at which samples of the analog signal are taken during an analog-to-digital conversion

DID YOU KNOW ?

Audio Frequency Range
The healthy human ear is sensitive to sound frequencies from about 20 Hz to 20 kHz. Thus, the sampling rate of 44.1 kHz is high enough for sampling digital sounds heard by humans.

 www.science.nelson.com

Answers

6. (a) 4
   (b) 8 Hz

8. 686 Hz

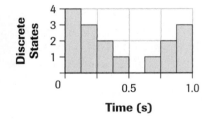

Figure 7

(or simply the *resolution*) is the maximum number of voltage states available for a digital signal. The number is always a power of 2, as illustrated in **Figure 6** for low values. A sampling resolution of 2^8 (256) works well for voice signals; a sampling resolution of 2^{16} (65 536) is sufficient for CDs and stereo systems; and a resolution of 2^{24} is used for digital videodiscs (DVDs). This latter value has almost 17 million possible voltage levels in the signal!

The **sampling rate** of an analog-to-digital conversion is the frequency at which samples of the analog signal are taken. Like other frequencies, it is measured in hertz (Hz), where 1 Hz = 1 cycle/s. Notice that the sampling rate for both signals in **Figure 6** is 10 Hz because there are 10 samples in 1.0 s. In general, the sampling rate is at least two times the maximum frequency of the maximum input analog signal. The sampling rate for digital phones may be 6 kHz, but for commercial voice transmission the standard sampling rate is 8 kHz, which gives a higher efficiency. For digital music and stereo systems, the standard sampling rate is 44.1 kHz; that is more than 44 000 samples of the waveform every second!

Digital signals have many other applications, including facsimile (fax) transmissions, image scanning, digital cameras, DVDs, CDs, video phone technology, and high-definition TV (HDTV). You will learn more about some of these applications in section 8.9 and in Unit 5, Communications Technology.

▶ **Practice**

Understanding Concepts

5. How does the smoothness of a digital signal compare with that of an analog signal? Draw a diagram to illustrate your answer.

6. (a) How many states are there in the signal in **Figure 7**?
   (b) What is the sampling rate? State how you found the answer.
   (c) Copy the diagram into your notebook and draw the signal's envelope.

7. **Figure 8** shows an analog signal. In your notebook, draw the corresponding digital signal with a sampling rate of 10 Hz and a sampling resolution of 4.

8. For a 343-Hz analog signal, what is the minimum sampling rate needed for a corresponding digital signal?

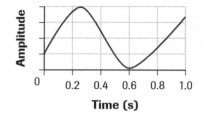

Figure 8

Applying Inquiry Skills

9. (a) No more than seven straight-line segments are needed to display a numerical digital readout. (This is called a "seven-segment display.") Draw diagrams of the numbers 0 to 9 to show this. Do not let the line segments touch.

   (b) How could you improve the resolution of the output? Draw a diagram of the number 6 to illustrate your answer.

Making Connections

10. Some people say that analog sound is smoother and more pleasant than digital sound. What is your opinion?

SUMMARY   *Analog and Digital Signals*

- An analog signal varies continuously with time. An example is a musical sound wave.

- A digital signal is information represented as on–off pulses, or ones and zeroes.

- An analog signal can be converted to a digital signal to copy, process, transmit, and store information quickly and accurately.

- A digital signal can be converted into an analog signal.

- The quality of a digital signal depends on the number of states and the sampling rate used when making it.

▶ *Section 8.8* Questions

Understanding Concepts

1. Is the electric current from an AC generator an analog signal or a digital signal? How can you tell?

2. How could you use a stairway and a wheelchair ramp to model the difference between analog and digital signals?

3. Which type of signal is easier to copy, analog or digital? Explain why.

4. Explain the difference between sampling resolution and sampling rate.

5. Copy **Figure 9** into your notebook, and draw the corresponding digital signal assuming there are five states and a sampling rate of 5 Hz.

Applying Inquiry Skills

6. In a lab demonstration, you have an electrical circuit with an electromagnetic coil. You also have an electronic circuit with a transistor–capacitor pair. Which circuit would you use for a digital signal and which for an analog signal? Explain why.

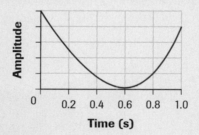

Figure 9

Making Connections

7. Instruments, meters, displays, appliances, and controls are often available in both analog and digital models. For example, bathroom weigh scales can be analog or digital. Give examples of devices available in both formats
   (a) for telling time
   (b) in the kitchen
   (c) in a physics lab
   (d) for voice communication

integrated circuit (IC) a combined electronic circuit with up to millions of diodes, transistors, resistors, and capacitors; it replaces numerous individual circuits

Integrated circuits have likely been the most significant development in the history of electronic technology. Designed to replace circuits with individual components, an **integrated circuit**, or **IC**, is one combined electronic circuit with up to millions of diodes, transistors, resistors, and capacitors. ("Integrate" means to bring parts together.) These components are etched or deposited onto *chips* or *wafers* of semiconductors, usually silicon (**Figure 1**). The chips are enclosed in cans or boxes with pins for connection to external components.

(a)

(b)

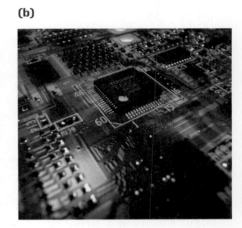

Figure 1
(a) The integrated circuit drastically reduced the size of personal computers (PCs). The process starts with a wafer of silicon, onto which millions of tiny circuit components can be manufactured.
(b) Examples of integrated circuits

Miniaturization allows personal computers to be very compact. If the notebook computer in **Figure 2** used old-fashioned vacuum tubes, it would be the size of a skyscraper and require thousands of dollars worth of electrical energy for a single day of operation.

(a)

(b)

Figure 2
(a) The pace of innovation in the computer industry is so rapid that new PCs, like this notebook model, are followed onto the market almost immediately by smaller, lighter models with greater storage capabilities and faster speeds.
(b) Using robotics to assemble microprocessors

An IC has so many parts and is so complex that it is often referred to as a "black box," that is, a device whose internal workings are unknown to the user. You don't need to learn the details of IC technology. However, understanding the advantages of integrated circuits over individual circuits is important:

- They are much more compact and can be manufactured to perform more functions.

- They perform functions at high speed because of the small distances between components.

- They require less electrical energy and power to operate and, thus, are cooler and more efficient.

- They are sealed, so they are protected and therefore more reliable, reducing down time for repairs.

- Maintenance is easier because a faulty IC can be replaced by a new one.

- They can be used in add-on circuit boards or smart cards.

- When produced in large numbers, they are inexpensive.

However, integrated circuits also have a limitation:

- They cannot be used with devices that produce a large amount of waste heat, such as large amplifiers and other high-power devices. These devices require vacuum tubes and power transistors as well as ICs.

Integrated circuits are a critical component in computer design. In a computer, the **central processing unit (CPU)** is the electronic device that controls the computer's operations. It consists of a microprocessor (the main IC or chip) and support circuits that contain memory and programming instructions. As you can see in **Figure 3**, these support circuits are the *random-access memory (RAM)* and the *read-only memory (ROM)*.

Capacitors are also used in the RAM chips in computers. In a RAM chip, information is stored as ones and zeroes. A single RAM chip contains millions of transistor–capacitor pairs, one of which is shown in **Figure 4**. The address line is used by the computer to locate a specific transistor–capacitor pair. The data line carries the data to be stored. An uncharged capacitor means that a "zero" has been stored. However, for the duration of one clock pulse, when the computer activates the address of one location, the transistor begins conducting the data signal current. This current transfers to the capacitor, charging it, which means that a "one" has been stored (called "write"). The transistor can also detect the charged state of the capacitor (called "read").

An important application that links ICs and digital signals is the modem. The word *modem* comes from its function as a "*mo*dulator–*dem*odulator." Consider, for example, what happens when you send a message over the Internet. The computer converts the input data into digital on–off (or one–zero) pulses. The modem converts (or *modulates*) these pulses into two signals that have different frequencies. These frequencies are carried through telephone systems. At the receiving end, another modem converts (or *demodulates*) the two frequencies back into on–off pulses, which the computer than interprets for the reader.

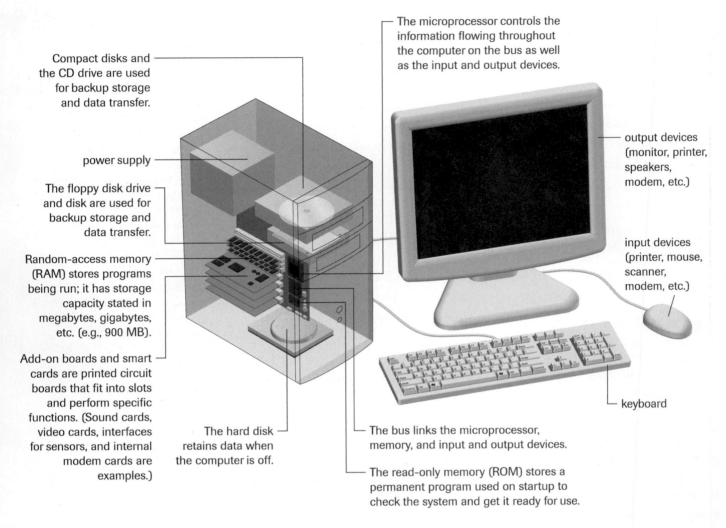

Compact disks and the CD drive are used for backup storage and data transfer.

The microprocessor controls the information flowing throughout the computer on the bus as well as the input and output devices.

power supply

The floppy disk drive and disk are used for backup storage and data transfer.

Random-access memory (RAM) stores programs being run; it has storage capacity stated in megabytes, gigabytes, etc. (e.g., 900 MB).

Add-on boards and smart cards are printed circuit boards that fit into slots and perform specific functions. (Sound cards, video cards, interfaces for sensors, and internal modem cards are examples.)

The hard disk retains data when the computer is off.

output devices (monitor, printer, speakers, modem, etc.)

input devices (printer, mouse, scanner, modem, etc.)

keyboard

The bus links the microprocessor, memory, and input and output devices.

The read-only memory (ROM) stores a permanent program used on startup to check the system and get it ready for use.

Figure 3
Typical components of a personal computer and their functions

address line

capacitor

data line

to data storage

transistor

Figure 4
A transistor–capacitor pair is part of a computer RAM chip.

The design of ever smaller microprocessors has resulted in powerful applications in small devices. Consider, for example, the features of the wristwatch shown in **Figure 5**. Although the watch is only about 6 cm × 5 cm × 2 cm, it can communicate with satellites and use the Global Positioning System (GPS) to pinpoint the user's location anywhere on Earth to within approximately 10 m. It can also be linked to a personal computer to plan and map routes stored in the watch itself. Many other miniature applications are available on the market.

Figure 5
The Casio PAT-2GP-1V satellite navigational watch

Notice the prefix "micro" in microprocessor. As a metric prefix, micro means 10^{-6}, or one one-millionth. However, when referring to microprocessors and microcomputers, the prefix is not a specific quantity; it simply means "very small." It means the same thing in the words microtechnology, micro-miniaturization, microcurrents, and so on. But the "micro" devices are rapidly becoming old technology. Each year, integrated circuits become smaller, so the next technology is called nanotechnology. (The SI prefix "nano" means 10^{-9}.) A glance at the list of metric prefixes in Appendix C, Table 2, reveals that as ICs become smaller, the corresponding names will be picotechnology, femtotechnology, and perhaps even attotechnology.

This section provides a very basic introduction to the topic of integrated circuits and computers. An entire book would be needed just to cover the main principles of these devices. Furthermore, the technology continually advances, so some of what is written here will be outdated by the time you read it. However, the basic concepts will remain the same.

▶ Practice

Understanding Concepts

1. For each variable listed, compare an IC with a circuit made from separate components:
   (a) amount of heat generated
   (b) speed of performance
   (c) ease of repairs

2. Name three devices with ICs that you or your friends use regularly.

3. Copy the following list of abbreviations into your notebook, and beside each one, write its meaning: IC; PC; CD; GPS; MB; GB; TB

4. Refer to the metric prefixes above. Assume that as the size of ICs decreases, the operational speed increases. If microtechnology corresponds to megahertz frequencies, state the frequencies that correspond to (a) nanotechnology and (b) picotechnology.

5. Convert the following measurements as indicated:
   (a) 0.85 GB = ? MB
   (b) 8.2×10^{12} Hz = ? THz
   (c) 4.5 ns = ? s

Making Connections

6. Research modems, and answer the following questions:
   (a) What electronic components are used to make them?
   (b) What function(s) do they serve?

7. Electronic digital pens that allow the user to store many written pages in memory use microtechnology. Find out more about this application, and describe briefly what you discover.

 www.science.nelson.com

Answers

5. (a) 850 MB
   (b) 8.2 THz
   (c) 4.5×10^{-9} s

SUMMARY Integrated Circuits

- Integrated circuits (ICs), which contain up to millions of microscopic electronic components, have many advantages due to their small size, low cost, reliability, and high operational speed.
- ICs form the main control system of the central processing unit (CPU) of a microcomputer.
- As technology advances, ICs become smaller and faster, and ways of using them increase.

▶ Section 8.9 Questions

Understanding Concepts

1. What does "integrated" refer to in integrated circuits?

2. Why is an IC sometimes called a "black box"?

3. How does the efficiency of an IC compare with the efficiency of an individual electronic circuit? Explain why this is so.

4. State three examples in a computer of
   (a) input devices
   (b) memory storage devices

5. Describe how the following parts of the human body can be used to model the parts of a computer: the heart; the brain (with the conscious and unconscious mind); the central nervous system; hands; ears; eyes; mouth.

6. Describe how a capacitor and a transistor can be used in an electronic circuit in a computer.

7. Convert the following measurements as indicated:
   (a) 1000 MB = ? GB
   (b) 2 million kilobytes = ? GB
   (c) 65 MHz = ? Hz

Making Connections

8. Research medical applications of IC technology. List the most significant uses for ICs that you discover. Some topics are cybernetic limb implants; pacemakers; therapeutic robots; medical scanners.

Key Understandings and Skills

8.1 Semiconductors

- A **semiconductor device** is an electronic component made of a **semiconductor** material.

- When the atoms of a **pure semiconductor** are stimulated with energy, **free electrons** migrate in one direction, leaving **holes** that migrate in the opposite direction.

- To increase the ability of a semiconductor to conduct electric current, the semiconductor can be **doped** with **impurities** that have either **donor atoms** (creating a negative, or *n*-type semiconductor) or **acceptor atoms** (creating a positive, or *p*-type semiconductor).

- Diodes can act as very fast switches when going from **forward bias** to **reverse bias** connections.

8.2 Diodes

- *P*-type and *n*-type semiconductors can be joined to make a **diode**. Diodes have several uses in electronic circuits, for example, as **light-emitting diodes**, **infrared light-emitting diodes**, and **photodiodes**, and in **rectifier circuits**.

8.3 Activity: Using Diodes

- A diode in good working order has a much lower resistance in forward bias than in reverse bias.

- Pairs of diodes can be arranged in different ways to control electrical circuits.

8.4 Transistors

- Semiconductors can also be joined to make a **transistor**, which comprises a **base** separating the **emitter** and the **collector**. These transistors are either *p-n-p* type or *n-p-n* type.

- Transistors are commonly used as amplifiers, in which a small input signal is increased to produce a large output signal.

- Semiconductor technology has many advantages over vacuum tube technology, which it has largely replaced.

8.5 Activity: Identifying Transistor Components

- The main parts of a transistor can be identified by inspecting the leads and the labels.

8.6 Capacitors

- A **capacitor** stores electric charge and thus energy on its two metal plates, which are separated by insulating material.

- **Capacitance**, measured in **farads** (F), is the ratio of the charge on a capacitor to the voltage across it. It is affected by the distance between the plates, and the surface area of the plates.

- Capacitors are found in most electronic devices. One common use is to smooth the output current in an AC-to-DC rectifier circuit.

8.7 Activity: Charging and Discharging a Capacitor

- An electrical circuit can be set up to observe and analyze the properties of a capacitor as it is being charged and discharged.

8.8 Analog and Digital Signals

- An **analog signal** varies continuously with time, so it is smooth.

- A **digital signal** is represented by the digits one and zero or as on–off pulses. It is a discrete-time signal.

- In an **analog-to-digital conversion**, analog signals are changed to digital signals to ease storage, reproduction, transmission, and processing. The digital signals can be changed back to analog signals in a **digital-to-analog conversion** to provide a smooth output.

- The quality of an analog-to-digital conversion depends on the **sampling resolution** and the **sampling rate** used.

8.9 Integrated Circuits

- Electronic components (diodes, transistors, resistors, and capacitors) are etched or deposited microscopically onto thin wafers to produce an **integrated circuit** (**IC**). When formed into a chip, an IC has many advantages in microtechnology.

- An IC chip is the brain of a computer's **central processing unit** (**CPU**), which controls the computer's operations. IC chips are also used in many other miniature devices, including smart cards and modems.

Key Terms

8.1
electronics
semiconductor
semiconductor device
pure semiconductor
free electron
hole
impurities
doping
donor atom
n-type semiconductor
p-type semiconductor
acceptor atom

8.2
diode
forward bias
reverse bias
rectifier circuit
light-emitting diode
   (LED)
infrared light-emitting
   diode (ILED)
photodiode

8.4
transistor
base
emitter
collector

8.6
capacitor
capacitance
farad

8.8
analog signal

digital signal
analog-to-digital
   conversion
digital-to-analog
   conversion
sampling resolution
sampling rate

8.9
integrated circuit (IC)
central processing unit
   (CPU)

Problems You Can Solve

8.1
- Distinguish between pure semiconductors and
  (a) electrical conductors and insulators and
  (b) doped semiconductors.

8.2
- Describe the design, operation, and uses of diodes.
- Explain the function of a diode in a control system.

8.3
- Use an ohmmeter to determine the resistance of a diode in forward bias and reverse bias.

8.4
- Describe the design, operation, and uses of transistors.
- Describe the advantages of using semiconductors rather than vacuum tubes in electronic circuits.

8.5
- Identify, by inspection, a transistor's emitter, base, and collector.

8.6
- Describe the design, operation, and uses of capacitors.

- Describe the properties of a capacitor as it is being charged and discharged, and draw voltage–time graphs to illustrate those properties.

8.7
- Draw voltage–time graphs to show the charging and discharging of a capacitor.

8.8
- Distinguish between analog and digital signals.
- Explain the advantages of analog-to-digital conversions.

8.9
- Describe the design and advantages of integrated circuits, and explain their importance in a computer's central processing unit.
- Describe how miniaturization has changed electronic technology.

▶ *MAKE* a summary

- Draw a variety of diagrams illustrating the use of as many of the components in **Figure 1** as possible. (You may use circuit diagrams.) For each diagram, state what the symbols mean, and explain what is occurring.
- Include as many key understandings and key terms from Chapter 8 as you can.

Figure 1

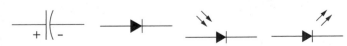

Write the numbers 1 to 10 in your notebook. Indicate beside each number whether the corresponding statement is true (T) or false (F). If it is false, write a corrected version.

1. A neutral donor atom has one more valence electron than an atom of a pure semiconductor material.

2. An *n*-type semiconductor is made by adding acceptor atom impurities to a semiconductor.

3. At 250 °C the current in a semiconductor is the same as it would be at −25 °C.

4. Light from an ILED is not visible.

5. In a photodiode, radiant energy is transformed into electrical energy.

6. The three circuits of a transistor are the base, emitter, and collector circuits.

7. In a parallel plate capacitor, the capacitance increases as the distance between the plates increases.

8. As a loudspeaker vibrates back and forth, it produces an analog sound signal.

9. An example of a sampling resolution is 512 Hz.

10. An integrated circuit is the main component of a computer's central processing unit.

Write the numbers 11 to 17 in your notebook. Beside each number, write the letter corresponding to the best choice.

11. The order of electrical conductivity from highest to lowest is
    (a) copper; doped semiconductor; pure semiconductor; vacuum
    (b) vacuum; pure semiconductor; doped semiconductor; copper
    (c) copper; pure semiconductor; doped semiconductor; vacuum
    (d) pure semiconductor; vacuum; doped semiconductor; copper

12. An impurity in a semiconductor
    (a) prevents any electric current in the semiconductor
    (b) increases the likelihood of a current
    (c) decreases the likelihood of a current
    (d) causes the material to crystallize

13. If the emitter of a transistor is a *p*-type semiconductor, then
    (a) the base and collector are both *n*-type semiconductors
    (b) the base is *p*-type and the collector is *n*-type
    (c) the base is *n*-type and the collector is *p*-type
    (d) the base can be either *p*-type or *n*-type, as long as the collector is *p*-type

14. In order to smooth the output signal of a diode rectifier circuit, the component added to the circuit is
    (a) a digital-to-analog converter
    (b) an analog-to-digital converter
    (c) an amplifier
    (d) a capacitor

15. In the circuit in **Figure 1**,
    (a) $I_E = I_C + I_B$
    (b) $I_C = I_B + I_E$
    (c) $I_B = I_E + I_C$
    (d) $I_E = I_C = I_B$

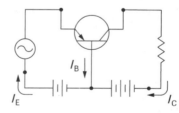

Figure 1

16. A frequency of 5.1×10^{15} Hz is equivalent to
    (a) 5.1×10^9 MHz
    (b) 5.1×10^6 GHz
    (c) 5.1×10^3 THz
    (d) all of the above

17. In an analog-to-digital converter, the appropriate sampling rate for a 60-Hz sound would be
    (a) 2^{60} Hz
    (b) 30 Hz
    (c) 60 Hz
    (d) 120 Hz

Understanding Concepts

1. How many free electrons does a perfect insulator have?

2. **Figure 1** shows the valence electron regions of four elements. Match each of the following descriptions to the correct element: donor element; acceptor element; electrical conductor; pure semiconductor.

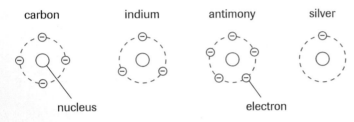

carbon   indium   antimony   silver

nucleus   electron

Figure 1

3. Create a table with the following titles, and compare *p*-type and *n*-type semiconductors: Type of Semiconductor; Typical Number of Valence Electrons; Possible Doping Element; Type of Atom; Majority Migration; Type of Charge.

4. When some photocopiers are first turned on, they cannot be used for several seconds (unlike some electronic devices that work almost immediately). Name two electronic components that could explain this wait time.

5. Explain why an electric current will flow in only one of the two circuits shown in **Figure 2**.

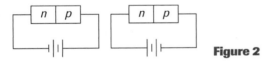

Figure 2

6. (a) What does the prefix "photo" mean in photodiode?
   (b) List other names for this device.
   (c) Describe the function of a photodiode, and write the corresponding energy-transformation equation.

7. Using a circuit diagram, describe an electronic system in which the electronic components control
   (a) a door opener
   (b) a security alarm

8. (a) Name the type of transistor shown in **Figure 3**.
   (b) Describe the function of this transistor.
   (c) Draw a voltage–time graph of the output signal.

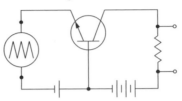

Figure 3

9. State how the maximum charge of a capacitor depends on (a) the plate separation and (b) the surface area of the plates.

10. Refer to the circuit in **Figure 4**.
    (a) Draw a graph of the voltage across the capacitor as a function of time after the switch is closed. (Numerical values are not required.)
    (b) Repeat (a) for the electric current in the circuit.
    (c) With the switch closed, will the battery wear out quickly? Explain your answer.

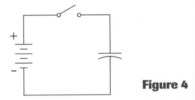

Figure 4

11. Refer to the circuit diagram in **Figure 5**.
    (a) Copy the diagram into your notebook with the switch in position 1. Describe what happens in the circuit. On the diagram, show the direction of the conventional current.
    (b) Repeat (a) after the switch has been changed from position 1 to position 2.

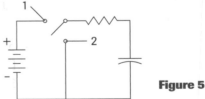

Figure 5

12. **Figure 6** shows a human nerve cell (or neuron) and an enlarged view of its axon, which carries electrical signals between cells. When at rest, the

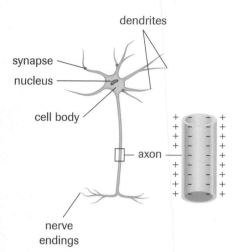

synapse
dendrites
nucleus
cell body
axon
nerve endings

Figure 6

axon has a net negative charge on the inside of the membrane and a net positive charge on the outside.

(a) The membrane is approximately 8 nm thick. Convert this measurement to millimetres and metres.

(b) The voltage across the axon membrane is -68 mV while at rest; when activated, it becomes $+29$ mV. What is the change in electric potential difference between the rest and action states?

(c) The axon's capacitance is 0.2 pF; convert this to microfarads and farads.

13. (a) Describe the features of an analog signal, and give an example of a device that operates with an analog circuit.
    (b) Repeat (a) for a digital signal and a digital circuit.

14. Signal A is a digital signal with a sampling rate of 10 Hz, and signal B is a digital signal with a sampling rate of 5 Hz. Which would more closely resemble an analog signal? Why?

15. Explain the advantages of using a digital signal to store and transmit data.

16. When a personal computer is first turned on, which operates first, the RAM or the ROM? Explain why. (If necessary, refer to **Figure 3** in section 8.9, page 400.)

17. Explain how the CPU and the IC of a micro-processor are related.

Applying Inquiry Skills

18. (a) Using symbols, draw a circuit diagram to explain the operation of a half-wave rectifier circuit connected to an AC generator.
    (b) Draw voltage–time graphs of the input and output in (a).
    (c) In a circuit diagram, show how to make the output in (a) smooth. Sketch a voltage–time graph of the new output.

19. Using symbols, draw a circuit diagram of a *p-n-p* transistor used to amplify the voltage signal from an AC generator. The load is a resistance with a voltmeter connected across it. Label the direction of the conventional current.

Making Connections

20. Describe an environmental concern related to the use of electronic devices, and describe ways of reducing that concern.

21. Two identical capacitors are tested because one may be faulty. After one capacitor is fully charged, it keeps its charge for a long time. The other loses its charge fairly quickly. Assume that the discharge is caused by charges leaking across the insulating material separating the plates. Draw a diagram to show how this could occur. (Start with a fully charged capacitor.)

22. In section 8.9, **Figure 3** (page 400), the bus is shown inside the computer. However, the bus can perform its function outside the computer. Give examples of systems where the bus is outside the computer. (Hint: What does "remote" mean?)

23. A Canadian satellite called *CubeSat* is a cube only 10 cm on each side. Designers call this a picosatellite. Other small satellites are called nanosatellites or microsatellites, depending on their masses and dimensions. Research the *CubeSat* program on the Internet.
    (a) What mass ranges apply to the names pico-, nano-, and microsatellites?
    (b) Look at ICs on board this satellite; note especially RAM and storage capabilities. Describe briefly what you discover.

 www.science.nelson.com

Building a Control Device

 Your teacher must check all designs before you assemble any equipment or circuits.

One characteristic common to the devices in **Figure 1** is that they are *control devices:* They respond to an input signal and perform a task, or create an output. Another common characteristic is that they are electrical or electronic devices.

In Option 1, you design, build, test, and analyze an electrical or electronic control device that accomplishes a specific task. In Option 2, you research the details of the design and operation of a control device, and then build a model to illustrate its operation. Both options apply the principles you learned in Chapters 7 and 8. The device or model you build will have at least one

Your completed task will be assessed according to the following criteria:

Process

- Draw up detailed plans and safety considerations for the design, tests, and modifications of the device or model.
- Choose appropriate research tools (especially for Option 2), such as books, magazines, and the Internet.
- Choose appropriate materials to construct the device or model.
- Appropriately and safely carry out the construction, tests, and modifications of the device or model.
- Analyze the process (as described in the Analysis).
- Evaluate the task (as described in the Evaluation).

Product

- Demonstrate an understanding of the relevant physics principles, laws, rules, and equations.
- Prepare a suitable research summary (Option 2).
- Submit a report containing the design plans for the device or model, as well as test results and calculations of the equivalent resistance, current, potential difference, and any other related quantities.
- Use terms, symbols, equations, and SI metric units correctly.
- Demonstrate that the final product works (for Option 1).

(a)

(c)

(b)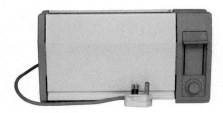

(d)

Figure 1
(a) A programmable thermostat turns the furnace on when the inside temperature drops below the preset value, and it turns the furnace off when the temperature rises to the preset value.
(b) When the toasting is complete, the automatic toaster pops the toast up.
(c) The smoke alarm sounds as soon as it detects smoke.
(d) On a highway in Kootenay National Park, British Columbia, an infrared sensor responds to the heat from an animal, such as an elk, a moose, or a deer, crossing a highway at night. A nearby sign lights up, warning motorists of the danger.

electrical circuit or one electronic circuit. If the circuit is electrical, the components will include a source, a load, connecting wires, and controls. A control can be a switch, a fuse, or circuit breaker. If the circuit is electronic, it can have common components, such as a diode, photocell, transistor, and/or capacitor. Tools (screwdrivers, pliers, wire strippers, crimping tools, etc.) and instruments (multimeter or voltmeter, ammeter, and ohmmeter) must be used safely in this project. The analysis involves electric current, potential difference, resistance, Ohm's law, and Kirchhoff's rules. Energy, power, and the operation of electronic components may also be analyzed.

The Task

Option 1 A Control Device

Your task is to design, build, test, modify, and analyze an electrical or electronic control device that can perform a specific function. For example, your control device could detect motion, indicate water level, turn on a fan or motor, or control an electrical circuit subject to temperature changes. Before starting the task, your group should decide the function of the device, the criteria chosen to evaluate the success of the device, and the necessary safety precautions. Then you should discuss the materials, tools, and instruments required with your teacher to be sure the task is safe and appropriate.

An alternative to designing an original device is to obtain a commercial kit that uses electrical or electronic circuits. In this case, you will build, test, modify, and analyze the device and evaluate your process.

Option 2 A Model Control Device

Your task is to research the design and operation of an electrical or electronic control device that performs a specific function. For example, your device could be an infrared motion detector, a water level control, an alarm system, a switch for a fan or motor, a programmable or mechanical thermostat, or an automatic toaster. You will then create a model of the device that illustrates the physics principles involved in its operation. To build the model, you can use materials that are inexpensive and easily and safely assembled. You will communicate (through diagrams or other means) how instruments are used to monitor or troubleshoot the components of the device.

Analysis

(a) What physics principles are applied in the design and use of your device?

(b) Describe the energy transformations that occur when using the device, and write the corresponding energy-transformation equations.

(c) Describe the function of each of the main circuit components in the device.

(d) How can you judge whether your device or model is successful?

(e) For what purpose can the device you designed or researched be used? What functions can it perform?

(f) What careers are related to the manufacture and use of the device?

(g) What safety precautions did you follow in building and testing your device or model?

(h) How could the process you used in this task be applied in the real world?

(i) List problems you had while building the device or model, and explain how you solved them.

Evaluation

(j) Evaluate the tools you used to construct the device (Option 1) or evaluate the resources you used in your research (Option 2).

(k) If you did this task again, how would you modify the process to obtain a better final product?

1. Each of the following is named after a scientist. Name the scientist.
   (a) the unit of electric current
   (b) the unit of electric resistance
   (c) the unit of electric potential difference
   (d) the unit of electrical power
   (e) the current and voltage rules of electrical circuits
   (f) the unit of capacitance

2. Write the letters (a) to (l) in your notebook. Beside each letter, write the word or term that corresponds to each of the following:
   (a) the particle in an atom's nucleus that has a neutral charge
   (b) the type of current in which the electrons reverse their direction periodically
   (c) the type of connection used to place an ammeter in a circuit
   (d) the energy per charge given by a source
   (e) a material that does not readily allow the flow of electric charges
   (f) the type of resistor connection in which the equivalent resistance exceeds any of the individual resistances
   (g) the outermost region of electrons in an atom
   (h) the type of semiconductor with extra holes
   (i) the type of circuit in which AC is converted to DC
   (j) an electronic device with three leads used to amplify a signal
   (k) a device used to store electric charge
   (l) a signal that varies continuously with time

3. Write the letters A to O in your notebook. Beside each letter, write the name of the component labelled in **Figure 1**.

Write the numbers 4 to 19 in your notebook. Indicate beside each number whether the corresponding statement is true (T) or false (F). If it is false, write a corrected version.

4. A DC cell is an example of a load in a DC circuit.

5. The symbol for ampere or amp is *I*.

6. In a circuit diagram with a DC source, electron flow is shown leaving the negative terminal of the source, and conventional current is shown leaving the positive terminal of the source.

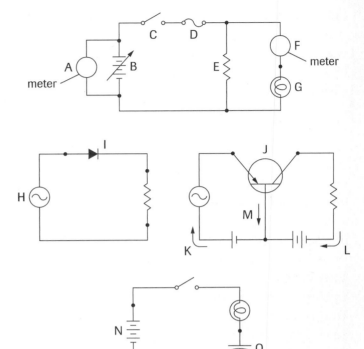

Figure 1

7. Moisture causes a person's electric resistance to increase.

8. The sum of the voltages across resistors connected in parallel equals the voltage across the source.

9. The equivalent resistance of two 4.0-Ω resistors connected in parallel is 8.0 Ω.

10. As more resistors are added in series to an electrical circuit, the equivalent resistance increases.

11. If two 1.5-V DC cells are connected in parallel, the total electric potential rise is 1.5 V.

12. The slope of the line on a graph of voltage (vertical axis) versus current equals the resistance.

13. A neutral acceptor atom has more valence electrons than an atom of a pure semiconductor material.

14. In a semiconductor material, the direction of the conventional current is the same as the direction of hole migration.

15. The purpose of a rectifier circuit is to convert AC into DC.

16. The "I" in ILED stands for "invisible."

17. The junction in a junction diode separates electrons and holes.

18. In a junction transistor, the emitter current equals the sum of the currents in the base and the collector.

19. In a parallel plate capacitor, the capacitance increases as the area of the plates increases.

Write the numbers 20 to 31 in your notebook. Beside each number, write the letter corresponding to the best choice.

20. An example of an electrical load is
    (a) 25 V
    (b) 25 A
    (c) 25 Ω
    (d) 25 kg

21. Volts divided by amps yields
    (a) joules
    (b) watts
    (c) ohms
    (d) seconds

22. A household circuit that operates an electric clothes dryer has these properties:
    (a) 240 V, DC
    (b) 120 V, DC
    (c) 120 V, AC
    (d) 240 V, AC

23. In **Figure 2**, $R_1 > R_2$. Therefore,
    (a) $V_1 = V_2 = V_3$
    (b) $V_1 > V_2 > V_3$
    (c) $V_1 > V_2 = V_3$
    (d) $V_1 < V_2 = V_3$

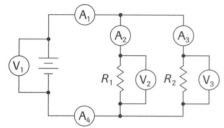

Figure 2

24. In **Figure 2**,
    (a) $A_1 > A_2 > A_3$
    (b) $A_1 > A_4$ and $A_2 < A_3$
    (c) $A_1 = A_4$ and $A_2 < A_3$
    (d) $A_1 = A_4$ and $A_2 > A_3$

25. In **Figure 2**,
    (a) $R_t > R_1$ and $R_t > R_2$
    (b) $R_t > R_1$ and $R_t < R_2$
    (c) $R_t = R_1 + R_2$
    (d) $R_t < R_1$ and $R_t < R_2$

26. The number of valence electrons in an atom of a pure semiconductor is
    (a) 2
    (b) 3
    (c) 4
    (d) 5

27. A semiconductor material is made into n-type by
    (a) injecting electrons
    (b) removing electrons
    (c) adding an acceptor impurity
    (d) adding a donor impurity

28. Which of the following is *not* an advantage of semiconductors over vacuum tubes:
    (a) lighter weight
    (b) lower operating voltage
    (c) ability to withstand high voltages
    (d) smaller size

29. As the spacing between the plates of a capacitor decreases, and all other factors remain the same,
    (a) The capacitance increases.
    (b) The capacitance decreases.
    (c) The capacitance stays the same.
    (d) The capacitor discharges because charges are forced away from the plates.

30. The number of states that work well for transmitting digital voice signals is
    (a) 256
    (b) 8
    (c) 4
    (d) 2

31. The abbreviations IC and CPU stand for, respectively,
    (a) integrated capacitors; collector photodiode unit
    (b) current/capacitance; capacitor, p-type, UV diode
    (c) integrated circuit; central processing unit
    (d) impurity collector; centi, pico, micro

Understanding Concepts

1. Name the particle that each of the following statements describes, and state each particle's charge:
   (a) It moves easily in electrical conductors.
   (b) It forms the central part of an atom.
   (c) It is part of an atom and has the same charge as a hole.

2. Name the type of material that has
   (a) free electrons
   (b) electrons tightly bound to the nucleus
   (c) valence electrons shared with adjacent atoms

3. Write the energy-transformation equation for each of the following:
   (a) An electric ceiling fan circulates air in a room.
   (b) The bright light in a multimedia projector is illuminated.
   (c) A rechargeable battery operates a pocket PC.
   (d) A car's battery operates the audio system speakers. (Start with the chemical potential energy stored in the fuel.)

4. State the function of each of the following components:
   (a) the connecting wires in an electrical circuit
   (b) the load in an electrical circuit
   (c) an adaptor used to operate a DC device in a home
   (d) a donor impurity
   (e) a photodiode
   (f) a transistor
   (g) a capacitor
   (h) an analog-to-digital converter
   (i) a central processing unit

5. What is the frequency of AC in North America?

6. **Figure 1** shows the colour codes of two resistors.
   (a) Interpret the colour codes, and state the resistance of each resistor.
   (b) Determine the equivalent resistance of these resistors if they are connected in series, and then in parallel.

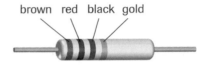

brown   red   black   gold

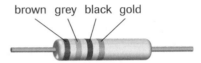

brown   grey   black   gold

Figure 1

7. Copy **Table 1** into your notebook and complete it. Wherever calculations are required, show the equation needed.

8. In **Figure 2**, $\Delta V_1 = 1.2$ V, $\Delta V_3 = 2.4$ V, $\Delta V_4 = 3.6$ V, and $R_2 = 18$ Ω. Calculate
   (a) ΔV_2 (c) R_t (total resistance)
   (b) I (d) R_1, R_3, and R_4

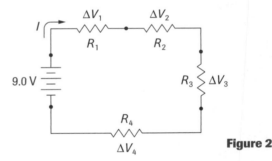

Figure 2

9. In **Figure 3**, $R_1 = 24$ Ω, $R_2 = 12$ Ω, and $I_3 = 0.25$ A. Determine
   (a) I_1, I_2, and I_t
   (b) R_3
   (c) R_t (total resistance)

Table 1   For Question 7

	Device	Current (A)	Voltage (V)	Resistance (Ω)	Power (W)	Type of Current	Useful Output
(a)	stereo	?	120	?	36	?	?
(b)	stove	25	240	?	?	?	?
(c)	toaster oven	15	?	?	1.8×10^3	?	?
(d)	?	?	120	?	5.0×10^2	?	air removal
(e)	?	2.0×10^{-3}	?	1.5×10^3	?	DC	calculations

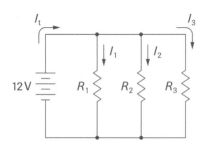

Figure 3

10. For each circuit in **Figure 4**, calculate the equivalent resistance, the total current, and the current through and voltage across each resistor.

(a)

$R_2 = 3.0\ \Omega$

$R_1 = 4.0\ \Omega$

6.0 V

$R_3 = 1.0\ \Omega$

(b)

$R_1 = 1.0\ \Omega$

4.5 V

$R_2 = 2.0\ \Omega$

$R_3 = 8.0\ \Omega$

$R_4 = 0.40\ \Omega$

Figure 4

11. Determine the maximum number of 200-W light bulbs that can be connected (in parallel) into a 120-V household circuit protected by a 15-A circuit breaker.

12. Calculate the power rating of an electronic toy that uses a 9.0-V battery and has a current of 150 mA.

13. Calculate the current, in milliamps, in a printer used in a 120-V circuit and rated at 3.4 W.

14. Calculate the electric potential of a 300-W windshield-wiper motor that has a maximum current of 25 A.

15. (a) Determine the number of megajoules in one kilowatt hour. (Hint: Apply the equation for energy in terms of power and time.)
    (b) If the cost of electricity is 2.5 ¢/MJ, what is the cost in cents per kilowatt hour?

16. Can semiconductors be called semi-insulators? Explain your answer.

17. (a) Under what conditions can a semiconductor become a conductor?
    (b) Name two materials in this category.

18. Using a diagram, explain how an electric current can be set up in a diode connected to a DC source.

19. How can a diode be used to protect an ammeter connected incorrectly in a circuit?

20. To compare semiconductor devices with vacuum tube devices, set up and complete a table with these titles:
    • Type of Device
    • Example of Use
    • Historical Facts
    • Safety Features or Concerns
    • Comparative Costs
    • Availability
    • Environmental Impact

21. **Figure 5** shows a familiar thermometer used to measure body temperature. It contains a component called a *thermistor*.
    (a) Look up "thermistor" in a dictionary and write the definition.

Figure 5

(b) Does the thermistor obey Ohm's law? Justify your answer.

(c) Should the thermistor be made of a conductor, an insulator, or a semiconductor? Explain why.

22. Refer to the circuit in **Figure 6**.
(a) Describe what happens to the brightness of the two bulbs after the switch is closed.
(b) Explain this statement: A capacitor in a branch of a DC circuit acts as an open switch in that branch.

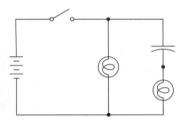

Figure 6

23. Refer to **Figure 7**. State what you would observe on the voltmeter in each of the following cases:
(a) Switch 1 is closed.
(b) Switch 1 is suddenly opened, and switch 2 is suddenly closed.
(c) Both switches 1 and 2 are closed.

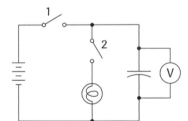

Figure 7

24. (a) Name one device that requires an analog circuit to function and one that uses a digital circuit.
(b) State two advantages of using digital signals for such applications as copying and processing data.

Applying Inquiry Skills

25. You are given a set of Christmas tree lights that are wired in series. You plug the set in, but it does not light up. Explain how you would use your troubleshooting skills to solve the problem. (You may use a multimeter.)

26. (a) Describe the steps you would take to determine whether a resistive component obeys Ohm's law. Using appropriate symbols, draw a circuit diagram of the experimental setup.
(b) What results would you observe if the device is nonohmic? Sketch the corresponding voltage–current graph.

27. The conducting metal plates in **Figure 8** are insulated from the base along which they can slide. A voltmeter is connected across the plates.
(a) Name the electronic component that the plates model.
(b) Once the plates are fully charged, what variable can be adjusted when using the apparatus?
(c) Make up a question for an investigation with this setup.
(d) If you were performing an investigation with this setup, what results would you expect to observe?

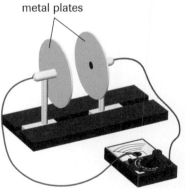

Figure 8   voltmeter

28. Describe how you would use electronic components in an electrical circuit to design a floor mat that acts as an alarm when stepped on. Using appropriate circuit symbols, draw a circuit diagram of your design.

29. **Figure 9** shows an analog signal.
(a) Using a sampling rate of 5 Hz and a sampling resolution of 8, draw a digital signal with an envelope that is the same as the analog signal.
(b) How could the sampling rate and/or the resolution be changed to make the digital signal smoother?

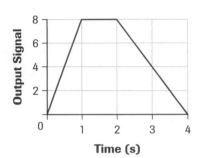

Figure 9

30. Explain how you would design one of the following electric devices:
    (a) a hair dryer with settings of 600 W, 900 W, and 1200 W
    (b) a curling iron for use in North America (120 V) and in other countries (240 V)

31. During a lab practical test, a teacher checks the electrical circuits before allowing the students to close the switch. Explain why the teacher will not approve either circuit shown in **Figure 10**. In each case, state two faults, and then draw the circuit correctly in your notebook.

(a)

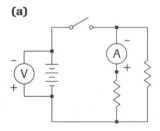

(b)

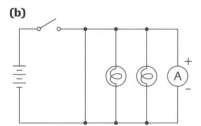

Figure 10

Making Connections

32. Explain the need for extra care in the safe use of electricity in hospital operating rooms.

33. "Smart textiles" are fabrics embedded with electronic components that accomplish a specific task. For example, a "smart blanket" has thin fibres that conduct current and generate thermal energy that can keep an ice fisher warm for hours. A "smart apron" uses integrated circuits and a built-in microphone to allow a person with a motor disability to operate kitchen appliances hands-free.

(a) You have been asked to design a wearable hooded electronic jacket. Describe the features you would like your jacket to have. Use your imagination! (Think of sights and sounds, as well as the ability to monitor variables such as heart rate, location, and distance moved.)

(b) Research smart textiles on the Internet. Briefly describe an interesting application you discover.

 www.science.nelson.com

34. Some electrical energy companies charge customers a very low rate if they can cut off the power at the company's option at any time. Is this a wise method of charging for electrical energy? Consider the point of view of both the buyer and the seller.

35. Estimate each of the following, showing your reasoning or calculations in each case:
    (a) the number of colour television sets in Canada
    (b) the average power rating of the televisions
    (c) the cost of the electrical energy consumed by these televisions in one year

36. Discuss the advantages and disadvantages of developing more uses of current electricity in our society.

37. Many people connect their personal computer system to an electrical outlet through a surge protector extension cord.
    (a) What is surge protection?
    (b) Why is it wise to have this type of protection for computers and other electronic devices?
    (c) Research surge protectors using a resource such as a reference book or the Internet. If necessary, change your answers in (a) and (b) above.
    (d) Based on your research, draw a diagram of a typical surge protector, and describe how the electrical and electronic components operate.

 www.science.nelson.com

Communications Technology

Canada is a world leader in the field of communications. The first domestic communications satellite was Canada's *Anik I,* launched into orbit around Earth in 1972. Additions to the *Anik* series have improved the area of coverage and the quality of radio, phone, and TV signals, and provide mobile phone coverage for automobiles, ships, and airplanes.

This unit examines sound and light energies used in communications and other technologies. Many modern careers are closely related to these technologies.

In this unit, you will study the scientific principles and technological applications of communications systems, and carry out experiments to investigate these principles and system components. You will also identify Canadian contributions to communications technology, and demonstrate awareness of the influence of communications technology on the global community.

In Chapter 9, you will study communication with sound, and in Chapter 10, you will study communication with light, leading to the Unit Performance Task, in which you design a communications system.

▶ Overall Expectations

In this unit, you will be able to

- understand the scientific principles and technological applications involved in the design, development, and operation of communications systems
- design and carry out experiments to investigate communications systems
- identify and describe Canadian contributions to communications technology, and demonstrate an awareness of the influence of communications technology on the global community
- identify and describe science- and technology-based careers related to communications technology

► **Prerequisites**

Concepts

- production of sound
- transmission of sound and light
- absorption of sound and light
- reflection of sound and light
- refraction of light
- plane, concave, and convex mirrors
- visible spectrum
- electromagnetic spectrum

Skills

- solve an equation with one unknown
- select and use appropriate vocabulary to communicate scientific ideas
- apply appropriate safety precautions required in a laboratory environment
- write lab reports for investigations

Knowledge and Understanding

1. (a) In general, what is needed to produce sounds?
   (b) Explain how sound travels from a source to your ears. (Use a diagram if needed.)

2. Can dogs hear sounds that humans cannot? Explain your answer.

3. Both sound energy and light energy can be absorbed by certain materials.
   (a) Describe the properties of a material that absorbs sound energy well, and name two examples of this type of material.
   (b) Repeat (a) for a material that absorbs light energy well.

4. The electromagnetic spectrum includes radio waves, infrared radiation, visible light, ultraviolet light, and X-rays.
   (a) What do all the waves of the electromagnetic spectrum have in common?
   (b) In what ways does visible light differ from the other parts of the electromagnetic spectrum?

5. Copy **Figure 1** into your notebook. In each case, indicate which way the light ray will travel when it passes through or reflects from the object. Give a reason for each choice.

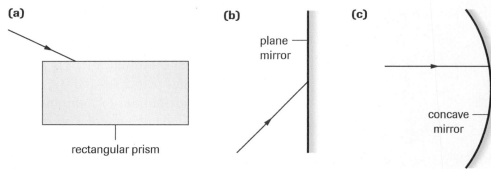

(a)

rectangular prism

(b)

plane mirror

(c)

concave mirror

Figure 1

6. Draw a circle about 10 cm in diameter in the centre of your notepaper. (You can use a protractor to draw the circle.)
   (a) If the circle represents Earth, show the approximate location of a passenger jet, a weather satellite, and the communications airship from **Figure 11,** page 503. (The ship's altitude is19.8 km.)
   (b) Using the same circle to represent Earth, where would a communications satellite be located?

Inquiry and Communication

7. Describe how you would experimentally determine the speed that sound travels in the air outside your school. Include any equation(s) you would apply.

8. **Figure 2** shows two waveforms. Measure the amplitude and wavelength of each wave.

(a)

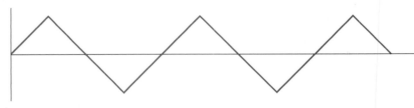

(b)

Figure 2

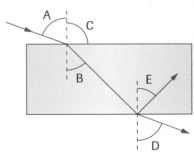

Figure 3

(a)

(b)

9. Use words and diagrams to describe and illustrate plane, concave, and convex shapes.

Math Skills

10. Measure angles A to E in **Figure 3**.

11. Determine the reciprocal of each of the following measurements, and write the answer to the correct number of significant digits:
    (a) 15 cycles/s
    (b) 0.024 s/cycle
    (c) 1.4×10^4 cycles/s
    (d) 8.3×10^{-4} s/cycle

Technical Skills and Safety

12. Describe the safety precautions you would follow when using
    (a) an electric wave generator to experiment with water waves in a shallow tank
    (b) a laser to demonstrate the way light behaves when it strikes a mirror

Making Connections

13. **Figure 4** shows three musical instruments.
    (a) To which family of musical instruments does each belong?
    (b) Does the frequency of these instruments depend on their size? Explain your answer.

14. Ultrasound and X-rays are two types of waves or energies used to provide medical images of the human body. Which one is not safe for imaging a fetus? Explain your answer.

15. Describe two uses of concave reflectors. (Think of both sound energy and light energy.)

Figure 4
(a) A violin and **(b)** a cello, and a double bass

Communication with Sound

Getting Started

When setting up a home theatre, one must take into consideration many of the physics principles related to sound energy to get the best results. The equipment needed to produce high-quality surround sound includes a DVD player, audio receivers with control functions, and speakers (**Figure 1**). Several speakers can be used to produce full, natural sounds that rival the sound quality at an entertainment megaplex. The home theatre demonstrates how much the technology of communication with sound has changed since people first began listening to music through a single, small speaker.

In this chapter, you will reinforce what you learn about the basic principles of vibrations, waves, and sound by doing experiments. Then you will combine your new knowledge with what you have learned about other physics principles, especially energy transformations and electronics, to explain many interesting applications of communication with sound.

Figure 1

1. Draw the side view of a playground swing pulled back and ready to be released. On your drawing show
   (a) the length of rope from the seat of the swing to the point where the rope is attached to the frame
   (b) the amplitude of the swing's vibration
   (c) one complete cycle of vibration of the swing

2. Summarize what you know about frequency and period (from Chapter 1) in a table with these titles: Quantity; Meaning; Symbol; Unit; Equation; Example. Express each equation in terms of the other variable.

3. How does the maximum frequency with which you can vibrate your hand compare with the maximum frequency with which you can vibrate your entire arm? Explain how the length of the hand and the length of the arm affect frequency.

4. As you fill a graduated cylinder with water, the pitch of the sound changes as the length of the air column decreases.
   (a) Does the pitch rise or fall?
   (b) What causes this effect?

5. Name a medical imaging technique that uses waves or sound.

6. When a crowd is cheering at a basketball game in the school gymnasium, the sound can be very loud and garbled. The sound quality is different in an auditorium with a cheering crowd of similar size. Explain the difference.

▶ TRY THIS activity   Can Pendulums Communicate?

Set up the apparatus shown in **Figure 2**, with the support stands about 50 cm apart and clamped securely to the lab bench. Connect the stands with a string, and make sure the string is taut. Suspend two pendulums of equal length (about 30 cm) and equal mass (50 g) from the string. Ensure the masses are securely attached to the string.

- Predict what will happen when you start one pendulum swinging. Then gently start the pendulum swinging. Watch carefully for several minutes. Record your observations.

- Shorten the string on one of the pendulums to about 20 cm, and repeat the above step.

(a) In what way can it be said that the two pendulums "communicate" with each other?

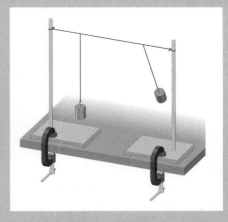

Figure 2
A double pendulum setup

(b) One of the observed motions is called a "forced vibration," and the other is called a "sympathetic vibration." Which motion is which? Explain your answer.

Figure 1
While the swing vibrates, this Cirque du Soleil acrobat performs a variety of acrobatic moves.

vibration the periodic motion of a particle or mechanical system; also called an oscillation

transverse vibration periodic motion that is perpendicular to the rest axis

longitudinal vibration periodic motion that is parallel to the rest axis

torsional vibration periodic motion in which the object or system twists around the rest axis

length (of a pendulum) the distance from its suspension point to the centre of the mass; symbol *l*

amplitude the maximum displacement of a vibration from its normal rest position; symbol *A*

cycle one complete vibration

Many motions can be considered vibrations. The swing carrying the acrobat in **Figure 1** vibrates back and forth. A plucked guitar string vibrates far more rapidly, producing sound. The wings of an insect vibrate, also producing a sound. After reaching the end of the cord, a bungee jumper moves up and down for a short while.

A **vibration**, the periodic or repeated motion of a particle or mechanical system, can be classified as one of three types:

- An object whose motion of vibration is *perpendicular* to its axis at the normal rest position undergoes a **transverse vibration**. An example is a child swinging on a swing (**Figure 2(a)**).

- An object whose motion of vibration is *parallel* to its axis at the rest position undergoes **longitudinal vibration**. An example is a coiled spring (**Figure 2(b)**).

- An object that *twists* around its axis at the rest position undergoes **torsional vibration**. For example, when a string supporting an object is twisted, the object turns, or vibrates, around and back (**Figure 2(c)**).

A simple pendulum is an excellent device for studying transverse vibrations. **Figure 3(a)** illustrates the mass and rest axis of a pendulum, as well as two distances that can be measured: *length* and *amplitude*. The **length** (*l*) of a pendulum is the distance from its suspension point to the centre of the mass. The **amplitude** (*A*) is the maximum horizontal displacement of the mass from its normal rest position. A **cycle** is one complete vibration of the pendulum (**Figure 3(b), (c)**).

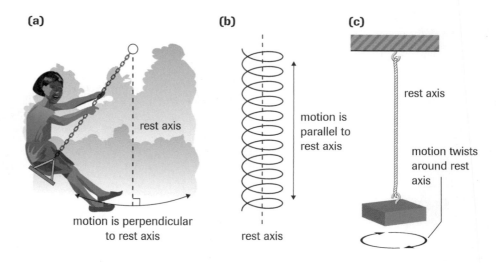

Figure 2
(a) Transverse vibration
(b) Longitudinal vibration
(c) Torsional vibration

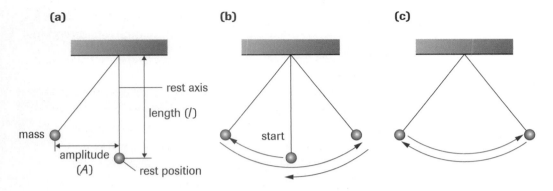

Figure 3
(a) Definitions
(b) One cycle
(c) One cycle is equivalent to one complete vibration.

▶ *SAMPLE* problem *1*
Calculating Distance Moved

A child is swinging on a swing with a constant amplitude of 1.2 m. Through what total horizontal distance does the child move in 3 cycles?

Solution

$A = 1.2$ m

In one cycle, the child moves 4 times as far as one amplitude, or
4×1.2 m $= 4.8$ m.
In 3 cycles, the child moves 3×4.8 m $= 14.4$ m.
The child moves 14.4 m in 3 cycles.

A mass attached to the end of a spring provides a simple model of longitudinal vibration. **Figure 4(a)** illustrates a spring and mass at rest. In **Figure 4(b)**, the mass has been raised to amplitude *A,* its maximum displacement from the rest position. Then the mass is released, and it drops to its lowest position (**Figure 4(c)**). The cycle is complete when the mass returns to its highest position (**Figure 4(d)**).

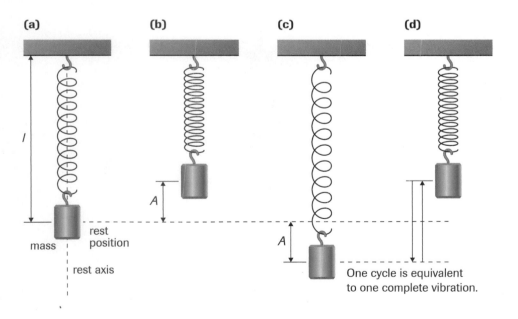

One cycle is equivalent to one complete vibration.

Figure 4
Longitudinal vibrations on a spring

Communication with Sound   **423**

Torsional vibrations are illustrated by the type of clock shown in **Figure 5**. The central shaft rotates first in one direction, then in the opposite direction, and so on, in uniform periods of time.

Figure 5
An "anniversary" clock demonstrates torsional vibration.

Answers

2. (b) 8.5 cm
   (c) 1.7 m
3. (b) 0.49 m
   (c) 3.5 cm

▶ *Practice*

Understanding Concepts

1. State the type of vibration demonstrated in each of the following:
   (a) A diving board vibrates momentarily after a diver jumps off.
   (b) A woodpecker pecks a tree.
   (c) The agitator in one model of washing machine turns back and forth.
   (d) The spring of the suspension system on a motorcycle vibrates as the bike travels over a rough road.

2. For the pendulum in **Figure 6**,
   (a) State the type of vibration.
   (b) Calculate the amplitude of vibration.
   (c) Calculate the total distance, in metres, that the mass moves in 5 cycles. (Assume that the amplitude remains constant.)

3. In **Figure 7(a)**, the mass on the spring is at rest; in **Figure 7(b)** and **(c)**, it is vibrating.
   (a) State the type of vibration.
   (b) Calculate the amplitude of vibration.
   (c) Calculate the total distance, in metres, the mass moves in 3.5 cycles. (Assume that the amplitude remains constant.)
   (d) Starting with the mass at the rest position, describe the energy transformations that occur up to the position in **Figure 7(c)**. Write the corresponding energy-transformation equation.

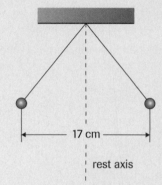

Figure 6
For question 2

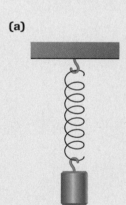

(a)

at rest

(b)

(c)

7.0 cm

Figure 7
For question 3

Frequency and Period of Vibration

A vibrating object has a *frequency* of vibration and a *period* of vibration. The **frequency** (f) is the number of cycles that occur in a specified amount of time:

$$f = \frac{\text{number of cycles}}{\text{total time}}$$

The SI unit of frequency is **hertz** (Hz), or cycles per second. Mathematically, this is the same as $\frac{1}{s}$, or s^{-1}.

The **period** (T) is the time required for one cycle of vibration:

$$T = \frac{\text{total time}}{\text{number of cycles}}$$

The SI unit of period is seconds per cycle, or simply seconds (s).

Note that frequency is measured in cycles per second and period in seconds per cycle: They are the inverse, or reciprocals, of each other. Thus,

$$T = \frac{1}{f} \text{ and } f = \frac{1}{T}$$

frequency the number of cycles per second; symbol f

hertz the SI unit of frequency; symbol Hz

period the number of seconds per cycle; symbol T

▶ *SAMPLE* problem *2*

Calculating Frequency and Period

A mass attached to the end of a spring vibrates vertically 15 times in 12 s. Calculate (a) the frequency and (b) the period of the vibration.

Solution

(a) number of cycles = 15 cycles
   total time = 12 s
   $f = ?$

$$f = \frac{\text{number of cycles}}{\text{total time}}$$

$$= \frac{15 \text{ cycles}}{12 \text{ s}}$$

$$= 1.25 \, \frac{\text{cycles}}{\text{s}}$$

$$f = 1.2 \text{ Hz}$$

The frequency of the vibrating spring is 1.2 Hz.

(b) Using the data in (a),
   $f = 1.25$ Hz (the unrounded value)
   $T = ?$

$$T = \frac{1}{f}$$

$$= \frac{1}{1.25 \text{ Hz}}$$

$$T = 0.80 \text{ s}$$

The period of the vibrating spring is 0.80 s.

in phase vibrating objects that have the same frequency and are at the same positions in their cycles at the same instant

out of phase vibrating objects that are at different positions in their cycles at the same instant

Two identical pendulums vibrate with the same frequency. They are said to be vibrating **in phase** if they are at the same positions in their cycles at the same instant. Two identical pendulums are vibrating **out of phase** if they are at different positions in their cycles at the same instant (**Figure 8**).

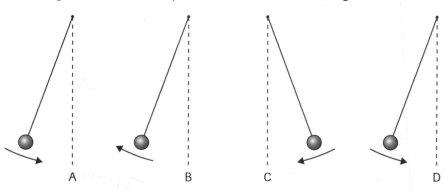

A   B   C   D

Figure 8
These pendulums all have the same frequency. Pendulums A and D are in phase; pendulums B and C are out of phase.

Answers

4. (a) 3.0×10^1 Hz; 3.3×10^{-2} s
   (b) 9.0×10^1 Hz; 1.1×10^{-2} s
   (c) 7.7 Hz to 11 Hz;
        0.13 s to 0.094 s
5. (a) 5.9 s
   (b) 2.5×10^{-7} s
   (c) 8.3×10^4 s
6. (a) 1.0×10^2 Hz
   (b) 5.0×10^7 Hz
   (c) 1.2×10^{-5} Hz

> ▶ *Practice*

Understanding Concepts

4. Calculate the frequency, in hertz, and the period, in seconds, for each situation described below. Assume two significant digits.
   (a) A movie projector displays 1800 frames each minute.
   (b) The horned sungem, a bird native to South Africa, has the fastest wingbeat of any bird, at 1800 beats in 20 s.
   (c) Most butterflies beat their wings between 460 and 640 times per minute. (In this case, find a range for the answer.)

5. Calculate the period of vibration, in seconds, for each of the following frequencies:
   (a) 0.17 Hz
   (b) 4.0 MHz
   (c) 1.2×10^{-5} Hz

6. Calculate the frequency, in hertz, for each of the following periods:
   (a) 0.010 s
   (b) 2.0×10^{-8} s
   (c) 1.0 d (one day)

7. Describe the movement of your arms and legs while you walk at a regular pace as in-phase or out-of-phase vibrations.

Applying Inquiry Skills

8. **Figure 9** shows a device called a mechanical metronome.
   (a) What type of vibration occurs in this device?
   (b) What is the function of a metronome? (If you need a hint, think of music lessons.)
   (c) Compare and contrast a mechanical metronome and a pendulum.
   (d) Describe how you would conduct an experiment to determine what factor affects the metronome's frequency of vibration. Include the apparatus you would require. If possible, carry out your experiment and describe what you discover.

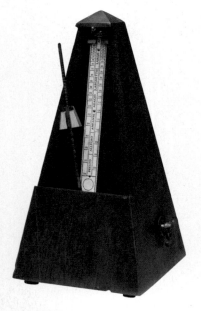

Figure 9
A mechanical metronome

Vibrations

- A vibration is a periodic motion of a particle or mechanical system that can be classified as transverse, longitudinal, or torsional, depending on the motion relative to the rest axis.

- Frequency of vibration, measured in hertz (Hz), is the number of cycles per second.

- Period, measured in seconds (s), is the number of seconds per cycle.

- Frequency and period are reciprocals of each other: $T = \dfrac{1}{f}$ and $f = \dfrac{1}{T}$.

▶ *Section 9.1 Questions*

Understanding Concepts

1. State the type of vibration in each of the following cases:
   (a) A tall tower sways in the wind.
   (b) A needle in a sewing machine vibrates up and down.

2. A pendulum moves 16 cm in one cycle. Calculate the amplitude of vibration.

3. Calculate the frequency and period in each of the following:
   (a) A child, while skipping, jumps off the ground 57 times in 95 s.
   (b) A pulse beats 27 times in 15 s.
   (c) A homeowner shovels snow at a rate of 15 shovelsful per minute.

4. The world record for pogo jumping is more than 122 000 jumps in 15 h and 26 min (**Figure 10**).
   (a) What type of vibration does the coiled spring of the pogo stick undergo?
   (b) Calculate the average period of vibration.
   (c) Calculate the average frequency of vibration.

5. The frequency of vibration of the water molecules in an object in a microwave oven is 2.0×10^{10} Hz. Calculate the period.

6. A sound wave in a steel rail has a period of 4.0×10^{-3} s. Calculate the frequency.

7. The caption for **Figure 8** says that pendulums B and C are out of phase. Name two other examples of pendulums in the diagram that are out of phase. Explain your reasoning.

Applying Inquiry Skills

8. (a) Describe how you would take measurements to determine the frequency and period of your steps when you are walking at a regular pace and when running at your fastest pace.
   (b) Name two major sources of error in making these measurements.

Figure 10

The Pendulum

This investigation is a controlled experiment in which you change one factor at a time to determine the effect on the dependent variable. The dependent variable in this investigation is the frequency of the pendulum. The factors you will change (the independent variables) are the amplitude of vibration, the mass of the pendulum bob, and the length of the pendulum. (To review controlled experiments, refer to Appendix A2.)

A major source of error in this investigation is using a process that requires two reactions: one student's reaction time in starting the stopwatch at the same instant that a second student releases the pendulum. A good way to reduce the effects of this source of error is to have one student set the pendulum swinging according to the instructions, then count backward (4, 3, 2, 1, 0), and on zero start the stopwatch. This process requires only one reaction.

Question

What happens to the frequency of a pendulum as its amplitude, mass, and length are changed, one at a time?

Predictions

(a) Predict, with an explanation, what will happen to the frequency of the pendulum when (i) the mass increases, but the length and amplitude remain constant; (ii) the amplitude increases, but the mass and length remain constant; and (iii) the length increases, but the mass and amplitude remain constant.

Materials

For each group of two or three students:
support stand (longer than 100 cm)
C clamp
buret clamp
split rubber stopper
string, around 110 cm in length
stopwatch
metre stick
3 masses (50 g, 100 g, and 200 g)

Inquiry Skills

○ Questioning	● Conducting	● Evaluating
● Predicting	● Recording	● Communicating
○ Planning	● Analyzing	● Synthesizing

For each student:
graph paper

 Use the C clamp to secure the support stand to the lab bench. Make sure the mass is securely attached to the string.

Procedure

1. Set up a data table based on **Table 1**. Fill in the first three columns according to the steps outlined below.

Table 1   Observations for Investigation 9.2

Length (cm)	Mass (g)	Amplitude (cm)	Time for 20 Cycles (s)	Frequency (Hz)
100	50	10		

2. Use the C clamp to secure the support stand to the lab bench as shown in **Figure 1**. Ensure there is enough clearance so that the swinging pendulum does not hit the clamp.

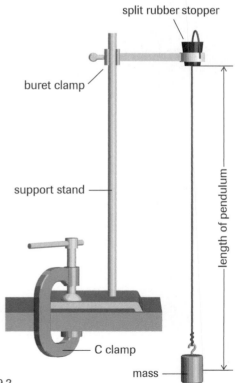

split rubber stopper

buret clamp

support stand

length of pendulum

C clamp

mass

Figure 1
Setup for Investigation 9.2

3. Attach a 50-g mass securely to one end of the string. Place the other end of the string into the split rubber stopper, adjust the pendulum length to 100 cm, and clamp the rubber stopper firmly. Be sure to measure the length to the centre of the mass.

4. Give the pendulum an amplitude of 10 cm, and measure the time required for 20 complete cycles. (Use the synchronized timing technique described in the introduction.) Repeat twice for accuracy, then calculate the frequency. Enter the measurements and calculations in your data table.

5. Repeat step 4 using amplitudes of 20 cm and 30 cm. Tabulate your data.

6. Determine the relationship between the frequency and the mass of a pendulum using a length of 100 cm, an amplitude of 10 cm, and masses of 100 g and 200 g. Be sure to measure the length to the centre of each mass. Tabulate the data, including the 50-g data that you already have.

7. Determine the relationship between the frequency and the length of a pendulum using an amplitude of 10 cm, a constant mass of 50 g, and lengths of 80 cm, 60 cm, 40 cm, and 20 cm. Tabulate your data, including the 50-g data that you already have.

Analysis

(b) With frequency as the dependent variable, plot graphs of frequency versus
- amplitude, for a length of 100 cm and a constant mass of 50 g
- mass, for a length of 100 cm and a constant amplitude of 10 cm
- length, for a constant amplitude of 10 cm and a constant mass of 50 g

(c) Which variables, if any, had little or no effect on the frequency of the pendulum?

(d) Which variables, if any, had the greatest effect on the frequency of the pendulum?

(e) In a short paragraph, summarize your findings on what happens to the frequency of a pendulum when the pendulum's amplitude, mass, and length are changed, one at a time.

Evaluation

(f) Evaluate your three predictions.

(g) Describe the main sources of error in this investigation, and explain how you reduced them to a minimum.

Synthesis

(h) Even though the displacement of the pendulum's swing became smaller during 20 cycles of vibration, the results of this investigation are still valid. Explain why.

Pulses on a Spring

The purpose of this investigation is to study the properties of pulses on a coiled spring.

If you hold a spring that is stretched out on the floor, and you flick your wrist to one side and back (**Figure 1**), you can create a pulse. A pulse is a disturbance caused by the transfer of energy from your hand to the spring. If you define one side as positive, then the other side is negative. You can then compare the phases of the pulses that you observe travelling through and reflecting off the far end of the spring. The far end is called "fixed" if it is held rigid, and "free" if it is held by a long string. The amplitude of the pulse (A) is the same as the amplitude of the vibration that causes the pulse.

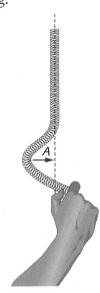

Figure 1
Creating a pulse on a spring

The distances given in the Procedure steps are intended for short springs. If long springs are used, the distances should be increased. Ask your teacher for recommended distances. Simulation software can be used for some of the steps, particularly steps 5, 6, and 7.

Questions

On what does the speed of a pulse along a spring depend?
How are pulses reflected from a fixed end of a spring?

Predictions

(a) Predict how you think the speed of a pulse along a spring depends on the amount that the spring is stretched when all other factors are kept the same.

(b) Predict how you think the speed of a pulse along a spring depends on the amplitude of the vibration that causes the pulse when all other factors are kept the same.

Inquiry Skills

○ Questioning ● Conducting ● Evaluating
● Predicting ● Recording ● Communicating
● Planning ● Analyzing ● Synthesizing

(c) Predict whether the pulse that reflects off the fixed end of a spring is in phase or out of phase with the initial pulse.

✋ **Do not overstretch the springs.**
Do not release a stretched spring.
Wear safety goggles.

Materials

For each group of three or four students:
spring (such as a Slinky toy)
piece of masking tape
piece of notepaper
stopwatch
metre stick

For each student:
safety goggles

Procedure

1. Put on your safety goggles. Attach the piece of masking tape to a coil as shown in **Figure 2(a)**. Stretch the spring along a smooth surface (the floor) to a length of 2.0 m. With one end of the spring held rigidly (i.e., fixed), use a rapid sideways jerk at the other end to create a transverse pulse. Watch the tape attached to the spring, and describe the motion of the coils of the spring by noting the motion of the tape.

2. Using the same setup as in step 1, push forward rapidly to produce a longitudinal pulse along the spring (**Figure 2(b)**). Describe the motion of the coils.

3. Stand a folded piece of notepaper on the floor close to the end of the spring (**Figure 2(c)**). Use the energy transferred by a transverse pulse to knock the paper over. Describe where the energy came from and how it was transmitted to the paper.

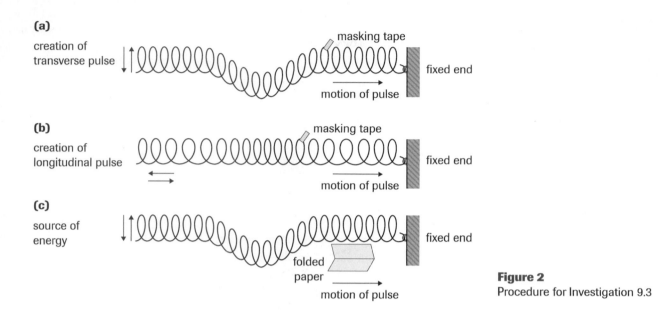

(a)

creation of transverse pulse

masking tape

fixed end

motion of pulse

(b)

creation of longitudinal pulse

masking tape

fixed end

motion of pulse

(c)

source of energy

fixed end

folded paper

motion of pulse

Figure 2
Procedure for Investigation 9.3

4. Hold one end of the spring, and send a transverse pulse toward the fixed end. Is the pulse that reflects off the fixed end in phase (on the same side) or out of phase (on the opposite side) compared with the original pulse? Illustrate your observation in a sketch.

5. Plan the measurements and calculations needed to determine whether the speed of the pulse depends on the tension in the spring. (The tension can be varied by changing the spring's extension.) Repeat the measurements at least three times for accuracy while trying to keep the amplitude constant. Record your measurements and calculations.

6. Plan the measurements and calculations needed to determine whether the speed of the pulse depends on the amplitude of the vibration that causes the pulse. Record your measurements and calculations.

7. If different types of springs are available, compare the speed of a transverse pulse along each of them. Record your measurements and calculations.

Analysis

(d) Write the energy-transformation equation that explains how the piece of paper was knocked over.

(e) Is the reflection in phase or out of phase for a pulse undergoing fixed-end reflection?

(f) State what happens to the speed of a pulse in a material under the following circumstances:
- The condition of the material changes. (For example, stretching a spring changes its tension.)
- The amplitude of the pulse increases.

(g) Does the speed of a pulse depend on the material through which the pulse travels? Use your observations in step 7 to explain your answer.

Evaluation

(h) Evaluate your three predictions.

(i) Describe the main sources of error in this investigation and how you reduced them to a minimum.

Synthesis

(j) How does friction affect the displacement of the travelling pulse from the rest axis?

Suppose you are holding one end of a rope, and the other end is caught loosely around a tree stump. Rather than walking over to the stump to untie the rope, you flick your wrist upward, producing a pulse (**Figure 1**). With just the right movement, perhaps the rope will become free of the stump.

Figure 1
Transferring energy with a pulse

If you continue vibrating the rope up and down, a *wave* will be created. A **wave** is a disturbance, caused by a vibration, that transfers energy over a distance. The vibration set up on the rope is transverse, so the wave is also transverse (**Figure 2(a)**). A transverse wave can be produced on a spring as well (**Figure 2(b)**). A longitudinal wave, which is produced by a longitudinal vibration, cannot be produced on a rope, but it can be produced on a coiled spring (**Figure 2(c)**).

wave a disturbance, caused by a vibration, that transfers energy over a distance

(a)

(b)

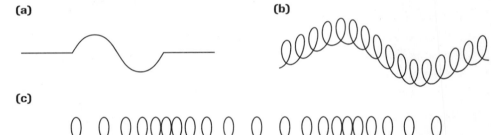

Figure 2
(a) Transverse wave on a rope
(b) Transverse wave on a spring
(c) Longitudinal wave on a spring

(c)

Whether a wave is transverse or longitudinal, its effect is the same: the transfer of energy. In Investigation 9.3, you used a pulse to knock over a piece of paper. The initial kinetic energy of your hand was transformed into elastic potential energy and kinetic energy in the spring coils. This energy was transferred along the spring. When the pulse reached the paper, some of its energy transferred to the paper, giving it kinetic energy. Keep in mind that it is the wave's energy that transfers, not the coils or particles of the material in which the wave travels.

A single cycle of a transverse wave is illustrated in **Figure 3(a)**. The two main parts of a transverse wave are the **crest** above the rest axis and the **trough** below it. The amplitude (A) is the maximum displacement from the rest axis. The **wavelength** (λ) is the distance between adjacent points on a wave that are in phase. (The symbol λ is the Greek letter lambda.) In **Figure 3(b)**, the wave is a **periodic wave**, which is produced by a source vibrating at a constant frequency. Wavelength in a periodic wave can be measured between any two consecutive points that are in phase.

crest the part of a transverse wave above the rest axis

trough the part of a transverse wave below the rest axis

wavelength the distance between adjacent points on a wave that are in phase; symbol λ

periodic wave a travelling wave produced by a source vibrating at a constant frequency

(a)

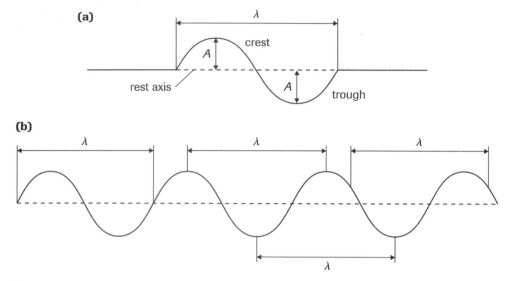

(b)

Figure 3
(a) The parts of a transverse wave
(b) Periodic wave showing examples of wavelength (λ)

▶ **TRY THIS** activity   *Generating Sine Waves*

A transverse periodic wave is drawn as a curve called a sine wave, as seen in **Figure 3(b)**. Generate your own sine wave using a computer software program or manually using the technique shown in **Figure 4**.

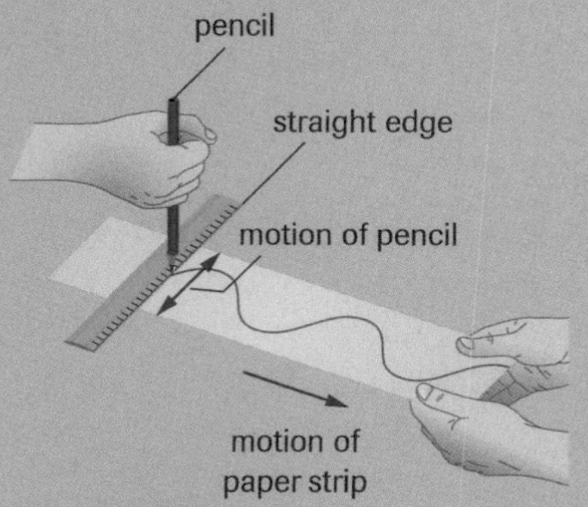

pencil

straight edge

motion of pencil

motion of paper strip

Figure 4
One student moves the pencil back and forth as smoothly as possible while a second student pulls the strip of paper at a constant speed.

Figure 5 shows that a longitudinal wave consists of a **compression**, where the particles are close together, and a **rarefaction**, where the particles are *rarefied*, or spread apart. Compressions and rarefactions are created where parts of the spring are displaced from their rest positions. The amplitude is the maximum displacement from the rest position. Again, wavelength can be measured between any two consecutive points that are in phase.

compression the part of a longitudinal wave where the particles are close together

rarefaction the part of a longitudinal wave where the particles are spread apart

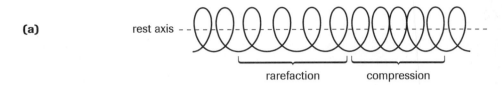

(a) rest axis

rarefaction   compression

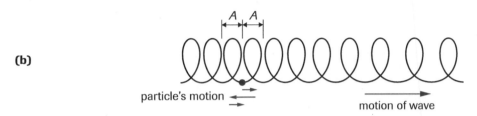

(b)

A A

particle's motion

motion of wave

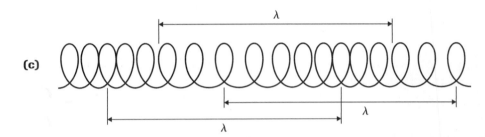

λ

(c)

λ

λ

Figure 5
(a) Compression and rarefaction
(b) The vibration and wave are parallel to the rest axis, and the amplitude (A) is measured from the rest position of a particle.
(c) Periodic wave showing examples of wavelength (λ)

▶ **Practice**

Understanding Concepts

1. How does the phase of a crest compare to the phase of a trough?

2. Measure the amplitude and wavelength of the periodic transverse wave in **Figure 3(b)**, page 433.

3. Measure the wavelength of the periodic longitudinal wave in **Figure 5(c)**.

4. A periodic wave drawn with a constant displacement is an "ideal wave."
   (a) What happens to the displacement of a real wave on a rope or a spring as the wave travels along the floor? Explain why.
   (b) Draw a sketch showing the difference between an ideal transverse periodic wave and a wave that is not ideal.

The Universal Wave Equation

To find the speed of a wave on a spring, we can use the equation for speed:
$\text{speed} = \dfrac{\text{distance}}{\text{time}}$ or $v = \dfrac{d}{t}$. From this basic equation we can derive an equation based on frequency and wavelength.

Consider **Figure 6(a)**, which shows a set of train cars travelling to the right. If each car is 18 m long and takes 4.0 s to pass point X, then the train's speed is

$$v = \frac{d}{t}$$

$$= \frac{18 \text{ m}}{4.0 \text{ s}}$$

$$v = 4.5 \text{ m/s}$$

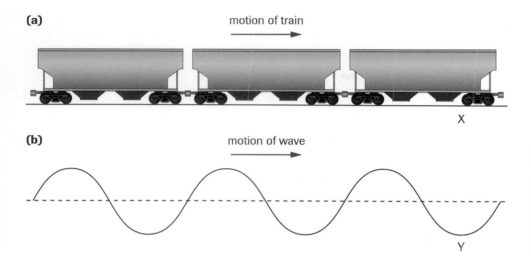

(a)

motion of train

X

(b)

motion of wave

Y

Figure 6
The speed of a periodic wave travelling past point Y can be determined using the same equation as was used to determine the speed of a train car travelling past point X.

Now compare the train cars to the periodic wave in **Figure 6(b)**. If each 18-m cycle was created in a period of 4.0 s, then the speed of the periodic wave past point Y is the ratio of the wavelength to the period of vibration of the source of the wave:

$$v = \frac{d}{t}$$

$$= \frac{\lambda}{T} \qquad \text{(because in time } t = T, \text{ a wave travels a distance } d = \lambda)$$

$$= \frac{18 \text{ m}}{4.0 \text{ s}}$$

$$v = 4.5 \text{ m/s}$$

This equation for speed, $v = \frac{\lambda}{T}$, may also be written $v = \left(\frac{1}{T}\right)\lambda$. Because $f = \frac{1}{T}$, we can write the equation for speed in terms of frequency and wavelength:

$$v = f\lambda$$

This equation is called the **universal wave equation**. The equation is considered universal because it applies to all waves, not just the mechanical ones described here.

universal wave equation the equation for the speed of a wave in terms of the frequency of its source and wavelength of the wave; $v = f\lambda$

▶ **SAMPLE** problem **1**

Calculating Wave Speed of Water

A wave machine, vibrating with a frequency of 4.0 Hz, generates a water wave of wavelength 2.4 m (**Figure 7**). What is the speed of the water wave?

Solution

f = 4.0 Hz

λ = 2.4 m

v = ?

$$v = f\lambda$$
$$= (4.0 \text{ Hz})(2.4 \text{ m})$$
$$= \left(4.0 \; \frac{\cancel{\text{cycles}}}{\text{s}}\right)\left(\frac{2.4 \text{ m}}{\cancel{\text{cycle}}}\right) \quad \text{(See the Learning Tip.)}$$
$$v = 9.6 \text{ m/s}$$

The speed of the wave is 9.6 m/s.

Figure 7
A wave pool

▶ **SAMPLE** problem **2**

Calculating Wave Speed on a Spring

Every 0.50 s, a periodic source produces a wave of wavelength 3.2 m on a spring. Calculate the speed of the waves.

Solution

λ = 3.2 m

T = 0.50 s

v = ?

$$v = \frac{\lambda}{T}$$
$$= \frac{3.2 \text{ m}}{0.50 \text{ s}}$$
$$v = 6.4 \text{ m/s}$$

The speed of the wave is 6.4 m/s.

LEARNING TIP

Unit Cancellation
If frequency is stated in cycles per second and wavelength in metres per cycle, the cycles cancel out, leaving metres per second, the correct unit of speed.

> **Practice**

Understanding Concepts

5. Calculate the speed of the following waves:
   (a) $f = 18$ Hz; $\lambda = 2.7$ m
   (b) $T = 4.5 \times 10^{-4}$ s; $\lambda = 9.0 \times 10^4$ m

6. Write an equation for each of the following:
   (a) f in terms of v and λ
   (b) T in terms of v and λ
   (c) λ in terms of v and f
   (d) λ in terms of v and T

7. Calculate the quantities indicated by question marks in **Table 1**. Show the equation required to solve each unknown.

Table 1   For Question 7

	Speed (m/s)	Frequency (Hz)	Period (s)	Wavelength (m)
(a)	3.0×10^5	?	—	1.5×10^2
(b)	0.60	—	?	0.12
(c)	850	25	—	?
(d)	2.0×10^8	—	4.2×10^{-7}	?

Applying Inquiry Skills

8. Describe two ways you can determine the speed of a transverse periodic wave on a long, strong rubber rope stretched out on the floor. State the measurements and calculations needed.

Making Connections

9. Does the equation $v = f\lambda$ apply to sound waves on another planet that has an atmosphere? How do you know?

Answers

5. (a) 49 m/s
   (b) 2.0×10^8 m/s
7. (a) 2.0×10^3 Hz
   (b) 0.20 s
   (c) 34 m
   (d) 84 m

SUMMARY   *Waves*

- A wave is a disturbance caused by a vibration; its effect is a transfer of energy.

- A transverse wave consists of crests and troughs.

- A longitudinal wave consists of compressions and rarefactions.

- A periodic wave is created by a regular, periodic vibration. Its wavelength can be measured between any two consecutive points that are in phase.

- Wave speed can be determined from the equation $v = \dfrac{\lambda}{T}$, as well as from the universal wave equation, $v = f\lambda$. The universal wave equation applies to all waves.

Understanding Concepts

1. How does the phase of a rarefaction compare to the phase of a compression?

2. Determine the wavelength in each of the following cases:
   (a) A crest is 4.1 cm long.
   (b) In a water tank, there are 5 parallel crests within a distance of 18 cm.
   (c) A rarefaction is 1.1 m long.

3. In each of the following cases, calculate the speed of the waves in metres per second:
   (a) $f = 2.1 \times 10^4$ Hz; $\lambda = 2.0 \times 10^5$ cm
   (b) $T = 2.0$ ms; $\lambda = 3.4$ km

4. A wave machine generates a 6.5-Hz wave in a water tank 45 cm long. If each part of the wave takes 0.50 s to travel the length of the tank, calculate the wavelength of the wave.

5. A 17-cm sound wave is moving at 3.4×10^2 m/s. Calculate the
   (a) frequency of the sound, in hertz and kilohertz
   (b) period of vibration of the source of the sound

Applying Inquiry Skills

6. Periodic waves are generated on a rubber rope with a source frequency of $f_1 = 1.5$ Hz, and later with a source frequency of $f_2 = 3.0$ Hz.

   (a) Predict which periodic wave, if either, will travel faster. Give a reason for your prediction.
   (b) How would you test your prediction experimentally? Indicate the measurements and calculations needed.

Making Connections

7. Waves on a spring on the floor lose energy as they travel. In your experience, do sound waves display the same property? Give evidence to support your answer.

8. Earthquakes produce *seismic waves*, which travel through Earth. *Primary (P) waves* are longitudinal; they can travel through both solids and liquids. *Secondary (S) waves* are transverse; they can travel through solids only. P waves travel at a speed of approximately 8.0 km/s, and S waves travel at a speed of approximately 4.5 km/s. Following an earthquake, vibrations are recorded at seismological stations around the world (**Figure 8**).
   (a) Determine how long P waves and S waves take to travel from an earthquake to a seismological station 2.4×10^3 km away. Express the answers in minutes.
   (b) Why do you think the transverse waves are called secondary waves?
   (c) By referring to **Figure 8**, explain how observing P waves and S waves helps scientists analyze the structure of Earth's interior.

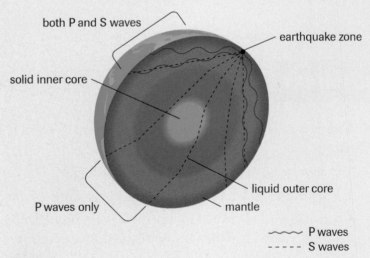

both P and S waves

earthquake zone

solid inner core

liquid outer core

P waves only

mantle

〜〜〜 P waves

- - - - - S waves

Figure 8

If a pulse moving in one direction in a single medium (or material) meets a pulse moving in the opposite direction in the same medium, the pulses interact with each other for an instant as they pass. This interaction is called wave *interference*. It is observed in transverse and longitudinal pulses and waves.

▶ TRY THIS activity   *Pulses Meeting Pulses*

To observe what happens when pulses travelling in opposite directions meet each other, you can use a computer simulation or a spring (such as a Slinky toy, see **Figure 1**) with a small piece of masking tape attached to a coil in the middle of the spring. First decide which side of the spring will be the crest side and which will be the trough side. You may need to repeat each step several times to be sure of the result.

• With a student at each end, stretch the spring an appropriate amount. As one student generates a crest, the other student generates a trough of equal size. Observe what happens before, during, and after the pulses meet at the middle of the spring.

• Observe what happens when two crests are generated at the same instant and meet at the middle of the spring.

• Observe what happens when two troughs are generated at the same instant and meet at the middle of the spring.

(a) Describe what happens to the displacement of a disturbance from the rest axis when

  • a crest meets a trough of equal amplitude

  • a crest meets a crest

  • a trough meets a trough

 Do not overstretch the springs.
Do not release a stretched spring.
Wear safety goggles.

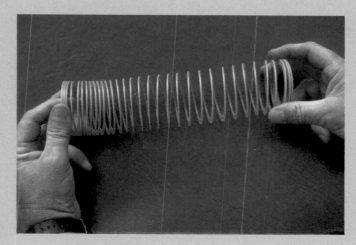

Figure 1
A Slinky toy

There are two types of interference: *constructive* and *destructive*. When pulses or waves meet, or interfere, and cause a reduction in displacement from the rest axis, the interference is called **destructive**. For transverse pulses, if a

destructive interference the reduction in displacement that results when pulses or waves meet

node a position of zero amplitude resulting from destructive interference of pulses or waves

crest meets a trough of equal amplitude and shape, their displacements cancel each other for an instant, creating a **node** of zero amplitude. Then the crest and trough continue in their original directions, as shown in **Figure 2**. For longitudinal pulses, destructive interference results when a compression meets a rarefaction.

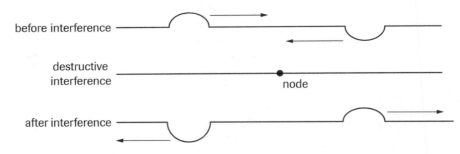

Figure 2
Destructive interference of transverse pulses

When pulses or waves interfere with a resulting larger displacement, the interference is called **constructive**. This happens with transverse pulses when a crest meets a crest, causing a **supercrest**, as shown in **Figure 3**, and when a trough meets a trough, causing a **supertrough**. For longitudinal pulses, constructive interference results when a compression meets a compression, and when a rarefaction meets a rarefaction.

constructive interference the increase in displacement that results when pulses or waves interfere

supercrest the result when a crest interferes with a crest

supertrough the result when a trough interferes with a trough

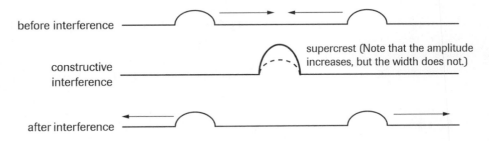

Figure 3
Constructive interference of transverse pulses

In **Figures 2** and **3**, the interference is shown at the instant the pulses overlap or become superimposed on one another. If we call displacements on one side of the rest axis positive, and displacements on the opposite side negative, then the superimposed displacement is simply the addition of the individual displacements. For example, displacements of $+1.0$ cm and -1.0 cm are added to produce a zero displacement. The principle of displacement addition is summarized in the *principle of superposition*:

Principle of Superposition
The resulting displacement of two interfering pulses or waves is the algebraic sum of the displacements of the individual pulses or waves.

This principle is especially useful for finding the resulting pattern when pulses that are unequal in size or shape interfere with one another. Note, however, that it applies only to displacements that are reasonable in size—large pulses that interfere may distort the material through which they are travelling.

To learn how to apply the principle of superposition in one dimension, study **Figure 4**, where coloured arrows are used to show displacement. In **Figure 4(a)**, two straight-line pulses interfere; in **Figure 4(b)**, two curved-line pulses interfere. In both cases, the resulting pulse (in the second last diagram) can also be called the resulting **waveform**. Displacements can be added electronically, and the resulting waveform can be displayed on a computer monitor or an oscilloscope.

waveform the instantaneous displacement of interfering waves

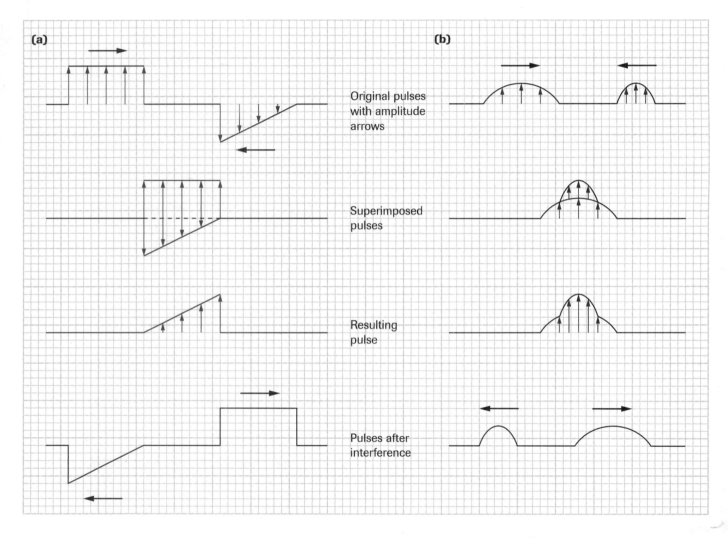

Figure 4
(a) Straight-line pulses
(b) Curved-line pulses

▶ *Practice*

Understanding Concepts

1. In each of the following, state whether the interference is constructive or destructive:
   (a) A large crest meets a small trough.
   (b) A supertrough is formed.
   (c) A small compression meets a large compression.

2. Using the principle of superposition, determine the resulting pulse when the pulses shown in **Figure 5** are superimposed on each other. (The point of overlap should be the horizontal midpoints of the pulses.)

Making Connections

3. The pulses on ropes, such as those in **Figures 2** and **3**, resemble analog signals. (Analog and digital signals were presented in section 8.8, pages 393 to 397.)
   (a) Draw a transverse crest in digital format.
   (b) Name at least one instrument or other device that probably uses the principle of superposition of digital signals.

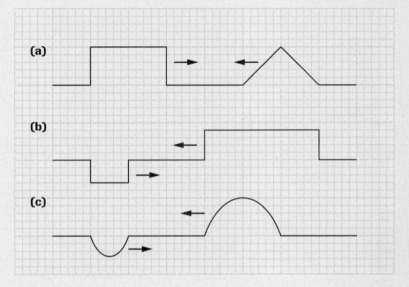

Figure 5

SUMMARY *Interference of Pulses and Waves*

- If the resulting displacement of two interfering waves or pulses is less than the individual displacements, the interference is destructive; if it is greater, the interference is constructive.

- The resulting displacement of two interfering pulses or waves is the algebraic sum of the displacements of the individual pulses or waves; this is called the principle of superposition.

Section 9.5 Questions

Understanding Concepts

1. **Figure 6** shows two sets of pulses approaching each other. In each case, draw the pattern that results when the two pulses meet and their centres coincide.

    (a)

    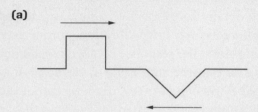

    (b)

    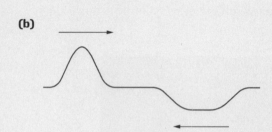

    Figure 6
    For question 1

2. Copy the pulses shown in **Figure 7** into your notebook. In each case, apply the principle of superposition to draw the resulting pattern at the instant shown.

    (a)

    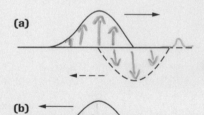

    (b)

    Figure 7

Making Connections

3. One application of the principle of superposition is noise-cancellation headphones. These electronic headphones are more effective than methods of protection that simply cover the ears. Pilots of some noisy military aircraft wear these headphones to protect their hearing. Some musicians also use this technology.

    (a) Based on the name of these headphones, do you think they make use of constructive or destructive interference? Explain your answer.

    (b) The noise sensors in a headphone receive the signal shown in **Figure 8**. Draw the signal that the electronic components must create to cancel the noise.

    Figure 8

    (c) Research noise-cancellation and noise-reduction headphones. Describe briefly what you discover about their operation and use.

    GO   www.science.nelson.com

resonant frequency the natural frequency of a vibrating object

mechanical resonance the maximum response due to the transfer of energy from one object to another in a mechanical system at the same natural frequency

Resonance in the Human Body
Experiments have shown that the body, as a whole, has a mechanical resonant frequency of about 6 Hz. The head has a frequency of between 13 Hz and 20 Hz, and the eyes, between 35 Hz and 75 Hz. Large-amplitude vibrations at any of these frequencies could irritate or even damage parts of the body. In the transportation and road construction industries, the effects of mechanical vibrations on the human body are an occupational hazard that must be reduced.

Splashing Soup
Resonant frequency explains why it is difficult to carry a bowl of soup or other liquid without spilling it. The frequency of the soup's motion approaches the frequency of the walking pace. This causes a relatively large amplitude in a relatively small bowl of soup.

sympathetic vibration the response of an object to another vibration with the same resonant frequency

forced vibration the response of an object to a vibration with a different frequency from the object's resonant frequency

An object can be vibrated most easily at its own natural frequency, known as its **resonant frequency**. At this frequency the amplitude is largest. Energy can be transferred in different ways to obtain this resonance. One way is to apply a small, repeated force, which causes a relatively large vibration. For mechanical vibrations, such as a vibrating pendulum, the response is called **mechanical resonance** and is maximum. For example, the amplitude of vibration of a playground swing can be increased by pushing at the correct instant in each cycle—the frequency of the application of the repeated force equals the resonant frequency of the swing, which generates a large amplitude (**Figure 1**).

Figure 1
A playground swing

A spectacular demonstration of mechanical resonance was the disastrous collapse of the Tacoma Narrows Bridge in the state of Washington in 1940 (**Figure 2**). The bridge was suspended by huge cables across a river. On a windy day four months after its official opening, the bridge began vibrating at its resonant frequency. At first, it vibrated as a transverse wave. Then, when one of the suspension cables loosened, the entire 850-m centre span length of the bridge underwent torsional vibrations. The vibrations were so severe that the bridge collapsed!

You observed a second way to transfer energy in the Chapter 9 introductory activity, page 421, in which the energy transferred from one pendulum to another of equal length. The second pendulum vibrated in resonance with the first pendulum, a response called **sympathetic vibration**.

Finally, energy can be transferred by *forced vibration*. For example, if soldiers march across a small bridge in unison, and the frequency of the soldiers' steps is near the resonant frequency of the bridge, the **forced vibration** could cause the bridge to collapse. To prevent this, soldiers are told to "break step" as they cross bridges. When they break step, the step frequencies are varied, and most no longer match the bridge's resonant frequency. As a result, the amplitude of vibration cannot build up. (You an saw example of a forced vibration in the Chapter 9 introductory activity, page 421 when the pendulums were of different lengths.)

(a)

(b)

(c)

(d)

Figure 2

(a) The Tacoma Narrows Bridge begins to vibrate.

(b) The centre span of the Bridge vibrates torsionally before collapsing.

(c) The bridge eventually collapsed as a result of the vibrations. (No injuries were reported because no one was on the bridge.)

(d) The Tacoma Narrows Bridge today. Notice the structural changes at the towers and in the size and design of the girders supporting the road.

Understanding Concepts

1. A playground swing with a child sitting on it has a resonant frequency of 0.25 Hz.
   (a) With what frequency and period must you push the child to build up a large amplitude of vibration?
   (b) What will happen if you try to push the child with a frequency of 1.5 Hz? (You can test your answer using a simple pendulum.)

2. Describe examples of mechanical resonance other than those mentioned in this section.

Applying Inquiry Skills

3. A long-stemmed glass can be made to resonate and produce a sound by rubbing a moist finger around the rim.
   (a) Describe an experiment to determine how the resonant frequency of the glass depends on the amount of water in the glass.
   (b) Predict the relationship in (a).
   (c) With your teacher's permission, try your experiment and describe what you discover.

Making Connections

4. If a car is stuck in the snow, how can you apply the principle of mechanical resonance to help get it out?

5. Describe how an athlete on a trampoline would apply the concept of resonance
   (a) to jump as high as possible
   (b) to reduce the vibration amplitude to a minimum before getting off the trampoline

Answers

1. (a) 0.25 Hz; 4.0 s

Standing Waves

If periodic transverse waves of equal wavelength and amplitude travel in opposite directions, for example, on a spring, a rope, or in a ripple tank, the waves interfere with each other and set up an obvious pattern. The pattern has loops made of supercrests and supertroughs that go up and down, and nodes that stand in the same position. The formation is called a *standing wave interference pattern*, or simply a **standing wave**.

You can easily observe standing waves in a one-dimensional medium: With one end of a rubber rope tied securely to a fixed support, such as a doorknob, send periodic waves toward the fixed end. (You learned in Investigation 9.3 that a pulse that strikes a fixed end reflects back out of phase.) Those waves will reflect back and interfere with the incoming, or incident, waves. This interference causes the nodes and loops. At the middle of each loop is an **antinode**, or position of maximum displacement from the rest axis. It is equal in magnitude to the amplitude of a supercrest or supertrough. Four patterns are shown in **Figure 3**. In each case, only one frequency produces the pattern. That frequency is the resonant frequency of the system. You can also observe standing waves in two dimensions as part of Activity 9.7.

standing wave a pattern of loops and nodes created by the interference of periodic waves of equal wavelength and amplitude

antinode a position of maximum displacement from the rest axis in a standing wave

The lowest resonant frequency of a system that produces a standing wave is called the **fundamental frequency**. The next resonance occurs with a frequency two times the fundamental frequency, then three times, and so on. In each case, the standing wave pattern occurs only at a specific resonant frequency that is a whole-number multiple of the fundamental frequency.

It is evident in **Figure 3** that the distance from one node to the next in a standing wave is half the wavelength that produced the pattern, or $\frac{1}{2}\lambda$. The distance between the centres of adjacent antinodes is also $\frac{1}{2}\lambda$.

fundamental frequency the lowest resonant frequency of a system that produces a standing wave

Tidal Nodes and Antinodes
Tidal action in the oceans and seas can set up standing wave patterns with very long wavelengths. A few places on Earth—Tahiti, for example, in the South Pacific Ocean—have no noticeable tide because they lie on a tidal node. Other places, such as the Bay of Fundy in eastern Canada, have very high tides because they lie on a tidal antinode.

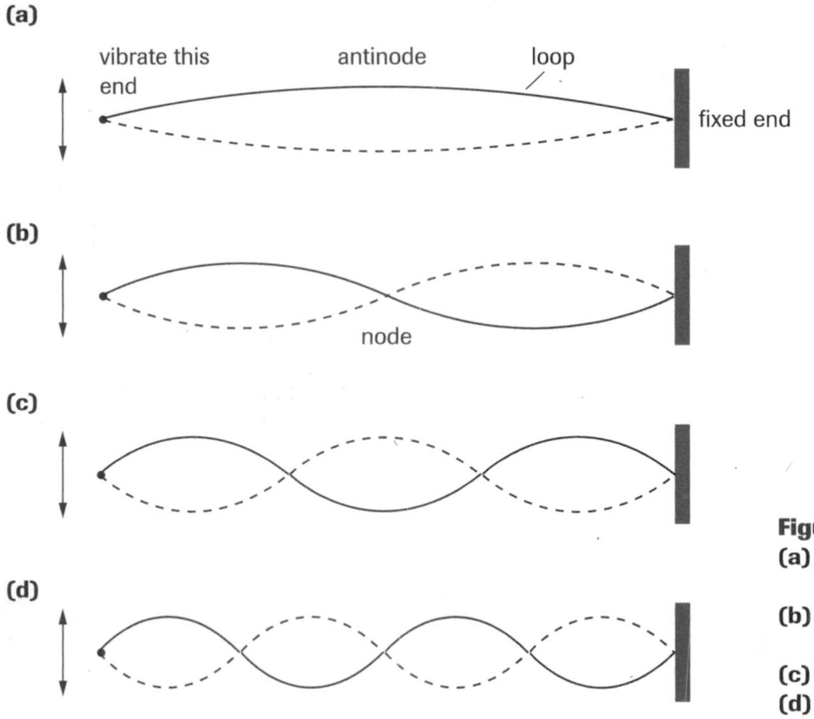

(a)

(b)

(c)

(d)

Figure 3
(a) Low frequency, long wavelength, zero nodes between ends
(b) Higher frequency, shorter wavelength, one node between ends
(c) Two nodes between ends
(d) Three nodes between ends

▶ **TRY THIS** activity ***Producing Standing Waves***

Using a rope, such as a skipping rope, try to create patterns as described and illustrated in **Figure 3**. Try producing standing waves with three, four, or even more nodes between the ends. Is it true that standing waves occur at specific frequencies only?

The Speed of Waves Forming a Standing Wave

A standing wave is produced on a 6.0-m rope using a 5.5-Hz source. If there are three antinodes between the ends, what is the speed of the waves that produced the pattern?

Solution

The standing wave resembles the one shown in **Figure 3(c)**, which also has three antinodes between the ends. The length of the standing wave pattern is 1.5λ.

$1.5\lambda = 6.0$ m

$\lambda = 4.0$ m

$f = 5.5$ Hz

$v = ?$

$$v = f\lambda$$
$$= (5.5 \text{ Hz})(4.0 \text{ m})$$
$$v = 22 \text{ m/s}.$$

The speed of the waves is 22 m/s.

The solution in Sample Problem 1 tells us that the motion in the rope consists of two equal, superimposed waves, one moving to the right and the other to the left. Each travels at 22 m/s, while the up–down movements occur at 5.5 Hz.

▶ **Practice**

Understanding Concepts

Answers

6. 4.0 m

7. (a) 0.40 Hz
   (b) 0.80 Hz
   (c) 1.6 Hz

6. Draw a scale diagram of a standing wave pattern on an 8.0-m rope with four antinodes between the ends. What is the wavelength of the waves that produced the pattern?

7. The speed of a wave on a 4.0-m rope is 3.2 m/s. What frequency of vibration is needed to produce a standing wave pattern with (a) one antinode, (b) two antinodes, and (c) four antinodes?

Applying Inquiry Skills

8. Two students want to determine whether standing waves can be produced on a spring stretched along the floor. Should they both try to generate waves of equal frequency to produce a steady pattern? If not, what should they do? Explain your answer.

Making Connections

9. How could a wind tunnel have helped prevent the Tacoma Narrows Bridge collapse? (Wind tunnels were discussed in section 6.1, page 280.)

SUMMARY *Mechanical Resonance and Standing Waves*

- Mechanical resonance is the transfer of energy from one object to another at the same resonant frequency.
- Sympathetic vibrations occur at the same resonant frequency, and forced vibrations occur at a different resonant frequency.
- Standing waves, a pattern of nodes and loops in a medium, are caused by interference and resonance.
- The distance between adjacent nodes or antinodes in a standing wave interference pattern is one-half the wavelength of the interfering waves.

 Section 9.6 *Questions*

Understanding Concepts

1. The toy illustrated in **Figure 4** can be made to vibrate with its own resonant frequency. Predict, with a reason, how the resonant frequency would change from **(a)** to **(b)** to **(c)**. If possible, test your prediction experimentally.

2. Standing waves are produced on a string by two waves travelling in opposite directions at 6.3 m/s. The distance between the second node and the sixth node is 84 cm.
   (a) Sketch the standing wave pattern from the second node to the sixth node.
   (b) Calculate the wavelength of the waves producing the pattern.
   (b) Calculate the frequency of the source of the waves.

3. Waves on a 2.0-m rope travel at 2.8 m/s. Determine the frequency needed to produce a standing wave with (a) one antinode, (b) two antinodes, and (c) three antinodes.

Applying Inquiry Skills

4. A mass is suspended from a spring and set into a vertical vibration. The mass–spring system has its own resonant frequency. (You learned about springs, Hooke's law, and elastic potential energy in Chapter 3.)
   (a) Make up a question for an investigation to determine the variables on which the frequency of the mass–spring system depends. (You should consider two main factors in your question.)
   (b) Predict an answer to your question in (a).

Making Connections

5. (a) What happens to the resonant frequency of a simple pendulum as the length decreases? Explain using principles of physics and previously established relationships.
   (b) Use your answer in (a) to explain why you are most likely to walk with your arms relaxed at your side but to run with your arms bent at the elbow.

6. Name two careers in which mechanical resonance could present a health hazard. In each case, explain the reason for the hazard.

(a)

(b)

(c)

Figure 4
Changing the resonant frequency

Observing Waves in Two Dimensions

Until now we have looked at pulses and waves in one dimension: a rope or spring's length. Water waves have two dimensions: length and width. Studying waves in two dimensions will help you understand sound waves, which travel through the air in three dimensions: length, width, and depth.

A ripple tank is a piece of equipment used to demonstrate two-dimensional waves in water. It is a raised, shallow tank with a glass bottom (**Figure 1**). For most demonstrations, the tank is level and contains water to a depth of between 5 mm and 15 mm. A light source held by a stand above the water allows the transverse water waves to be easily seen on a screen beneath the tank. Each crest acts like a magnifying glass, focusing the light to produce a bright region beneath the tank. Each trough spreads the light out, producing a dark region. The bright and dark regions appear on the screen. In a ripple tank, a periodic wave with straight wavefronts is produced by a motor connected to a straight bar, shown in **Figure 2**. A periodic wave with circular wavefronts is produced by a motor connected to a spherical source.

The properties of water waves can also be observed in videos of ripple tank demonstrations or using simulation software. All these methods of observing waves explore the same questions.

Materials

For a class demonstration:
ripple tank and related apparatus (light source, motor, connecting wires, construction paper)
source of straight wavefronts
two point sources
ruler or metre stick
water
stopwatch
barriers (as described in the Procedure steps)
metric ruler
protractor

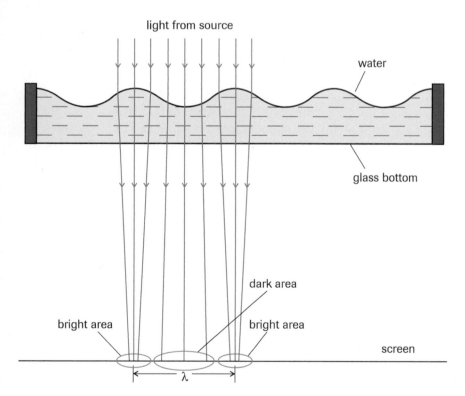

Figure 1
Bright lines appear on the screen where light rays converge.

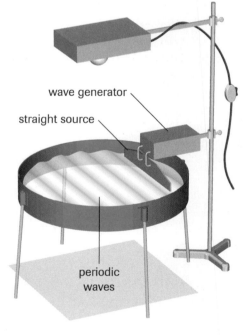

Figure 2
A wave with straight wavefronts

The light source and wave generator are electrical; all wiring must be taped away from the water.

The lab will be darkened for this demonstration; therefore, all bags, books, and other belongings must be kept out of the aisles and away from the exits.

Secure the legs of the ripple tank before filling it.

Ensure that all tank attachments (e.g., the light source and the wave generator) are properly secured.

Unplug the tank and accessories by pulling on the plug, not on the cord.

Procedure

1. Add water to a depth of 5 mm. Devise a way to determine the speed of a straight wavefront that travels from one end of the tank to another.

2. Add water to a depth of 15 mm, and determine the speed of a straight wavefront.

3. Send a periodic wave with straight wavefronts toward a straight barrier at an angle to the surface of the barrier (**Figure 3**). Draw the incoming, or incident, and reflected wavefronts. Measure the angle between the incident wavefront and the reflecting surface. Measure the angle between the reflected wavefront and the reflecting surface.

4. Send a periodic wave with straight wavefronts toward a concave parabolic reflector (**Figure 4**). Locate the position where the reflected wavefronts converge. Label this position *focal point* (*F*). Measure and label the *focal length* (*f*), which is the distance from the focal point to the reflecting surface.

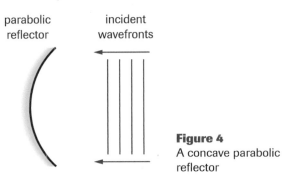

Figure 4
A concave parabolic reflector

5. Send a periodic wave with straight wavefronts from deep water to shallow water that is above a flat barrier (**Figure 5**). Draw a diagram showing what happens to the wavelength of the periodic wave as it travels from one medium (deep water) into a second one (shallow water).

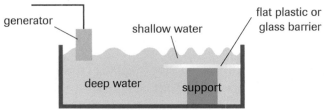

Figure 5
An area of shallow water can be created by placing a flat barrier on a support in the deep water.

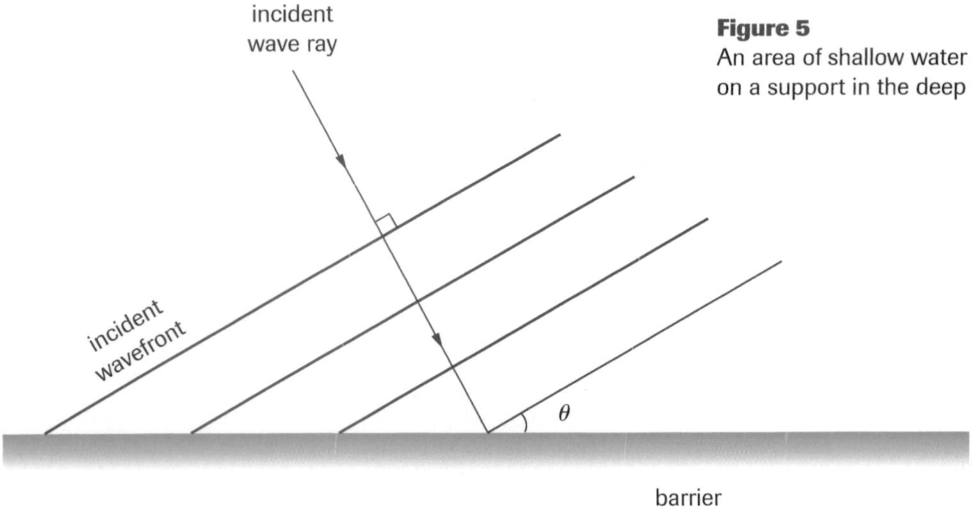

Figure 3
The wavefronts reflect off the barrier.

6. Send a periodic wave with straight wavefronts from deep water to shallow water at an angle to the interface between the two media (**Figure 6**). Draw a diagram showing how the waves *refract*, or bend, as they enter the second medium. Measure the angle between the incident wavefront and the interface between the media, and the angle between the refracted wavefront and the interface.

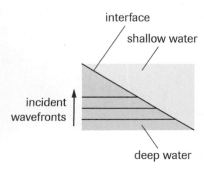

Figure 6
The wavefronts travel from one medium into another at an angle.

7. (a) Aim a periodic wave with straight wavefronts of low frequency (long wavelength) toward an opening between two barriers (**Figure 7**). Draw a diagram showing the *diffraction* of the wave (bending within the same medium) as it travels through the opening.
   (b) Aim a periodic wave with straight wavefronts of higher frequency toward the same opening. Draw a diagram showing how the diffraction changes when the wavelength changes.

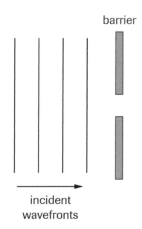

Figure 7
Observing the diffraction of a wave in water

8. (a) Aim a periodic wave with straight wavefronts of low frequency toward a straight barrier. Adjust the position of the barrier so that a standing wave pattern occurs between it and the source. Plan how to measure the wavelength of the waves.
   (b) Repeat (a) using a higher source frequency.

9. Replace the source of straight wavefronts with two sources of circular wavefronts.
   (a) Adjust the two point sources so that when they vibrate, both sources move up and down at the same time (i.e., they are in phase with each other). Adjust the frequency of the motor to a low frequency. Observe how the circular wavefronts produced by the two sources interfere with each other as they move through the water (**Figure 8**).
   (b) Predict the effect of an increase in the frequency of the sources on the pattern. Increase the frequency, and describe what you observe.

Analysis

(a) Compare the speed of a wave in deep water and in shallow water.

(b) How does the angle of an incident wavefront compare to the angle of the corresponding reflected wavefront that strikes a barrier?

(c) Describe the difference(s) between the focal point and focal length of a parabolic reflector.

(d) What happens to the speed of periodic waves as they travel from deep water to shallow water? (Hint: You can apply the universal wave equation to the wavelength of the waves as they travelled at a constant frequency into the shallow water in step 5.)

(e) As periodic waves travel from deep water into shallow water at an angle to the interface, how does the angle in the shallow water compare to the angle in the deep water? What would happen if the waves travelled from shallow water into deeper water?

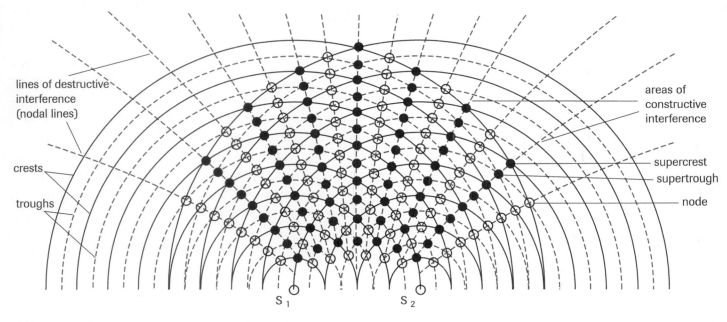

Figure 8
Two sources, S_1 and S_2, produce in-phase periodic circular waves that interfere. The pattern has *nodal lines* (lines of destructive interference), and supercrests and supertroughs (areas of constructive interference).

(f) How does the diffraction of periodic waves in water depend on the wavelength of the periodic waves?

(g) How does the spacing of nodal lines in a two-source interference pattern depend on the wavelength of the periodic circular waves?

(h) What are the advantages of using videos or software programs over ripple tanks to observe the properties of water waves? What are the disadvantages?

Synthesis

(i) Both refraction and diffraction involve the bending of waves. How do these properties differ?

(j) The parabolic reflector used in this activity has a concave shape. Should this type of reflector be called a converging or a diverging reflector? Explain why.

Figure 1
Visible vibrations of guitar strings

We use different words to describe the sounds we hear: Leaves rustle, lions roar, babies cry, birds chirp, corks pop, orchestras crescendo—the list is long. The energy that produces sound of all types originates from vibrating objects. Some vibrations that make sound are visible. If you strike a low-frequency tuning fork or pluck a guitar string (**Figure 1**), the vibrations of the object can be seen. Similarly, if you watch the low-frequency woofer of a loudspeaker system, you can see it vibrating. However, some vibrations are not visible. For example, when you speak, parts of your throat vibrate. When you make a whistling sound by blowing over an empty pop bottle, the air molecules in the bottle vibrate invisibly to produce sound. Vibrations are the source of *all* sound energy, whether they are visible or invisible. The definitions and concepts related to vibrations presented earlier in the chapter also apply to vibrations that cause sound.

> ▶ **TRY THIS** *activity* ***Vibrating Tuning Forks***
>
> For this activity, you will need a low-frequency tuning fork, a plastic beaker, water, a rubber hammer, and a suspended pith ball.
>
> **Do not use a glass beaker for this activity; it could break.**
>
> (a) Strike the tuning fork with a rubber hammer, and touch the prongs to the surface of the water in a plastic beaker. Describe what happens and why.
>
> (b) Touch a vibrating tuning fork to a suspended pith ball. Describe what happens and why.

Unlike light energy, sound cannot travel through a vacuum; it must be transmitted through a medium. Sound travels by means of longitudinal waves. Consider the tuning fork in **Figure 2**. Tuning forks are constructed so that striking one prong from the side causes both prongs to move together (**Figure 2(a)**). This causes the compression of the air molecules between the prongs and the rarefaction of the air molecules outside the prongs at the same instant (**Figure 2(b)**). At the next instant, shown in **Figure 2(c)**, the prongs move apart, causing rarefaction between the prongs and compression outside them. The air molecules vibrate back and forth at the same frequency as the tuning fork. They transfer the sound energy away from the source through the air to the listener by colliding with each other (**Figure 2(d)**).

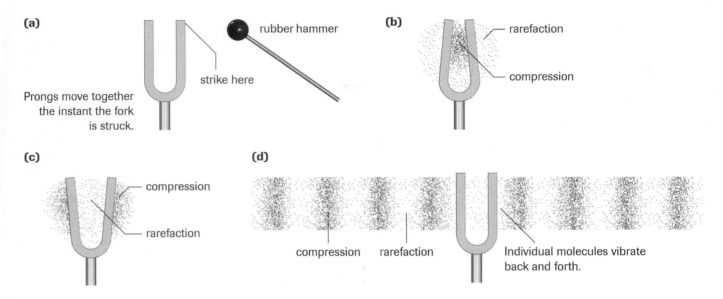

(a)

rubber hammer

strike here

Prongs move together
the instant the fork
is struck.

(b)

rarefaction

compression

(c)

compression

rarefaction

(d)

compression   rarefaction

Individual molecules vibrate
back and forth.

Figure 2
(a) Tuning forks are designed so that the two prongs approach each other when struck.
(b) Prongs coming together
(c) Prongs spreading apart
(d) Longitudinal waves from a tuning fork

▶ *Practice*

Understanding Concepts

1. When each of the following produces sound, what specific part or substance vibrates?
   (a) banjo   (c) drum
   (b) coach's whistle   (d) crackling fire

2. You are speaking to a friend who is about one metre from you. When you speak, air molecules near you vibrate. Do those molecules reach your friend's ears? Explain your answer.

3. Measure the wavelengths of the longitudinal waves in **Figure 2(d)**.

Applying Inquiry Skills

4. An electric bell is placed inside a sealed jar, which is connected to a vacuum pump (**Figure 3**). The bell produces sound when a vibrating arm strikes a gong.
   (a) Predict what you would observe as the vacuum pump removes air from the jar.
   (b) What does this demonstration show?
   (c) If possible, observe the demonstration, and comment on your prediction.

Making Connections

5. There is no air on the Moon. How do you think astronauts on the Moon can hear each other?

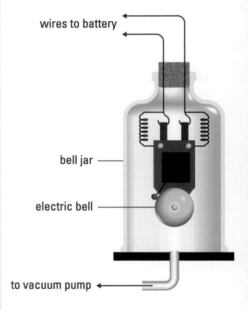

wires to battery

bell jar

electric bell

to vacuum pump

Figure 3
An electric bell in a vacuum

Interference of Sound Waves

Sound is transmitted in air by waves, so sound waves exhibit the same properties as all waves, including interference. Under normal conditions, interference can be difficult to observe. However, it is easy to demonstrate using a tuning fork.

A top view of the longitudinal waves travelling outward from the prongs of a tuning fork is shown in **Figure 4**. The waves travel outward not only from the sides of the prongs, but also from between the prongs. The sounds in these four directions are relatively loud. However, whenever there is a compression from the sides of the prongs, a rarefaction occurs between the prongs. Thus, at each of the four corners of the prongs, a compression always meets a rarefaction. This results in interference, causing a soft sound (a nodal line) at the four corners.

Now consider what happens when two sounds of nearly the same frequency are heard together. In **Figure 5**, two tuning forks are mounted on wooden boxes, or sounding boards, which undergo sympathetic vibrations, thus increasing the loudness of the sounds. The forks have the same resonant frequency (e.g., 256 Hz). However, the frequency of one has been lowered slightly by placing an elastic band around one prong. When a rubber hammer strikes each fork, the result is a series of loud and soft sounds. In this case, we have both constructive and destructive interference. The series of loud and soft sounds that are produced by the interference of two nearly identical frequencies is called the production of **beats**. **Figure 6** shows how the principle of superposition is applied in adding two waves to obtain beats. The resulting interference pattern illustrates the cause of the loud and soft sounds.

The number of beats heard per second is called the **beat frequency** and is measured in hertz (Hz). It is found by subtracting the lower frequency from the higher frequency.

beats the series of loud and soft sounds produced by the interference of sounds of two nearly identical frequencies

beat frequency the number of beats heard per second; measured in hertz (Hz)

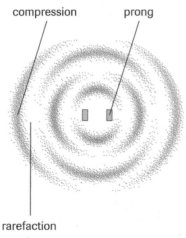

compression · prong

rarefaction

Figure 4
Top view of a tuning fork, showing longitudinal sound waves spreading outward from the prongs

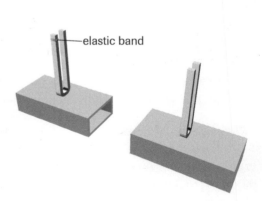

elastic band

Figure 5
The elastic band slightly lowers the frequency of one of two identical tuning forks.

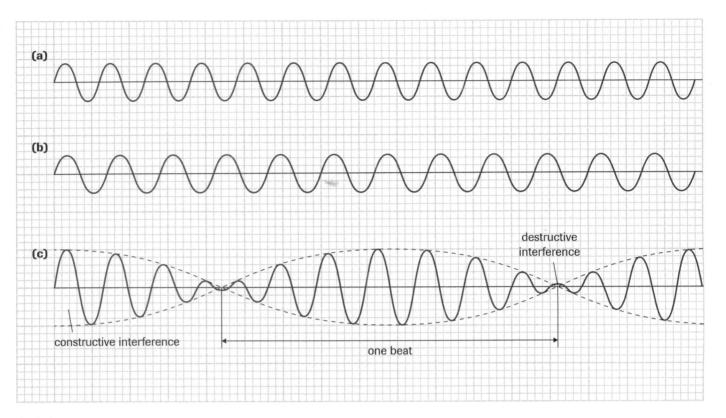

Figure 6
(a) Wave X: $\lambda = 12$ mm, $A = 5.0$ mm
(b) Wave Y: $\lambda = 14$ mm, $A = 5.0$ mm
(c) X + Y

▶ **SAMPLE** problem **1**

Beat Frequency

Two 384-Hz tuning forks are sounded together, and no beats are produced. Then a metal clip is attached to a prong of one fork, and again the forks are sounded together. This time a beat frequency of 4 Hz is produced. What is the new frequency of the fork with the clip?

Solution
Adding extra mass to a tuning fork reduces the resonant frequency, so the frequency must be 384 Hz − 4 Hz = 380 Hz. Thus, the new frequency is 380 Hz.

DID YOU KNOW ?

Human Detection of Beat Frequencies
The human ear can only detect beat frequencies less than 7 Hz.

Beats are used to tune musical instruments, for example, pianos, violins, and guitars. A note on the instrument is sounded at the same time as a tuning fork of the desired frequency. The tension in the string is then adjusted until the beat frequency decreases to zero, at which time the frequencies are identical.

As a final example of sound interference, recall the two-source interference pattern in water in Activity 9.7, **Figure 8**, page 453. Imagine that the two sources of periodic circular wavefronts are replaced by two loudspeakers set

CAREER CONNECTION

Combine your love of music with skill as a musical instrument repair technician or fine craftsperson building precision instruments.

Communication with Sound **457**

up in your classroom. With sound waves of the same frequency emitted from the speakers, a three-dimensional interference pattern can be observed in the room. (This demonstration is sometimes difficult to hear because of sound reflections in the room.)

▶ TRY THIS activity · *Demonstrating Sound Interference*

Demonstrate sound interference by trying these activities:

- Strike a tuning fork with a rubber hammer, and then hold the fork vertically a few centimetres from your ear. As you slowly rotate the fork, have your partner help you locate the areas of loud and soft sounds.

- As a class demonstration, attach a rubber band or metal clip to one of two mounted tuning forks. Tap with a rubber hammer and listen for beat frequencies. Try changing the location of the rubber band or clip to see whether the change in position affects the interference.

- As a class demonstration, set up a two-source interference pattern in the classroom using identical sources, for example, two speakers (**Figure 7**). As you walk slowly across the room, keep one ear facing the speakers. Listen carefully to the sounds from the sources and try to detect the nodal lines. Predict the effect of increasing the frequency of the two sources. Test your prediction.

 If using speakers, comply with all safety requirements for electrical appliances.

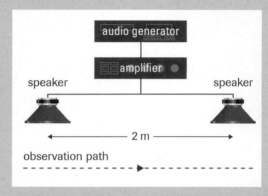

Figure 7
Listening to sounds from two identical sources

Answers

7. (a) 2 Hz
   (b) 6 Hz
   (c) 5 Hz

8. 2 Hz; 6 Hz; 7 Hz; 9 Hz; 13 Hz; 15 Hz

▶ Practice

Understanding Concepts

6. What evidence suggests that sound energy in air travels by means of longitudinal waves?

7. State the beat frequency when the following pairs of frequencies are heard together:
   (a) 202 Hz, 200 Hz
   (b) 341 Hz, 347 Hz
   (c) 1003 Hz, 998 Hz

8. Four tuning forks, with frequencies of 512 Hz, 518 Hz, 505 Hz, and 503 Hz, are available. List all possible beat frequencies when any two forks are sounded together.

Applying Inquiry Skills

9. Describe how you could determine experimentally whether a note produced by a musical instrument is in tune (i.e., at the correct frequency).

Resonance in Sound

You have probably heard the high-pitched squeal produced by a moist finger moving around the rim of a long-stemmed glass. This squeal is a resonant frequency that can be changed by adding water to the glass. If a nearby sound with the same frequency is loud enough, the glass may shatter (**Figure 8**).

Sound resonance, like mechanical resonance, results when a small force produces a relatively large natural vibration. For example, when a tuning fork is struck, it vibrates at its own resonant frequency. A short tuning fork has a high frequency and a long tuning fork has a low frequency, just like short and long pendulums.

Figure 8
This long-stemmed glass shattered when it vibrated in sympathy with a source of the same resonant frequency and a large amplitude of vibration.

> ▶ **TRY THIS** activity **Sound Resonance**

Obtain two identical low-frequency tuning forks. Strike one fork and hold it close to the other, as in **Figure 9**. After about 15 s, stop the first fork from vibrating. Listen closely to the second one. Repeat the procedure using tuning forks of different frequencies.

(a) Describe what you observe.

(b) Relate your observations to sympathetic and forced vibrations.

(c) Explain how the sound energy is transferred from one fork to another.

 Do not strike the forks on a hard surface. Use a rubber or soft surface.

Sometimes, it is necessary to prevent an object from vibrating with maximum amplitude at its own resonant frequency or frequencies. A good loudspeaker, for example, is designed in such a way that its many resonant frequencies are controlled. With this arrangement, no single frequency is dominant. Otherwise, the sound output for some frequencies would be louder than for other frequencies for similar input levels.

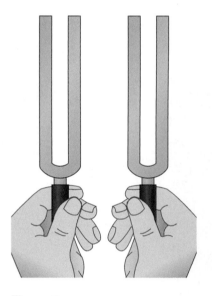

Figure 9
Using sympathetic vibrations to observe sound resonance

> ▶ **Practice**

Understanding Concepts

10. (a) What happens to the resonant frequency of a source of sound energy when the length of the source increases?
    (b) On what facts do you base your answer in (a)?

Making Connections

11. If an acoustic piano is available, depress the right (sustaining) pedal to free all the strings in the piano. Sing a single note loudly into the piano and listen for the sound of the strings that vibrate. Explain what you hear.

audible range the range of frequencies that an ear can detect

ultrasonic sound sound with a frequency higher than the normal audible range of a human

Table 1   Audible Ranges

Animal	Audible Range (Hz)
human	20–20 000
dog	15–50 000
cat	60–65 000
bat	1000–120 000
porpoise	150–150 000
robin	250–210 000

echolocation the process of using reflected sounds to determine distances

sonar sound navigation and ranging

Human Audible Range and Ultrasonic Sounds

Our ears are very sensitive organs that react to a wide range of frequencies. With an audio frequency generator connected to an appropriate loudspeaker, your teacher can check the frequency range of the hearing of everyone in the class. All the frequencies you can hear make up your **audible range**. The audible range of most students is between 20 Hz and 20 kHz. The ear is most sensitive to frequencies between about 1000 Hz and 5000 Hz. People who expose their ears to loud sounds for long periods of time may reduce their audible range and hearing sensitivity. In general, the aging process also decreases the audible range in the higher frequencies.

Sounds with frequencies higher than 20 kHz are called **ultrasonic**. (*Ultra* is Latin for "higher than.") Dogs' ears are sensitive to frequencies that are higher than humans can hear, so dog whistles are made to produce ultrasonic frequencies. Dogs are not the only animals that hear ultrasonic sounds; the audible ranges of several creatures are listed in **Table 1**.

Some animals navigate and hunt using ultrasonic sounds. Some types of bats, for example, emit high-frequency sounds that are reflected off objects. The reflected sounds return to the bat and allow it to judge what is nearby. This enables the bat to navigate in the dark.

The use of the reflection of ultrasonic sounds in water is called **echolocation**. It is used to determine the depth of water below a ship or to locate reefs, submarines, or schools of fish. Ultrasonic sounds are sent out in pulses from the ship, as shown in **Figure 10**. They reflect off the object and return to the ship. An instrument measures the time taken for the signal to return; then the equation for speed ($v = d/t$) is used to find either the speed of the sound or the distance it travelled. This method of detection is also called **sonar**, which stands for **s**ound **n**avigation **a**nd **r**anging.

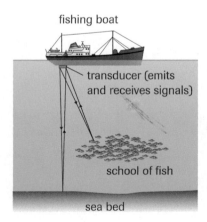

Figure 10
An emitting–receiving device called a hydrophone sends ultrasonic signals and receives the reflected signals.

> ▶ *Practice*

Understanding Concepts

12. Can humans hear all the sound frequencies that dogs can hear? Explain your answer by referring to **Table 1**.

13. Why are ultrasonic sounds used to view a baby in a womb rather than X rays?

14. The speed of sound in a freshwater lake is 1500 m/s. Ultrasonic sound sent from the surface of the water to the bottom of the lake returns in 0.20 s. How deep is the lake?

Ultrasonic sounds have many applications. In industry they help locate flaws in welded joints by revealing breaks in uniformity. They are also used to drill small holes in glass and steel, and to clean electronic parts of watches and other instruments. Medical uses include detecting brain damage and cancers, making glasses for people who are blind, and studying of the growth of infants in the womb (**Figure 11**). For medical applications, ultrasonic sounds are less dangerous than high-energy X rays. In the home, ultrasonic sounds are used in television remote control units.

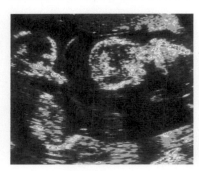

Figure 11
Ultrasonic image of twins in the womb

Loudness of Sounds

Our ears are sensitive to a wide range of loudness (i.e., intensity), which is sometimes referred to as the "volume" of sound. To characterize loudness of sounds, we refer to the **intensity level** of a sound, measured in decibels. The decibel, symbol dB, is one-tenth of a bel, symbol B. This unit is named in honour of Alexander Graham Bell (1847–1922), who invented the telephone. **Table 2** lists the intensity levels in decibels of various sounds.

> ▶ **TRY THIS** *activity*   *Testing Loudness*
>
> Obtain a sound level meter (a decibel meter), and use it to measure the sound levels of music from a car stereo system. Start at low loudness levels of sounds, and then increase the level to the value you normally use. Record the readings, and compare them to the values listed in **Table 2**.
>
>   **Do not sustain loud sounds; they may damage your eardrums.**

CAREER CONNECTION

As a radiation or ultrasound medical imaging technician, you assist doctors in the diagnosis of problems or disease. With further training, you can move into more complex technologies, such as CT (computed tomography) and MRI (magnetic resonance imaging). There are employment opportunities in hospitals, clinics, and in industry.

GO www.science.nelson.com

intensity level represents the loudness of a sound; unit dB

Table 2 Intensity Levels of Sound

Intensity Level (dB)	Example
0	near the threshold of hearing
10	empty church on a quiet street
20	average whisper, at 1 m
30	library reading room
40	inside a car, engine running
50	quiet restaurant
60	conversation, at 1 m
70	factory machinery in use
80	noisy street corner
90	loud stereo in average-size room
100	rock concert
110	jet taking off, at 60 m
120	threshold of feeling
130	threshold of pain

How Can the Effects of Loud Sounds Be Reduced?

Exposure to loud sounds can increase stress and blood pressure, cause hearing loss, reduce rest and sleep, and affect learning. The effects of the loud sounds depend on the intensity of the sounds, the length of exposure time, and how often the listener is exposed to them.

To explore the effects of loud sounds in our society and means of reducing them, your class can share the responsibility of researching the following areas:

• workplace noise (construction, demolition, manufacturing, mining, etc.) (**Figure 12**)

• environmental noise (airplanes, highways, yard machines, sirens, etc.)

• household noise (appliances, barking dogs, nearby traffic, leaf blowers, etc.)

• recreational noise (movies, car stereos, concerts, fireworks, balloon pops, hunting, etc.)

• rural noise (farm machinery, power tools, grain dryers, etc.)

After choosing an area of concern, each group can explore answers to some or all of the questions listed below.

1. What are the current standards (in decibels, dB) for the workplace and other locations?

2. How do loudness standards compare in different jurisdictions in Canada and in other countries?

3. How do recommended maximum loudness levels compare for short duration times and long duration times?

4. How can sound levels be reduced at the source?

5. What methods are available to reduce the effects of loud sounds on people living near busy highways or airports?

6. How effective are various ear protection devices (such as earplugs and noise-reduction earphones) in reducing hearing loss?

7. Whose responsibility is it to care about the sound environment?

8. What careers are available in the area of sound safety?

9. Should maximum loudness standards be implemented at concerts and movie theatres?

10. What other questions would you propose to address this issue?

 www.science.nelson.com

Take a Stand

Based on answers to the above questions from your research, create a position paper to suggest ways of addressing the issue of exposure to loud sounds. The paper can be a poster board, an essay, a letter, a Web page, a video, or an audio presentation. Your final report should contain your research notes, your answers to the questions, and your position paper.

Evaluation

(a) Evaluate the usefulness of the Web sites you used as references.

(b) How do your opinions about problems and solutions concerning loud sounds compare with those of your peers?

Figure 12

SUMMARY	*Sound Waves*

- Sound energy is produced by vibrations and transmitted through air by longitudinal waves.
- Beats are a series of loud and soft sounds produced when two sound waves of nearly identical frequencies interfere; they can be used to tune a musical instrument.
- Beat frequency, or the number of beats per second, is found by subtracting the lower frequency from the higher one.
- Sound resonance can be observed when one tuning fork vibrates in sympathy with another of the same resonant frequency.
- Ultrasonic sounds are above the human audible range of 20 Hz to 20 kHz. They are used for medical imaging, industrial applications, and navigation and ranging.
- Intensity level, in decibels (dB), is a measure of the loudness of a sound.

LEARNING *TIP*

Intensity and Intensity Level
Sound *intensity* is a measure of the power of a sound wavefront covering a unit area; it is measured in W/m^2. The range of sound intensities of the human ear is from 10^{-12} W/m^2 to more than 10^1 W/m^2, which is a cumbersome set of values to deal with. Therefore, scientists prefer to use a ratio scale, with the intensity calculated as a ratio to 10^{-12} W/m^2. (The intensity of earthquakes is also measured using a ratio scale.) This scale gives the intensity level, with the following equalities:

$0 B = 0 dB = 10^{-12} W/m^2$
$4 B = 40 dB = 10^{-8} W/m^2$
$8 B = 80 dB = 10^{-4} W/m^2$
$12 B = 120 dB = 10^0 W/m^2$

▶ *Section 9.8 Questions*

Understanding Concepts

1. Name two examples of the production of sound in which the vibrations that cause the sound are (a) visible and (b) invisible.

2. An explosion occurs at an oil refinery. Several seconds later, windows in an adjacent neighbourhood shatter. Why do the windows shatter?

3. Can sound travel in outer space? Explain your answer.

4. Three tuning forks, with frequencies of 384 Hz, 381 Hz, and 379 Hz, are available.
   (a) What are all the possible beat frequencies when any two forks are sounded together?
   (b) Describe two ways you could obtain a beat frequency of 4 Hz.

5. A figurine is placed on the flat top of a stereo speaker. Certain sounds from the speaker cause the figurine to vibrate noticeably. Why does that happen?

6. Describe, with a diagram, how the reflection of waves is applied in sonar technology.

Applying Inquiry Skills

7. (a) Predict how the sounds from a ruler vibrated over the edge of a desk depend on the projection of the ruler beyond the edge.
   (b) Test your prediction experimentally.

8. An audio frequency generator can create sounds identical to the sounds from a tuning fork. However, the hand dial on the generator has come loose, so it does not indicate the correct frequency. How could you apply the principle of beats to set the dial so you could tighten it back in place?

Making Connections

9. Some entertainers have damaged their hearing through prolonged exposure to loud music. They are trying to persuade music enthusiasts to protect their hearing by monitoring and reducing their own exposure. Create a cartoon, poem, short story, poster, or other communication to advertise the risks and reduction methods.

How would you describe the difference between noise and music? You might say that noise is unpleasant and annoying, and music is pleasant and harmonious. However, this description depends largely on individual judgment—it is subjective. There is a scientific difference between the two. In this section, you will apply the knowledge of vibrations, waves, and sound that you gained earlier in the chapter to the analysis of the scientific difference between noise and music.

Suppose you hear the middle A note equally loudly from a piano, a violin, and a trumpet. The three sounds have equal frequency (440.0 Hz) and loudness. However, each note sounds different because each instrument has a unique quality.

quality the pleasantness of a sound related to the waveform of the sound

The *quality* of musical sounds can differ greatly. For example, a novice music student may create sound of poor quality on a musical instrument, whereas an experienced player using the same instrument will produce sound of high quality. A portable radio produces sound of poor quality compared with an expensive sound system.

harmonics the frequencies that make up a sound

Scientifically, the **quality** of a sound refers to how musical or pleasing the sound is to the human ear—as the quality relates to the shape of the sound waves. The shape of a sound wave depends on the wave's *harmonic* structure. **Harmonics** consist of the **fundamental frequency** of a musical sound, as well as the frequencies that are whole-number multiples of the fundamental frequency. The quality of a sound depends on the number and relative loudness of the harmonics that make up the sound.

fundamental frequency the lowest frequency of all the harmonic frequencies of a sound; also called the first harmonic

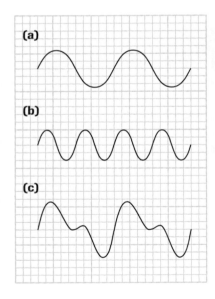

Figure 1
(a) Fundamental frequency (f)
(b) The frequency is $2f$, and the amplitude is 80% of the amplitude in (a).
(c) Frequencies f and $2f$ are added.

▶ **TRY THIS** activity **Demonstrating Harmonics**

You can hear the effects of harmonic structure by plucking the string of an acoustic guitar or other stringed instrument. Start by gently plucking a string at the exact middle, and listen carefully to the sound produced. Then gently pluck the string at other locations, such as one-quarter, one-eighth, and then one-sixteenth of the distance from one end. Compare the sounds heard.

For example, assume that an instrument produces a fundamental frequency (f), also called the first harmonic, of 110.0 Hz, and that it has a second harmonic of $2f$ (220.0 Hz) with 80% of the amplitude of f. By themselves, these sounds are "pure," and they sound boring, like the sound from a mounted tuning fork. However, when the sounds are superimposed, the resulting waveform is more complex and more pleasant. **Figure 1** shows the individual frequencies and their addition using the principle of superposition. Likewise, adding more harmonics creates even more complex sounds that can become more pleasant. Experiments with people listening to sounds show that the ratios of the frequencies of higher harmonics to lower harmonics must remain simple, such as 2:1 and 3:2. Adding too many harmonics changes the ratios; for example, the ratio 16:15 decreases the pleasantness of the sounds.

▶ **TRY THIS** activity   *Comparing Waveforms*

Connect a sound sensor to a computer interface, or a microphone–amplifier system to an oscilloscope. View the waveform traces of varying qualities of musical sounds as you listen to them. Draw diagrams of the waveforms you observe.

▶ **Practice**

Understanding Concepts

1. Scientifically, on what does the quality of a musical sound depend?

2. The note middle A ($f = 440.0$ Hz) is struck on a piano. Determine the frequency of the following:
   (a) second harmonic
   (b) third harmonic
   (c) fifth harmonic

3. A string has a length of 48 cm and a fundamental frequency of 330 Hz.
   (a) Draw sketches of the string vibrating at the first harmonic and the second harmonic. Label the sketches.
   (b) Determine the frequency of the string in the second harmonic.
   (c) What is the distance between nodes on the string vibrating in the second harmonic?

Answers

2. (a) 880.0 Hz
   (b) 1.320×10^3 Hz
   (c) 2.200×10^3 Hz

3. (b) 660 Hz
   (c) 24 cm

Stringed Instruments

Stringed instruments consist of two main parts: the vibrator and the resonator. The **vibrator** is the string, and the **resonator** is the case, box, or sounding board on which the string is mounted. A string by itself does not give a loud sound or even a necessarily pleasing one. It must be attached to a resonator, so that forced vibrations improve the loudness and quality of the sound. Even a tuning fork has a louder and better sound if its handle touches an object, such as a desk, wall, or resonance box.

Stringed instruments can be played by plucking, striking, or bowing them. The quality of sound is different in each case. The quality also depends on where the string is plucked, struck, or bowed. For example, a string plucked gently in the middle produces a strong fundamental frequency or first harmonic (**Figure 2(a)**). A string plucked at a position one-quarter of its length from one end vibrates in several modes (**Figure 2(b)**). In this case, the second harmonic is added to the first harmonic, changing the sound quality. **Figure 2(c)** shows the third harmonic superimposed on the first harmonic. This also results in a different quality of sound.

vibrator the string of a stringed instrument

resonator the case, box, or sounding board of a stringed instrument

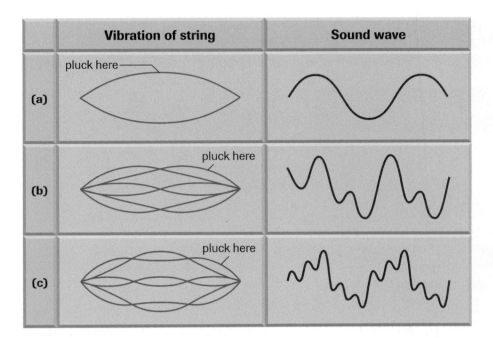

Vibration of string	Sound wave
(a) pluck here	
(b) pluck here	
(c) pluck here	

Figure 2
(a) First harmonic
(b) Second harmonic superimposed on the first harmonic
(c) Third harmonic superimposed on the first harmonic

Figure 3
A harp

Figure 4
The strings and hammers inside a piano

Stringed instruments that are usually plucked include the banjo, guitar, mandolin, ukulele, and harp (**Figure 3**). The harp, with 46 strings, is a complex instrument. It consists of a hollow soundboard (the resonator), a vertical pillar, and a curved neck. The strings are stretched between the pins on the soundboard and the pegs on the curved neck. A pedal mechanism enables the player to raise the frequency of individual strings. The other stringed instruments that are plucked resemble one another. They have four to eight strings, as well as metal cross pieces called frets to guide the placement of the fingers. The thick wires under low tension create the lower notes, while the thin wires under greater tension create the higher notes.

The best-known stringed instrument that is struck is the piano. A piano key is connected by a system of levers to a hammer, which strikes the string (or, in some cases, multiple strings) to produce a note (**Figure 4**). A modern piano has 88 notes, with a frequency range of 27.5 Hz to 4186 Hz. The short, high-tension wires produce high-pitched notes, and the long, thick wires produce low-pitched notes. The sounds from the strings are increased in loudness and quality by the wooden sounding board of the piano.

Stringed instruments that are usually bowed belong to the violin family. This family consists of the violin (**Figure 5**), viola, cello, and double bass. One side of each bow consists of dozens of fine fibres that are rubbed with rosin to increase the friction when stroked across a string. Each instrument has four strings and wooden sounding boards at the front and back of the case. Unlike the members of the guitar family, the members of the violin family have no frets. Thus, their frequency can be changed gradually rather than in steps. (The trombone is another instrument that is capable of changing frequency gradually.) The double bass, the largest member of the violin family, produces low-frequency sounds.

> ### *Practice*

Understanding Concepts

4. A fishing line under tension can be made to vibrate, producing a sound. But that sound has poor musical quality. Explain why.

5. Describe how the low-frequency wires of a piano differ from the high-frequency wires.

6. A note of frequency 440 Hz is played on a violin string. Describe ways in which the quality of the sound can be changed.

Wind Instruments

All wind instruments contain columns of vibrating air molecules. Following the pattern you have observed for all vibrating objects, the large instruments create low-frequency sounds, and the small instruments create high-frequency sounds.

In some wind instruments, like the pipe organ in **Figure 6**, the length of each air column is fixed. However, in most wind instruments, such as the trumpet or trombone, the length of the air column can be changed.

To cause the air molecules to vibrate, something else must vibrate first. There are four general mechanisms for forcing air molecules to vibrate in wind instruments:

- In *air reed instruments,* air is blown across or through an opening. The moving air sets up turbulence inside the column of the instrument. Examples of air reed instruments are the pipe organ, flute, piccolo, recorder, and fife. The flute and piccolo have keys that are pressed to change the length of the air column. The recorder and fife have side holes

Figure 5
The violin, the smallest instrument in the violin family, produces high-frequency sounds.

Figure 6
The pipe organ of Notre Dame Basilica, Montreal

that must be covered with the fingers to change the length of the air column and, thus, control the pitch.

- In *single-reed instruments,* moving air sets a single reed vibrating, which in turn sets the air in the instrument vibrating. Examples of these instruments are the saxophone, clarinet, and bagpipe. In the bagpipe, the reeds are located in the four drone pipes attached to the bag, not in the mouth pipe. Again, the length of the air column is changed by holding down keys or covering side holes in single-reed instruments.

- In *double-reed instruments,* moving air forces a set of two reeds to vibrate against each other. This causes air in the instrument to vibrate. Examples are the oboe, English horn, and bassoon. Keys are pressed to alter the length of the air column.

- In *lip reed instruments,* also called *brass instruments,* the player's lips function as a double reed. They vibrate, causing the air in the instrument to vibrate. None of the air escapes through side holes, as in other wind instruments; rather, the sound waves must travel all the way through a brass instrument. Examples of such instruments are the bugle, trombone, trumpet, French horn, and tuba. The length of the air column is changed either by pressing valves or keys that add extra tubing to the instrument, or, in the case of the trombone, by sliding the U-tube.

The quality of sound from wind instruments is determined by such factors as the construction of the instrument and the experience of the player. However, just as with stringed instruments, it also depends on the harmonics produced by the instrument.

▶ TRY THIS activity   *An Air Column Concert*

In a large group, design and carry out an activity using empty plastic pop bottles as air reed instruments to play a musical tune. Each bottle can contain water at a different level and can be "played" by blowing across the bottle opening. Bottles can be tuned in various ways; for example, a portable synthesizer may be used as a reference.

▶ Practice

Understanding Concepts

7. What vibrates to create sound in a column of air?

8. Describe the main factors that affect the frequency of the sounds from wind instruments.

Making Connections

9. From each of the pairs of instruments listed below, choose the one that has the higher range of pitches. (It will be helpful to discuss in class the size of the instruments.)
   (a) piccolo, flute
   (b) bassoon, English horn
   (c) oboe, English horn
   (d) tuba, trumpet

SUMMARY   *The Quality of Musical Sounds*

- Scientifically, the quality of a sound depends on the harmonic structure of the sound waves.

- A pleasant sound is created when multiples of the fundamental frequency are superimposed on the fundamental frequency.

- The quality of sound from a stringed instrument is enhanced by the use of a resonator and by creating complex vibrations on the string.

- Wind instruments are classified according to how air molecules are forced to vibrate: air reed instruments, single-reed instruments, double-reed instruments, and lip reed (or brass) instruments.

▶ *Section 9.9 Questions*

Understanding Concepts

1. **Figure 7** shows four sets of waveforms observed on an oscilloscope screen.
   (a) Match the waveforms to these sources of sound: a pure sound; the addition of *f* and *2f*; noise; the addition of *f* and *8f*.
   (b) In each case, describe the quality of the sound you would hear.

2. A sound of frequency 480.0 Hz is the second harmonic of a set of frequencies.
   (a) What is the fundamental frequency?
   (b) What is the third harmonic?
   (c) When the first and second harmonics are sounded together, what happens to the resulting waveform and quality of the sound? Explain why.

3. Compare the ways in which air molecules are forced to vibrate in air reed and lip reed instruments. Give two examples of each type of instrument.

4. A 256-Hz tuning fork is mounted on a wooden resonance board. Assuming that the speed of the sound in the room is 345 m/s, determine the wavelength of the sound from the tuning fork.

Applying Inquiry Skills

5. A flexible toy tube open at both ends is 60 cm long and 3 cm in diameter. When one end is held rigidly by hand and the tube is twirled in a large circle, air rushes through the tube creating a sound.
   (a) What will you hear as the speed of twirling increases?
   (b) What would happen to the sound if the person twirling the tube covered one end?
   (c) If this type of tube is available, try twirling it safely. Do the sounds change in frequency gradually or in distinct jumps? How does this relate to the concept of resonance?

Making Connections

6. Each pair of notes listed below is sounded at the same time on a piano. The ratios of the frequencies are given in brackets. Which pairs sound relatively pleasant? Explain your choices.
   (a) A_5, 880.0 Hz and A_4, 440.0 Hz (2:1)
   (b) A_{sharp} or B_{flat}, 466.2 Hz and A_4, 440.0 Hz (16:15)
   (c) E_5, 659.3 Hz and A_4, 440.0 Hz (3:2)

(a)

(b)

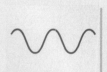

(c)

(d)

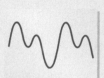

Figure 7

Sound communications technology has changed the way music is produced, transmitted, reproduced, and heard. Electric instruments, electronic amplifiers, electronic synthesizers, and computer controls are just a sampling of the many examples of communications technologies. Here we take a brief look at a small number of these applications.

An important component of any electric or electronic instrument is the **transducer**, which is an energy-transforming device. The energy transformation depends on the application, as you will see shortly.

transducer a device used to transform energy in an electrical or electronic system

Electric and Electronic Musical Instruments

Electric amplifier systems are made of four main parts: a *source* of sound, a *microphone,* an *amplifier,* and a *loudspeaker.* Both the microphone and the loudspeaker are transducers. At hockey and football games, for example, the announcer's voice creates sound energy that falls on the microphone. The microphone transforms sound energy into electrical energy. This energy is amplified and causes vibrations in a loudspeaker. The loudspeaker transforms electrical energy into sound energy, reproducing the original sound with an increased loudness.

Many musical instruments discussed in previous sections can be included in an electric amplification system by adding a microphone, amplifier, and loudspeaker. Stringed instruments, which normally give out low amounts of power, are often amplified. A microphone is attached directly to the body of the instrument. In some cases, the design of the instrument is altered. Most electric guitars, for example, have a solid body rather than the hollow body of acoustic guitars (**Figure 1**).

Loudspeakers play an important role in sound quality in an electric amplifier system. A single loudspeaker does not have the same frequency range as our ears, so a set of two or three must be used to give both quality and frequency range. **Table 1** lists details of the three common sizes of loudspeakers used in electric amplifier systems.

Figure 1
An electric guitar

Table 1   Details of Loudspeakers

Name	Approximate Diameter (cm)	Frequency Range (Hz)	Wavelength Range (cm)
woofer (low-range)	25–40	25–1000	34–1400
squawker (mid-range)	10–20	1000–10 000	3.4–34
tweeter (high-range)	4–8	3000–20 000	1.7–11

The final column of **Table 1** shows that the sound waves from the tweeter have much shorter wavelengths than those from the woofer. Long wavelengths are diffracted easily through doorways and around furniture and people. However, the short waves from a tweeter are not diffracted around large objects, so their sound tends to be directional. As a result, the listener must be in front of the tweeter to get the full sensation of its sound, especially in the very high frequency range. (Recall what you observed about the diffraction of water waves in Activity 9.7.)

> ▶ **TRY THIS** activity
>
> ## *"Unplugged" versus "Plugged"*
>
> Use an electric instrument, such as an electric guitar, to produce sounds with the electricity on and then off. Describe the differences between the sounds, especially differences in quality.
>
> **Comply with all safety requirements for electrical appliances.**

In contrast with stringed and wind instruments, a musical synthesizer is an electronic instrument that produces vibrations using electronic circuits with amplifiers (**Figure 2**). A synthesizer consists of four main parts:

- The *oscillator* creates the vibrations.
- The *filter circuit* selects the frequencies that are sent to the mixing circuit.
- The *mixing circuit* adds various frequencies together to produce the final signal.
- The *amplifier* and *speaker system* make the sound loud enough to be heard. (As in an electric system, the speaker is a transducer.)

Synthesizers can vary the shape of the sound waves they produce; as a result, they can imitate the sound of almost any musical instrument. The basic shapes of the waves used to create examples of more complex waves are shown in **Figure 3**.

Figure 2
A portable electronic synthesizer

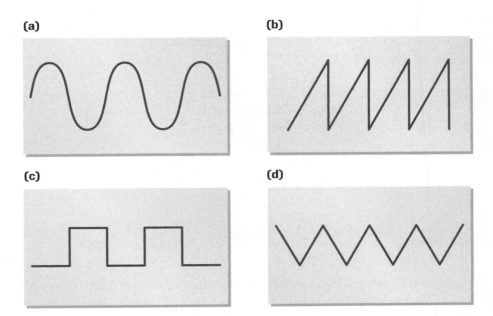

Figure 3
(a) Sine wave
(b) Saw-tooth wave
(c) Square wave
(d) Triangular wave

Synthesizers can also control the attack and decay patterns of a sound. The *attack* occurs when the sound is first heard. It may be *sudden, delayed,* or *overshot,* as illustrated in **Figure 4(a)**. After the attack, the sound is often sustained, which means its envelope is kept constant. This is followed by the *decay* as the sound comes to an end. Decay may be *slow, fast,* or *irregular,* as shown in **Figure 4(b)**.

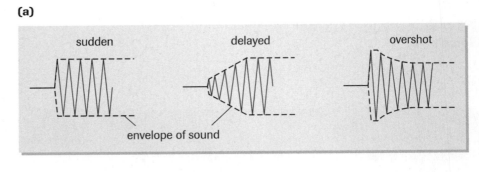

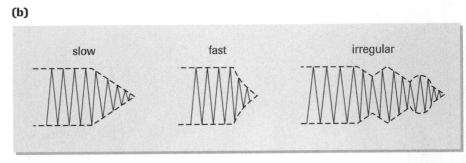

Figure 4
Control patterns
(a) Growth patterns
(b) Decay patterns

▶ TRY THIS activity   *Creating Electronic Music*

Listen to the sounds produced by an audio frequency generator that creates sine waves, square waves, and triangular waves, all at a constant frequency. Describe the differences in the sounds, and try to explain why the differences exist. Observe a demonstration of a musical synthesizer, if available.

✋ **Comply with all safety requirements for electrical appliances.**

▶ Practice

Understanding Concepts

1. Describe the energy transformations in a microphone–amplifier–loudspeaker system. Write the corresponding energy-transformation equation.

2. **Figure 5** shows a top view of a speaker system with a woofer and a tweeter facing right. Compare the sounds heard by observers at A, B, and C.

3. Describe the differences between an electric amplifier system and a synthesizer.

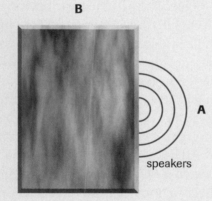

Figure 5

Applying Inquiry Skills

4. A synthesizer produces a constant-frequency triangular waveform sound with a delayed attack, then a sustained amplitude, and finally a slow decay. Draw a diagram of the waveform of the sound.

Making Connections

5. Bursts of loud bass sounds on a car speaker system require more power than is available from a car's battery. Large capacitors, called *diamond-like capacitors* or *carbon ultra capacitors,* provide enough power for short time intervals to play the bass sounds. Research this technology on the Internet.
   (a) What is the current capacitance of these capacitors? (Recall from Chapter 8 that capacitance is measured in farads, F.)
   (b) Describe features of these capacitors, including their dimensions, maximum current, cost, and other variables.
   (c) What other technology benefits from the use of high-farad capacitors?

 www.science.nelson.com

Virtual Music: Using Computers to Generate Music

In this chapter, you learned that waves are caused by vibrations, and that sound waves interfere with other sound waves according to the principle of superposition. In Chapter 8, you studied analog and digital circuits, and how digital signals can be controlled. These concepts are joined in the production of electronic keyboards, musical synthesizers, and other modern musical instruments.

A standard way of linking several digital instruments together under the control of computers is called **Musical Instrument Digital Interface**, or **MIDI**. With MIDI, a single electronic instrument can be used to construct music that sounds like a whole orchestra. MIDI files contain digital instructions for the music, which are transmitted to an output circuit, such as a speaker circuit.

Figure 6 shows the basic operation of MIDI. Start at the input, and follow through to the output to see how music is created, transmitted, controlled, and heard in a MIDI system. The composer can give instructions for notes to play, how long to play them, and with what attack, decay, and intensity. This allows the composer to create the virtual music of an entire orchestra and change the keys, tempo, or instruments without needing a musician. A major advantage of the MIDI system is that it is now standardized so that software programs, hardware devices, and interfaces can be interchanged or shared.

MIDI can create many other sounds. For example, it can produce the sounds of thunder, dogs barking, frogs chirping, or screaming (for special effects in movies). MIDI creates these virtual sounds at a fraction of the cost of creating most real sounds.

Musical Instrument Digital Interface (MIDI) a system of electronic devices that can create music

Three examples of input control
Data can be entered on a computer keyboard. On-screen choices can be made using a mouse. A MIDI instrument, such as a musical keyboard, can be played. Striking a computer key harder increases the time interval of the signal.

Examples of output options
A laser printer capable of printing music fonts can create music manuscripts. Speakers allow the music to be heard.

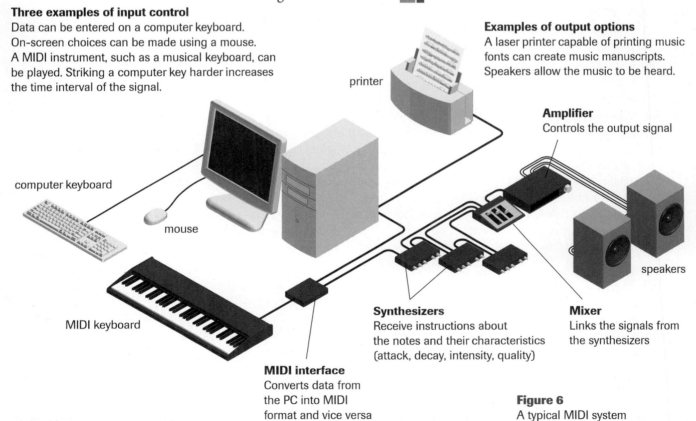

printer

Amplifier
Controls the output signal

computer keyboard

mouse

speakers

MIDI keyboard

Synthesizers
Receive instructions about the notes and their characteristics (attack, decay, intensity, quality)

Mixer
Links the signals from the synthesizers

MIDI interface
Converts data from the PC into MIDI format and vice versa

Figure 6
A typical MIDI system

> ## Practice

Understanding Concepts

6. Why can MIDI music be called virtual music?

7. What are the advantages of a standardized MIDI format?

8. An electronic keyboard is used to produce trumpet sounds. How can the loudness of the sounds be controlled by the keyboard user?

Making Connections

9. Would you expect the signals in the MIDI interface to be analog or digital? Explain why. (Analog and digital signals were described in section 8.8.)

SUMMARY   *Sound Communications Technology*

- Sounds from electric musical systems are amplified and heard through loudspeakers.

- Sounds from electronic instruments are generated by an oscillator, amplified by electronic circuits, and then sent to loudspeakers.

- Musical Instrument Digital Interface (MIDI) is a computerized electronic system used to create and manipulate complex musical sounds.

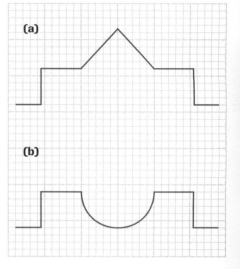

(a)

(b)

Figure 7

> ## Section 9.10 Questions

Understanding Concepts

1. (a) How can an acoustic guitar be converted into an electric guitar? (Include the term *transducer* in your answer.)
   (b) What are the advantages of doing this? Are there disadvantages?

2. (a) Which wavelengths of sounds tend to be most directional?
   (b) Explain why these sounds are called directional. (Use the concept of diffraction in your answer.)

3. (a) What is an attack pattern?
   (b) Name and draw three examples of attack patterns using saw-tooth waves.

4. MIDI technology combines concepts of sound, electricity, and electronics.
   (a) Explain the function of a synthesizer in a MIDI system.
   (b) What are the energy transformations that occur to produce an output of sound energy? (Include the name of any device that is a transducer.)
   (c) What other output is possible?

Applying Inquiry Skills

5. **Figure 7** shows two waveforms. Each is generated by the addition of two separate waveforms using a synthesizer. Draw one complete wavelength of the two waveforms used to create each waveform shown.

6. A synthesizer produces a constant-frequency sine waveform sound with an overshoot attack, then a sustained amplitude, and finally an irregular decay. Draw a diagram of the waveform of the sound, showing the resulting envelope.

Making Connections

7. A stereo speaker must be connected to the amplifier in a specific pattern. (The red and black colours distinguish the positive and negative terminals.) If the polarities of the wires from the amplifier to one of the two speakers are reversed, destructive interference results. How can this happen?

8. Describe the advantages and disadvantages of electronic music. Show that you realize that music is both objective and subjective.

Key Understandings

9.1 Vibrations

- **Transverse vibrations** are perpendicular to the rest axis, **longitudinal vibrations** are parallel to the rest axis, and **torsional vibrations** rotate around the rest axis.

- **Frequency** (in **hertz**) and **period** (in seconds) are reciprocals of each other.

9.2 Investigation: The Pendulum

- A controlled experiment can be performed to determine how the frequency of a simple pendulum depends on the pendulum's mass, amplitude, and length.

9.3 Investigation: Pulses on a Spring

- Tests can be performed to determine the properties of pulses travelling along a spring, such as the factors that affect speed, the effect of the pulse, and the phase of reflected pulses.

9.4 Waves

- A **wave** is a disturbance that transfers energy; a **periodic wave** is produced by a source vibrating at a constant frequency; its **wavelength** is the distance between any two adjacent points that are in phase.

- A transverse wave consists of **crests** and **troughs**; a longitudinal wave consists of **compressions** and **rarefactions**.

- The **universal wave equation** gives the speed of a periodic wave in terms of the frequency and wavelength.

9.5 Interference of Pulses and Waves

- The **principle of superposition** can be applied to show how **destructive interference** of periodic waves produces **nodes**, and how **constructive interference** produces **supercrests** and **supertroughs**.

9.6 Mechanical Resonance and Standing Waves

- **Mechanical resonance** is the maximum response of the transfer of energy at the **resonant frequency** of an object.

- A **sympathetic vibration** results when the input energy is at the resonant frequency, and a **forced vibration** results when the input energy is at a different frequency.

- A **standing wave** is caused by the interference of two waves of the same amplitude and wavelength moving in opposite directions. It results in a series of **nodes** and **antinodes**. The nodes are half a wavelength apart.

9.7 Activity: Observing Waves in Two Dimensions

- Ripple tank demonstrations and simulations reveal the properties of two-dimensional waves, such as reflection, refraction, diffraction, and interference.

9.8 Sound Waves

- Sound waves are produced by vibrations and are transmitted through a medium by longitudinal waves.

- When two nearly identical frequencies are sounded together, **beats** are produced. These alternating loud and soft sounds can be used to tune musical instruments.

- The human **audible range** is 20 Hz to 20 kHz. Sounds of higher frequency are called **ultrasonic**.

- The **intensity level**, in decibels (dB), is a measure of the loudness of a sound.

- The effect of loud sounds in our environment is an important issue.

9.9 The Quality of Musical Sounds

- The scientific **quality** of musical sounds depends on their **harmonics**.

- Stringed instruments and wind instruments are classified according to their method of producing sounds.

9.10 Sound Communications Technology

- A **transducer** transforms energy from one form into another; an example is a loudspeaker that transforms electrical energy into sound energy.

- Electric and electronic musical instruments use various components to create sounds.

- A composer can create and control complex sounds using a **MIDI** system.

Key Terms

9.1
vibration
transverse vibration
longitudinal vibration
torsional vibration
length
amplitude
cycle
frequency
hertz
period
in phase
out of phase

9.4
wave
crest
trough
wavelength
periodic wave
compression
rarefaction
universal wave equation

9.5
destructive interference
node
constructive interference
supercrest
supertrough

principle of superposition
waveform

9.6
resonant frequency
mechanical resonance
sympathetic vibration
forced vibration
standing wave
antinode
fundamental frequency

9.8
beats
beat frequency
audible range

ultrasonic sound
echolocation
sonar
intensity level

9.9
quality
harmonics
fundamental frequency
vibrator
resonator

9.10
transducer
Musical Instrument
   Digital Interface (MIDI)

Key Equations

9.1

- $f = \dfrac{1}{T}$

- $T = \dfrac{1}{f}$

9.4

- $v = f\lambda$

- $v = \dfrac{\lambda}{T}$

Problems You Can Solve

9.1

- Define, describe the properties, and give examples of transverse, longitudinal, and torsional vibrations.
- Given the number of vibrations during a time interval, determine the frequency and period of vibration.
- Given either period or frequency, determine the other quantity.

9.2

- State what happens to the frequency of a simple pendulum when the mass, amplitude, and length are increased, one at a time.

9.3

- Describe how the speed of a pulse on a spring depends on the tension in the spring.

- State what a pulse or wave accomplishes.
- Given the diagram of an incident pulse, draw the diagram of the pulse that reflects off a fixed end.

9.4

- Describe examples of the transfer of energy by waves.
- Given any two of speed, frequency, and wavelength, determine the third quantity.

9.5

- Draw and label diagrams illustrating the destructive and constructive interference of transverse waves.

9.6

- Describe and write the energy-transformation equations for sympathetic vibrations.

- Given either the distance between nodes on a standing wave or the wavelength of the wave, determine the other quantity.

9.7

- Draw diagrams of periodic water waves showing how they reflect off straight and parabolic barriers, refract as they travel from a faster medium to a slower one, and diffract through openings.
- Recognize the pattern and components of a two-source interference pattern of water waves.

9.8

- Describe how sound energy is transmitted in air from the source to the listener.
- Describe examples of sound interference and sound resonance.
- Identify the frequencies and uses of sounds that are audible and ultrasonic.
- Assess the risk of prolonged exposure to loud sounds, and describe ways of avoiding the risk.

9.9

- Given one harmonic frequency of a sound, determine other harmonic frequencies.
- Describe how air molecules are made to vibrate in the four categories of wind instruments.
- Describe how to improve the quality of sound from stringed and wind instruments.

9.10

- Name and describe the function of the main parts of electric and electronic instruments.
- Given a diagram of a MIDI system, describe the functions of the main components.

▶ **MAKE** a summary

Draw a large diagram of an entertainment venue for your school that includes the ideas presented in this chapter.

- Begin with an appropriate setting that has a stage (e.g., an auditorium). Add an orchestra that has stringed, wind, electric, and electronic instruments.
- Add a microphone–amplifier–loudspeaker system.
- Add a sound engineer who uses an oscilloscope to analyze the sounds produced.
- Draw examples of instrument designs to show that you understand vibrations, waves, standing waves, resonance, and interference.
- Add a listener, and show how sound energy is transmitted to the listener.
- In your design, include as many of the concepts, skills, and key words from this chapter as possible.

Write the numbers 1 to 9 in your notebook. Indicate beside each number whether the corresponding statement is true (T) or false (F). If it is false, write a corrected version.

1. A torsional vibration results when an object twists around its rest axis.

2. As the amplitude of vibration of a simple pendulum increases, the frequency of vibration decreases.

3. As the tension in a spring increases, the speed of a pulse along the coil increases.

4. The speed of periodic waves on water increases as the frequency of the source of waves increases.

5. When a trough meets a trough, destructive interference produces a supertrough.

6. The amount of diffraction of waves through an opening increases as the wavelength of the waves increases.

7. In a two-source, in-phase interference pattern, the distance between the nodal lines increases as the frequency of the source increases.

8. The fundamental frequency of a guitar string increases as the length of the string decreases.

9. The loudspeaker that best generates high frequencies is called the tweeter.

Write the numbers 10 to 17 in your notebook. Beside each number, write the letter corresponding to the best choice.

10. A simple pendulum of length 50 cm vibrates with an amplitude of 4.0 cm. In two cycles, the pendulum bob moves
    (a) 8.0 cm   (c) 32 cm
    (b) 16 cm   (d) 100 cm

11. For a transverse wave, one wavelength can be measured
    (a) from one crest to the next
    (b) from one trough to the next
    (c) from the beginning of the crest to the end of the adjacent trough
    (d) all of the above

12. As the frequency of a source increases, the wavelength of a periodic transverse wave

    (a) increases, but the speed of the wave remains constant
    (b) increases, and the speed of the wave increases
    (c) decreases, but the speed of the wave remains constant
    (d) decreases, and the speed of the wave decreases

13. Waves of wavelength 12 cm and amplitude 3.0 cm are used to create a standing wave. The distance from one node to the next in the pattern is
    (a) 3.0 cm   (c) 12 cm
    (b) 6.0 cm   (d) 24 cm

14. When waves pass through an opening,
    (a) the amount of diffraction increases as the wavelength increases
    (b) the amount of diffraction decreases as the wavelength increases
    (c) the amount of diffraction remains constant as the wavelength increases
    (d) no diffraction occurs because diffraction occurs only around barriers

15. Two tuning forks, one 340 Hz and the other 344 Hz, are sounded together. The beat frequency produced is
    (a) 340 Hz   (c) 684 Hz
    (b) 344 Hz   (d) 4 Hz

16. For humans, an example of an ultrasonic sound is
    (a) 30 000 Hz   (c) 30 Hz
    (b) 3000 Hz   (d) 3 Hz

17. If f is the fundamental frequency of the sound of a vibrating string, then the waveform in **Figure 1** results from the addition of
    (a) f and f (c) f and $3f$
    (b) f and $2f$ (d) f and $4f$

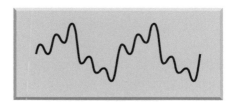

Figure 1

NEL   An interactive version of the quiz is available online.
GO www.science.nelson.com

Communication with Sound   **479**

Understanding Concepts

1. The mass of a simple pendulum moves a total horizontal distance of 14 cm in one cycle.
   (a) State the type of vibration the pendulum undergoes.
   (b) Calculate the pendulum's amplitude of vibration.
   (c) What happens to the frequency of the pendulum as the amplitude decreases?
   (d) What happens to the energy of the mass as the amplitude decreases? Does this agree with the law of conservation of energy?

2. Calculate the period, in seconds, and frequency, in hertz, of each of the following motions:
   (a) A clothes dryer drum rotates 25 times in 18 s.
   (b) In 11 s, a chipmunk grooms its front paws 33 times.

3. Assume that the device in **Figure 1** is connected to a motor and set into a constant high-speed rotation. Then a stiff piece of paper is held against the toothed disks, one level at a time. Describe the sounds produced, and explain why they differ from one another.

Figure 1

4. State the relationship (if any) between the following pairs of variables:
   (a) period and frequency of a vibration
   (b) frequency and wavelength of a periodic wave in a single medium
   (c) amplitude and speed of a pulse in a single medium

5. Each diagram in **Figure 2** shows a transverse pulse travelling along a rope toward a fixed end. Draw a diagram showing the reflected pulse in each case.

(a)

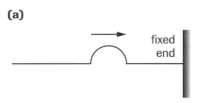

(b)

Figure 2

6. A 5.2×10^2-Hz sonar signal travels through water with a wavelength of 2.9 m. Calculate the speed of the signal in water.

7. Ocean waves that measure 12 m from crest to adjacent crest pass by a fixed point every 2.0 s. Calculate the speed of the waves.

8. **Figure 3** shows a vibrating mass hanging at the end of a spring. Its up and down frequency of vibration is 0.80 Hz, and it moves 16 cm in one complete cycle. A pen connected to the mass creates marks on a paper pulled to the right with a constant speed of 0.20 m/s.
   (a) What type of vibration does the spring undergo?
   (b) Sketch the pattern that the pen creates on the paper. Label the amplitude.
   (c) Calculate the wavelength of the wave you drew in (a).

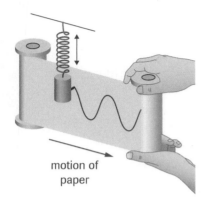

motion of
paper

Figure 3

9. State the conditions necessary to produce
   (a) constructive sound interference
   (b) destructive sound interference

10. A spring is stretched 4.5 m along the floor and is fixed at the far end. Periodic transverse waves of period 0.25 s are generated from the near end toward the far end. Each wave returns to the source after 0.50 s.
    (a) Calculate the speed of the waves along the spring.
    (b) Calculate the wavelength of the waves.
    (c) If a standing wave pattern is produced, how many nodes and antinodes are there between the ends of the spring?

11. A 6.0-m rope is used to produce standing waves. Draw a diagram of the standing wave pattern produced by waves of wavelength (a) 12 m, (b) 6.0 m, and (c) 3.0 m.

12. Define, in your own words, mechanical resonance.

13. Define sympathetic vibration, and include an example.

14. Water waves with straight wavefronts of low frequency are sent toward an opening between two barriers.
    (a) What happens to the waves as they pass through the opening?
    (b) Describe what happens as the frequency is gradually increased.

15. Describe how sound energy is transferred from its source to the listener through air.

16. Is it easier to identify the direction from which a sound is coming if it has a high pitch or a low pitch? Assume equal sensitivity in both ears. Explain your reasoning. (Hint: Consider the diffraction of sound waves.)

17. What beat frequencies are possible when these tuning forks are used in pairs: 256 Hz, 249 Hz, and 251 Hz?

18. A 440-Hz tuning fork is sounded together with the A-string on a guitar, and a beat frequency of 3 Hz is heard. Then an elastic band is wrapped tightly around one prong of the tuning fork, and a new beat frequency of 2 Hz is heard.
    (a) Determine the frequency of the guitar string.
    (b) What should be done to tune the string to 440 Hz?
    (c) Explain how your answer in (b) is an application of the interference of waves to sound communications.

19. The fundamental frequency of a string is 220.0 Hz. Determine the frequency of the
    (a) second harmonic
    (b) fourth harmonic

20. A pulse is sent from a ship to the floor of the ocean 420 m below; 0.60 s later the reflected pulse is received at the ship. Calculate the speed of the sound in the water.

21. The Ontario Science Centre in Toronto has many hands-on displays related to sound communications, one of which is shown in **Figure 4**. Students can communicate with other students at an identical reflector on the far side of the room by talking softly and listening.
    (a) Sketch the room and show the two reflectors used in this display. Name the type of reflector.
    (b) Explain how sounds produced a long distance away can be heard, even if the sounds are not loud.

Figure 4

22. The sound waveform displayed on an oscilloscope in **Figure 5** was produced by a tuning fork. Copy the waveform into your notebook, and draw a diagram of each of the following:
    (a) a louder sound wave with the same frequency
    (b) a sound wave with a higher frequency but the same loudness
    (c) a sound wave with the same frequency but created by a musical instrument with a different quality of sound

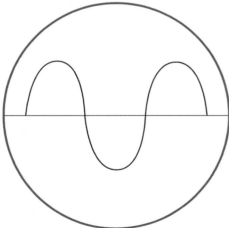

Figure 5

23. Describe how the quality of a musical sound is affected by the harmonic structure of the sound. Apply the principle of superposition.

24. An electric guitar is connected to a microphone, an amplifier, and a loudspeaker. You strum a string. Describe the resulting energy transformations and the sound energy transmissions. Start from the motion of the player's hand and end at the listeners' ears; indicate which components are transducers.

25. List four methods of forcing air to vibrate in wind instruments. Name an instrument that illustrates each method.

Applying Inquiry Skills

26. **Figure 6** shows a strobe photograph of a swinging pendulum.
    (a) Does the photograph show a half cycle or more than a half cycle? Describe evidence that supports your answer.
    (b) Describe how you can tell where the kinetic energy of the pendulum mass is greatest.
    (c) What data would you need in order to calculate an approximate value of the maximum speed of the mass?

Figure 6

27. (a) Describe how you would create a musical instrument made of bars that makes at least six notes that correspond to notes on a piano scale. The bars can be metal, hard plastic, or wood. (You may have seen toy instruments like this.)
    (b) How would you test your instrument to determine whether it is in tune?

28. Listen to the sound produced by a small loudspeaker connected to an audio frequency generator. Now cut a hole the size of the speaker in a large board or piece of cardboard. Hold the speaker at the hole and listen again. Use the concepts studied in this chapter to explain what you hear.

Making Connections

29. (a) How does an animal's size relate to its audible range? (You can refer to **Table 1**, section 9.8, page 460.)

 (b) Relate your answer in (a) to the relationship between the size and frequency of tuning forks.

30. A *tsunami* is a fast-moving ocean wave that results after an underwater earthquake or volcanic eruption. In the deep ocean, the wavelength might be more than 250 km, the amplitude only about 5 m, and the speed up to 800 km/h. The wave might pass under a ship and not even be noticed, but it can strike a shore with an amplitude of perhaps 30 m and do severe damage. How can a wave that seems harmless at sea do such damage onshore?

31. Describe practical ways to reduce noise pollution at a busy intersection in a city.

32. Each button on a touch-tone phone has a distinct sound. The sounds are combinations of pure tones of specific frequencies arranged according to rows and columns, as shown in **Figure 7**. For example, all the buttons in the first row produce a 697-Hz sound, and all the buttons in the first column produce a 1209-Hz sound. Therefore, each button produces two frequencies when pushed. When a button is pushed, the pair of frequencies is transmitted to the central phone control office, where they activate electronic circuits to complete the call.

 (a) State the pair of frequencies produced by each of the following buttons: 2; 6; 7; #.

 (b) Calculate the periods of vibration of sounds produced by button 9.

697 Hz
770 Hz
852 Hz
941 Hz

1209 Hz   1477 Hz
1336 Hz

Figure 7
Each button on a touch-tone phone produces two frequencies.

Communication with Light

- define and explain the concepts and units related to communications technology involving light and other electromagnetic waves

- explain and illustrate how the reflection of waves is used in communications technology

- using Snell's law, explain and predict the refraction of electromagnetic waves

- describe and illustrate total internal reflection, and explain its significance in communications systems

- analyze and describe the energy transformations and transmissions that occur in communications systems involving light energy

- experimentally verify Snell's law and identify the conditions required for total internal reflection

- investigate the reflection and refraction of light

- design and construct a simple communications system, and demonstrate the operation of its major components

- evaluate models of a communications system

- describe and evaluate Canadian contributions to communications science and technology involving electromagnetic radiation

- assess the risks and benefits to society and the environment of introducing a particular technology from the communications industry

Getting Started

Light and other electromagnetic waves form the basis of many communications systems. This is due in part to the fact that light can be a messenger: When light is transmitted and received, it carries information. For example, a laser is an integral part of a CD player. Electromagnetic waves are useful in communications systems for another reason: They exhibit wave properties. In this chapter, you will discover how light and other electromagnetic waves can reflect, refract, interfere, and diffract, and how these properties are used in communications systems.

You will also learn how the combination of light and electronics provides fascinating and fun communications systems. For example, light-emitting diodes allow the Ford GloCar (**Figure 1**) to communicate with its surroundings in an intelligent way. If another vehicle comes too close, the light intensity of the diodes increases to warn of danger. The driver can also change the appearance of the car so that it either stands out or blends in. In addition to learning about the application of light-emitting diodes in this chapter, you will learn how light and electronics can be combined to create state-of-the-art technologies such as charge-coupled devices (CCDs) in digital cameras.

💡 REFLECT on your learning ▼

1. Compare what you have learned about sound waves with what you already know about light waves. Use the following questions as a guide:
   (a) Can both sound waves and light waves travel in a vacuum?
   (b) How does the speed of light in air compare with the speed of sound in air?

2. (a) What is meant by the refraction of light?
   (b) Where have you observed it?

3. (a) Give an example of an object or material that is (i) transparent, (ii) translucent, and (iii) opaque.
   (b) Which of the materials in (a) absorbs light energy most readily? Explain your answer.

4. Compare and contrast AM and FM radio signals.

5. What are some uses of lasers?

Figure 1
This Ford GloCar uses electronic circuits to create light that "communicates" with its surroundings. The car can be seen at night from all directions, reducing the chances of accidents, especially at intersections.

▶ **TRY THIS** *activity*   *Prisms and Light*

A prism is a transparent block of acrylic or glass used in light experiments (**Figure 2**). Obtain two 45°-90°-45° prisms and a source of light rays, such as a ray box. Plan steps to answer the following questions, and use a diagram to illustrate your answers:

(a) How does a single prism cause light to refract?

(b) How does light reflect off the inside and outside of the prism?

(c) How does the prism cause light to split into its spectral colours?

(d) How can two prisms be used to focus light to a bright area?

 Do not allow the ray box cord to hang where someone might walk into it.

Unplug the ray box by pulling on the plug, not on the electrical cord.

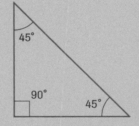

Figure 2
A transparent 45°-90°-45° prism

Be careful when touching the ray box after use; it may be hot.

Do not touch the ray box light bulb or look directly into the light.

Turn off the ray box when not in use.

Allow the ray box to cool before putting it away.

Communication with Light   **485**

diffraction the bending of waves as they travel around objects or through openings

Figure 1
A diffraction pattern produced by light in air striking a razor blade

Figure 2
Light from a bright, white source passing through two narrow slits produced this interference pattern.

visible spectrum the set of colours visible to the human eye

Light is a form of radiant energy that allows us to see. It has several properties in common with other forms of radiant energy, as you learned in Chapters 3 and 4. In this section, we explore the properties of radiant energy, how radiant energy is transmitted and absorbed, and how it can be controlled in order to transmit signals in communications systems.

Properties of Electromagnetic Waves

Once light is emitted from a source, it travels in the form of waves. Light displays the properties of waves presented in Chapter 9. One wave property is **diffraction**, which is the bending of waves as they travel around objects or through openings. **Figure 1** shows a diffraction pattern observed when light strikes an object that has sharp edges.

A second example of a wave property of light is interference. Recall the two-source interference pattern observed with water waves (Activity 9.7) and with sound waves (section 9.8). **Figure 2** shows a pattern of bright and dark regions produced when white light is viewed through two parallel slits. The pattern becomes more interesting if the light is viewed through several sets of fine parallel slits. You will discover more evidence of the wave nature of light as the chapter progresses.

Notice that portions of the pattern in **Figure 2** have the colours of the rainbow. The pattern was created by white light, so this observation provides evidence that white light comprises several colours. The set of colours visible when white light is dispersed or separated, for example, by a prism (**Figure 3**), is called the **visible spectrum**. The six main colours of this spectrum are red, orange, yellow, green, blue, and violet.

The visible spectrum is only a small portion of the range of radiant energies of electromagnetic waves. Other radiant energies are radio waves, microwaves, infrared radiation (IR), ultraviolet (UV) radiation, X rays, and gamma rays.

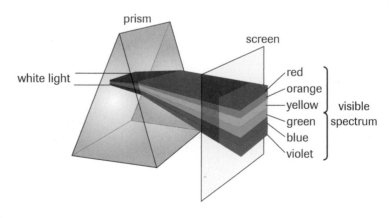

Figure 3
White light that strikes a prism at an angle is separated into the spectral colours of the rainbow.

▶ **TRY THIS** activity *Diffraction and Interference of Light*

Your teacher will set up demonstrations that provide evidence of the wave nature of light.

(a) Observe the diffraction pattern as light from a laser passes through a narrow opening with sharp, parallel edges. How does the pattern change as the opening becomes narrower? What does the width of the opening reveal about the wavelength of light waves?

(b) Observe the interference pattern created when light from a laser passes through a set of two parallel narrow slits. Try another set of slits with a different separation. Describe the effect on the pattern when the distance between the slits decreases.

(c) Observe a white light source with a vertical filament (**Figure 4**) through a plastic sheet with many fine parallel lines (this is called a diffraction grating). Draw a diagram of what you observe.

 Do not direct the laser beam toward anyone's face.

Be sure that the laser beam does not reflect off a shiny object.

Figure 4
A white light source with a straight filament

These forms of energy are invisible, but they share some properties with visible light. The entire range of radiant energies, including visible light, is called the **electromagnetic (EM) spectrum.**

One property of all EM radiation is that it can travel in a vacuum; no particles are needed for its transmission. (Recall section 3.1, page 128.) This is different from both sound waves and water waves, which must be transmitted in a medium (particles). Another important feature of EM waves is that they all travel at the same speed in a vacuum, 3.00×10^8 m/s. At this speed, light takes about 1.3 s to travel from Earth to the Moon and about 8.0 min to reach Earth from the Sun. Light from the nearest star beyond the solar system, Proxima, in the Alpha Centauri system, takes more than 4 years to reach us. Light from the nearest large galaxy, Andromeda, takes more than 2 million years to reach Earth.

Although all types of electromagnetic radiation share some features, each portion of the EM spectrum has its own wavelength and frequency (**Figure 5**).

electromagnetic (EM) spectrum
the range of radiant energies

DID YOU KNOW❓

Black Light
The Famous People Players are renowned for their innovative "black light" entertainment. Black light is ultraviolet (UV) radiation, which is invisible. However, when UV radiation strikes certain substances, the particles in the substance gain energy and re-emit some of that energy as visible light. The Famous People Players use such substances in their costumes, makeup, and props when they mimic famous entertainers.

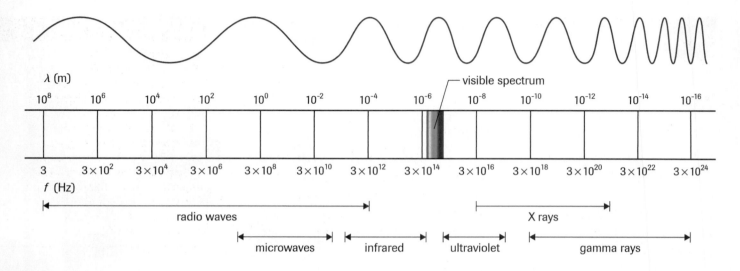

λ (m)

| 10^8 | 10^6 | 10^4 | 10^2 | 10^0 | 10^{-2} | 10^{-4} | 10^{-6} | 10^{-8} | 10^{-10} | 10^{-12} | 10^{-14} | 10^{-16} |

visible spectrum

| 3 | 3×10^2 | 3×10^4 | 3×10^6 | 3×10^8 | 3×10^{10} | 3×10^{12} | 3×10^{14} | 3×10^{16} | 3×10^{18} | 3×10^{20} | 3×10^{22} | 3×10^{24} |

f (Hz)

radio waves

X rays

microwaves   infrared   ultraviolet   gamma rays

Figure 5
Parts of the electromagnetic spectrum and their uses. (Waves are not drawn to scale.)

A clue to understanding how radiant energy travels is found in the term "electromagnetic wave." Each wave has a vibrating electric component and a vibrating magnetic component (**Figure 6**). When an EM wave is emitted from a source, the electric component continues to produce the magnetic component, and the magnetic component continues to produce the electric component. These interactions keep the wave vibrating and travelling until it meets an obstruction.

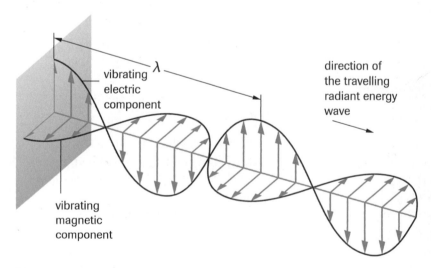

vibrating electric component

λ

direction of the travelling radiant energy wave

vibrating magnetic component

Figure 6
The electric and magnetic components of an EM wave are transverse; they are perpendicular to each other and perpendicular to the wave's direction of travel.

Notice in **Figure 6** that an EM wave is drawn as a transverse wave. Like other waves, it has a wavelength and a frequency. The frequency corresponds to the frequency of the vibrating particles of the source. (The particles are charged particles, such as electrons and protons.) Thus, the universal wave equation, $v = f\lambda$, applies to radiant energy. Using the symbol c to represent the speed of light and other radiant energy in a vacuum, the equation can also be written as

$$c = f\lambda$$

where $c = 3.00 \times 10^8$ m/s.

Since $f = \dfrac{1}{T}$, this equation is equivalent to $c = \dfrac{\lambda}{T}$ or $v = \dfrac{\lambda}{T}$.

The equation $v = f\lambda$ applies to light transmitted in air, glass, plastic, and other media, as well as in a vacuum. However, the speed depends on the medium, as you will discover in section 10.4. It follows from the equations $v = f\lambda$ and $c = f\lambda$ that, as the frequency of the source increases, the wavelength of the wave decreases. (You can see this pattern in the diagram of the EM spectrum in **Figure 5**.)

Of course, EM waves cannot be transmitted through all materials. Some materials absorb EM waves, some reflect them, others scatter them; many display all these properties at the same time. Materials can be classified according to what happens to light when it strikes them. **Transparent materials**, such as clear glass and shallow water, allow light to be transmitted easily. A clear image can be seen through these materials. **Translucent materials**, such as waxed paper and frosted glass, allow the transmission of some light but scatter it as well, so no clear image can be seen through them. **Opaque materials**, such as concrete and wood, allow no light to pass through; all the light is absorbed and/or reflected, and no image is seen through them.

An important application of the absorption of EM waves is solar cell technology. As you learned in Chapter 4, solar cells have a fairly low efficiency. One of the technical problems with these cells is that they can only transform radiant energy into electrical energy if the EM waves have a frequency above a certain level. Thus, today's solar cells only absorb the waves at the higher frequency end of the visible spectrum, particularly violet. Research continues to improve the efficiency of these devices by using materials that absorb radiant energy across a wider range of frequencies.

transparent material a substance in which light can be easily transmitted, so a clear image is seen through it

translucent material a substance in which some light is transmitted and some is scattered, so no clear image is seen through it

opaque material a substance that absorbs and/or reflects all the light that strikes it, so no image is seen through it

▸ *Practice*

Understanding Concepts

1. Under normal conditions, we do not notice the diffraction of light. What condition allows us to observe this property?

2. Based on what you learned about a two-source interference pattern in water, what would you call the two dark lines in the light diffraction pattern in **Figure 2**, page 486?

3. Compare and contrast visible light with the other components of the EM spectrum.

4. Based on the information in **Figure 5**, state the relationship between the wavelengths of EM waves and the frequencies of their sources.

5. **Figure 7** shows white light reflected off the thousands of grooves on a CD. Describe how this pattern resembles the interference pattern seen through a diffraction grating.

Figure 7
Visible spectra observed from the reflection of white light off CDs

6. The average distance from Earth's surface to the Moon's surface is 3.84×10^{8} m. Calculate the length of time it takes a beam of laser light to travel from Earth to the Moon and back again.

7. Calculate the wavelength of each of the following in a vacuum or air:
   (a) an X ray with a frequency of 2.00×10^{18} Hz
   (b) the Ryerson University radio station in Toronto, CKLN, with a radio frequency of 88.1 MHz
   (c) an EM wave with a period of 1.67×10^{-2} s

8. When hydrogen gas in an enclosed tube is exposed to a high voltage, it emits light colours of distinct wavelengths.
   (a) Calculate the frequency of an emitted light that has a wavelength of 410 nm.
   (b) According to **Figure 8**, what is the colour of this light?

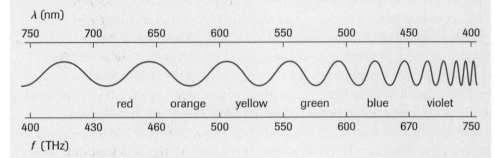

Figure 8
Approximate wavelength and frequency ranges for visible light. (Waves are not drawn to scale.)

9. Name two examples not mentioned in this section of each of the following types of materials:
   (a) transparent
   (b) translucent
   (c) opaque

10. (a) How does the speed of sound in air compare with the speed of light in air?
    (b) Use your answer in (a) to explain the order in which you see lightning and hear thunder during a storm.

Applying Inquiry Skills

11. (a) Predict what you would observe if you aimed red light at the equilateral prism in **Figure 3**, page 486, instead of white light.
    (b) Repeat (a) for blue light.
    (c) With your teacher's permission, try it. Comment on your predictions.

12. On a clear day you can create your own visible spectrum. With your back to the Sun, spray a mist of water into the air from a garden hose. Change the angle of spray until you see a spectrum. Draw a diagram of the arrangement you used.

Making Connections

13. Describe as many situations as you can in which you have seen a visible spectrum.

14. The photograph in **Figure 9** was taken with a crosshair filter, also called a starburst filter, placed over the camera's lens. The light source shining on the gemstones is a small, bright point source.
(a) Draw a diagram showing how you think the pattern would appear if the source were a vertical light source (i.e., a showcase bulb).
(b) What property of light causes this type of pattern?
(c) How would a photographer use this pattern for special effects?

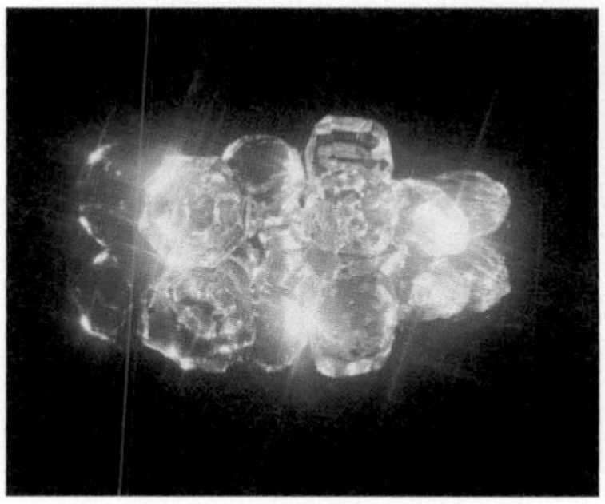

Figure 9
A camera filter with fine cross lines produced the pattern of light surrounding these gemstones.

Modulating EM Waves

When listening to the radio, you can choose an AM station or an FM station. AM stands for *amplitude modulation* (i.e., control), and FM stands for *frequency modulation*. These are the two ways—amplitude and frequency—that radio signals are modulated, or controlled. In both cases, the audio signal, which has been transformed into an EM wave, is combined with an EM carrier wave to produce an output EM wave.

With **amplitude modulation**, the output EM wave has a constant frequency, and the wave amplitude varies according to the waveform of the audio signal, as illustrated in **Figure 10**. Notice that the left part of the diagram shows the unmodulated EM carrier wave. When the EM audio signal is superimposed on the EM carrier wave, interference occurs and the waves "add" according to the principle of superposition. The resulting wave is the modulated output wave. The frequency of the carrier wave is the frequency marked on the tuning scale of the AM receiver. Different frequencies are assigned to different radio stations.

amplitude modulation (AM) the method of controlling radio signals by varying the amplitude of the EM carrier wave, which has a constant frequency

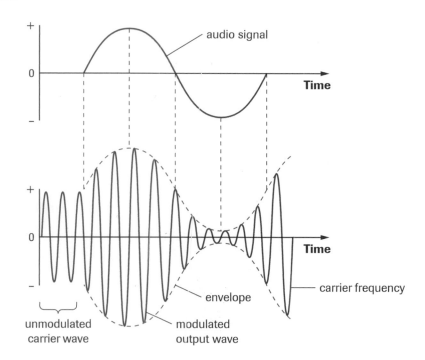

Figure 10
With amplitude modulation (AM), the amplitude of the output EM wave varies according to the audio signal.

frequency modulation (FM) the method of controlling radio signals by varying the frequency of the EM carrier wave, which maintains a constant amplitude

With **frequency modulation**, the carrier amplitude remains constant, and the carrier frequency is modulated according to the EM audio waveform, as illustrated in **Figure 11**. Again, the left side of the diagram shows the unmodulated carrier wave; its frequency is the frequency assigned to the FM radio station. It is that frequency that is modulated to become the output EM wave. When FM is used for transmitting TV signals, both audio and video signals are carried by the EM signal.

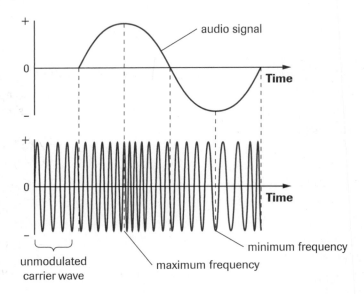

Figure 11
With frequency modulation (FM), the frequency of the output EM wave varies according to the audio signal.

> **Practice**

Understanding Concepts

15. The Channel 7 TV station uses a carrier frequency of 175 MHz.
    (a) Calculate the wavelength of this signal.
    (b) Is this AM or FM transmission?

16. Why do AM radio waves travel better around hills than FM radio waves?

SUMMARY   *Light and Electromagnetic Waves*

- Radiant energy exhibits wave properties, for example, diffraction and interference.

- The visible spectrum is the set of radiant energy colours we can see; it is a small part of the entire electromagnetic (EM) spectrum.

- The radiant energies of the EM spectrum travel in a vacuum at the speed $c = 3.00 \times 10^8$ m/s.

- Electromagnetic waves obey the universal wave equation, $v = f\lambda$ (or $c = f\lambda$ for EM waves in a vacuum). The universal wave equation can also be written as $v = \dfrac{\lambda}{T}$ (or $c = \dfrac{\lambda}{T}$).

- Materials can be classified as transparent, translucent, or opaque, depending on how light and other EM waves react when in contact with the materials.

- EM waves act as carrier waves for transmission of AM and FM radio signals.

▶ **Section 10.1 Questions**

Understanding Concepts

1. Describe two phenomena that illustrate the wave nature of light.

2. Place the following EM waves in order from lowest to highest wavelength: blue light; X rays; microwaves; infrared radiation; orange light.

3. Make a list of all the types of EM waves you have experienced in the past year.

4. What evidence exists that white light comprises several colours?

5. State the relationship between the frequencies of EM waves and their wavelengths.

6. If light from the Sun takes 5.00×10^2 s to reach us, what is the radius of Earth's orbit? (Assume that the orbit is circular.)

7. The data in **Table 1** represent EM waves travelling in air. Copy the table into your notebook and complete it. Show the equation used in each calculation.

8. Compare and contrast the transmission of signals using AM and FM.

Applying Inquiry Skills

9. A window screen has two sets of fine parallel wires, one vertical and the other horizontal.
   (a) Predict the pattern you would observe if you looked through a window screen at a bright source of white light, such as a street light. Illustrate the pattern in a sketch. (Hint: Extend the pattern observed using vertical slits to one that uses both vertical and horizontal slits.)
   (b) Try the activity. If the pattern differs from your prediction, draw the observed pattern.
   (c) What property or properties of light does this activity demonstrate?

Making Connections

10. What type of clothing would be most appropriate on a hot, sunny day? Explain your answer.

Table 1   For Question 7

	Frequency, f (Hz)	Wavelength, λ (m)	Period, T (s)	Type of EM Wave
(a)	?	6.00×10^{-15}	?	?
(b)	9.00×10^{15}	?	?	?
(c)	?	?	5.00×10^{-4}	?

Reflection of Light

In this investigation, you will view images in plane and curved mirrors, and investigate how light reflects off them.

Even if you have performed experiments with mirrors before, a review of the relevant terminology will refresh your memory. The diagrams in **Figure 1(a)** to **(d)** show the shapes of the four major types of mirrors you will be investigating; **Figure 1(e)**

Inquiry Skills

○ Questioning	● Conducting	● Evaluating
● Predicting	● Recording	● Communicating
● Planning	● Analyzing	● Synthesizing

shows a light ray box used to direct rays of light; this is needed to discover where they travel after they reflect off mirrors.

Light travels in straight lines as long as it remains in a single medium, so its transmission can be represented by rays. **Figure 2** reviews how to draw a normal line from a surface so that the angle between the ray and the normal can be measured.

(a)

(c)

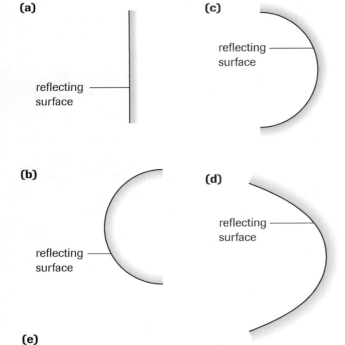

(b)

(d)

(e)

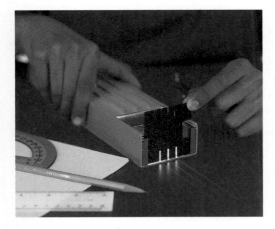

Figure 1
(a) Plane mirror
(b) Convex (diverging) mirror
(c) Circular concave (converging) mirror
(d) Parabolic concave mirror
(e) Ray box with a triple-slit window

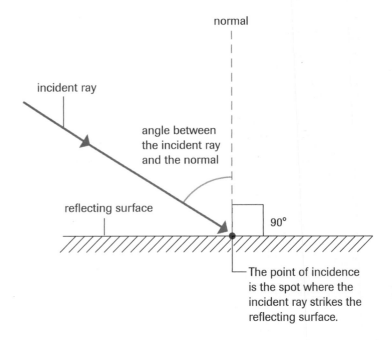

Figure 2
The angle of a light ray can be measured to the normal drawn on the diagram.

Questions

What are the properties of the images seen in plane, convex, and concave mirrors?

How do light rays reflect off plane, convex, concave, and parabolic mirrors?

Predictions

(a) Predict which type of mirror will produce an image that is (i) larger than the object; (ii) always smaller than the object; (iii) always the same size as the object.

(b) For a light ray that reflects off a mirror, predict how the angle of the reflected ray compares with the angle of the incident ray.

(c) Predict which type of mirror will cause parallel light rays to (i) converge, or come together, when they reflect and (ii) diverge, or spread apart, when they reflect.

Materials

For each student:
metric ruler
protractor

For each group of three or four students:
viewing mirrors (plane, concave, and convex)
mirrors for use with a ray box (plane, convex, circular concave, parabolic concave)
ray box
single-slit and multiple-slit windows

 Handle the mirrors carefully; broken pieces are very sharp.

Do not touch the ray box light bulb or look directly into the light.

Turn off the ray box when it is not in use.

Be careful when touching the ray box after use; it may be hot.

Unplug the ray box by pulling on the plug, not on the electrical cord.

Do not allow the ray box electrical cords to hang in areas used as walkways.

Procedure

1. Have your partner hold a plane viewing mirror so that you can see your image from a near distance, and then from farther away. State the attitude of the image (upright or inverted) and the size of the image relative to the object (larger, smaller, or the same size). Record your observations, then trade responsibilities so that your partner can make observations.

2. Repeat step 1 using a convex viewing mirror.

3. Repeat step 1 using a concave viewing mirror.

4. Using a ruler, draw a straight line on a piece of paper. Place the reflecting surface of the plane mirror to be used with a ray box along the line, as shown in **Figure 2**. Aim an incident ray from the ray box at the mirror. Use small dots to mark the incident and reflected rays. Remove the ray box and mirror. Using a ruler, draw the rays. Use a protractor to draw a normal from the point where the rays meet at the mirror. Label and measure the angles between the rays and the normal.

5. Repeat step 4 using two new diagrams and distinctly different angles.

6. Place the convex mirror flat on your paper, and draw the outline of its reflecting surface. Use the ray box with the multiple-slit window to aim parallel rays of light toward the mirror so that the middle ray reflects back onto itself (**Figure 3**). Draw the incident and reflected rays. Remove the mirror, and use your ruler to extend the reflected rays as broken lines behind the mirror until they meet. Label the point of intersection F, and measure the distance *f* from F to the reflecting surface.

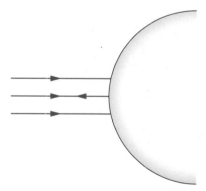

Figure 3
When aiming light rays toward a convex mirror, be sure that the rays are parallel to each other to obtain accurate measurements.

7. Place the circular concave mirror flat on your paper, and draw the outline of its reflecting surface. Use the ray box with the multiple-slit window to aim parallel rays of light toward the mirror so that the middle ray reflects back onto itself (**Figure 4**). Draw the incident and reflected rays. Label the point of intersection F, and measure the distance *f* from F to the reflecting surface.

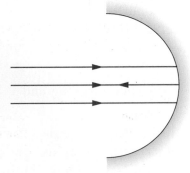

Figure 4
Aiming light rays toward a concave mirror

8. Plan and carry out an experimental step to determine why a parabolic concave mirror focuses light to a single focal point better than a circular concave mirror. (Hint: With each mirror, use several lights rays parallel to each other.)

Analysis

(d) State the size and attitude of the image in each of the following viewing mirrors: plane mirror; convex mirror; close to a concave mirror; far from a concave mirror.

(e) How does the angle between the incident ray and the normal compare with the angle between the reflected ray and the normal?

(f) Compose definitions for focal point (F) and focal length (*f*) for a convex mirror.

(g) Compose definitions for focal point (F) and focal length (*f*) for a concave mirror.

(h) Based on your observations, why is a convex mirror called a "diverging mirror," and why is a concave mirror called a "converging mirror"?

(i) What experimental evidence shows the advantage of a parabolic concave mirror over a circular concave mirror?

Evaluation

(j) How accurate were your predictions?

(k) Describe the main sources of error in this investigation and how you tried to reduce them.

Synthesis

(l) If you were designing a concave mirror for use with a flashlight bulb, would you recommend a spherical or a parabolic reflector? (A spherical mirror is the three-dimensional version of a circular or two-dimensional mirror.) Explain why, using a diagram to illustrate where you would place the bulb and where the light rays would go.

(m) Automobile headlights are concave reflectors. Where should the bulbs be placed for high-beam and low-beam adjustments?

If you look around the room, almost everything you can see is visible because of the reflection of light. Light is produced by a source, strikes the objects around you, and reflects off those objects. Then some of the reflected light travels to your eyes. Many of the reflection examples presented in this section involve visible light. Some examples involve other EM waves used in communications systems.

Plane Mirror Reflection

The reflection of light off a smooth, shiny surface that allows us to see an image is called **regular reflection**, for example, light reflected off shiny metals, the surface of still water, and mirrors. **Figure 1(a)** shows that parallel rays of light striking a regular surface are still parallel after reflecting. **Figure 1(b)** shows the **angle of incidence**, which is the angle between the incident ray and the normal, and the **angle of reflection**, which is the angle between the reflected ray and the normal. (Recall from your studies of normal forces in Chapter 1 that "normal" means perpendicular to the surface.)

The experimental observations of light reflecting off regular surfaces are summarized in the *laws of reflection:*

regular reflection the reflection of light off a smooth, shiny surface; allows us to see an image

angle of incidence the angle between the incident ray and the normal; symbol θ_i

angle of reflection the angle between the reflected ray and the normal; symbol θ_r

diffuse reflection the reflection of light off a rough surface; does not produce an image

---

Laws of Reflection

The angle of incidence equals the angle of reflection ($\theta_i = \theta_r$).

The incident ray, normal, and reflected ray all lie in the same plane.

On rough surfaces, the reflection of light is irregular. This type of reflection is called **diffuse reflection**; it does not allow us to see an image (**Figure 2**). Painters take advantage of diffuse reflection to reduce unwanted glare. For example, instead of using paint with gloss and semigloss finishes, which increase reflection and glare, they use paint with mat and flat finishes to provide diffuse reflection. You can probably see examples of diffuse reflection in your classroom.

Plane mirror reflection is used in many applications beyond simply viewing a reflection, for example:

- If the space in an optometrist's office is limited, the patient is directed to look at a plane mirror from a distance of 3.0 m to see the image of the eye chart. With this arrangement, the chart appears to be 6.0 m away. This is because the distance of the image behind the mirror to the mirror equals the distance of the object to the mirror.

- Security windows, also called one-way mirrors, act like mirrors on the side where the lighting is bright. However, the security guard, located in a darker room, can see through the window.

(a)

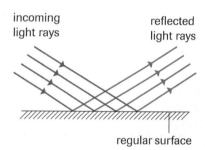

(b)

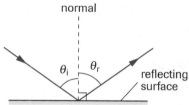

Figure 1
(a) Regular reflection
(b) Measuring the angle of incidence (θ_i) and the angle of reflection (θ_r)

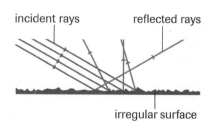

Figure 2
Irregular surfaces produce diffuse reflection. For each ray of light, $\theta_i = \theta_r$ but no image will be seen.

Figure 3
Multiple images are observed when two plane mirrors are nearly parallel and face each other.

- Plane mirrors placed at angles to each other produce multiple images, which provide entertainment (**Figure 3**).
- A corner mirror with three sides that are perpendicular to each other was installed on the Moon; scientists use reflected laser light pulses from the Moon to measure the Earth–Moon distance to the closest 15 cm.

▶ **TRY THIS** activity   *Multiple Images*

An equation to determine the number of images formed by two mirrors at specific angles to each other is $n = \dfrac{360°}{a} - 1$, where n is the number of images, and a is the angle between the plane mirrors. Design and perform an experiment to determine whether this equation works for angles 180°, 90°, 60°, and 45°.

 Handle glass carefully.

Answers
2. (a) 48 cm
   (b) 96 cm

▶ **Practice**

Understanding Concepts

1. State the angle of reflection if the angle of incidence is (a) 23°, (b) 78°, and (c) 0°.

2. A 160-cm tall student is standing 48 cm from a long plane mirror. What is the distance from the
   (a) image to the mirror?
   (b) student to the image?

3. Which type of paint, flat or glossy, produces more glare? Explain why.

4. Describe the circumstances in which a window in your home might act as a one-way mirror when viewed (a) from the inside and (b) from the outside.

Applying Inquiry Skills

5. How could a basketball, a stairway, and a smooth floor be used to demonstrate the difference between regular and diffuse reflection? (Do not try this; the bounce could be dangerous.)

6. In your notebook, draw a straight line to represent a plane mirror. Draw and label a normal. Then add incident rays with angles of incidence of 32° and 85°. Draw the reflected rays. Label all angles, and show all ray directions.

Making Connections

7. In some public areas, a plane mirror is installed with the mirror angled away from the wall. What is the purpose of this arrangement? Draw a sketch to show the arrangement. Include a person using the mirror.

Curved Mirror Reflection

Curved mirrors are classified according to how they reflect light rays from a distant source. A **diverging mirror** causes the light rays to diverge, or spread apart. Its reflecting surface is convex (curved outward), as in **Figure 4**. Parallel light rays that reflect off a diverging mirror never meet, but if they were extended behind the mirror they would meet at a point called the **focal point** of the mirror. In a diverging mirror, the focal point is called *virtual* because the rays only appear to come from there. The distance from the focal point to the reflecting surface is called the **focal length** (*f*) of the mirror.

A diverging mirror can be used to survey a large area because its curved surface reflects light to our eyes from a large portion of the surroundings. Images are always upright and smaller than the objects viewed, no matter where the objects are located. **Figure 5(a)** shows why a diverging mirror has a wider area of view than a plane mirror. This property is applied in the surveillance mirrors in stores and side-view mirrors on many vehicles (**Figure 5(b)**).

A **converging mirror** causes reflected light rays to converge, or come together. Its reflecting surface is concave (curved inward), as shown in **Figure 6(a)**. If the mirror is circular or spherical in shape, the parallel incident rays close to the middle of the mirror reflect and meet at the focal point. However, rays that are closer to the outer edge of the mirror reflect inside the focal point, as shown in the diagram. (This is a property of the mirror's shape because the rays are still obeying the first law of reflection.) The result is a distorted image. To overcome this problem, a *parabolic mirror* is used. A **parabolic mirror** is a mirror in the shape of a paraboloid, which is

diverging mirror a curved mirror that causes the reflected light to spread outward

focal point in a diverging mirror, the point from which parallel incident rays appear to reflect; in a converging mirror, the point where parallel incident rays reflect and meet

focal length the distance from the focal point to the reflecting surface of a curved mirror; symbol *f*

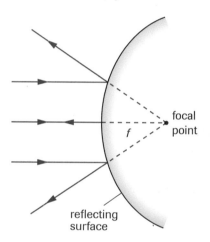

Figure 4
The focal point and focal length of a diverging mirror can be found by aiming parallel light rays toward the mirror.

converging mirror a curved mirror that causes the reflected light to come together

parabolic mirror a curved mirror in the shape of a paraboloid

(a)

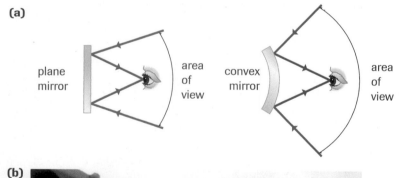

(b)

Figure 5
(a) The reflection in a diverging mirror gives a larger area of view than a reflection in a plane mirror of similar size.
(b) A diverging mirror on the front of a school bus allows the driver to see children both beside and in front of the bus. If a diverging mirror is used to see behind the bus, the driver must be careful because the objects are actually closer than their images suggest they are.

(a)

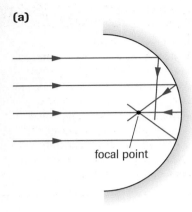

focal point

(b)

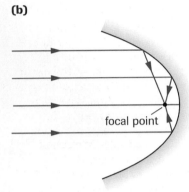

focal point

Figure 6

(a) Only the parallel rays that strike close to the middle of the circular or spherical concave mirror reflect to the focal point. Rays near the outer part of the mirror reflect inside the focal point.

(b) A parabolic concave mirror: All the parallel incident rays hitting the parabolic concave mirror reflect to a common focal point.

the three-dimensional version of a parabola. As shown in **Figure 6(b)**, even if the parallel rays are close to the outer edge of the mirror, they reflect to the focal point.

Parabolic concave mirrors have many applications that use visible light as well as other parts of the EM spectrum. In **Figure 7(a)**, a flashlight bulb placed near the focal point of the mirror behind it emits rays in all directions. The rays that reflect off the parabolic mirror are nearly parallel and form a bright light beam. In **Figure 7(b)**, a reflecting telescope gathers light from distant objects such as stars and galaxies. The parabolic mirror combines with a plane mirror and the eyepiece lens to bring the object being viewed into focus. The largest telescopes, including space telescopes, use this design. **Figure 7(c)** shows a microwave communications tower with parabolic reflectors that send and receive signals, such as TV signals in the microwave range. Radio telescope

(a)

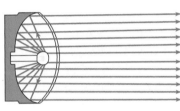

(b)

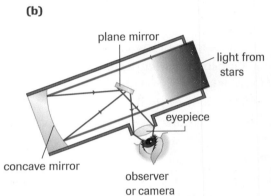

plane mirror

light from stars

eyepiece

concave mirror

observer or camera

(c)

(d)

Figure 7

(a) In this flashlight, the light bulb filament is near the focal point of the mirror behind it. The rays that reflect form a beam of light.

(b) A reflecting telescope produces an image that can be viewed directly, photographed, or recorded digitally.

(c) The microwaves sent from and received at this tower have longer wavelengths than visible light. Find the focal points of the reflectors.

(d) The dish on this radio telescope is 46 m in diameter. Radio waves have longer wavelengths than microwaves, so the dish needs to be considerably larger. The location of the focal point is obvious.

dishes are also parabolic (**Figure 7(d)**). These telescopes collect long-wavelength radio waves emitted from objects in distant parts of the universe.

> ▶ *Practice*

Understanding Concepts

8. Explain why a convex mirror has a virtual focal point.

9. (a) Rank plane, diverging, and converging mirrors of similar size according to their area of view when you look into them. Which type of mirror requires you to stand directly in front of it to see a clear image?
   (b) Describe as many situations as possible in which the mirror with the largest area of view is used.

Applying Inquiry Skills

10. Draw a diagram showing how you could boil water in a large metal pot using solar energy, with a curved mirror acting as the solar collector. Include incident and reflected rays in your diagram.

11. Describe how you would determine experimentally whether the laws of reflection apply to curved mirrors. With your teacher's permission, try your experimental design.

Making Connections

12. Plane mirrors are usually back-surfaced, but the small mirrors used by dentists are front-surfaced.
   (a) Draw sketches to compare the reflecting surfaces in these two types of mirrors.
   (b) Name one advantage and one disadvantage of a front-surfaced mirror compared with a back-surfaced mirror. In your answer, include the mirrors used by dentists as an example.

LEARNING TIP

Light Reversibility
Figure 7(a) shows an example of the reversibility of light rays. You discovered experimentally that light rays parallel to each other reflect through the focal point. If the rays are reversed, that is, they start from the focal point and strike the mirror, they reflect parallel to each other.

DID YOU KNOW ?

A Parabolic Liquid Mirror
Researchers at the University of British Columbia have built a reflecting telescope whose mirror is a liquid. The 6-m mirror consists of a thin layer of liquid mercury, which is highly reflective. The mirror stays horizontal to prevent spillage and spins at a rate of 7 rotations per minute. The spinning forces the normally flat liquid surface into a parabolic shape. Because the telescope is aimed vertically, only a small portion of the sky can be seen. Thus, images photographed on successive nights are combined by computer to obtain a larger image.

Satellite Technology

Canada has long been a world leader in satellite technology. This is due directly to the large size of our country and the need to communicate over large distances. Most Canadians live in the southern part of the country, within a few hundred kilometres of the border with the United States. But many others are scattered over vast areas, often far from any city or town. People who live and work in these remote areas rely on satellites to keep in contact with the rest of the country and the world. Satellite technology uses parabolic reflectors on Earth and on the satellites to send and receive EM signals (**Figure 8**).

In 1962, Canada became the third country to send a satellite into space. This satellite, *Alouette 1*, was used for scientific research: to study particles in Earth's upper atmosphere. *Alouette 1* was an example of a satellite in *low-Earth orbit*, which is an orbit just above Earth's atmosphere, at an altitude of 200 km–1000 km. Low-Earth orbits may be around the equator, around the poles, or at any angle between the equator and the poles. At speeds of approximately 28 000 km/h, low-Earth-orbit satellites take only about 1.5 h to orbit Earth.

CAREER CONNECTION

With the current digital revolution, the field of telecommunications has become very exciting, especially in the area of wireless products. Specialists in this field design, service, and sell high-tech electronic products. Of equal importance to this field are the telecommunications specialists needed to install and maintain the actual communications systems.

 www.science.nelson.com

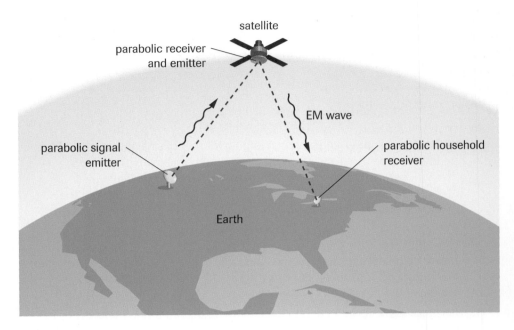

satellite

parabolic receiver
and emitter

EM wave

parabolic signal
emitter

parabolic household
receiver

Earth

Figure 8
The parabolic dishes used for sending and receiving EM signals have the emitter–receiver located at the focal point of the parabolic dish. Household dishes have only the receiver.

Figure 9
This modern satellite dish is less than 50 cm in diameter. Compare this diameter with that of older satellite dishes.

Satellites in low-Earth orbits are not used for TV communications. Imagine a television satellite dish, such as the one shown in **Figure 9**, trying to receive signals from such a satellite. The dish would have to follow the satellite as it raced across the sky. Then, when the satellite disappears below the horizon, all signals would be lost. For a satellite dish to receive signals continuously, the satellite must remain in the same location above Earth's surface.

> **TRY THIS** activity   **Dish Hunt**

Look at several satellite dishes in your area, and answer these questions:

• What size, shape, and design are the dishes?

• Which way are they pointed?

• Are they all pointed in exactly the same direction?

Research uses of satellites dishes. Explain why the dishes have the characteristics you observed. (Refer to shapes of curved reflectors, focal length, focal point, and wavelengths of waves.)

 www.science.nelson.com

The orbit of such a satellite is called a *geosynchronous orbit*. "Geo" means Earth, and "synchronous" means taking place at the same rate. Since it takes Earth 24 h to rotate once on its own axis, a satellite in geosynchronous orbit must take 24 h to orbit Earth. From the ground, it appears as if the satellite is not moving at all. The easiest place to control such an orbit is directly above Earth's equator, at an altitude of 36 000 km above sea level, which is much higher than the altitude of low-orbit satellites. To keep its position at this altitude, the satellite must travel at a speed of 11 060 km/h (**Figure 10**).

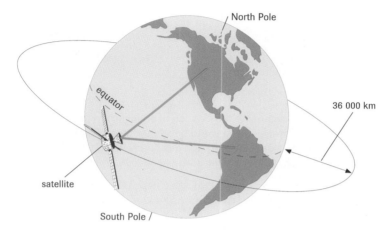

Figure 10
All geosynchronous orbits lie above the equator. A satellite orbiting the equator receives signals from Earth-based transmitters and sends signals back to Earth-based receivers. (The scale for Earth is larger than the scale for the orbit.)

The first Canadian domestic communications satellite was *Anik A1*, launched into geosynchronous orbit in 1972. One of the most recent technological advances in the *Anik* series is a satellite that sends signals to and receives signals from mobile telephones in cars, trucks, ships, and airplanes. Similar satellites provide radio service to remote areas of Canada.

Some satellites that are not used for TV communications travel in orbits somewhat lower than geosynchronous orbits. For example, satellites used for search-and-rescue operations travel in 12-hour orbits approximately 20 000 km above Earth's surface. When a ship or an aircraft is lost or at risk, an onboard device can be activated to start transmitting signals. The satellites receive the signals and send them back to Earth. The signals reveal the location of the vessel so that rescuers can try to reach it. A network of 24 of these satellites forms the *Global Positioning System (GPS)*.

The GPS can determine the position of an object on Earth's surface to within approximately 10 m. The boat in **Figure 12** has a computer-controlled GPS receiver that detects signals from each of three satellites. These signals help determine the distance between the boat and the satellite using the speed of the signal and the time it takes for the signal to reach the boat. If the boat were able to receive a signal from a fourth satellite, the boat's altitude above sea level could also be determined. The cost of a GPS receiver has fallen to the point where it is offered as a feature in some cars, boats, and airplanes. It is also used by hikers in remote areas.

DID YOU KNOW?

Communications Airships
Floating airships are an alternative to satellite technology for two-way Internet communications. A Canadian company, 21st Century Airships, is a leader in the design of these airships. These spherical airships float at an altitude of approximately 19.8 km, about twice the cruising altitude of commercial aircraft. They are highly manoeuvrable, use solar energy to operate slow-moving propellers, and are easily retrieved for repair and maintenance.

Figure 11
The airships will provide communication links in North America at a significantly lower cost than that provided by satellite systems.

 www.science.nelson.com

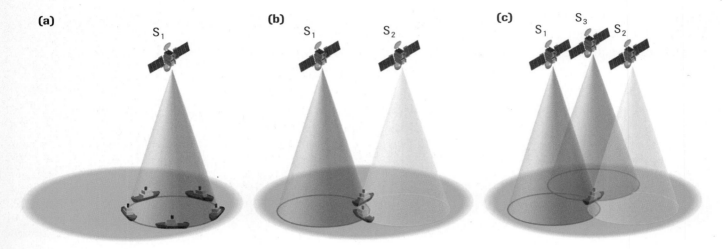

(a) S_1

(b) S_1 S_2

(c) S_1 S_3 S_2

Figure 12
GPS satellites can be used to determine the location of an object, in this case a boat. **(a)** Using only one satellite, the boat's location is known to be somewhere along the circumference of a circle. **(b)** Using two satellites simultaneously, the location is found to be at one of two intersection locations. **(c)** Using three satellites simultaneously, the intersection of three circles is the exact location of the boat.

The future of Canadian satellite communications is exciting. The *Anik F1* satellite, launched in 2000, was the first in the *Anik F* series (**Figure 13**). These satellites will provide various communications applications for many years, for example, multimedia services such as tele-medicine, tele-learning, tele-working, e-commerce, and high-speed Internet. Canada is also an active partner in communications multimedia projects with the European Space Agency.

Figure 13
The *Anik F1* being tested in an anechoic chamber, where all the walls, the floor, and the ceiling are lined with a type of sound-absorbing material. Anechoic chambers are used to measure the sound power of machines and the frequency responses of microphones and loudspeakers.

▶ **Practice**

Understanding Concepts

13. How does the expression "Necessity is the mother of invention" apply to Canadian communications satellite technology?

14. Assuming that Earth's radius is 6.4×10^3 km, calculate the speed in kilometres per hour of each of the following. (The circumference of a circle is $C = 2\pi r$.)
  (a) A geosynchronous satellite at an altitude 3.6×10^4 km orbits Earth in 24 h.
  (b) A satellite at an altitude of 2.0×10^2 km orbits Earth in 1.5 h.
  (c) Compare these speeds with those given in the text.

15. What is the GPS? What is its function?

Applying Inquiry Skills

16. (a) Cup one hand so it has the shape of a receiver–transmitter dish on an orbiting satellite. From a standing position and with your arm with the cupped hand stretched out, move in such a way as to model a geosynchronous orbit of your hand (the satellite) around your head (Earth). Describe the motion.
  (b) With a partner, create a way to model a satellite in low-Earth orbit. Describe the motion.

Answers

14. (a) 1.1×10^4 km/h
    (b) 2.8×10^4 km/h

SUMMARY   *Reflection of Electromagnetic Waves*

- Light reflecting off smooth, shiny surfaces produces regular reflection; light reflecting off irregular surfaces produces diffuse reflection.

- The laws of reflection state that the angles of incidence and reflection are equal ($\theta_i = \theta_r$), and that the incident ray, the normal, and the reflected ray lie in the same plane.

- Diverging mirrors are convex in shape; converging mirrors are concave in shape. In both cases, the focal length is the distance from the focal point to the reflecting surface.

- Plane, diverging, and converging mirrors have many applications.

- Parabolic mirrors overcome distortion effects of circular or spherical mirrors.

- Canada's contributions to satellite technology are many, including low-Earth-orbit satellites used for analyzing Earth's atmosphere and geosynchronous satellites used for communications.

Understanding Concepts

1. State the angle of reflection in each of the following cases:
   (a) The angle of incidence is 23.0°.
   (b) The angle between the incident ray and the reflected ray is 90.0°.
   (c) The incident ray is perpendicular to the mirror.

2. Optometrists view the inside of a patient's eye by using an ophthalmoscope. One version of this instrument, illustrated in **Figure 14**, has a small but bright light source that directs light to a small mirror. The light travels from the mirror to the eye, and the physician views the illuminated eye through a small opening in an opaque cover. Copy the top part of the diagram, that is, the light reflecting off the mirror, into your notebook, and measure the angles of incidence and reflection.

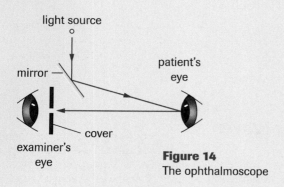

Figure 14
The ophthalmoscope

3. A person is moving toward a plane mirror at a speed of 20 cm/s.
   (a) At what speed is the person's image behind the mirror approaching the mirror?
   (b) At what speed are the person and the image approaching each other?

4. After a snowfall, a driver prints the word "snow" in the snow on a car's rear window. What does the driver see when looking into the plane rear-view mirror from the driver's seat?

5. Explain why many types of light bulb are frosted.

6. Draw a sketch showing how you would use several small plane mirrors to make a large parabolic concave mirror.

7. What is the relationship between the size of a parabolic reflector used for communications purposes and the wavelength of the EM waves it reflects? (If necessary, refer to the EM spectrum in **Figure 5**, page 488 of section 10.1, and **Figure 7**, page 500.)

8. Explain the disadvantage of using a satellite in low-Earth orbit as a TV communications satellite.

9. Set up a table to compare satellites in low-Earth orbits with satellites in geosynchronous orbits. Include the altitude, period, speed, and use of each type of satellite.

10. What is the minimum number of satellites needed to pinpoint a location on Earth's surface using the GPS? Explain why.

Applying Inquiry Skills

11. (a) Draw and label a diagram of a metal tablespoon to show how it can act as a model of both diverging and converging mirrors.
    (b) Describe how you would hold the spoon to view an image of your eye that is (i) small and inverted, (ii) small and upright, and (iii) large and upright. Check your answers experimentally.

12. Using your protractor, draw a semicircle in your notebook. The semicircle is a model of circular diverging and converging mirrors used in ray box experiments.
    (a) Label the diverging and converging sides of the semicircle.
    (b) Draw incident and reflected rays on the diverging side, and locate the focal point.
    (c) Draw incident and reflected rays on the converging side. Show the disadvantage of this type of mirror.

Making Connections

13. On the diverging mirror attached to a car's front passenger door is a warning: "Objects in mirror are closer than they appear."
    (a) What does the statement mean?
    (b) How could the statement be clarified?

14. For safety reasons, a single mirror is to be installed near the exit of an underground parking garage.
    (a) What type of mirror would you install? Explain why.
    (b) Draw a sketch showing the design of the installation.

15. (a) Look up the prefix "tele" in a dictionary, and write down its origin and meaning.
    (b) Relate your answer in (a) to applications such as tele-medicine and tele-learning.

Have you ever noticed a distorted view of your legs as you walk in clear, waist-deep water? Or have you ever wondered how a highway can appear to have a pool of water on the road ahead on a hot, dry day (**Figure 1**)? These effects are caused by the bending of light. This bending of light as it travels at an angle from one medium to another is called **refraction**.

Figure 1
The shimmering image of water ahead of you is a result of the refraction of light in air.

refraction the bending of light as it travels at an angle from one medium to another of different density

▶ TRY THIS activity   The Effects of Refraction

For this activity, you will need a coin, an opaque evaporating dish, a small beaker, and some water. Place the coin in the middle of the base of the evaporating dish, and position your eyes at a level where you just miss seeing the coin (**Figure 2**). Slowly add water to the dish without moving the coin, and observe the results. Explain your observations, given that your eyes "believe" that light travels in straight lines.

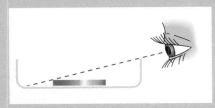

Figure 2
The line of sight before water is added

🤚 Do not use a chipped glass beaker.

Index of Refraction and Snell's Law

When light travels from air into a block of glass at an angle other than 90° to the surface, it refracts toward the normal, as shown in **Figure 3(a)**. (Some of the light is reflected off the surface of the block, but we will ignore that.) The reason the light is refracted as shown is that its speed decreases when it enters the glass from the air. This change in speed is illustrated in **Figure 3(b)** by a set of wheels travelling from a paved surface into sand. One wheel of the set reaches the sand first and slows down. The other wheel is still on the pavement and continues travelling at the original speed. The result is that the set of

(a)

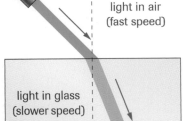

normal

light in air
(fast speed)

light in glass
(slower speed)

(b)

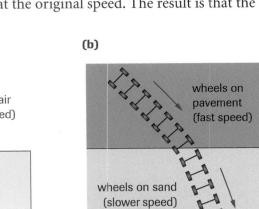

wheels on
pavement
(fast speed)

wheels on sand
(slower speed)

Figure 3
(a) Light is refracted as it travels from one medium to another of different density.
(b) The set of wheels travelling from a paved surface into sand illustrates how the change in direction results.

wheels changes its direction of motion. When light travels in the opposite direction, that is, from the glass toward the air at an angle other than 90° to the surface, it refracts away from the normal. This occurs as the light that emerges into the air increases its speed.

Light travels at different speeds in different transparent media. Thus, when light is refracted upon entering one medium from another, the amount of refraction depends on the relative speeds of light in the two materials. For example, light is refracted more when entering glass from air at a given angle than when entering water from air at the same angle. We therefore say that glass has a greater *optical density* than water. Greater optical density refers to the fact that light travels more slowly in the glass. (Optical density is not directly related to physical density, mass, or volume.)

The ratio of the speed of light in a vacuum (c) to the speed of light in a given medium (v) is called the **index of refraction**, n, of that medium. Thus,

index of refraction the ratio of the speed of light in a vacuum to the speed of light in a given medium; symbol n

$$n = \frac{c}{v}$$

The units for the ratio of the speeds cancel out, so the index of refraction has no units. **Table 1** lists the speed of light in various media and the indexes of refraction for those media. Notice that as the speed of light in various media decreases, the index of refraction increases.

Table 1   The Speed of Light and Indexes of Refraction*

Medium	Speed of Light in Medium (m/s)	Index of Refraction, n
air (at 101.3 kPa)	3.00×10^8	1.00
ice	2.31×10^8	1.30
water	2.26×10^8	1.33
ethyl alcohol	2.21×10^8	1.36
benzene	2.00×10^8	1.50
acrylic	2.00×10^8	1.50
glass		
quartz	2.05×10^8	1.46
crown glass	1.97×10^8	1.52
light flint	1.90×10^8	1.58
heavy flint	1.82×10^8	1.65
zircon (a gemstone)	1.58×10^8	1.90
diamond	1.24×10^8	2.42

* Values are average values for white light and are given for 0 °C.

► **SAMPLE** problem **1**

Using Speeds to Determine Index of Refraction

Light travels in glycerin at a speed of 2.04×10^8 m/s. Calculate the index of refraction of glycerin.

Solution

$c = 3.00 \times 10^8$ m/s

$v = 2.04 \times 10^8$ m/s

$n = ?$

$$n = \frac{c}{v}$$

$$= \frac{3.00 \times 10^8 \text{ m/s}}{2.04 \times 10^8 \text{ m/s}}$$

$$n = 1.47$$

The index of refraction of glycerin is 1.47.

DID YOU KNOW ?

Skillful Fish
Certain fish have adapted very
well to the effects of the refraction
of light when looking upward from
water toward air. For example,
some types of fish can shoot a
high-pressure jet of water from the
mouth into the air toward a
stationary prey, such as an insect,
located up to 3 m above the
surface of the water. The fish takes
aim from beneath the surface,
where refraction of light must be
considered. When the fish scores a
"hit," the startled prey falls into the
water, and the fish scoops it up.

In the seventeenth century, Willebrord Snell (1591−1626), a Dutch mathematician, analyzed refraction experimentally and mathematically using incident and refracted rays and angles (**Figure 4**). His analysis led to a way of calculating the index of refraction of a transparent medium by finding the ratio of the sine of the angle of incidence to the sine of the angle of refraction. This ratio is constant for any pair of materials, and is equal to the index of refraction or the ratio of speeds. The discovery is summarized in a statement that honours Snell's contribution, called *Snell's law of refraction*.

Snell's Law of Refraction
The ratio of the sine of the
angle of incidence to the sine
of the angle of refraction is
constant and is equal to the
index of refraction. In
equation form,

$$n = \frac{\sin \theta_i}{\sin \theta_R}$$

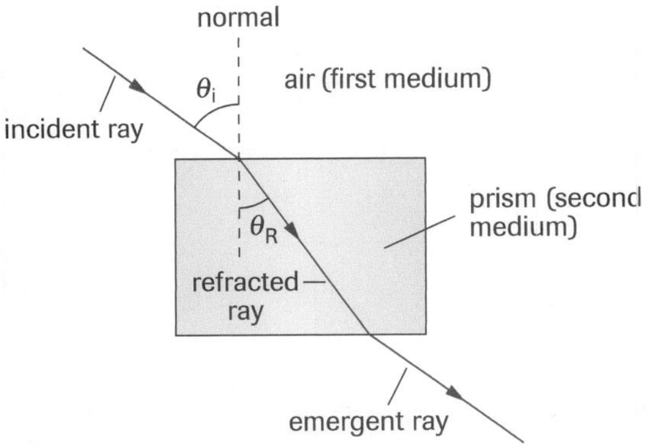

Figure 4
To analyze refraction mathematically,
the angle of incidence (θ_i) and the
angle of refraction (θ_R) must be known.

It is difficult to find the speed of light in a diamond crystal or plastic prism by investigating it directly in a physics lab. However, Snell's law of refraction can be applied to allow us to find the speed because the index of refraction equals both the ratio of the speeds and the ratio of the sines of the angles:

$$n = \frac{c}{v} = \frac{\sin \theta_i}{\sin \theta_R}$$

Using Angles to Calculate Index of Refraction and Speed of Light

In a refraction experiment, light travels from air into acrylic such that the angle of incidence is 50.0° and the angle of refraction is 30.7°. Calculate

(a) the index of refraction of the acrylic

(b) the speed of light in the acrylic

Solution

(a) $\theta_i = 50.0°$

$\theta_R = 30.7°$

$n = ?$

$$n = \frac{\sin \theta_i}{\sin \theta_R}$$

$$= \frac{\sin 50.0°}{\sin 30.7°}$$

$$n = 1.50$$

The index of refraction of the acrylic is 1.50.

(b) $c = 3.00 \times 10^8$ m/s

$n = 1.50$

$v = ?$

$$n = \frac{c}{v}$$

Solving for v:

$$v = \frac{c}{n}$$

$$= \frac{3.00 \times 10^8 \text{ m/s}}{1.50}$$

$$v = 2.00 \times 10^8 \text{ m/s}$$

The speed of light in the acrylic is 2.00×10^8 m/s.

▶ **Practice**

Understanding Concepts

1. A ray of light in air enters glass.

   (a) What happens to the speed and direction of the light ray if the angle of incidence is 0°?

   (b) What happens to the speed and direction if the angle of incidence is greater than 0°?

2. Calculate the index of refraction for light travelling from air into a medium in which the speed of light is (a) 2.40×10^8 m/s and (b) 1.80×10^8 m/s.

3. Calculate the speed of light in media with indexes of refraction of (a) 1.63 and (b) 1.20.

Answers

2. (a) 1.25

  (b) 1.67

3. (a) 1.84×10^8 m/s

  (b) 2.50×10^8 m/s

4. Light rays in air are aimed at an angle of 45° to the surfaces of two different substances, ice and zircon, both of which are listed in **Table 1**, page 508. In which substance is the amount of refraction greater? Explain your answer.

5. A ray of light is aimed from air into three different materials, X, Y, and Z, such that the angle of incidence in each case is 56.0°.
   (a) Determine the index of refraction of X, Y, and Z, given that the angles of refraction are 26.0°, 40.0°, and 30.0°, respectively.
   (b) Using **Table 1**, determine the likely identity of materials X, Y, and Z.

6. **Figure 5** shows a ray of light travelling from air into a liquid. Determine the speed of light in the liquid.

7. According to **Table 1**, which type of glass is high-index glass? What is its index of refraction?

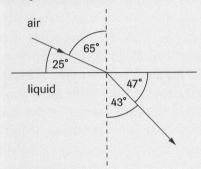

Figure 5

8. Two swimmers, G and B, stand at the edge of a clear lake and look at a rock in the water (**Figure 6**). B says that if he stands on the rock, the water will be at his waist. G disagrees because she thinks the rock is too far below the surface.
   (a) According to the diagram, who is right?
   (b) Why was the other swimmer wrong? (Use a diagram with light rays to explain your answer.)

Figure 6

Applying Inquiry Skills

9. Place a solid object such as a stick, pencil, or ruler in a water-filled beaker or other clear container. Observe the object from various directions including above and below, and explain what you observe.

Total Internal Reflection

For some communications applications, such as fibre optics, it is desirable to have light reflecting completely inside the material. This can happen only when the material is surrounded by material of lower optical density. The effect, called *total internal reflection*, can be explained by considering refraction.

You have learned that as light travels from a material of higher optical density (such as acrylic) into air, it is refracted away from the normal. This means that the angle of refraction in air, θ_R, is greater than the angle of incidence, θ_i, in the more dense medium. This observation is shown in **Figure 7(a)** for light travelling from acrylic into air.

In **Figure 7(b)**, the angle of incidence in the glass has increased, and the angle of refraction in the air is almost 90°. When this happens, the white light splits up into the colours of the rainbow. Also, some light is internally reflected; in other words, some light is reflected off the inside surface of the glass.

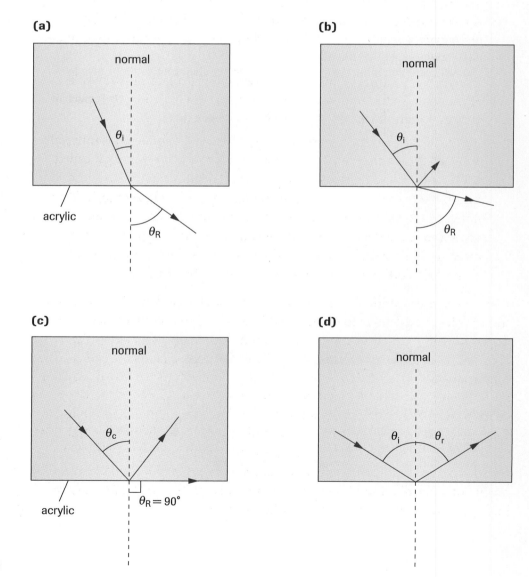

(a)

(b)

(c)

(d)

Figure 7
(a) The angle of refraction, θ_R, is greater than θ_i but less than 90°.
(b) The angle of refraction, θ_R, is approaching 90°.
(c) The angle of refraction, θ_R, is perpendicular to the normal.
(d) Total internal reflection, $\theta_i = \theta_R$

As the angle of incidence inside the glass increases, less light refracts into the air and more reflects inside the glass. At a certain angle, the refracted light disappears along the surface of the glass, and all the light reflects internally. The minimum angle of incidence in the glass or other medium that results in *total internal reflection* is called the **critical angle**, θ_c, as in **Figure 7(c)**. **Total internal reflection** is the reflection of light in a transparent medium that occurs for angles of incidence in the medium equal to or greater than the critical angle, as illustrated in **Figure 7(d)**. (You can observe refraction and total internal reflection and determine the critical angle in different media in Investigation 10.5.)

Figure 8 illustrates four applications of total internal reflection. Note that prisms—not mirrors—are used. Plane mirrors tarnish easily and do not last as long as prisms. They also absorb more light energy that prisms. When mirrors are either inconvenient or unsatisfactory for reflection, prisms are used.

- **Figure 8(a)** shows a prism periscope, a device used in submarines to view the seascape above the surface of the water while the submarine remains submerged. This periscope has two prisms that reflect light internally.

- A bicycle reflector, shown in **Figure 8(b)**, uses prisms to reflect light. Light from a vehicle behind the bicycle strikes the reflector and bounces back, alerting the driver.

- Internally reflecting prisms are also used in binoculars, illustrated in **Figure 8(c)**. Without prisms, the binoculars would have to be longer to give an upright image under the same magnification.

- **Figure 8(d)** shows solid plastic tubing in which a light beam is internally reflected every time it strikes an inside surface. This technology, called **fibre optics**, is used in the transmission of telephone, television, and Internet signals on laser beams in solid, thin fibres (**Figure 8(e)**). Fibre optics is also valuable in industry and medicine. For example, an arthroscope is a device with thin, flexible tubes linked to video or still cameras. It is used to view internal parts of machines or human bodies. As a result, cutting or major surgery can often be avoided.

Have you ever wondered why diamonds sparkle more than glass or zircon stones? The critical angle for a diamond is 24.4°. Light rays that enter a diamond will be totally internally reflected if they strike a surface on the diamond at an angle greater than 24.4°. Keep in mind that a diamond is cut in such a way that when there is even a very slight motion relative to an observer, the light enters and exits from a different surface. Thus, the chances are very good that light entering a diamond will be totally internally reflected many times before it exits. The result is a sparkling effect as the diamond is tilted in the light (**Figure 9**). Fake diamonds, often made of the gemstone zircon, have a critical angle that exceeds 30°, so the sparkling effect is not as noticeable. The quality of a fine diamond is determined not only by its purity (lack of flaws), but also by the craftsmanship of its polished faces.

critical angle the minimum angle of incidence of light inside a transparent medium that results in total internal reflection; symbol θ_c

total internal reflection the reflection of light that strikes the interior surface of a transparent medium at angles equal to or greater than the critical angle

fibre optics the study of the transmission of light in solid, transparent fibres

(a)

(b)

(c)

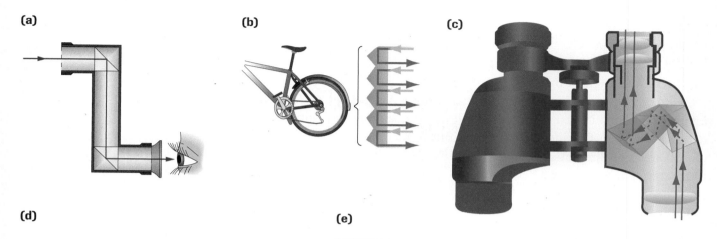

(d)

incoming ray

outgoing ray

Figure 8
(a) Prism periscope
(b) Bicycle reflector
(c) Prism binoculars
(d) Light in a solid glass or plastic fibre follows the fibre even around corners.
(e) Laser light transmitted through thin, solid fibres carries many more telephone calls at once, with much less energy loss, than is possible with numerous thick copper wires.

(e)

Figure 9
The sparkle of a diamond is produced by the refraction and total internal reflection of light in the diamond.

▸ **Practice**

Understanding Concepts

10. Name the two conditions required for total internal reflection.

11. The fibres used in fibre optics are surrounded by a thin, transparent film.
  (a) Should the film have a lower or higher optical density than the central fibre? Explain your answer.
  (b) What is the function of the film?

12. Would it be possible to make a prism periscope like the one shown in **Figure 8(a)** using a material whose critical angle is 48°? Explain your answer.

Applying Inquiry Skills

13. (a) Describe the demonstration in **Figure 10**.
  (b) What arrangement must be made to ensure that the demonstration works properly?

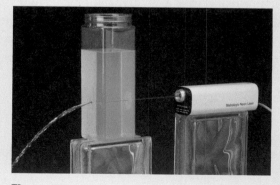

Figure 10

| SUMMARY | *Refraction and Total Internal Reflection* |

- Refraction is the bending of light as it travels at an angle from one medium into another in which the speed of light differs.

- The index of refraction of a medium is the ratio of the speed of light in a vacuum to the speed of light in the medium, $n = \dfrac{c}{v}$.

- Snell's law states that the index of refraction equals the ratio of the sine of the angle of incidence to the sine of the angle of refraction; $n = \dfrac{\sin \theta_i}{\sin \theta_R}$.

- To achieve total internal reflection, a transparent medium must be surrounded by one that is less optically dense, and the angle of incidence in the medium must be equal to or greater than the critical angle.

- The critical angle is the minimum angle of incidence of light inside a transparent medium that results in total internal reflection.

- Total internal reflection has useful applications, for example, in fibre optics.

▶ **Section 10.4 Questions**

Understanding Concepts

1. (a) Under what condition(s) does light refract when it travels from one medium into another?
   (b) Why does this refraction occur?

2. In each case, determine the index of refraction of the medium:
   (a) Light travels at a speed of 2.24×10^8 m/s in the medium.
   (b) Light in air strikes a medium at an angle of incidence of 60.0° and refracts in the medium at an angle of 30.0°.

3. In each case, determine the speed of light in the medium:
   (a) $n = 1.52$
   (b) θ_i (in air) $= 45.0°$; $\theta_R = 25.9°$

4. Describe two advantages of using fibre optics for transmitting communications signals rather than wire cables.

Applying Inquiry Skills

5. Describe how you would determine the speed of light in a clear liquid in the lab. Include the materials, procedure steps, and calculations needed.

Making Connections

6. Zircon gemstones are often used instead of diamonds in jewellery. The critical angle for light in a zircon gemstone is 31.8°.
   (a) Which gemstone can sparkle more? Why?
   (b) Why is diamond generally more expensive than zircon?

7. Much of the development of communications technology occurred on the Atlantic coast of North America. Why was the development of wireless communications important for the shipping and fishing industries, and for national security?

Refraction and Total Internal Reflection

In this investigation, you will apply Snell's law (presented in section 10.4) to determine the index of refraction of a solid and a liquid. You will explore the conditions required for light to be totally internally reflected, and measure the critical angle in a solid and a liquid.

Ray diagrams for the refraction of light resemble those for the reflection of light. **Figure 1** shows a typical labelled ray diagram of light that is refracted twice, first as it leaves the air and enters a rectangular prism, then as it leaves the prism and returns to the air. The normals are drawn perpendicular to the surfaces from which the rays enter and leave the prism. The angles of incidence (θ_i), refraction (θ_R), and emergence (θ_e) are measured from the normal to the ray.

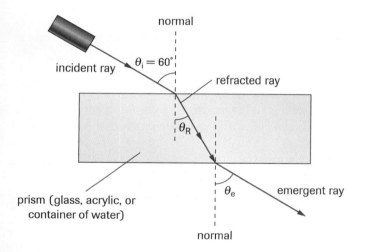

Figure 1
A labelled ray diagram of the refraction of light into and out of a rectangular prism

LEARNING TIP

Sines of Angles
Some of the calculations in this investigation require you to find the sines of angles using your calculator. Be sure your calculator is in the degree mode. To review angles and applying the sines of angles, refer to Appendix A1.

Inquiry Skills

○ Questioning	● Conducting	● Evaluating
● Predicting	● Recording	● Communicating
○ Planning	● Analyzing	○ Synthesizing

Questions

How can Snell's law be applied experimentally to determine the index of refraction of a medium? How can the critical angle of light in a medium be determined experimentally?

Predictions

(a) Predict an answer to each Question above.

Materials

For each student:
metric ruler
protractor

For each group of two or three students:

For Part A

ray box with single-slit window
rectangular solid prism (glass or acrylic)
polar graph paper
thin, transparent rectangular dish to hold water
water

For Part B

ray box with single-slit and double-slit windows
semicircular solid prism (glass or plastic)
semicircular plastic dish to hold water
water
two triangular solid prisms with angles of 45°, 45°, and 90°

 Handle glass components carefully; cut glass is very sharp.

Prevent water from spilling.

Do not use chipped glass prisms.

Do not touch the ray box light bulb or look directly into the light.

Turn off the ray box when not in use.

Do not allow the ray box cord to hang where someone might walk into it.

Be careful when touching the ray box after use; it may be hot.

Unplug the ray box by pulling on the plug, not on the electrical cord.

Allow the ray box to cool before putting it away.

Procedure

Part A

1. Set up a data table based on **Table 1**.

2. Place the solid prism on a piece of paper and draw its outline. Remove the prism so that you can draw a normal (broken line) and an incident ray (solid line) with $\theta_i = 60°$, as shown in **Figure 1**. Label the angle and lines.

3. Place the prism back on the diagram, and aim a single ray from the ray box along the incident ray. Draw the ray that emerges on the opposite side of the prism, then remove the prism and draw the entire path of the light. Draw a second normal at the surface where the light emerges from the prism. Measure and label the angle of refraction (θ_R) in the prism and the angle of emergence back into the air (θ_e). Enter the measured values in your data table.

4. Repeat steps 2 and 3 using water in a transparent container.

Part B

5. Place the semicircular solid prism on a piece of polar graph paper so that the middle of the flat edge of the prism is at the centre of the paper. Draw the outline of the prism. Aim a single light ray ($\theta_i = 30°$) from the curved side of the prism directly toward the *middle* of the flat edge, as shown in **Figure 2(a)**. Complete your drawing, including the normal and all rays and angles.

6. Using the setup from step 5, slowly move the ray box to increase the angle of incidence until

rainbow colours appear in the air. Then move the box slightly farther, until the light in the air just disappears, as in **Figure 2(b)**. Mark the rays, remove the prism, and measure the angles of incidence and reflection inside the prism. Have your teacher check your values.

7. Using the same setup, determine what happens to a light ray aimed so that the angle of incidence inside the prism is greater than the angle found in step 6. Describe what you observe.

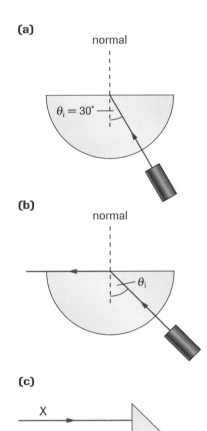

Figure 2
For step 5

Table 1 Data for Part A

Steps	Material	Angle of Incidence, θ_i	Angle of Refraction, θ_R	Angle of Emergence, θ_e	sin θ_i	sin θ_R	$\dfrac{\sin \theta_i}{\sin \theta_R}$
2, 3	acrylic	60°					
4	water	60°					

8. Repeat steps 5 to 7 using water in a semicircular plastic dish.

9. A prism periscope is an application of total internal reflection. To observe how light travels in a prism periscope, set up the two triangular prisms as in **Figure 2(c)**. Aim two rays, X and Y, as shown, and draw the paths they follow. Determine whether the final emergent rays are upright or inverted when compared to the incident rays.

Analysis

(b) Use your calculator to calculate the sine of each angle of incidence and refraction in Part A. Enter the data in your table. Then calculate the ratio $\dfrac{\sin \theta_i}{\sin \theta_R}$ for each ray to three significant digits. Enter the data in your table. For light entering acrylic or water, how does this ratio compare? (The ratio is equal to the index of refraction of the medium.)

(c) When light travels at an angle from a medium of low optical density (such as air) to one of higher optical density (such as water), is it refracted away from or toward the normal?

(d) When light travels at an angle from a medium of high optical density to one of lower optical density, is it refracted away from or toward the normal?

(e) How does the internal angle beyond which light is totally internally reflected in acrylic compare to the corresponding angle in water?

Evaluation

(f) How good were your predictions?

(g) Refer to **Table 1** in section 10.4, page 508, for the accepted indexes of refraction of acrylic and water. In each case, determine the percent error of your experimental value. To review experimental error, refer to Appendix A1.

(h) Describe the major sources of error in this investigation, and explain how you tried to minimize them.

Several examples of communications technology have been presented in this text. This section features a few more applications. Some of them involve lasers, and others link EM waves with electronics. You can also explore and evaluate a communications technology in Investigation 10.7.

Laser Technology

Most sources of light emit light as a result of the spontaneous acceleration of particles. The light energies have many different values, so the emitted light has many different wavelengths (**Figure 1(a)**). A **laser** is different: It emits light with only one wavelength (or a controlled set of discrete wavelengths), and the waves are in phase so the crests and troughs move along together, as illustrated in **Figure 1(b)**. The word *laser* stands for *l*ight *a*mplification by the *s*timulated *e*mission of *r*adiation.

Because of the single-wavelength, in-phase nature of laser light, lasers have unique characteristics that allow them to be used in numerous applications:

- Light from lasers spreads very little as it travels. The fine, straight laser light provides precise alignment during the construction of bridges, roads, tunnels, and skyscrapers.
- Lasers help survey rugged terrain that is difficult or impossible to approach.
- They are used in conjunction with prisms, lenses, and music for entertainment at concerts and spectacular indoor and outdoor laser shows.
- Lasers are improving the field of communications: Telephone and other messages are transmitted along optical fibres using the total internal reflection of laser light.
- They can be controlled to emit intense beams of light. This is useful in industry to drill fine holes and also in surgery (**Figure 2**).

laser a technology that emits light in phase and with one wavelength; acronym for *l*ight *a*mplification by the *s*timulated *e*mission of *r*adiation

(a)

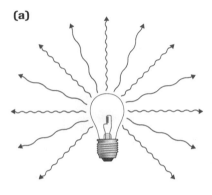

(b)

Figure 1
(a) White light from an incandescent source consists of waves of many wavelengths.
(b) Light from a laser consists of waves that are in phase and have a single wavelength.

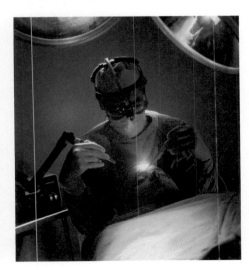

Figure 2
Two applications of lasers are **(a)** cutting steel and **(b)** treating skin conditions.

- In police work, lasers can help "see" fingerprints previously impossible to detect.
- Lasers are used in compact disc (CD) technology (**Figure 3**). Laser light does not damage the CD, so the disc lasts much longer than older technologies, such as vinyl records and magnetic tape. Digital videodiscs (DVDs) operate in a similar fashion.

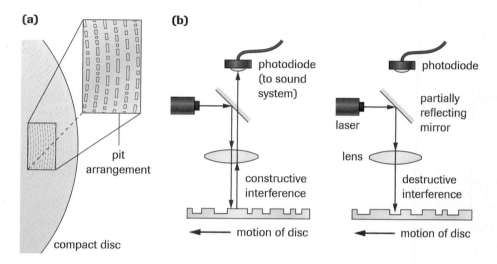

(a)

(b)

Figure 3
(a) Information is encoded onto the disc in the form of tiny pits arranged in a continuous data track.
(b) As the disc spins, the laser light that has been focused by the lens reflects off the raised parts; this reflected light is in phase, causing constructive interference and "on" signals. The reflection off the pits is out of phase, causing destructive interference and "off" signals. A photodiode receives the resulting on–off signals and changes them into electric signals to produce sound.

The electronics and computer industries are changing rapidly, in part because of the use of lasers. Tiny semiconductor lasers, no larger than a grain of salt, are combined with other optical devices such as lenses and prisms to produce an optical computer. Such computers, which operate on tiny pulses of light rather than electricity, perform operations thousands of times faster than electronic digital computers.

Holography is another exciting application of the laser. **Holography** is the process of making a three-dimensional photograph on a two-dimensional surface using the interference of laser light. In one holography technique, a laser beam is split into two parts called the *object beam* and the *reference beam*, as shown in **Figure 4(a)**. The reference beam goes to the photographic film. The object beam strikes the object to be photographed and is then reflected to the holographic film. At the film, the reference and object beams interfere, creating an interference pattern called a *hologram*. (The word "hologram" is derived from the Greek *holos*, meaning "whole," and *gramma*, meaning "something written or recorded.") When this pattern is illuminated by laser light identical to the reference beam, the image seen is a strikingly faithful reproduction of the original object (**Figure 4(b)**). Holograms produced by

holography the process of making a three-dimensional image on a two-dimensional surface using the interference of laser light

(a) 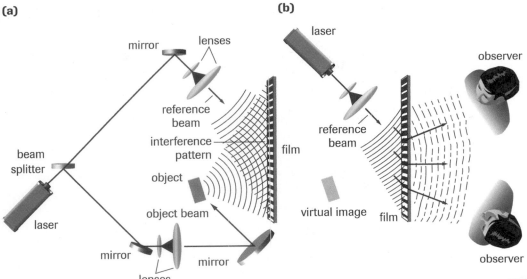 **(b)**

Figure 4
(a) Making a hologram
(b) Viewing a hologram

other techniques can be viewed in white light. These holograms are used as a security feature on many credit cards. A hologram is shown in **Figure 5**.

There are numerous uses for holography, and more are being developed. A common use may be found at your local supermarket checkout, where a laser scanner reflects light from a product's Universal Product Code (**Figure 6**). Light from the laser is aimed toward a rotating disc in the product scanner. This disc has 21 separate holograms, each of which directs a beam of light in a slightly different direction. Only one beam needs to reflect off the bar code to a photodetector. The photodetector transforms the optical signal into electrical signals, so that the product's name and price appear on the cash register screen and are printed on your receipt. At the same instant, the purchase information also goes to update the store's inventory.

Figure 5
Holographic artists can achieve dramatic effects.

Figure 6
The Universal Product Code is the bar code found on most packaged products.

▶ **TRY THIS** activity

A Simple Communications System

In a group, design a simple communications system, and describe how its main components operate. One example is an intercom system for an apartment entrance; another is a laser show, where low-power laser light reflects off a small mirror attached to a loudspeaker (**Figure 7**). Have your teacher approve your design, then build the system and demonstrate how it works.

 If the activity you design uses a laser, do not allow the beam to reflect toward anyone's eyes.

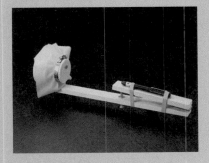

Figure 7
A sound and light show demonstration

Understanding Concepts

1. In what ways is laser light different from white light?

2. Describe the way in which holography is an example of an application of the interference of light.

3. Is the operation of a CD player digital or analog? Explain your answer.

Applying Inquiry Skills

4. Use a magnifying glass to view holograms on credit cards and paper money. Describe what you observe.

Making Connections

5. Holography has many practical applications.
   (a) What is the purpose of a hologram on a credit card?
   (b) Find other examples of the applications of holography using resources of your choice. Describe one example.

 www.science.nelson.com

Case Study Canadian Contributions to Early Communications

What do the following communications technologies have in common?

- first long-distance telephone call
- first Morse code radio signals transferred across an ocean
- first person to transmit a voice message using radio waves

The answer is that they are all part of Canadian communications history.

Alexander Graham Bell (1847–1922), working in his laboratory in Brantford, Ontario, developed the telephone (**Figure 8**). He made the first long-distance telephone call in 1876 from Paris, Ontario, to Toronto, a distance of about 110 km. That phone call was carried through wires.

Figure 8
Much of Alexander Graham Bell's research was done in his lab in Brantford, Ontario.

A 10-year-old boy named Reginald Fessenden (1866–1932) saw Bell at work in his lab. He became interested in the new technology and in creating his own inventions. Fessenden was born in Quebec but spent his youth in Ontario. As a young adult, he developed solutions to several problems related to electrical and communications devices. He was especially interested in long-distance communication using radio waves rather than wires. Like many people, he was excited by the first transatlantic radio message, received on the coast of Newfoundland in 1901 in the form of Morse code. The original signal, sent from the coast of England, was picked up by a radio receiver on Signal Hill just at the entrance to the harbour at St. John's (**Figure 9**).

Figure 9
Cabot Tower is located on Signal Hill outside St. John's, Newfoundland. For centuries, it served as the communications link between land and sea in the St. John's harbour.

Italian inventor Guglielmo Marconi (1874–1937) had designed the technology to send and receive these radio signals. At the time, most scientists believed that EM waves such as radio waves would not be able to travel more than 2000 km across an ocean. They reasoned that, even if you stand on top of the highest tower, you can't see another tower hundreds of kilometres away because of Earth's curvature. However, Marconi believed that radio waves would follow Earth's curvature. It was his radio receiver that picked up the radio signal sent from England, proving he was right.

To understand how these signals can cross the Atlantic Ocean, look at **Figure 10**, which shows Earth's atmosphere. As radio waves sent from one side of the ocean travel through the atmosphere, some refract into space. However, those that strike the part of the atmosphere called the ionosphere at a large enough angle reflect back toward Earth. These reflected signals can be received thousands of kilometres away.

Unlike many scientists of his time, Fessenden wanted to send voice and music messages as well as Morse code signals. It took six years of dedicated experimentation to create a working design. In 1906, Fessenden became the first person to broadcast a voice message using radio waves. His message was sent from Brant River, Massachusetts, to ships at sea.

DID YOU KNOW?

Reflection at Night
Radio waves constantly reflect off the upper part of the ionosphere, but the phenomenon is more commonly observed at night. Solar energy creates many more ions (charged particles) in the lower part of the ionosphere during the day than at night. The ions tend to absorb the radio waves, reducing the daytime reflection from higher in the ionosphere.

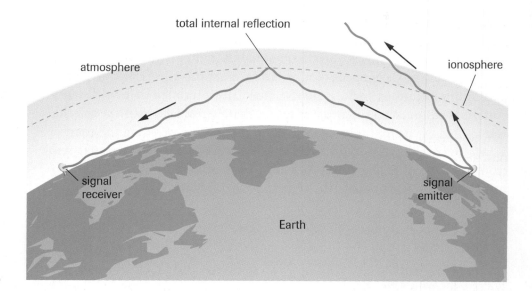

Figure 10
Radio waves can be reflected in the atmosphere if the waves strike the ionosphere at a great enough angle.

▶ **TRY THIS** activity

Signal Communications

Early communications sent from Signal Hill to ships approaching the harbour used hand signals, flags, and sound signals from cannons. At airports today, ground crew use arm signals to direct aircraft to a safe stop at the terminal gates. In a group, you can create your own hand or arm communications system to solve the problem outlined below.

Your task is to design a nonverbal communications system as follows: Map out a course that includes start and finish lines, at least 6 m apart, with several obstacles between them. One student uses visual signals to guide a second student from start to finish with the minimum number of moves. The second student must not know the destination. The signals allowed are as follows:

- one step forward or backward
- one step to the left or to the right
- size of step (either large or small)

No voice communication is allowed, so both the sender and receiver must memorize the agreed signals beforehand.

▶ **Practice**

Understanding Concepts

6. (a) Explain what is meant by wireless communication. How is it achieved?
   (b) Evaluate its importance to people who live in urban and remote areas of Canada.

Making Connections

7. On that famous day in 1901, when the first Morse code message was received in Newfoundland, Marconi flew a kite on Signal Hill with a wire antenna attached. How would this have helped his experiment?

8. Find out more about one of the inventors featured in this case study. Write a brief report using information that has not been presented here.

 www.science.nelson.com

Digital Imaging

One of the most widespread applications linking light and electronics is the *charge-coupled device* used in digital cameras and other technologies. A **charge-coupled device (CCD)** is a semiconductor chip used to convert light into electrical energy.

Digital imaging technology is different from traditional film photography. In film cameras, light strikes the film, which causes a chemical reaction that is recorded on the film. Once the film is developed using another chemical process, the film is set permanently.

In a digital camera, light enters the camera lens and passes through red, green, and blue filters. Then it strikes the CCDs, which are silicon semiconductors (**Figure 11(a)**). Numerous CCDs are arranged in an array forming small sections called *pixels*. Some cameras have 6 million pixels or more. When light hits the semiconductors, electrons are released, just as they are in a solar cell. The released electrons accumulate and become trapped in the pixels until a voltage change in the electrodes allows them to be transferred and recorded.

The accumulated charge is an analog signal based on light intensity that is transmitted to an analog-to-digital converter (**Figure 11(b)**). (Refer to section 8.8 for details about this technology.) The digital output consists of arrays of zeroes and ones that together represent the picture taken by the camera. The signals are stored on a disk, which allows the image to be viewed immediately, sent to a printer, stored on a computer, or manipulated to create changes as desired.

CCDs are used in digital camcorders, TV cameras, and electronic scanners, as well as in still cameras. Astronomers also find CCD technology especially useful for obtaining images of the universe because light can be gathered from very dim, faraway objects over periods of seconds or minutes rather than hours or days. They sometimes use CCDs with semiconductors sensitive to EM waves beyond the visible spectrum to study the universe. Some images help astronomers study the origin of the universe (**Figure 12**).

charge-coupled device (CCD)
a semiconductor chip that converts light into electrical energy

(a)

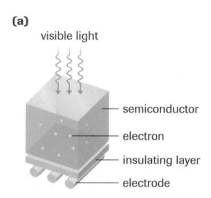

(b)

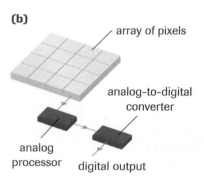

Figure 11
(a) The basic design of a CCD
(b) Pixels are linked to an analog-to-digital converter to create digital signals.

Figure 12
Light from the Cat's Eye Nebula began travelling billions of years ago and was captured for this image using CCD technology.

▶ **Practice**

Understanding Concepts

9. Describe the difference in the energy transformations that occur in film and digital photography.

10. Do CCDs create digital signals? Explain your answer.

11. Digital camcorders can obtain images in much dimmer lighting conditions than film recorders. Explain why.

12. Is the semiconductor in a CCD array doped with donor or acceptor atoms? How can you tell?

Flat-Panel Screens

Colour TVs, computer monitors, and video screens come in all shapes and sizes (**Figure 13**):

- hand-held video games with colour screens
- large, conventional television sets found in many homes
- TVs the size of a wristwatch
- flat-panel TVs that hang on a wall like a framed painting
- visors used in virtual reality headsets
- screens on personal digital assistants, on which the user can write with a special pen or a finger

In a conventional colour TV, the inside surface of the screen contains thousands of individual pixels in the shape of bars or dots. Arranged in sets of three, the pixels are made of phosphors that emit one of three colours (red, green, or blue) when hit by high-energy electrons. Three electron guns, one for each type of phosphor, are located near the back of the TV tube (**Figure 14**). At controlled instants, they send electrons toward the pixels on the screen. Each pixel struck by electrons gains energy, then emits that energy in the form of visible light.

TVs as thin as a few centimetres are called *flat-panel* TVs. They offer some advantages over conventional electron-gun TVs: They occupy a small space, they weigh much less, and they operate with much less electrical energy. However, at present, they are comparatively expensive.

One type of flat-panel TV uses liquid crystal diodes (LCDs) to control the light. In this type of TV, light comes from behind the screen and then passes through the screen to the front. On the way through the screen, however, the

(a)

(b)

Figure 13
(a) Hand-held video game
(b) Flat-panel TV screen

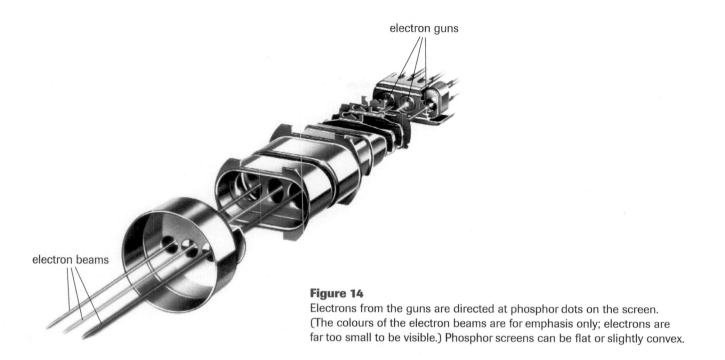

Figure 14
Electrons from the guns are directed at phosphor dots on the screen. (The colours of the electron beams are for emphasis only; electrons are far too small to be visible.) Phosphor screens can be flat or slightly convex.

light can be controlled by more than a million pixels. These pixels have electronic switches that either block the light or let it pass. Light that passes through the pixels then passes through one of three colours of filter: red, green, or blue. The pixels, electronic switches, and filters are all sandwiched between two flat glass plates (**Figure 15**).

TVs with LCD screens are safer than electron-gun TVs because they can operate at a lower voltage. Also, they require less electrical energy. Because they are very versatile, they can be used in the following ways:

- Small TV sets placed in airplanes, automobiles, buses, and trains provide travellers with entertainment, travel schedules, and weather reports.

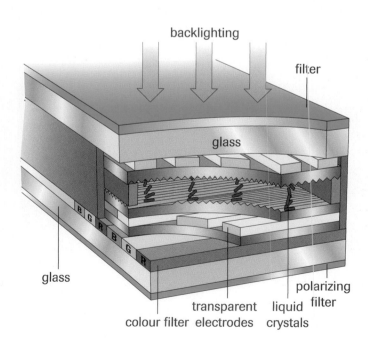

Figure 15
The basic structure of a flat-panel TV made with liquid crystals

- Screens can provide maps connected to the GPS in cars.

- Screens used for advertising can be fitted into almost any space.

- Portable notepads can be used for banking, paying bills, booking airplane flights, and communicating.

LCDs could make TVs more common and accessible than they already are. Some outcomes may not be good things:

- Children will be able to watch more TV and see more advertising than they already do.

- What will happen to the old TVs? The average North American home has more than two TV sets.

- Computers can be linked to your TV to produce interactive TV. However, when you are watching, someone may be watching you, collecting information on what you watch, read, and buy.

> **Practice**

Understanding Concepts

13. Starting with electrical energy, describe the energy transformations that occur to produce the light emitted from an electron-gun TV. Write the energy-transformation equation.

14. What are LCD screens?

15. Compare and contrast LCD screens and electron-gun TVs.

> **EXPLORE** an issue

Should Flat-Panel TVs Replace Conventional TVs?

Decision-Making Skills

- ○ Define the Issue
- ● Defend the Position
- ● Analyze the Issue
- ● Identify Alternatives
- ● Research
- ● Evaluate

Find out more about the benefits, drawbacks, and alternatives of flat-panel TVs in order to evaluate them. Decide on the criteria your group will use to evaluate the types of TVs available and their uses. Then create a position paper to defend your opinion regarding whether flat-panel TVs should replace conventional TVs. The paper can be presented as a poster board, an essay, a letter, a Web page, a video, a PowerPoint presentation, or an audio presentation. Your final report should contain your

research notes, your answers to Practice questions 13 to 15, and your position paper.

Evaluation

(a) Evaluate the usefulness of the references you researched.

(b) Do all students or groups in your class agree on this issue? Explain your answer.

SUMMARY — Communications and Electromagnetic Waves

- The special properties of laser light enable it to be used for many applications, for example, playing CDs and transmitting data through optical fibres.

- Canadian technology has made valuable contributions to the communications industry.

- Digital imaging uses charge-coupled devices (CCDs) to convert light energy into electrical energy; the electrical energy is changed to digital signals, which can be stored and manipulated.

- Flat-panel TVs use liquid crystal diodes (LCDs) with colour filters to produce colour images.

▶ Section 10.6 Questions

Understanding Concepts

1. Describe, with an example, how lasers are applied in communications technologies.

2. The helium–neon laser used in physics classrooms has a wavelength of 633 nm.
   (a) Express the wavelength in metres.
   (b) Calculate the frequency of the source of this radiation.

3. Write the energy-transformation equation for
   (a) film photography
   (b) digital photography

4. Compare and contrast film photography and digital photography.

5. Compare the colours of the filters or phosphors in digital camera technology, electron-gun TVs, and LCD screens.

Applying Inquiry Skills

6. Describe how you would use two laser beams to ensure that the height of bottles travelling along a conveyor belt is always the same. Draw a sketch to illustrate your design.

Making Connections

7. The study of the interrelationship between light and electronics is called *photonics*. Photonics is a fast-growing technology in Canada, with many career opportunities. For example, a college graduate in photonics can qualify for positions in maintenance, repair, and technical sales involving photo imaging, fibre optics, electronic printing, consumer electronics, and telecommunications. Visit a few of the many Web sites available, and make a list of careers in photonics of interest to you.

 www.science.nelson.com

10.7 Investigation

Analyzing and Evaluating a Communications Technology

Inquiry Skills
- Questioning
- Predicting
- Planning
- Conducting
- Recording
- Analyzing
- Evaluating
- Communicating
- Synthesizing

In this investigation, you will choose a communications technology that interests you and analyze available models of that technology using your own criteria. Based on your analysis and evaluation, you will recommend the model you think is best.

Some of the many communications technologies to choose from are listed below. You may think of other examples as well.

- cell phone
- home entertainment system
- computer system
- satellite versus cable TV service
- digital camera
- video camera
- flat-panel technology
- night-vision goggles
- infrared (IR) scanners
- black light theatre
- AM radio
- FM radio
- CD player
- DVD player
- pagers
- music kiosks
- laser shows
- IMAX versus regular movie theatre
- holography

Question

(a) Make up your own questions regarding your choice of technology and your criteria for evaluation.

Predictions

(b) Predict an answer to your question. Give reasons for your prediction.

Materials

(c) Make a list of the resources you intend to use, and add to the list as your research progresses.

Procedure

1. Refer to Appendixes A3 and A4 to brainstorm ideas about the factors you want to compare. List the factors, and describe how you intend to evaluate each one. For example, if the factor is economic, you can compare the initial costs and the operating costs over a period of time. If the factor is career possibilities, you can determine which careers are available locally and which are not. Have your teacher approve your list.

2. Research two or three models of your chosen communications technology. Create a portfolio of your research.

3. Combine your evaluation criteria with your research portfolio to summarize the advantages and disadvantages of the models.

4. Decide which model of the technology you would recommend. Create a report to support your recommendation.

Analysis

(d) Which physics principles apply to the type of communications technology you researched?

(e) What were the three most significant criteria you used in your evaluation?

(f) Describe the effect that the technology has had or could have on your life.

(g) Name three careers that relate to the manufacture and use of the technology.

Evaluation

(h) Evaluate the resources you used.

(i) If you did this investigation again, what changes would you make in the procedure? Why?

Synthesis

(j) What life skills have you learned while performing this investigation that will help you become a wise consumer?

Key Understandings

10.1 Light and Electromagnetic Waves

- The **visible spectrum** is the band of radiant energy colours visible to the human eye; it is part of the **electromagnetic (EM) spectrum**.

- Radiant energies travel in a vacuum at the speed $c = 3.00 \times 10^8$ m/s, obey the universal wave equation, $c = f\lambda$ or $v = f\lambda$, and display wave properties.

- Materials are **transparent**, **translucent**, or **opaque**.

- **Amplitude modulation (AM)** and **frequency modulation (FM)** are two ways of transferring communications signals using EM as carrier signals.

10.2 Investigation: Reflection of Light

- Image characteristics differ in plane, convex, and concave mirrors.

- The focal point and focal length of curved mirrors can be found experimentally.

10.3 Reflection of Electromagnetic Waves

- **Regular reflection** results when light reflects off smooth, shiny surfaces; **diffuse reflection** results when light reflects off irregular surfaces.

- The **laws of reflection** state that the angles of incidence and reflection are equal, and that the incident ray, the normal, and the reflected ray lie in the same plane.

- The **focal length** of a **diverging mirror** or a **converging mirror** is the distance from the **focal point** to the reflecting surface.

- **Parabolic mirrors** overcome the distortion effects of circular and spherical mirrors.

- Satellites have many applications.

10.4 Refraction and Total Internal Reflection

- When light travels at an angle from one medium into another in which the speed of light differs, the light is **refracted**.

- The **index of refraction** of a medium is the ratio of the speed of light in a vacuum to the speed of light in the medium.

- **Snell's law of refraction** states that the index of refraction equals the ratio of the sine of the angle of incidence to the sine of the angle of refraction.

- **Total internal reflection** of light occurs in a transparent material that is surrounded by one that is less optically dense when the angle of incidence of light in the material is greater than the **critical angle**.

10.5 Investigation: Refraction and Total Internal Reflection

- Light rays can be used to determine the index of refraction of a medium and the critical angle.

10.6 Communications and Electromagnetic Waves

- **Laser** light has a single wavelength and is in phase; this permits many applications, for example, **holography** and CD technology.

- Digital cameras use **charge-coupled devices (CCDs)** to transform light energy into electrical energy.

- Flat-panel TVs use liquid crystal diodes with colour filters to produce colour images.

10.7 Investigation: Analyzing and Evaluating a Communications Technology

- Various models of a communications technology can be analyzed and evaluated.

Key Terms

10.1
diffraction
visible spectrum
electromagnetic (EM)
 spectrum
transparent material
translucent material
opaque material
amplitude modulation
 (AM)

frequency modulation
 (FM)

10.3
regular reflection
angle of incidence
angle of reflection
laws of reflection
diffuse reflection
diverging mirror

focal point
focal length
converging mirror
parabolic mirror

10.4
refraction
index of refraction
Snell's law of refraction
critical angle

total internal reflection
fibre optics

10.6
laser
holography
charge-coupled device
 (CCD)

Key Equations

10.1

- $v = f\lambda$
- $c = f\lambda$

- $v = \dfrac{\lambda}{T}$ (or $c = \dfrac{\lambda}{T}$)

10.3

- $\theta_i = \theta_r$

10.4

- $n = \dfrac{c}{v} = \dfrac{\sin \theta_i}{\sin \theta_R}$

Problems You Can Solve

10.1

- Describe how the visible spectrum differs from the other portions of the EM spectrum.
- Apply the universal wave equation to determine either the frequency or the wavelength of an EM wave.
- Describe how radio signals are transmitted using AM and FM signals.

10.2

- Draw diagrams showing the reflection of light rays off a plane mirror, and determine and compare the angle of incidence and the angle of reflection.
- Use parallel rays to locate the focal point of convex, concave circular, and concave parabolic mirrors, and measure their focal lengths.

10.3

- Distinguish between diverging and converging mirrors, and describe uses for each type of mirror.
- Describe the advantage of a converging parabolic mirror over a converging circular or spherical mirror.
- Distinguish between satellites in low-Earth orbits and those in geosynchronous orbits, and state applications for each type of satellite.

10.4

- Given the angle of incidence in air and the angle of refraction in a medium, determine the index of refraction of the medium.
- Describe the conditions required for total internal reflection in a medium.
- Describe how total internal reflection is applied in the field of communications.

10.5

- State which way light refracts relative to the normal when it travels at an angle from air into a more optically dense material or in the reverse direction.

- Describe how to experimentally determine the critical angle in a transparent medium.

10.6

- Describe how laser light differs from ordinary white light, and describe examples of how laser light is used in communications technology.
- Describe how light and electronics are combined in the operation of a digital camera.

10.7

- List criteria that can be used to analyze and evaluate a communications technology.

▶ *MAKE* a summary

Sketch seven different devices to summarize this chapter:

- plane mirror* (Show the laws of reflection and applications.)
- diverging mirror* (Show experimental measurements, advantages, and applications.)
- converging circular mirror* (Show experimental observations and disadvantages.)
- converging parabolic mirror* (Show the advantages and applications in the visible spectrum and other parts of the EM spectrum, including satellite communications systems.)
- rectangular prism (Show Snell's law of refraction.)
- semicircular prism (Show the critical angle and total internal reflection.)
- optical fibre (Show applications of the use of a laser and total internal reflection.)

In your diagrams, include as many key understandings, skills, terms, and equations as possible.

*You can draw the mirrors used in ray box investigations to keep the diagrams as simple as possible.

Write the numbers 1 to 10 in your notebook. Indicate beside each number whether the corresponding statement is true (T) or false (F). If it is false, write a corrected version.

1. Interference and diffraction are wave properties exhibited by light.

2. Light can be transmitted in a vacuum, but radio waves cannot.

3. Regular reflection is the type of reflection that occurs off your hand when you hold your hand flat.

4. A diverging mirror has a convex shape.

5. A converging circular mirror is widely used because a converging parabolic mirror does not provide a clear focal point.

6. When light travels from a less optically dense medium into a more optically dense one, it refracts toward the normal.

7. As the speed of light in a material decreases, the index of refraction of that material increases.

8. If the critical angle in a medium is 24°, then all the rays at an angle of incidence less than 24° will totally reflect inside the medium.

9. A satellite in geosynchronous orbit must maintain a speed of zero in order to remain above the same location on Earth 24 h a day.

10. A laser hologram is made by using a reference beam that interferes with an object beam.

Write the numbers 11 to 17 in your notebook. Beside each number, write the letter corresponding to the best choice.

11. Infrared radiation
    (a) is visible to the human eye
    (b) has frequencies higher than those of the visible spectrum
    (c) has wavelengths shorter than those of the visible spectrum
    (d) has none of the above qualities

12. The order of EM waves from high frequency to low frequency is
    (a) IR; UV; X rays
    (b) UV; X rays; IR
    (c) X rays; UV; IR
    (d) IR; X rays; UV

13. In a plane mirror, the angle of incidence is
    (a) the angle between the incident ray and the reflected ray
    (b) larger than the angle of reflection
    (c) smaller than the angle of reflection
    (d) none of the above

14. For a light ray travelling from air into glass, if the angle of incidence is 30°, then the angle of refraction
    (a) $= 30°$
    (b) $< 30°$
    (c) $> 30°$
    (d) does not exist because the ray totally internally reflects

15. In **Figure 1**, the angle of incidence in the air and the angle of refraction in the glass are, respectively,
    (a) 1 and 3   (c) 2 and 4
    (b) 2 and 3   (d) 1 and 4

16. In **Figure 1**, the angle of incidence in the glass and the angle of emergence in the air are, respectively,
    (a) 6 and 7   (c) 5 and 7
    (b) 6 and 8   (d) 5 and 8

17. The phenomenon in Earth's atmosphere that explains how long-distance radio communication was possible before the use of satellites is
    (a) reflection   (c) interference
    (b) diffraction   (d) refraction

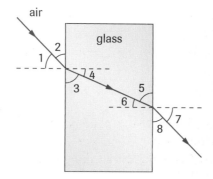

Figure 1

Understanding Concepts

1. Describe evidence from this chapter that supports the notion that light exhibits wave properties.

2. List the colours of the visible spectrum in the order of lowest to highest wavelength.

3. (a) What is the speed of light in a vacuum?
   (b) At that speed, how far can light travel in one microsecond?

4. The frequency of one of the bright colours emitted by hydrogen gas in a high-voltage tube is 4.57×10^{14} Hz.
   (a) Calculate the wavelength of this light.
   (b) What colour is the light? (Refer to the visible spectrum in **Figure 8**, page 490.)

5. A certain laser produces EM waves of wavelength of 2.0×10^{-10} m. Are the waves visible? (Refer to the diagram of the visible spectrum.)

6. (a) What causes the image created by the outside edge of a converging mirror to appear distorted?
   (b) What type of mirror can be used to correct this problem?

7. State the characteristics of a substance that prevents glare caused by reflected light.

8. Is it possible for a material to be transparent to some EM waves but opaque to others? Give at least one example to explain your answer.

9. What type of mirror can provide a wide-angled area of view? Use a sketch to illustrate how the mirror works.

10. For each mirror in **Figure 1**, state
    (a) the type of mirror
    (b) the size of the image compared to the object
    (c) one application of that type of mirror

Figure 1

11. The angle of incidence of a ray entering a rectangular prism is 0°. What is the angle of (a) refraction and (b) emergence?

12. Explain why light is refracted when it travels from air into water at an angle of incidence other than 0°.

13. Calculate the index of refraction of ocean water where the speed of light is 2.17×10^8 m/s.

14. A person is spearfishing from the edge of a pond. Draw a diagram to show where the person should aim the spear in order to strike a fish beneath the surface.

15. Calculate the speed of light in a material with an index of refraction of (a) 1.55 and (b) 1.09.

16. Draw a labelled diagram of a 3.0 cm $\times$ 5.0 cm solid rectangular prism that has a ray of light from air striking its long edge, such that $\theta_i = 50.0°$ and $\theta_R = 30.0°$.
    (a) What is the angle of emergence? Draw the emergent ray in your diagram.
    (b) Calculate the index of refraction of the material.
    (c) Calculate the speed of light in the prism.
    (d) Use **Table 1**, page 508, to determine a likely identity of the substance.

17. (a) Define "critical angle."
    (b) What is the relationship between the optical density of a substance and the size of the critical angle within it for light travelling toward air?

18. Explain the unique properties of laser light and why it is important.

19. Explain why digital photography is preferred by astronomers.

20. Explain why it is possible for a radio to receive signals from the radio station emitter even though there are obstacles between the emitter and the receiver.

Applying Inquiry Skills

21. A right-angled periscope with a single prism can be used to see around corners. Draw a diagram showing how you would design such a periscope by applying the principle of internal reflection.

22. Assume that you are given a sample of an unknown transparent liquid.
    (a) How would you find the speed of light in the liquid?
    (b) How would you use your experimental data to determine the identity of the liquid?

23. Place a glass or plastic prism in a beaker, and slowly add glycerin until the prism is covered. View the prism from various directions and explain what you observe.

24. **Figure 2** shows one method of heating water using solar energy.
    (a) Describe ways in which this method is an application of topics in this chapter.
    (b) Describe the energy transformations that occur in this method of heating water. Write the energy-transformation equation.
    (c) How would you determine experimentally the best position for the black hose for heating purposes?

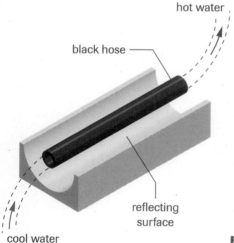

Figure 2

Making Connections

25. Is the image of the ambulance in **Figure 3** a direct view or the view from the rear-view mirror of a car ahead of the ambulance? Give evidence to support your answer.

26. Choose your favourite AM and FM radio stations. For each station, name or calculate the following:
    (a) its call letters

Figure 3

    (b) the frequency (in hertz)
    (c) the wavelength of the output waves

27. (a) List three Canadian contributions to communications science and technology.
    (b) Choose one of the contributions named in (a), and evaluate its importance to the technologies you use today.

28. List the criteria you would use to evaluate satellite dish technologies. Indicate which technical criteria are most important to you and why.

29. The swallowable, 3-cm long capsule in **Figure 4** contains a digital camera and takes about 2 h to move through the digestive tract. As it moves, it takes several images per second.
    (a) Describe how this technology applies the principles of light and electronics.
    (b) The images obtained by the camera can be transmitted to a receiver and then downloaded to a computer. Where would you place the receiver? What portion of the EM spectrum would this technology use to transmit the signals in your design? Can the waves pass safely through the human body? Explain your answers.

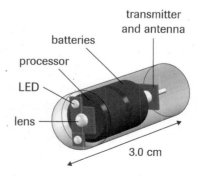

Figure 4

PERFORMANCE TASK

Assessment

Your completed task will be assessed according to the following criteria:

Process

- Draw up detailed plans and safety considerations of the design, tests, and modifications of the system or model.
- Choose appropriate research tools, such as books, magazines, and the Internet (especially for Option 2).
- Choose appropriate materials to construct the system or model.
- Appropriately and safely carry out the construction, tests, and modifications of the system or model.
- Analyze the process (as described in the Analysis).
- Evaluate the task (as described in the Evaluation).

Product

- Demonstrate an understanding of the relevant physics principles, laws, rules, and equations.
- Prepare a suitable research summary (Option 2).
- Submit a report containing the design plans for the system or model, as well as test results and analysis of properties such as the reflection, refraction, transmission, absorption, and/or interference of waves.
- Use terms, symbols, equations, and SI metric units correctly.
- Demonstrate that the final system works (for Option 1).

Design and Build a Communications System

For people with normal hearing and vision, communication using sound and light is straightforward. However, for people with impaired hearing or vision (or both), communication is much more challenging. Deaf and blind people can try to experience sounds and sights by spending time in a Snoezelen room (**Figure 1**). A Snoezelen room is filled with sensory-perception devices that stimulate whatever sense is challenged. Snoezelen facilities were developed in Holland in the late 1970s and have been built in hundreds of locations around the world, including Canada. Several Web sites are devoted to this type of communication.

 www.science.nelson.com

Figure 1
In a Snoezelen room, people experience sensations related to all five senses. Here, a child is experiencing sound by touching a device that vibrates. The word *Snoezelen* comes from two Dutch words that mean "to explore" and "to relax."

In Option 1 of this task, you will design, build, test, and analyze a communications system based on sound, light, or both. In Option 2, you will research the details of the design and operation of a communications system, and then build a model to illustrate its operation. Both options apply the principles you learned in Chapters 9 and 10. The system or model you build will use at least one energy transformation related to sound energy or electromagnetic energy. It will be operated with mechanical, electrical, or electronic components. Construction requires the safe use of tools and measuring instruments. The analysis demonstrates an understanding of the properties of waves, such as frequency, period, wavelength, type of wave, reflection, refraction, transmission, absorption, and interference.

 Before you build your system, have your teacher check your design to ensure that it is safe.

The Task

Option 1   Building a Communications System

Your task is to design, build, test, modify, and analyze a communications system using any or all of the following forms of energy: sound, ultrasound, visible light, or infrared radiation. For example, you can build an intercom system, a warning system (with a sound signal or a light signal), or a device designed for use in a Snoezelen room. Before starting the task, your group should decide the system's function, the criteria chosen to evaluate its success, and the necessary safety precautions. Have your teacher approve the materials, tools, and instruments.

An alternative to designing an original system is to build a communications system using a commercial kit or components salvaged from kits no longer in use. If you choose this alternative, you will build, test, modify, and analyze the system, and evaluate your process.

Option 2   A Model Communications System

Your task is to research the design and operation of a communications system that performs a specific function. Some examples are a device used in a Snoezelen facility, a virtual sound system, a remote control device that uses infrared radiation, a fish finder, an automatic door opener, an automatic light switch, a radar device that measures the speed of a baseball thrown by a pitcher (**Figure 2**), and a communications system that uses a satellite or an airship like the one featured on page 503. You will

then create a model of the system to illustrate the physics principles that are applied in its operation. To build the model, you can use materials that are inexpensive and easily and safely assembled. You will communicate (through diagrams or other means) how the components of the system work.

Analysis

(a) What physics principles are applied in the design and use of your system?

(b) Describe the energy transformations in the system, and write the corresponding energy-transformation equations.

(c) Describe the function of each of the main components of the system.

(d) How can you judge whether your system or model was successful?

(e) For what purpose can the system you designed or researched be used? What functions can it perform?

(f) What careers are related to the manufacture and use of your system?

(g) What safety precautions did you follow in building and testing your system or model?

(h) How could the process you used in this task be applied in business or industry?

(i) List problems you had while building the system or model, and explain how you solved them.

Evaluation

(j) Evaluate the tools you used in constructing the system (Option 1), or evaluate the resources you used in your research (Option 2).

(k) If you were to repeat this task, how would you modify the process to obtain a better final product?

Figure 2

1. Write the letters A to N in your notebook. Beside each letter, write the word or phrase that corresponds to the labels in **Figure 1**.

2. Write the letters (a) to (h) in your notebook. Beside each letter, write the word or phrase that best completes the blank(s).
   (a) The type of vibration experienced by a simple pendulum is called ____?____, which means that the motion of the pendulum is ____?____ to the rest axis.
   (b) When a crest meets a trough of equal size, the result is a(n) ____?____. This type of interference is called ____?____.
   (c) A compression travelling outward from a tuning fork is followed by a(n) ____?____.
   (d) During sound interference, a node is produced when ____?____.
   (e) The unit of sound intensity level is named after ____?____.
   (f) The bending of light as it passes from one material into another is called ____?____. The reason this bending occurs is ____?____.
   (g) The set of colours visible to the human eye is called the ____?____. The colours, in order of highest to lowest wavelength, are ____?____.
   (h) The visible spectrum is part of a larger set of waves called the ____?____. Two examples of waves whose frequencies exceed the frequency of visible light are ____?____.

3. Write the letters (a) to (h) in your notebook. Beside each letter, write the letter from A to O that corresponds to each of the terms listed below.
   (a) antinode
   (b) compression
   (c) audible range of a bat
   (d) normal
   (e) critical angle
   (f) hologram
   (g) diffraction
   (h) transmitted first voice message using radio waves

Title: A

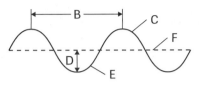

Title: G

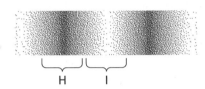

Title: J

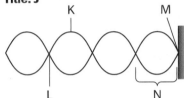

Figure 1

A   part of a longitudinal wave
B   part of a transverse wave
C   position of maximum amplitude in a standing wave
D   position of minimum amplitude in a standing wave
E   20 Hz to 20 kHz
F   1000 Hz to 120 kHz
G   another word for plane
H   another word for perpendicular
I   maximum angle of incidence in air
J   angle beyond which light reflects inside a medium
K   bending of light as it travels from one medium into another
L   bending of light as it passes through a narrow opening
M   a three-dimensional laser image
N   Marconi
O   Fessenden

Write the numbers 4 to 15 in your notebook. Indicate beside each number whether the corresponding statement is true (T) or false (F). If it is false, write a corrected version.

4. As the length of a simple pendulum increases, the frequency of vibration decreases.

5. As the amplitude of the pulse on a coiled spring increases, the speed of a pulse along the coil increases.

6. The amount of diffraction of waves through an opening increases as the frequency of the waves increases.

7. In a two-source interference pattern, the distance between the nodal lines increases as the wavelength of the waves increases.

8. A rarefaction is the part of a transverse wave where the particles are close together.

9. All sound is produced by vibrating objects.

10. The prefix "ultra" means "lower than."

11. Diffraction is a wave property exhibited by both sound and light.

12. The speed of EM radiation is constant at 3.00×10^8 m/s in all media.

13. An incident ray aimed along the normal to a mirror reflects back onto itself.

14. If the angle of incidence in air is 44° and the angle of refraction in a medium is 22°, the index of refraction of the medium is 2.0.

15. Alexander Graham Bell designed the original technology to send radio signals across the Atlantic Ocean.

Write the numbers 16 to 22 in your notebook. Beside each number, write the letter corresponding to the best choice.

16. When a crest travelling in one direction on a rope meets a crest travelling in the opposite direction, the result is
    (a) a standing wave
    (b) constructive diffraction
    (c) a loop
    (d) constructive interference

17. If the frequency of a source of waves in a medium changes from f to $2f$, the wavelength of the waves
    (a) changes from λ to 2λ
    (b) changes from λ to $\lambda/2$
    (c) changes from λ to $\lambda/4$
    (d) remains the same because the speed in the medium is constant

18. If the fundamental frequency of a standing wave is 100.0 Hz, the frequency of the second harmonic is
    (a) 200.0 Hz   (c) 400.0 Hz
    (b) 300.0 Hz   (d) none of these

19. In **Figure 2,** where the critical angle is 45°,
    (a) Ray A undergoes total internal reflection, and ray B emerges into the air.
    (b) Ray B undergoes total internal reflection, and ray A emerges into the air.
    (c) Both rays undergo total internal reflection.
    (d) Both rays undergo partial reflection in the medium and partial emergence in the air.

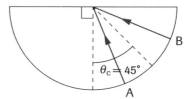

Figure 2

20. The parabolic reflectors that receive radio waves from distant stars and galaxies are much larger than TV satellite dishes because of this property of the radio waves:
    (a) high frequencies   (c) long wavelength
    (b) high energies   (d) high speed

21. Visible light is not used to transmit signals to and from TV communications satellites because
    (a) it does not allow viewing at night
    (b) it requires huge, land-based satellite dishesbecause of its long wavelengths
    (c) it cannot travel through clouds and other obstructions
    (d) cannot travel in the vacuum of space

22. The phenomenon that explains how holograms are made is
    (a) total internal reflection   (c) interference
    (b) diffraction   (d) refraction

Understanding Concepts

1. State the type of vibration that results in each of the following cases:
   (a) A pile driver pounds a metal post repeatedly to force it into the ground.
   (b) A camper rotates a stick between her hands to create a spark to light a campfire.
   (c) A bird feeder hanging from a tree vibrates after a bird flies away.

2. A tuning fork vibrates 88 times in 0.20 s.
   (a) Calculate the frequency and period of vibration of the tuning fork.
   (b) Describe how the sound from the tuning fork is transmitted to your ears.

3. State the relationship (if any) between the following pairs of variables:
   (a) period and length of a pendulum
   (b) mass and frequency of a pendulum
   (c) wavelength and period of a periodic wave

4. The frequency of a mechanical metronome can be altered by moving the mass. (See **Figure 9**, page 426, for a photograph of this device.) Should a music student move the mass closer to or farther from the pivot point to increase the metronome's frequency of vibration?

5. Each diagram in **Figure 1** shows an incident pulse travelling toward one end of a rope. Draw a diagram showing the reflected pulse in each case.

(a)

(b)

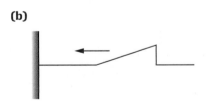

Figure 1

6. A sound generator emits a note with wavelength 0.672 m. Using 344 m/s as the speed of sound in air, calculate the frequency of the note.

7. Kingston Collegiate Vocational Institute has a radio station: CKVI at 91.9 MHz.
   (a) Calculate the wavelength of the radio waves from this station.
   (b) Is this an AM or FM station? Explain how you know.

8. A sound wave in a steel rail has a period of 4.10×10^{-3} s and a speed of 5.03 km/s. Calculate the wave's frequency and wavelength.

9. (a) Describe the sound heard when beats are produced.
   (b) Explain how beats can be used to tune a musical instrument.

10. An 8.0-m rope is used to produce the standing wave in **Figure 2**.
    (a) What is the wavelength of the waves?
    (b) Draw a diagram of the standing wave pattern on the same rope if the wavelength of the waves is 4.0 m.

Figure 2

11. The distance between nodes of a standing wave is 0.38 m, and the frequency of the source producing the wave by reflection is 88 Hz. Find the speed of the wave.

12. In 2000, on the opening day of a millennium bridge across the Thames River in London, England, thousands of runners attempted to cross. The bridge experienced unusually large vibrations. Explain the cause.

13. Provide evidence that sound cannot travel in a vacuum.

14. Explain what vibrates to produce the sound originating from each of the following:
    (a) acoustic guitar
    (b) pipe organ
    (c) stereo system

15. State the frequency range of
    (a) human hearing
    (b) ultrasonic sounds

16. (a) Compare the visible range of frequencies for humans to the audible range of frequencies.
    (b) Compare the visible range of light wavelengths for humans to the audible range of sound wavelengths. (Hint: To determine the audible wavelengths, you can apply the universal wave equation and use 344 m/s as the speed of sound.)
    (c) Use the wavelength ranges to explain why demonstrations of diffraction involve different dimensions for sound and light.

17. Describe two examples of sound resonance.

18. Ultrasonic sound is transmitted from a ship to locate a school of fish. The speed of sound in the ocean is 1.44 km/s, and the reflection of the sound reaches the ship 0.124 s after it is sent. How far is the school of fish from the ship?

19. A 900.0-Hz note is the third harmonic of a sound of a stringed instrument. Determine the first, second, and fourth harmonics.

20. What is the relationship between the frequency and length of a vibrating guitar string?

21. A note with a fundamental frequency of 392.0 Hz is played on a synthesizer. Describe ways in which the quality of the sound emitted can be altered.

22. A synthesizer produces a constant-frequency sound composed of a saw-tooth wave with a sudden attack, then a sustained amplitude, and finally a fast decay. Draw a diagram of the waveform of the sound.

23. Describe the energy transformations that occur in the operation of an electric amplifier system at a sports stadium when an announcer gives the name of a player who has scored.

24. Compare and contrast visible light with the other parts of the electromagnetic spectrum.

25. The standard metre is close to 1.65×10^6 wavelengths of a certain light emitted by krypton-86 atoms in a vacuum. Find the wavelength, frequency, and colour of this light.

26. In **Figure 3**, is light travelling more slowly in A or B? Which substance has the lower optical density?

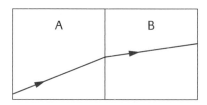

Figure 3

27. For the situation shown in **Figure 4**, determine
    (a) the index of refraction of the prism by applying Snell's law of refraction
    (b) the speed of light in the prism

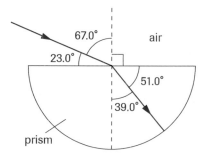

Figure 4

28. Explain how the principle of total internal reflection is applied in communications systems using long glass fibres.

Applying Inquiry Skills

29. How would you use a cork in a sink to demonstrate that a water wave transfers energy without transferring water molecules? Describe what you would expect to observe.

30. Two combs with the teeth arrangements shown in **Figure 5** are moved closer together until their teeth start to overlap. When the combs are viewed against a light background, the pattern observed is a model for the production of beats.
    (a) Predict the number of "beats" per centimetre observed on the left and on the right.

Figure 5

(b) Try the demonstration. If necessary, revise your answer in (a).

31. Use a rubber hammer to strike a tuning fork; then press the handle of the fork gently against the surface of a desk and other surfaces.
    (a) What happens to the loudness and quality of the sound when the tuning fork is in contact with the desk and other surfaces?
    (b) Relate your answer in (a) to the playing of stringed instruments.

32. Describe what apparatus you would use to demonstrate the characteristics of
    (a) mechanical waves in one dimension
    (b) mechanical waves in two dimensions
    (c) sound waves in three dimensions

33. Describe how you would experimentally determine the speed of light in a clear liquid.

34. (a) What shape of reflector is used in satellite communications systems?
    (b) How would you demonstrate the advantage of using this shape?

35. A solar cell sensitive to infrared (IR) can be connected to an oscilloscope to "see" the IR signals emitted by a remote control device, such as the ones used to operate a TV, VCR, CD, or DVD. The waveform of the Play button on one model of VCR is shown in **Figure 6**. The first wide peak is a setup signal; it is followed by 12 peaks representing the digits zero and one. **Table 1** gives the IR codes of this example, as well as others.

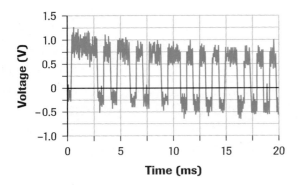

Figure 6

Table 1 IR Codes for Various Sony Devices

Device	Button	IR Code
VCR	1	000 000 001 000
	8	111 000 001 000
	9	000 100 001 000
	Play	010 110 001 000
TV	1	000 000 010 000
	8	111 000 010 000
	9	?
CD player	1	?
	8	?
	9	000 100 010 001

(a) Copy the table into your notebook. Look at the pattern, then complete the missing information.
(b) Study the pattern in **Figure 6** and determine how it relates to the code for the Sony VCR Play button. Then draw a corresponding graph for the Sony CD player's button 8.

Making Connections

36. Should a tourist sensitive to motion sickness choose an upper or lower deck on a cruise ship? Explain your answer. (Hint: Relate this decision to question 5.)

37. Explain why the Canadian satellite industry is an important part of communications science and technology.

38. Night-vision goggles are often used in peacekeeping and military operations (**Figure 7**). The green colour comes from the type of phosphors used on the screen in the goggles. Research night-vision goggles.
    (a) Explain the use of transducers and an amplifier in this technology.
    (b) How does the principle of operation differ from the one used to obtain an IR image?
    (c) What are the advantages and disadvantages of night-vision goggles?

 www.science.nelson.com

Figure 7

39. The use of sonar by the U.S. Navy is controversial. The Navy performs underwater sonar tests with signals as loud as 215 dB. However, at 180 dB, a whale's ears can explode. Dozens of whales and dolphins have been found stranded on beaches along the Atlantic and Pacific coasts after the Navy performed tests.
    (a) How does the intensity level of the sonar compare with the threshold of pain for human hearing?
    (b) Research the "Surveillance Towed Array Sensor System" (or SURTASS LFA). Describe what you discover about its purpose and results.
    (c) Do you think the U.S. Navy should continue to use this system? Justify your response.

     www.science.nelson.com

40. To check airline passengers for high fever, some airports use infrared (IR) scanners.
    (a) Describe the principle of using an IR scanner to determine body temperature.
    (b) List advantages and disadvantages of using an IR scanner at an airport.
    (c) Make up a short questionnaire to determine public opinion about the use of IR scanners to take body temperatures at airports.

41. Bar code technology is used in grocery stores to make up a customer's bill at the checkout counter and keep track of the store's inventory. Similar technology is used at self-check counters in many stores. Some stores even provide handheld price scanners, which allow customers to verify the price of an item and enter data for each item placed in their grocery cart. Describe advantages and disadvantages of this technology.

42. The technology of recording sounds in nature, such as bird songs, was improved by a Canadian invention called the "Dan Gibson microphone," named after its inventor. Canadian companies applying this technology are world leaders in making nature and environmental sound recordings and wildlife movies. Research this technology.
    (a) Describe the principle of operation of the Dan Gibson microphone.
    (b) What difficulties do sound production people face in obtaining natural sounds? How are these difficulties overcome?
    (c) Evaluate the contributions to communications of at least one of the following companies: Earth noise; Somerset Entertainment (formerly Solitudes Ltd.); Holborne Distributing Company.

     www.science.nelson.com

43. (a) Pager technology is continually being upgraded to provide new communications features for customers. Describe features of a pager that you would like to see.
    (b) List criteria for evaluating pagers in order to choose the one that provides the greatest number of the features you described in (a).

44. Both radio and television are communications technologies that have had profound effects on society. Create a list of the positive and negative contributions of these technologies.

contents

Appendices

A1 **Math Skills**

Significant Digits

Two types of quantities are used in science: exact values and measurements. Exact values include defined quantities (e.g., 1 kg = 1000 g) and counted values (e.g., 23 students in a classroom). Measurements, however, are not exact because they always include some degree of uncertainty.

In any measurement, the *significant digits* are the digits that are known reliably, or for certain. The single last digit is considered estimated or uncertain, but is also included in the count of significant digits. Thus, if the width of a piece of paper is measured as 21.6 cm, there are three significant digits in the measurement, and the last digit (6) is estimated or uncertain.

The following rules are used to determine whether a digit is significant in a measurement:

- All nonzero digits are significant: 345.6 N has four significant digits.

- In a measurement with a decimal point, zeroes placed before other digits are not significant; for example, the measurement 0.0056 m has two significant digits.

- Zeroes placed between other digits are always significant; for example, the measurement 7003 s has four significant digits.

- Zeroes placed after nonzero digits after a decimal are significant; for example, the measurements 9.100 km and 802.0 kg each have four significant digits.

- Scientific notation (see below) is used to indicate whether zeroes at the end of a measurement are significant; the measurement 4.50×10^7 km has three significant digits, and the measurement 4.500×10^7 km has four significant digits. The same number written as 45 000 000 km has at least two significant digits, but the total number is unknown unless the measurement is written in scientific notation. (An exception to this last statement is the following: if the number of significant digits can be assessed by inspection.

For example, a reading of 1250 km on a car's odometer has four significant digits.)

Rounding Off Calculated Answers

When measurements made in scientific experiments or given in problems are used in calculations, the final answer must take into consideration the number of significant digits of each measurement, and may have to be rounded off according to the following rules:

- When adding or subtracting measured quantities, the final answer should have no more than one estimated digit; in other words, the answer should be rounded off to the least number of decimals in the original measurements.

- When multiplying or dividing measured quantities, the final answer should have the same number of significant digits as the original measurement with the least number of significant digits.

Example

A piece of paper is 48.5 cm long, 8.44 cm wide, and 0.095 mm thick.

(a) Calculate the perimeter of the piece of paper.

(b) Calculate the volume of the piece of paper.

Solution

(a) L = 48.5 cm (The 5 is estimated.)
 w = 8.44 cm (The last 4 is estimated.)
 P = ?

$$P = L + L + w + w$$
$$= 48.5 \text{ cm} + 48.5 \text{ cm} + 8.44 \text{ cm} + 8.44 \text{ cm}$$
$$P = 113.88 \text{ cm}$$

According to the rule for adding and subtracting quantities, the answer must be rounded off to only one estimated digit. Thus, the perimeter is 113.9 cm.

(b) $h = 0.095$ mm $= 9.5 \times 10^{-3}$ cm (two significant digits)

$V = ?$

$$V = Lwh$$
$$= (48.5 \text{ cm})(8.44 \text{ cm})(9.5 \times 10^{-3} \text{ cm})$$
$$= 3.88873 \text{ cm}^3$$
$$V = 3.9 \text{ cm}^3$$

According to the rule for multiplying or dividing quantities, the answer is rounded off to two significant digits.

Other rules must be taken into consideration in some situations. Suppose that after calculations are complete, the answer to a problem must be rounded off to three significant digits. Apply the following rules of rounding:

- If the first digit to be dropped is 4 or less, the preceding digit is not changed; for example, 8.674 is rounded to 8.67.

- If the first digit to be dropped is 5, the preceding digit is increased by 1; for example, 8.675 123 is rounded up to 8.68.

- If the first digit to be dropped is a lone 5 or a 5 followed by zeroes, the preceding digit is not changed if it is even, but is increased if it is odd. For example, 8.675 is rounded up to 8.68, but 8.665 is rounded down to 8.66. (This rule exists to avoid accumulated error that would occur if the 5 were always rounded up. It is followed in this text, but not in all situations. For example, calculators and some computer software programs do not follow it. This rule is not crucial to your success in solving problems.)

When solving multistep problems, round-off error occurs if you use the rounded-off answer from the first part of the question in subsequent parts. Thus, when making calculations, record all the digits or store them in your calculator until the final answer is determined, and then round off the final answer to the correct number of significant digits. For example, in a multistep sample problem, the answer for part (a) is written to the correct number of significant digits, but all the digits of the answer are used to solve part (b).

Scientific Notation

Extremely large and extremely small numbers are awkward to write in common decimal notation. Furthermore, they do not always convey the number of significant digits of a measured quantity. In these cases, we can change the metric prefix before the unit of measurement so that the number falls between 0.1 and 1000; for example, 0.000 000 906 kg can be expressed as 0.906 mg. However, a prefix change is not always possible, either because an appropriate prefix does not exist or because it is essential to use a particular unit of measurement. In these cases, it is best to use *scientific notation*, also called standard form. Scientific notation expresses a number by writing it in the form $a \times 10^n$, where $1 \le |a| < 10$, and the digits in the coefficient a are all significant. For example, the magnitude of the acceleration due to gravity at a particular location is 9.79 m/s^2, and the speed of light in a vacuum is $2.997\ 924\ 58 \times 10^8$ m/s. Using this notation, calculations are easier.

The following rules are applied when performing mathematical operations:

- For addition and subtraction of numbers in scientific notation: Change all the factors to a common factor—the same power of 10—and add or subtract the numbers.

Example

$$1.234 \times 10^5 + 4.2 \times 10^4 = 1.234 \times 10^5 + 0.42 \times 10^5$$
$$= (1.234 + 0.42) \times 10^5$$
$$= 1.654 \times 10^5$$

This answer is rounded off to 1.65×10^5, so the answer has only one estimated digit, in this case two digits after the decimal.

- For multiplication and division of numbers in scientific notation: Multiply or divide the coefficients, add or subtract the exponents, and express the result in scientific notation.

Example

$$\left(1.36 \times 10^4 \frac{\text{kg}}{\text{m}^3}\right)\left(3.76 \times 10^3 \text{ m}^3\right) = 5.11 \times 10^7 \text{ kg}$$

When working with exponents, recall the following rules:

$$x^a \times x^b = x^{a+b}$$

$$\frac{x^a}{x^b} = x^{a-b}$$

On many calculators, scientific notation is entered using the EXP or the EE key. This key includes the "×10" from the scientific notation, so you need only enter the exponent. For example, to enter 6.51×10^{-4}, press 6.51 EXP +/− 4.

Error Analysis in Experimentation

In experiments involving measurement, there is always some degree of uncertainty. This uncertainty can be attributed to the instrument used, the experimental procedure, the theory related to the experiment, and/or the experimenter.

In all experiments involving measurements, the measurements and calculations should be recorded to the correct number of significant digits. However, a formal report of an experiment involving measurements may include an analysis of uncertainty, percent uncertainty, and percent error or percent difference.

Uncertainty is the amount by which a measurement may deviate from an average of several readings of the same measurement. This uncertainty can be estimated, so it is called the *estimated uncertainty*. Often it is assumed to be plus or minus half of the smallest division of the scale on the instrument; for example, the estimated uncertainty of 15.8 cm is ±0.05 cm or ±0.5 mm. (Uncertainty can also be called possible error. Thus, estimated uncertainty is estimated possible error.)

Percent error can be found only if it is possible to compare an experimental value with that of the most commonly accepted value. The equation is

$$\% \text{ error} = \frac{\text{measured value} - \text{accepted value}}{\text{accepted value}} \times 100\%$$

Percent difference is useful for comparing measurements when the true measurement is not known or for comparing an experimental value to a predicted value. The equation is

$$\% \text{ difference} = \frac{|\text{difference in values}|}{\text{average of values}} \times 100\%$$

Accuracy is a comparison of how close a measured value is to the true or accepted value. An accurate measurement has a low uncertainty. *Precision* is an indication of the smallest unit provided by an instrument. A highly precise instrument provides several significant digits. Accuracy and precision are compared in **Figure 1**.

Figure 1
The dart throws are **(a)** precise and accurate, **(b)** precise but not accurate, and **(c)** neither precise nor accurate.

Random error occurs in measurements when the last significant digit is estimated. Random error results from variation about an average value. One way to reduce random error is to take the average of several readings.

Parallax is the apparent shift in an object's position when the observer's position changes. This source of error can be reduced by looking straight at an instrument or dial.

Systematic error results from a consistent problem with a measuring device or the person using it. Such errors are reduced by adding or subtracting the known error, calibrating the instrument, or performing a more complex investigation.

Measuring Angles

Angles are commonly measured in degrees (°) using a protractor. (If you push the DRG or DEG button on your calculator, you will notice that angles can also be measured in radians (RAD) and gradients (GRAD).

Only degrees are used in this textbook.) **Figure 2** illustrates the following facts and measurements:

- There are 360° in a circle, 180° in a straight line, and 90° in a right angle.
- The origin position of a protractor is the reference point from which you measure any angle.
- Angles can be measured from the horizontal, from the vertical, and between two intersecting lines.
- A normal line is perpendicular to a surface.

Reading Analog and Digital Meters

Meters can be analog or digital. An analog meter has a needle that usually moves from zero at the left to a maximum value at the right. If there is only one scale, it is relatively easy to read the value, as in the ammeter

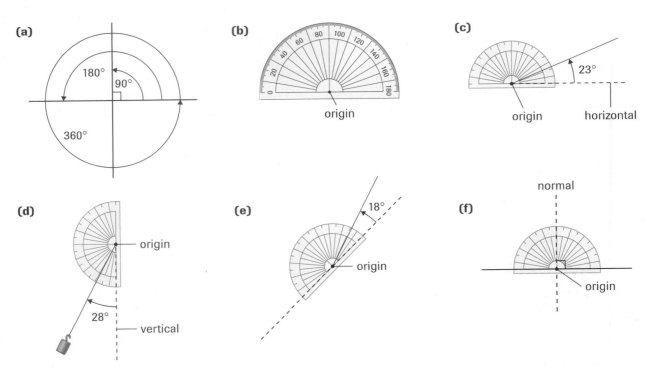

Figure 2
(a) Degrees in a circle, a straight line, and a right angle
(b) Using a protractor to measure an angle in degrees
(c) An angle measured from the horizontal
(d) An angle measured from the vertical
(e) The angle between two intersecting lines
(f) A normal line

reading in **Figure 3(a)**. If the meter has two or more scales as selected by a switch or terminal, then the value must be read on the scale that corresponds to the switch or terminal chosen. In **Figure 3(b)**, if the switch is on the 1 A setting, the reading is 0.72 A, but if the switch is on the 5 A setting, the reading is 3.6 A.

(a)   **(b)**

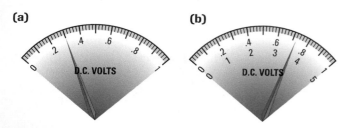

Figure 3

A digital meter often has several options, so the reading corresponds to the option selected by the switch. For example, the digital multimeter shown in **Figure 4** is set on 2 V, which means that the maximum expected voltage is 2 V. To measure voltage, the black wire is connected to the common terminal (COM), which corresponds to the negative terminal; the red wire is connected to the voltage-resistance (V/Ω) terminal, corresponding to the positive terminal.

Figure 4
A digital multimeter

Mathematical Equations

Several mathematical equations from the fields of geometry and trigonometry are used in this textbook.

Geometry

For a rectangle of length L and width w, the perimeter (P) and the area (A) are

$$P = 2L + 2w$$

$$A = Lw$$

For a triangle of base b and altitude h, the area is

$$A = \frac{1}{2}bh$$

For a circle of radius r, the circumference (C) and the area are

$$C = 2\pi r$$

$$A = \pi r^2$$

For a right circular cylinder of height h and radius r, the volume is

$$V = \pi r^2 h$$

For a regular solid of length L, width w, and height h, the volume is

$$V = Lwh$$

Trigonometry

Three common trigonometric functions are sine (sin), cosine (cos), and tangent (tan). These functions are based on similar triangles—the ratios of the corresponding sides are equal. For example, in **Figure 5**, triangle ABC is similar to triangle ADE. Using "opp" for opposite, "adj" for adjacent, and "hyp" for hypotenuse,

$$\sin A = \frac{\text{opp}}{\text{hyp}} = \frac{BC}{AB} = \frac{DE}{AD}$$

$$\cos A = \frac{\text{adj}}{\text{hyp}} = \frac{AC}{AB} = \frac{AE}{AD}$$

$$\tan A = \frac{\text{opp}}{\text{adj}} = \frac{BC}{AC} = \frac{DE}{AE}$$

You can use your calculator to check these equations. The first example is

$$\sin A = \sin 37° = 0.601815 = 0.60$$

$$\sin A = \frac{BC}{AB}$$

$$= \frac{3.0 \text{ cm}}{5.0 \text{ cm}}$$

$$\sin A = 0.60$$

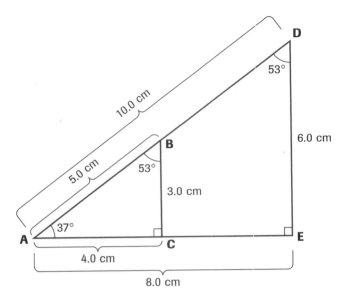

Figure 5
Similar triangles used to illustrate the sin, cos, and tan of angle *A*

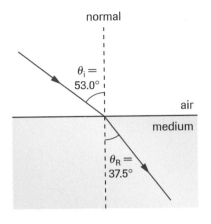

Figure 6
Applying Snell's law

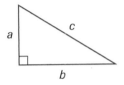

Figure 7
The law of Pythagoras

In the field of optics, trigonometry is used to determine the index of refraction (*n*) of a medium by using Snell's law of refraction:

$$n = \frac{\sin \theta_i}{\sin \theta_R}$$

where   *n* = index of refraction,

    θ_i = angle of incidence (in air), and

    θ_R = angle of refraction (in the medium).

In **Figure 6**,

$$n = \frac{\sin \theta_i}{\sin \theta_R}$$

$$= \frac{\sin 53.0°}{\sin 37.5°}$$

$$= \frac{0.798\ 635\ 5}{0.608\ 761\ 4}$$

$$n = 1.31$$

In this textbook, Snell's law is usually applied with three significant digits.

Law of Pythagoras: For the right-angled triangle in **Figure 7**, $c^2 = a^2 + b^2$, where *c* is the hypotenuse, and *a* and *b* are the other sides.

Unit Analysis

The process of using units to analyze a problem or an equation is called *unit analysis*. Unit analysis is a tool used to determine whether an equation has been written correctly and to convert units. For example, we can use unit analysis to determine whether the following equation is valid:

$$E_{g\ (top)} = E_{K\ (bottom)} \text{ for an object that falls from rest}$$

$$mg = \frac{1}{2}mv^2$$

The usual technique is to put the units in square brackets and ignore numbers like the $\frac{1}{2}$ in the equation. The square brackets mean that we are dealing with units only:

$$[kg]\left[\frac{m}{s^2}\right] = [kg]\left[\frac{m}{s}\right]^2$$

$$[kg]\left[\frac{m}{s^2}\right] = [kg]\left[\frac{m^2}{s^2}\right]$$

This equation is not valid because the units on the right-hand side do not equal the units on the left-hand side. The correct equation is

$$mgh = \frac{1}{2}mv^2$$

We can ignore the number $\frac{1}{2}$ because it has no dimensions. Dimensionless quantities include

- all plain numbers (4, π, etc.)
- counted quantities (12 people, 5 cars, etc.)
- angles (although angles have units)
- cycles
- trigonometric functions

Analyzing Experimental Data

Controlled physics experiments are conducted to determine the relationship between variables. The experimental data can be analyzed in a variety of ways to determine how the dependent variable depends on the independent variable(s). Often the relationship can be expressed in a proportionality statement or shown on a graph.

The statement of how one quantity varies in relation to another is called a *proportionality statement*. (It can also be called a *variation statement*.) Typical proportionality statements and their corresponding graphs are shown in **Figure 8**.

Graphing

When graphing experimental results,

- Place the *independent variable*, the one controlled by the experimenter, on the horizontal axis.
- Place the *dependent variable*, the quantity you are trying to find (it depends on the independent variable) on the vertical axis.
- Choose scales that occupy as large a portion of the graph paper as possible.
- Give the graph a title.
- Label the axes with the names of the quantities being plotted and their units.

Points on a line graph are obtained by plotting ordered pairs obtained from the experiment. If the points plotted appear to line up fairly closely, a single straight line of best fit should be drawn. Once a straight line is obtained on a graph, the slope of the line can be calculated using either of these equations:

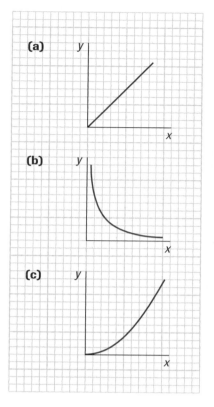

Figure 8
(a) Direct proportion: $y \propto x$
(b) Inverse proportion: $y \propto \dfrac{1}{x}$
(c) Square proportion: $y \propto x^2$

$$\text{slope} = \frac{\text{rise}}{\text{run}} \quad \text{or} \quad m = \frac{\Delta y}{\Delta x}$$

where Δy is the change in the value plotted on the vertical axis, and Δx is the corresponding change in the value on the horizontal axis. When calculating the slope of a line on a graph, always include units because they represent the meaning of the slope.

A graph of data can also be used to illustrate examples of interpolation and extrapolation, shown in **Figure 9**. *Interpolation* is the process of estimating values of the dependent variable *between* the plotted points. *Extrapolation* is the process of estimating values *beyond* measurements made in the experiment. Extrapolation is possible only if we assume that the line continues in a predictable fashion on the graph. In a rubber band experiment, for example, extrapolation would not be possible if the band were stretched to the point of breaking.

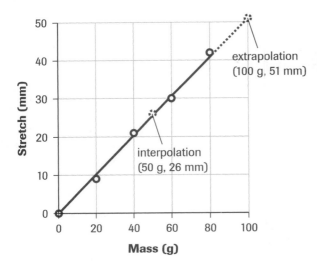

Experimental data:

Mass (g)	0	20	40	60	80
Stretch (mm)	0	9	21	30	42

Figure 9
A graph of the stretch of a rubber band suspended vertically as a function of the mass added to the end of the band

Graphing on a Spreadsheet

A *spreadsheet* is a computer program that can be used to create a table of data and then a graph of the data. It is composed of cells represented by a column letter (A, B, C, etc.) and a row number (1, 2, 3, etc.); thus, A3 and C1 are examples of cells (**Figure 10**). Each cell can hold a number, a label, or an equation.

Follow these steps to create a graph of data gathered in an investigation:

- Enter the data in the spreadsheet.
- Select the graph type (line, column, bar, or pie) that will display the data in the most appropriate form.
- Select the data from the spreadsheet that will be plotted on the graph.
- Enter the remaining graph information, such as the graph title, axes titles, and units.
- Indicate where the graph is to be located in the spreadsheet. It can be included on the same sheet as the data or on a new sheet.

	A	**B**	**C**	**D**	**E**
1	A1	B1	C1		
2	A2	B2			
3	A3				

Figure 10
Spreadsheet cells

A2 Planning an Investigation

To answer questions, solve mysteries, and explain events, we investigate using scientific inquiry. The methods used in scientific inquiry depend, to a large degree, on the purpose of the inquiry.

Controlled Experiments

In a *controlled experiment*, an independent variable is changed to determine its effect on a second dependent variable. All other variables are controlled or kept constant. Controlled experiments are performed when the purpose of the inquiry is to create, test, or use a scientific concept.

The common components of controlled experiments are outlined below. *Even though the components are presented in a step-by-step fashion, there are normally many cycles through the steps during an actual experiment.*

Stating the Purpose

Every scientific investigation has a purpose; for example,

- to develop a scientific concept (a theory, law, generalization, or definition)
- to test a scientific concept
- to determine a scientific constant or
- to test an experimental design, a procedure, or a skill

Determine which of these is the purpose of your investigation. Then write the purpose in your lab report.

Asking the Question

Your question forms the basis for your investigation: The investigation is designed to answer the question. Controlled experiments are about relationships, so the question could be about the effects on variable A when variable B is changed.

Predicting

A prediction is a tentative answer to the question you are investigating. In the prediction, you state what outcome you expect from your experiment and why.

Designing the Investigation

The design of a controlled experiment identifies how you plan to manipulate the independent variable, measure the response of the dependent variable, and control all the other variables to answer your question. It is a summary of your plan for the experiment.

Gathering, Recording, and Organizing Observations

There are many ways to gather and record observations during your investigation. It is helpful to plan ahead and think about what data you will need to answer the question and how best to record them. This helps to clarify your thinking about the original question, the variables, the number of trials, the procedure, the materials, and your skills. It will also help you organize your evidence for easier analysis.

Analyzing the Observations

After thoroughly analyzing your observations, you may have sufficient and appropriate evidence to enable you to answer the original question.

Evaluating the Evidence and the Prediction

At this stage of the investigation, you evaluate the processes that you followed to plan and perform the investigation.

You will also evaluate the outcome of the investigation, which involves evaluating any prediction you made. You must identify and take into account any sources of error and uncertainty in your measurements.

Finally, compare the answer you predicted with the answer generated by analyzing the evidence.

Reporting on the Investigation

In your report, you should describe your planning process and procedure clearly so that anyone reading it can repeat the experiment exactly as you performed it. Also, you should report your observations, your analysis, and your evaluation of the experiment accurately and honestly.

A3 Exploring an Issue

An issue is defined as a problem that has at least two possible solutions rather than a single answer. There can be many positions, generally determined by the values that an individual or a society holds, on a single issue. Which solution is "best" is a matter of opinion; ideally, the solution that is chosen is the one that is best for society as a whole.

The common process involved in the decision-making process is outlined below. *Even though the process is presented in a step-by-step fashion, you may go through several cycles before deciding you are ready to defend a decision.*

Defining the Issue

The first step in understanding an issue is to explain why it is an issue, describe the problems associated with the issue, and identify the individuals or groups, called stakeholders, involved in the issue. You could brainstorm the following questions to research the issue: Who? What? Where? When? Why? How? Gather background information on the issue by clarifying facts and concepts, and identifying relevant features or characteristics of the problem.

Identifying Alternatives/Positions

Examine the issue and think of as many solutions as you can. At this point, it does not matter if the solutions seem unrealistic. To analyze the alternative solutions, you should examine the issue from a variety of points of view. Stakeholders may bring different viewpoints to an issue, and these may influence their position on the issue. Brainstorm or speculate on how different stakeholders would feel about your alternatives. Points of view that stakeholders may consider are listed in **Table 1**.

Researching the Issue

Compose a research question that helps to limit, narrow, or define the issue. Then develop a plan to identify and find reliable and relevant sources of information. Outline the stages of your information search: gathering, sorting, evaluating, selecting, and integrating relevant information. You could use a flow chart, concept map, or other graphic organizer to

Table 1   Possible Points of View on an Issue

cultural	focused on customs and practices of a particular group
environmental	focused on effects on natural processes and other living things
economic	focused on the production, distribution, and consumption of wealth
educational	focused on the effects on learning
emotional	focused on feelings and emotions
aesthetic	focused on what is artistic, tasteful, beautiful
moral/ethical	focused on what is good/bad, right/wrong
legal	focused on rights and responsibilities
spiritual	focused on the effects on personal beliefs
political	focused on the aims of an identifiable group or party
scientific	focused on logic or the results of relevant inquiry
social	focused on effects on human relationships, the community
technological	focused on the use of machines and processes

outline the stages of your information search. Gather information from many sources, including newspapers, magazines, scientific journals, the Internet, and the library.

Analyzing the Issue

In this stage, you will analyze the issue to determine where you stand. First, you should establish criteria for evaluating your research to make sure it is relevant and significant. You can then evaluate your sources, determine what assumptions may have been made, and assess whether you have enough information to make your decision.

Follow these five steps:

1. List your criteria for determining the relevance and significance of the data you have gathered.

2. Evaluate the sources of information.

3. Identify and determine what assumptions have been made. Challenge unsupported evidence.

4. Determine any relationships associated with the issue, for example, is the issue of the same concern to a person living in a rural area as a person living in a city?

5. Evaluate the alternative solutions, possibly by conducting a risk–benefit analysis.

Defending the Decision

After analyzing your information, you can answer your research question and take an informed position on the issue. You should be able to defend your solution choice in an appropriate format: debate, class discussion, speech, position paper, multimedia presentation (e.g., computer slide show), brochure, poster, or video.

Your position on the issue must be justified using your research information. You should be able to defend your position to people with different points of view. In preparing for your defence, ask yourself the following questions:

- Do I have supporting evidence from a variety of sources?
- Can I state my position clearly?
- Do I have solid arguments (with solid evidence) supporting my position?
- Have I considered arguments against my position and identified their faults?
- Have I analyzed the strong and weak points of each point of view?

Evaluating the Process

The final phase of decision making includes evaluating the decision the group reached, the process used to reach the decision, and the part you played in decision making. After a decision has been reached, carefully examine the thinking that led to the decision. The following questions will help:

- What was my initial point of view on the issue? How has my point of view changed since I first began to explore the issue?
- How did we make our decision? What process did we use? What steps did we follow?

- In what ways does our decision resolve the issue?
- What are the likely short- and long-term effects of our decision?
- To what extent am I satisfied with our decision?
- What reasons would I give to explain our decision?
- If we had to make this decision again, what would I do differently?

A Risk–Benefit Analysis Model

Risk–benefit analysis is a tool used to organize and analyze information gathered in research. A thorough analysis of the risks and benefits associated with each alternative solution can help you decide on the best alternative.

- Research as many aspects of the proposal as possible. Look at it from different perspectives.
- Collect as much evidence as you can, including reasonable projections of likely outcomes if the proposal is adopted.
- Classify every individual potential result as being either a benefit or a risk.
- Quantify the size of the potential benefit or risk (perhaps as a dollar figure, the number of lives affected, or on a scale of 1 to 5).
- Estimate the probability (percentage) of that event occurring.
- By multiplying the size of a benefit (or risk) by the probability of its happening, you can assign a significance value for each potential result.
- Total the significance values of all the potential risks and all the potential benefits. Then compare the sums to help you decide whether to accept the proposed action.

Note that although you should try to be objective in your assessment, your beliefs will have an effect on the outcome—two people, even if using the same information and the same tools, could come to a different conclusion about the balance of risk and benefit for any proposed solution to an issue.

A4 Technological Problem Solving

There is a difference between the scientific process and the technological problem-solving process. The goal of science is to understand the natural world. The goal of technological problem solving is to develop or revise a product or a process in response to a human need. The product or process must fulfill its function, but, in contrast with scientific problem solving, it is not essential to understand why or how it works. Technological solutions are evaluated based on such criteria as simplicity, reliability, efficiency, cost, and environmental and political effects.

Although the process of technological problem solving is presented in a step-by-step fashion, there are normally many cycles through the steps in any problem-solving attempt.

Defining the Problem

This process involves recognizing and identifying the need for a technological solution. You need to clearly state both the question(s) that you want to investigate and the criteria you will use as guidelines to solve the problem and to evaluate your solution. In any design, some criteria may be more important than others. For example, if the product solution measures accurately and is economical, but is not safe, then it is clearly unacceptable.

Identifying Possible Solutions

Use your knowledge and experience to propose possible solutions. Creativity is also important in suggesting novel solutions.

You should generate as many ideas as possible about the function of your solution and about potential designs. During brainstorming, the goal is to generate many ideas without judging them. They can be evaluated and accepted or rejected later.

To visualize the suggested solutions it is helpful to draw sketches. Sketches are often better than verbal descriptions to communicate an idea.

Planning

Planning is the heart of the entire process. Your plan will outline your processes, identify potential sources of information and materials, and establish evaluation criteria.

Seven types of resources are generally used in developing technological solutions to problems: people, information, materials, tools, energy, cost, and time.

Constructing/Testing Solutions

In this phase, you will construct and test your prototype using systematic trial and error. Try to manipulate only one variable at a time. Use failures to help make decisions before your next trial. You may also complete a cost–benefit analysis on the prototype.

To help you decide on the best solution, you can rate each suggested solution on each of the design criteria using a five-point rating scale, with 1 being poor, 2 fair, 3 good, 4 very good, and 5 excellent. You can then compare your proposed solutions by totalling the scores.

Once you have made the choice from among all the possible solutions, you need to make and then test a prototype. While making the prototype you may need to experiment with the characteristics of different components. A model, on a smaller scale, might help you decide whether the product will be functional. The test of your prototype should answer three basic questions:

- Does the prototype solve the problem?
- Does it satisfy the design criteria?
- Are there any unanticipated problems with the design?

If these questions cannot be answered to your satisfaction, you may have to modify the design or select another solution.

Presenting the Preferred Solution

In presenting your solution, you will communicate your solution, identify potential applications, and put your solution to use.

Once the prototype has been made and tested, the best presentation of the solution is a demonstration of its use—a test under actual conditions. This demonstration can also serve as another test of the design. Any feedback should be considered for future redesign. Remember that no solution should be considered the absolute final solution.

Evaluating the Solution and Process

The technological problem-solving process is cyclical. At this stage, evaluating your solution and the process you used to arrive at your solution may lead to a revision of the solution.

Evaluation is not restricted to the final step; however, it is important to evaluate the final product using the criteria decided earlier, and to evaluate the processes used while arriving at the solution. The following questions will help:

- To what degree does the final product meet the design criteria?
- Did you have to make any compromises in the design? If so, are there ways to minimize the effects of the compromises?
- Are there other solutions that deserve consideration?
- How did your group work as a team?

A5 Lab Reports

Lab reports are prepared after an investigation is completed. To ensure that you can accurately describe the investigation, it is important to keep thorough and accurate records of your activities as you carry out the investigation. A sample lab report appears at the end of this section.

Investigators use a similar format in their final reports or lab books, although the headings and order may vary. Your lab book or report should reflect the type of scientific inquiry that you used in the investigation and should be based on the following headings, as appropriate:

Title

At the beginning of your report, write the number and title of your investigation. In this course, the title is given, but if you are designing your own investigation, create a title based on what the investigation is about. Include the date the investigation was done and the names of all lab partners (if you worked as a team).

Purpose

State the purpose of the investigation. Why are you doing this investigation? (In this textbook, the purpose is stated in the introductory part of each investigation.)

Question

This is the question that you attempted to answer in the investigation. If it is appropriate to do so, state the question in terms of independent and dependent variables.

Prediction

A prediction is a tentative answer to the question you are investigating. In the prediction you state what outcome you expect from your experiment with a reason.

Experimental Design

When you design your own investigation, you provide the experimental design. This is a brief general overview (one to three sentences) of what was done. If your investigation involved independent, dependent, and controlled variables, list them. Identify any control or control group that was used in the investigation.

Materials

This is a detailed list of all materials used, including sizes and quantities where appropriate. Be sure to include safety equipment and any special precautions when using the equipment or performing the investigation. Draw a diagram to show any complicated setup of apparatus.

Procedure

Describe, in detailed, numbered steps, the procedure you followed in carrying out your investigation. Include steps to clean up and dispose of waste materials.

Observations

This includes all qualitative and quantitative observations that you made. Be as precise as appropriate when describing quantitative observations, include any unexpected observations, and present your information in a form that is easily understood. If you have only a few observations, this could be a list; for controlled experiments and for many observations, a table will be more appropriate.

Analysis

Interpret your observations and present the evidence in the form of tables, graphs, or illustrations, each with a title. Include any calculations, the results of which can be shown in a table. Make statements about any patterns or trends you observed. Conclude the analysis with a statement based only on the evidence you have gathered, answering the question that initiated the investigation.

Evaluation

The evaluation is your judgment about the quality of observations obtained and about the validity of the prediction. This section can be divided into two parts:

- Did your observations provide reliable and valid evidence that let you answer the question? Are you confident enough in the evidence to use it to evaluate any prediction you made?

- Was the prediction you made before the investigation supported or falsified by the evidence?

The following questions should help you through the process of evaluation:

1. Were you able to answer the question using your experimental design? Are there any obvious flaws in the design? What other designs (better or worse) are available? As far as you know, is this design the best available in terms of controls,

efficiency, and cost? How confident are you in the chosen design?

You may sum up your conclusions about the design in a statement like this: "The experimental design [name or describe in a few words] is judged to be adequate/inadequate because …"

2. Were the steps that you used in the laboratory correctly sequenced and adequate to gather sufficient evidence? What improvements could be made to the procedure? What steps, if not done correctly, would have significantly affected the results?

Sum up your conclusions about the procedure in a statement like this: "The procedure is judged to be adequate/inadequate because …"

3. Which specialized skills, if any, might have the greatest effect on the experimental results? Was the evidence from repeated trials reasonably similar? Can the measurements be made more precisely?

Sum up your conclusions in a statement like this: "The technological skills are judged to be adequate/inadequate because …"

4. You should now be ready to sum up your evaluation of the experiment. Based on uncertainties and errors you have identified, what would be an acceptable percent difference for this experiment (1%, 5%, or 10%)?

State your confidence level in a summary statement: "Based upon my evaluation of the experiment, I am not certain/I am moderately certain/I am very certain of my experimental results. The major sources of uncertainty or error are …"

5. If appropriate, calculate the percent difference for your experiment:

$$\% \text{ difference} = \frac{|\text{difference in values}|}{\text{average of values}} \times 100\%$$

Sum up your evaluation of the results in a statement like this: "The results are judged to be acceptable/unacceptable because…"

6. Evaluate your prediction(s). Sum up your evaluation of the prediction(s):

"The predictions are judged to be acceptable/unacceptable because…"

Synthesis

You can synthesize your knowledge and understanding in the following ways:

- Relate what you discovered in the experiment to theories and concepts studied previously.
- Apply your observations and conclusions to practical situations.

Investigation: The Effect of Increasing Net Force on an Object's Acceleration

Submitted by Linh Yu, Pardeep Khan, and Colleen Wilson

Conducted on Sept. 30, 2009

Purpose

The purpose of this investigation was to determine quantitatively how the acceleration of an object depends on the net force applied to the object.

Question

What is the mathematical relationship between the net force applied to an object and the acceleration of the object?

Prediction

It seems logical that as the net force on an object increases, the acceleration should increase directly. Mathematically, we predict that the acceleration is directly proportional to the net force, or $\vec{a} \propto \vec{F}_{net}$.

Experimental Design

A ticker-tape timer and a tape were used to measure the time and distance moved as different forces were applied to a cart. The net force was the independent variable, and the acceleration was the dependent variable. The mass of the cart system was the controlled variable. To keep the mass of the system constant, we moved two 100-g masses, one at a time, from the cart to the overhang, as shown in **Figure 1**. Our teacher showed us that the equation used to calculate the magnitude of the acceleration of an object starting from rest is $a = \dfrac{2\Delta d}{(\Delta t)^2}$.

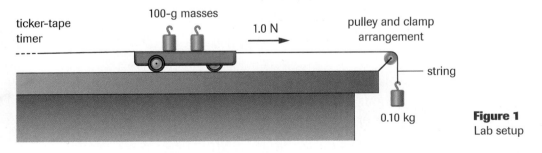

Figure 1
Lab setup

Sample lab report

Materials

cart	two clamps
three 100-g masses	ticker-tape timer and tape
force scale	masking tape
string (1.0 m long)	metre stick
pulley	graph paper

Procedure

1. Using the force scale, we measured the magnitude of the force on each 100-g mass.

2. We tested the ticker-tape timer to be sure it was in proper working order.

3. We set up an observation table to record the measured data.

4. We used a clamp to attach the pulley to the end of the lab bench. We attached one end of the string to a 100-g mass and the other end to the front of the cart using masking tape. We also used masking tape to secure two 100-g masses to the top of the cart, as shown in **Figure 1**.

5. We used masking tape to attach an 80-cm long strip of timer tape to the rear of the cart. Then we fed the tape through the ticker-tape timer, which was clamped to the bench.

6. With the suspended mass at its highest position and the cart held steady, we turned on the timer and released the cart. We stopped the cart just as the mass reached the floor.

7. We moved one 100-g mass from the cart and suspended it with the other mass. We then repeated steps 5 and 6.

8. We moved the final 100-g mass from the cart and suspended it with the first two masses. We then repeated steps 5 and 6.

9. We counted the spaces between the dots on each tape and measured each corresponding distance. We recorded the data in the observation table.

10. We put away the apparatus and cleaned up the lab area.

Observations

The measurements we made are summarized in **Table 1**.

Table 1 Measurements in the Acceleration Investigation

Mass Causing Acceleration (g)	Number of Time Intervals	Δd (m)
100	94	0.684
200	67	0.682
300	55	0.688

Analysis

Using the observations, we calculated the time interval of each trial and the magnitude of the acceleration of the cart system in each case. Sample calculations are shown for the first trial.

$$\Delta t = 94 \text{ intervals} \times \frac{1 \text{ s}}{60 \text{ intervals}} = 1.57 \text{ s}$$

$$a = \frac{2\Delta d}{(\Delta t)^2}$$

$$= \frac{2(0.684 \text{ m})}{(1.57 \text{ s})^2}$$

$$a = 0.555 \text{ m/s}^2$$

The results of the calculations are shown in **Table 2**. (Only magnitudes are shown.)

Table 2   Calculated Accelerations

Trial	F_{net} (N)	Δt (s)	Δd (m)	a (m/s²)
1	1.0	1.57	0.684	0.555
2	2.0	1.12	0.682	1.09
3	3.0	0.917	0.688	1.64

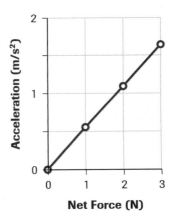

Figure 2

The acceleration–force graph in **Figure 2** shows the relationship between the dependent variable (the acceleration) and the independent variable (the net force). It is evident that the acceleration is directly proportional to the net force.

Evaluation

The experimental design of the investigation is judged to be adequate because the question was clearly answered and there are no obvious flaws. Even before the calculations were made, it was obvious that increasing the force applied to the cart system caused an increase in the acceleration.

The procedure is judged to be adequate because it produced sufficient data to obtain an obvious pattern on the graph. The procedure could be improved if more trials with each applied force were obtained.

The technical skills are judged to be adequate because the line of best fit on the graph joined the data points very closely.

There were some sources of error in the investigation:

- We assumed that the force of gravity on the overhanging mass (1.0 N, 2.0 N, and 3.0 N) was the net force in each case. This assumption does not take into consideration the friction of the cart on the lab bench or the string over the pulley. The cart's wheels appeared to have little friction. We tried to reduce the friction with the pulley by using smooth string.

- It was difficult to count the dots made by the timer, especially at the start of the motion where the dots were close together. We tried to reduce this source of error by holding the cart very still before releasing it.

- Parallax error could have occurred in measuring the distance from the starting dot to the finishing dot. We tried to reduce this error by looking straight at the metre stick.

Our prediction is judged to be acceptable because the resulting graph coincides with the mathematical relationship predicted.

Synthesis
Based on the results of this investigation, it is evident that to increase the acceleration of a vehicle, a greater net force and a lower friction are required.

A6 Self- and Peer Evaluation

You can make up a checklist using the following questions as a guide for self-evaluation or peer evaluation.

Applying Skills and Procedures:

- Have you (or you peer) applied previously learned skills and procedures in the new context?

Connecting Science and Technology to the Real World:

- Have you (or your peer) made connections between classroom/lab learning and familiar and unfamiliar contexts?

Understanding the Impact of Science and Technology in the World:

- Have you (or your peer) made an attempt to change your behaviour or the behaviour of others in considering social, political, environmental, and/or economic issues?
- Have you (or your peer) considered several points of view?

A7 Research Skills

General Research

Thanks to the Internet, we can now access more information than at any other time in history. However, you must know how to gather information—from *all* sources—efficiently and how to assess its credibility before you can make effective use of it.

Collecting Information

The following tips will help you collect information:

- Before you begin your research, list the most important words associated with your subject so that you can search for related topics.
- Brainstorm a list of possible resources. Consider all the sources of information available. Rank your list, starting with the most useful resource.

- Search and collect information from a variety of resources.
- Ask yourself: "Do I understand what this resource is telling me?"
- Make sure that the resource is current by checking the publication date.
- Consider the source of the information. From what perspective is it written? Is it likely to be biased?
- Keep organized notes or files while doing your research.
- Keep a complete list of the resources you used so that you can quickly find the source again if you need to. The list will also help you make a bibliography when writing your report.

- Review your notes. After your research, you may want to alter your original position or hypothesis or take your research in a slightly different direction.

Assessing the Credibility of Information Sources

Understanding and evaluating the work of others is an important part of research. When you do research, you may access information from the Internet, textbooks, magazines, chat lines, television, radio, and other sources.

PERCS

A useful framework for evaluating the credibility of information gathered from different sources is the PERCS checklist (**Table 1**). This framework, developed at Central Park East Secondary School in New York City, New York, uses a series of questions to critically assess information and arguments concerning an issue. You can use these questions to evaluate the information you have collected.

Internet Research

There is a huge variety of information on the Internet (facts, opinions, stories, interpretations, and statistics) created for many purposes (to inform, to persuade, to sell, to present a viewpoint, and to create or change an attitude or belief). However, anybody can post just about anything on the Internet without any proof of authenticity. Therefore, it is crucial that you critically evaluate the material you find on the Internet.

Evaluating Your Sources

Anyone with access to a server can put material on the Internet; there are almost no controls on what people choose to write and publish. It is your job as a researcher to evaluate what you find to determine whether it suits your needs. Keep the following questions in mind as you search to help you determine the quality of an Internet resource. The greater the number of questions answered "yes," the more likely that the source is of high quality.

Authority: Is it clear who is sponsoring the site? Does it appear to be permanent or part of a permanent organization? Is there a way of verifying the

Table 1 The PERCS Checklist

Perspective	From whose viewpoint are we seeing, reading, or hearing?
	From what angle or perspective?
Evidence	How do we know what we know?
	What is the evidence and how reliable is it?
Relevance	So what?
	What does it matter?
	What does it mean?
	Who cares?
Connections	How are things, events, or people connected to each other?
	What is the cause and what is the effect?
	How do they "fit" together?
Supposition	What if…?
	Could things be otherwise?
	What are or were the alternatives?
	Suppose things were different.

legitimacy of the page's sponsors? For example, is there a valid phone number or address posted? Is it clear who developed and wrote the material? Are that person's qualifications for writing on this topic stated?

Purpose: Is there information available describing the purpose of the sponsoring organization or individual? Is the point of view stated or otherwise obvious? Is the intended audience obvious?

Accuracy: Are the sources for factual information given so they can be verified? Is it clear who has the responsibility for the accuracy of the information presented? If statistical data are presented in graphs or charts, are they clearly labelled?

Objectivity: Is the site provided as a public service? Does it present a balance of views? If there is advertising on the page, is it clearly separated from the information?

Currency: Are there dates on the page to indicate when the page was written, first placed online, and last revised or edited? Are there any indications that the material is updated frequently? If the information is published in print in different editions, is it clear what edition the page is from? If the material is from a work that is out of print, has an effort been made to update the material?

Coverage: Is there an indication that the page has been completed and is not still under construction? If there is a print equivalent to the site, is there clear indication of whether the entire work or only a portion of it is available on the Internet?

Ease of Use: Is the information logically organized so that you can navigate easily and find relevant information? Is there a table of contents or index? Is it visually appealing and easy to read? Do the graphics help you understand the information presented?

Search Strategy

Before you begin searching the Internet, think about the information you are searching for. What is your topic? What are the key concepts in your question? What words would best describe your subject? Try to be as precise as possible. Are there other ways that you can express these key concepts? When you have answered these questions, you will have a list of search terms. Be willing to add to and subtract from your list as you evaluate what you have found to see if it is relevant and useful. (As you search, think about the questions posed earlier to determine the quality of the source.)

The primary ways of searching the Internet are as follows:

- *Search engines* (e.g., www.google.com, www.lycos.ca, www.altavistacanada.com) use key words that describe the subject you are researching.

- *Meta search engines* (e.g., www.go2net.com/index.html, www.search.com, www.infind.com) search several search engines at once.

- *Subject gateway and databases* (e.g., www.looksmart.com, www.yahoo.com, http://infomine.ucr.edu, www.searchdatabase.com) provide an organized list of Internet sites, divided into subject areas.

- *E-mail* and *discussion lists* put you in touch with individuals interested in your research topic.

Search Results

Once you have done a search, you will be confronted with a list of Internet sites and a number of "matches" for your search. There is often some information to help you decide which pages to look at in detail. You can always refine your search to reduce the number of "matches" you receive. There are several ways of refining or improving your search:

- Use more search terms to get fewer, more relevant records.

- Use fewer search terms to get more records.

- Search for phrases by enclosing search terms in quotation marks (e.g., "alexander graham bell").

- Choose search engines that allow you to refine your search results (e.g., AltaVista).

- Limit your searches to Canadian sites by using local search engines, or limit your searches to sites with the .ca at the end of their domain name.

- Use Boolean operators: + (an essential term) and − (a term that should be excluded).

Every site on the Internet has a unique address, or URL (Universal Resource Locator). Looking at the URL can help you decide if a page is useful. The URL sometimes tells you the name of the organization hosting the site and can indicate, by a tilde (~) in the URL, that you are viewing a personal page. The address includes a domain name, which also contains clues about the organization hosting the site (**Table 2**). Some organizations are likely to provide more reliable information than others. For example, the URL www.ec.gc.ca/ is the home page for Environment Canada; "ec.gc.ca" is the domain name—a reliable source.

Table 2   Codes

URL Code	Organization Information
com or co	commercial
edu or ac	educational
org	nonprofit organization
net	networking service providers
mil	military
gov	government
int	international organizations
ca, au	country code, e.g., Canada, Australia

B1 **Safety Conventions and Symbols**

Although every effort is undertaken to make the science experience a safe one, inherent risks are associated with some scientific investigations. These risks are generally associated with the materials and equipment used, and the disregard of safety instructions that accompany investigations and activities. However, there may also be risks associated with the location of the investigation, whether in the science laboratory, at home, or outdoors. Most of these risks pose no more danger than one would normally experience in everyday life. With an awareness of the possible hazards, knowledge of the rules, appropriate behaviour, and a little common sense, these risks can be practically eliminated.

Remember, you share the responsibility not only for your own safety, but also for the safety of those around you. Always alert the teacher in case of an accident.

In this text, equipment and procedures that are potentially hazardous are accompanied by the symbol and safety instructions.

WHMIS Symbols and HHPS

The Workplace Hazardous Materials Information System (WHMIS) provides workers and students with complete and accurate information regarding hazardous products. All chemical products supplied to schools, businesses, and industries must contain standardized labels and be accompanied by Material Safety Data Sheets (MSDS) providing detailed information about the product. Clear, standardized labelling is an important component of WHMIS (**Table 1**). These labels must be present on the product's original container or be added to other containers if the product is transferred.

The Canadian Hazardous Products Act requires manufacturers of consumer products containing chemicals to include a symbol specifying both the nature of the primary hazard and the degree of this hazard. In addition, any secondary hazards, first aid treatment, storage, and disposal must be noted. Household Hazardous Product Symbols (HHPS) are used to show the hazard and the degree of the hazard by the type of border surrounding the illustration (**Figure 1**).

	CORROSIVE
	This material can burn your skin and eyes. If you swallow it, it will damage your throat and stomach.
	FLAMMABLE
	This product or the gas (or vapour) from it can catch fire quickly. Keep this product away from heat, flames, and sparks.
	EXPLOSIVE
	Container will explode if it is heated or if a hole is punched in it. Metal or plastic can fly out and hurt your eyes and other parts of your body.
	POISON
	If you swallow or lick this product, you could become very sick or die. Some products with this symbol on the label can hurt you even if you breathe (or inhale) them.

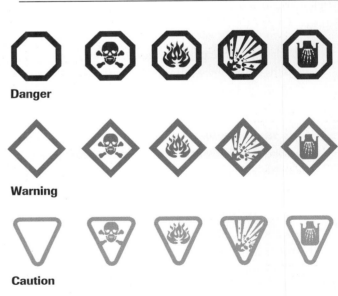

Figure 1
Hazardous household product symbols

Table 1 Workplace Hazardous Materials Information System (WHMIS)

Class and Type of Compounds	WHMIS Symbol	Risks	Precautions
Class A: Compressed Gas Material that is normally gaseous and kept in a pressurized container		• could explode due to pressure • could explode if heated or dropped • possible hazard from both the force of explosion and the release of contents	• ensure container is always secured • store in designated areas • do not drop or allow to fall
Class B: Flammable and Combustible Materials Materials that will continue to burn after being exposed to a flame or other ignition source		• may ignite spontaneously • may release flammable products if allowed to degrade or when exposed to water	• store in properly designated areas • work in well-ventilated areas • avoid heating • avoid sparks and flames • ensure that electrical sources are safe
Class C: Oxidizing Materials Materials that can cause other materials to burn or support combustion		• can cause skin or eye burns • increases fire and explosion hazards • may cause combustibles to explode or react violently	• store away from combustibles • wear body, hand, face, and eye protection • store in proper container that will not rust or oxidize
Class D: Toxic Materials Immediate and Severe Poisons and potentially fatal materials that cause immediate and severe harm		• may be fatal if ingested or inhaled • may be absorbed through the skin • small volumes have a toxic effect	• avoid breathing dust or vapours • avoid contact with skin or eyes • wear protective clothing, and face and eye protection • work in well-ventilated areas and wear breathing protection
Class D: Toxic Materials Long Term Concealed Materials that have a harmful effect after repeated exposures or over a long period		• may cause death or permanent injury • may cause birth defects or sterility • may cause cancer • may be sensitizers causing allergies	• wear appropriate personal protection • work in a well-ventilated area • store in appropriate designated areas • avoid direct contact • use hand, body, face, and eye protection • ensure respiratory and body protection is appropriate for the specific hazard
Class D: Biohazardous Infectious Materials Infectious agents or a biological toxin causing a serious disease or death		• may cause anaphylactic shock • includes viruses, yeasts, moulds, bacteria, and parasites that affect humans • includes fluids containing toxic products • includes cellular components	• special training is required to handle materials • work in designated biological areas with appropriate engineering controls • avoid forming aerosols • avoid breathing vapours • avoid contamination of people and/or area • store in special designated areas
Class E: Corrosive Materials Materials that react with metals and living tissue		• eye and skin irritation on exposure • severe burns/tissue damage on longer exposure • lung damage if inhaled • may cause blindness if it contacts eyes • environmental damage from fumes	• wear body, hand, face, and eye protection • use breathing apparatus • ensure protective equipment is appropriate • work in a well-ventilated area • avoid all direct body contact • use appropriate storage containers and ensure proper nonventing closures
Class F: Dangerously Reactive Materials Materials that may have unexpected reactions		• may react with water • may be chemically unstable • may explode if exposed to shock or heat • may release toxic or flammable vapours • may vigorously polymerize • may burn unexpectedly	• handle with care avoiding vibration, shocks, and sudden temperature changes • store in appropriate containers • ensure storage containers are sealed • store and work in designated areas

Electrical Certification

Certification markings are very important for your safety. They show that prescribed tests have been performed to ensure that electrical equipment and appliances sold in Ontario are manufactured to a rigid set of standards. The certification marks used in Ontario are as follows:

 CSA Canadian Standards Association

 Electrical Safety Authority

 Warnock Hersey

 UL Underwriters Laboratories

 ULC Underwriters Laboratories of Canada

 Met

 Entela

 O-TL

 Ontario Hydro

B2 Safety in the Laboratory

General Safety Rules

Safety in the laboratory is an attitude and a habit more than it is a set of rules. It is easier to prevent accidents than to deal with the consequences of an accident. Most of the following rules are common sense:

- Do not enter a laboratory unless a teacher or other supervisor is present, or you have permission to do so.
- Familiarize yourself with your school's safety regulations.
- Make your teacher aware of any allergies or other health problems you may have.
- Wear eye protection, lab aprons or coats, and gloves when appropriate.
- Wear closed shoes (not sandals) when working in the laboratory.
- Place your books and bags away from the work area. Keep your work area clear of all materials except those that you will use in the investigation.
- Do not chew gum, eat, or drink in the laboratory. Food should not be stored in refrigerators in laboratories.
- Know the location of MSDS information, exits, and all safety equipment, such as the fire blanket, fire extinguisher, and eyewash station.
- Use stands, clamps, and holders to secure any potentially dangerous or fragile equipment that could be tipped over.
- Avoid sudden or rapid motion in the laboratory that may interfere with someone carrying or working with chemicals or using sharp instruments.
- Never engage in horseplay or practical jokes in the laboratory.
- Ask for assistance when you are not sure how to do a procedural step.
- When heating a test tube over a laboratory burner, use a test-tube holder and a spurt cap.

Holding the test tube at an angle, facing away from you and others, gently move the test tube backward and forward through the flame.

- Never attempt any unauthorized experiments.
- Never work in a crowded area or alone in the laboratory.
- Clean up all spills, even spills of water, immediately.
- Always wash your hands with soap and water before or after you leave the laboratory. Definitely wash your hands before you touch any food.
- Do not forget safety procedures when you leave the laboratory. Accidents can also happen outdoors, at home, and at work.

Eye and Face Safety

- Always wear approved eye protection in a laboratory, no matter how simple or safe the task appears to be. Keep the safety glasses over your eyes, not on top of your head. For certain experiments, full face protection may be necessary.
- Never look directly into the opening of flasks or test tubes.
- If, in spite of all precautions, you get a solution in your eye, quickly use the eyewash or nearest running water. Continue to rinse the eye with water for at least 15 minutes. This is a very long time—have someone time you. Unless you have a plumbed eyewash system, you will also need assistance in refilling the eyewash container. Have another student inform your teacher of the accident. The injured eye should be examined by a doctor.
- If you must wear contact lenses in the laboratory, be extra careful; whether or not you wear contact lenses, do not touch your eyes without first washing your hands. If you do wear contact lenses, make sure that your teacher is aware of it. Carry your lens case and a pair of glasses with you.

- If a piece of glass or other foreign object enters your eye, seek immediate medical attention.

- Do not stare directly at any bright source of light (e.g., a burning magnesium ribbon, lasers, or the Sun). You will not feel any pain if your retina is being damaged by intense radiation. You cannot rely on the sensation of pain to protect you.

- Be careful when working with lasers; be aware that a reflected laser beam can act like a direct beam on the eye.

Handling Glassware Safely

- Never use glassware that is cracked or chipped. Give such glassware to your teacher or dispose of it as directed. Do not put the item back into circulation.

- Never pick up broken glassware with your fingers. Use a broom and dustpan.

- Do not put broken glassware into garbage containers. Dispose of glass fragments in special containers marked "Broken Glass."

- Heat glassware only if it is approved for heating. Check with your teacher before heating any glassware.

- If you cut yourself, inform your teacher immediately.

- If you need to insert glass tubing or a thermometer into a rubber stopper, get a cork borer of a suitable size. Insert the borer in the hole of the rubber stopper starting from the small end of the stopper. Once the borer is pushed all the way through the hole, insert the tubing or thermometer through the borer. Ease the borer out of the hole, leaving the tubing or thermometer inside. To remove the tubing or thermometer from the stopper, push the borer from the small end through the stopper until it shows from the other end. Ease the tubing or thermometer out of the borer.

- Protect your hands with heavy gloves or several layers of cloth before inserting glass into rubber stoppers.

- Be very careful when cleaning glassware. There is an increased risk of breakage from dropping when the glassware is wet and slippery.

Using Sharp Instruments Safely

- Make sure your instruments are sharp. Dull cutting instruments require more pressure than sharp instruments and are therefore much more likely to slip.

- Select the appropriate instrument for the task. Never use a knife when scissors would work best.

- Always cut away from yourself and others.

- If you cut yourself, inform your teacher immediately and get appropriate first aid.

- Be careful when working with wire cutters or wood saws. Use a cutting board where needed.

Heat and Fire Safety

- In a laboratory where burners or hot plates are being used, never pick up a glass object without first checking the temperature by lightly and quickly touching the item, or by placing your hand near, but not touching, it. Glass items that have been heated stay hot for a long time, but do not appear to be hot. Metal items such as ring stands and hot plates can also cause burns; take care when touching them.

- Do not use a laboratory burner near wooden shelves, flammable liquids, or any other item that is combustible.

- Before using a laboratory burner, make sure that long hair is always tied back. Do not wear loose clothing (wide long sleeves should be tied back or rolled up).

- Never look down the barrel of a laboratory burner.

- Always pick up a burner by the base, never by the barrel.

- Never leave a lighted Bunsen burner unattended.

- If you burn yourself, *immediately* run cold water gently over the burned area or immerse the burned area in cold water and inform your teacher.

- Make sure that heating equipment, such as a burner, hot plate, or electrical equipment, is secure on the bench and clamped in place when necessary.

- Always assume that hot plates and electric heaters are hot and use protective gloves when handling.

- Keep a clear workplace when performing experiments with heat.

- Remember to include a "cooling" time in your experiment plan; do not put away hot equipment.

- Very small fires in a container may be extinguished by covering the container with a wet paper towel or ceramic square.

- For larger fires, inform the teacher and follow the teacher's instructions for using fire extinguishers, blankets and alarms, and for evacuation. Do not attempt to deal with a fire by yourself.

- If anyone's clothes or hair catch fire, tell the person to drop to the floor and roll. Then use a fire blanket to help smother the flames.

Electrical Safety

- Water or wet hands should never be used near electrical equipment.

- Do not operate electrical equipment near running water or a large container of water.

- Check the condition of electrical equipment. Do not use if wires or plugs are damaged.

- Make sure that electrical cords are not placed where someone could trip over them.

- When unplugging equipment, remove the plug gently from the socket. Do not pull on the cord.

- When using variable power supplies, start at low voltage and increase slowly.

Waste Disposal

Waste disposal at school, at home, and at work is a societal issue. Some laboratory waste can be washed down the drain or, if it is in solid form, placed in ordinary garbage containers. However, some waste must be treated more carefully. It is your responsibility to follow procedures and dispose of waste in the safest possible manner according to the teacher's instructions.

First Aid

The following guidelines apply if an injury, such as a burn, cut, chemical spill, ingestion, inhalation, or splash in eyes, is to yourself or to one of your classmates:

- If an injury occurs, inform your teacher immediately.

- Know the location of the first aid kit, fire blanket, eyewash station, and shower, and be familiar with the contents/operation.

- If you have ingested or inhaled a hazardous substance, inform your teacher immediately. The MSDS will give information about the first aid requirements for the substance in question. Contact the Poison Control Centre in your area.

- If the injury is from a burn, immediately immerse the affected area in cold water. This will reduce the temperature and prevent further tissue damage.

- If a classmate's injury has rendered him/her unconscious, notify the teacher immediately. The teacher will perform CPR if necessary. Do not administer CPR unless under specific instructions from the teacher. You can assist by keeping the person warm and reassured.

Table 1 Système International (SI) Base Units of Measurement

Quantity	Quantity Symbol	SI Base Unit	Unit Symbol
length	$L, l, h, d, w, r, \lambda, \vec{\Delta d}$	metre	m
mass	m	kilogram	kg
time	t	second	s
electric current	I	ampere	A
thermodynamic temperature	T	kelvin	K
amount of substance	n	mole	mol
luminous intensity	l_v	candela	cd

Table 2 Metric Prefixes

Prefix	Abbreviation	Meaning
exa	E	10^{18}
peta	P	10^{15}
tera	T	10^{12}
giga	G	10^{9}
mega	M	10^{6}
kilo	k	10^{3}
hecto	h	10^{2}
deca	da	10^{1}
standard unit		10^{0}
deci	d	10^{-1}
centi	c	10^{-2}
milli	m	10^{-3}
micro	μ	10^{-6}
nano	n	10^{-9}
pico	p	10^{-12}
femto	f	10^{-15}
atto	a	10^{-18}

Table 3 Some SI Derived Units

Quantity	Symbol	Unit	Unit Symbol	SI Base Unit
acceleration	$\vec{a}$	metre per second per second	m/s^2	m/s^2
area	A	square metre	m^2	m^2
density	ρ, D	kilogram per cubic metre	kg/m^3	kg/m^3
electric potential	V	volt	V	$kg \cdot m^2/A \cdot s^3$
electric resistance	R	ohm	Ω	$kg \cdot m^2/A^2 \cdot s^3$
energy	E	joule	J	$kg \cdot m^2/s^2$
force	$\vec{F}$	newton	N	$kg \cdot m/s^2$
frequency	f	hertz	Hz	s^{-1}
heat	Q	joule	J	$kg \cdot m^2/s^2$
period	T	second	s	s
power	P	watt	W	$kg \cdot m^2/s^3$
pressure	p	newton per square metre	N/m^2	$kg/m \cdot s^2$
speed	v	metre per second	m/s	m/s
velocity	$\vec{v}$	metre per second	m/s	m/s
volume	V	cubic metre	m^3	m^3
weight	$\vec{F}_W$	newton	N	$kg \cdot m/s^2$
work	W	joule	J	$kg \cdot m^2/s^2$

Table 4 Measurement Conversions

Length
1 inch (in) = 2.54 cm
1 foot (ft) = 0.3048 m
1 mile (mi) = 5280 ft = 1.609 km
1 nautical mile = 1.151 mi = 6077 ft = 1.852 km
1 light-year = 9.46×10^{15} m

Area
1 m^2 = 10^4 cm^2 = 10.46 ft^2 = 1550 in^2
1 ft^2 = 144 in^2 = 9.29×10^{-2} m^2 = 929 cm^2
1 hectare (ha) = 10^4 m^2 = 2.471 acre
1 acre = 4.365×10^4 ft^2

Speed
1 mi/h = 1.609 km/h = 1.467 ft/s = 0.4470 m/s
1 km/h = 0.6214 mi/h = 0.2778 m/s = 0.9113 ft/s
1 knot = 1 nautical mi/h = 0.5144 m/s

Volume and Capacity
1 L = 1000 mL = 1000 cm^3 = 10^{-3} m^3 = 0.03531 ft^3
1 imperial gallon = 1.201 U.S. gallons = 4.546×10^{-3} m^3 = 0.1606 ft^3

Time
1 day (d) = 24 h = 1.44×10^3 min = 8.64×10^4 s
1 yr = 365.24 d = 3.156×10^7 s

Mass
1 ounce = 28.35 g
1 slug = 14.59 kg
1 kg = 1000 g = 6.852×10^{-2} slug
1 metric tonne (t) = 1000 kg
1 imperial ton = 2000 lb = 907.2 kg
1 kg = 2.2 lb (where $\lvert \vec{g} \rvert$ = 9.8 N/kg)

Force and Pressure
1 lb = 4.448 N
1 N = 10^5 dyne = 0.2248 lb
1 Pa = 1 N/m^2 = 1.45×10^{-4} lb/in^2
1 lb/in^2 = 6.895 kPa
1 atmosphere (atm) = 1.013×10^5 Pa = 1.013 bar = 14.70 lb/in^2 = 760 torr = 76 cm Hg (at 0 °C and $\lvert \vec{g} \rvert$ = 9.8 N/kg)

Energy and Power
1 J = 10^7 ergs = 0.7376 ft·lb
1 kcal = 4186 J
1 British thermal unit (Btu) = 1055 J
1 kW·h = 3.6×10^6 J = 860 kcal
1 horsepower (hp) = 745.7 W = 550 ft·lb/s
1 W = 0.7376 ft·lb/s

Table 5 Hydraulic and Pneumatic Circuit Symbols

Transmission Lines

———————	fluid transmission line
———▶	liquid (in direction of arrow)
———▷	gas (in direction of arrow)
- - - - - - - -	drain line
+ +	lines crossing
+ +	lines joining
	return line to reservoir

Rotary Devices

◯	rotary device (motor, pump, or compressor)
◉	rotary device with rotation direction
Ⓜ	electric motor or power source
⬤	pump (for liquids)
◐	compressor (for gases)

Storage Devices

▢	reservoir for liquid (atmospheric pressure)
▭	receiver for gas (pressurized)
▯	accumulator (storage cylinder)
▤	accumulator (spring-loaded)
▽	accumulator (pressurized)
▯	accumulator (weighted)

Filters

◇	fluid conditioner
◈	filter or strainer
◈	lubricator
◈	filter with drain

Cylinders

▭	cylinder
▭	cylinder, single-acting
▭	cylinder, double-acting

Gauges

◔	pressure gauge
◍	temperature gauge

Actuators

⌇	spring
⊡	solenoid (electrical)
⊟	manual control
◁	pressure
⌇⊟	spring and solenoid

Table 5 *(continued)*

Valves

Symbol	Description	Symbol	Description
	valve, one position or infinite positions between off and on	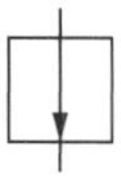	infinite position, normally open
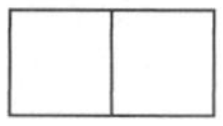	2-position valve	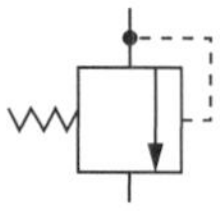	pressure relief valve, spring-controlled
	3-position valve		
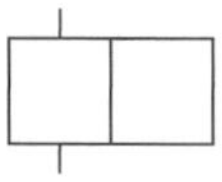	2-position, 2-port (2-way) valve	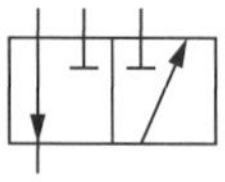	2-position, 3-port, normally open
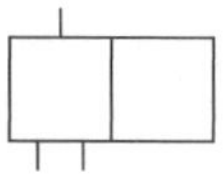	2-position, 3-port (3-way) valve	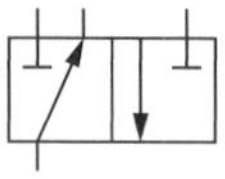	2-position, 3-port, normally closed
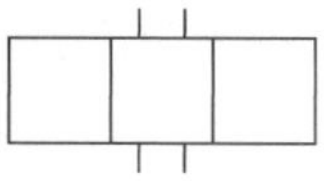	3-position, 4-port or 5-port (4-way) valve	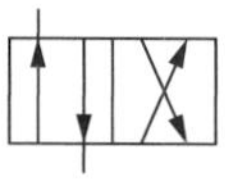	2-position, 4-port, normal
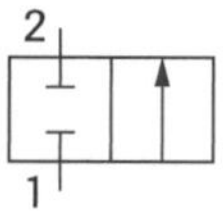	2-position, 2-port, flow blocked at start of cycle; flow from port 1 to port 2 when valve is shifted	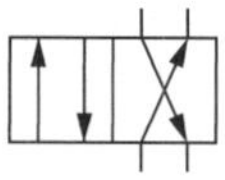	2-position, 4-port, activated
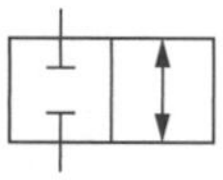	2-position, 2-port, flow blocked at start of cycle; flow in either direction when valve is shifted	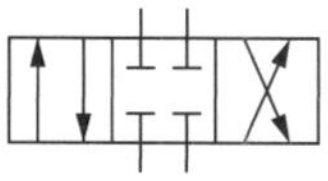	3-position, 4-port, normal position (closed)
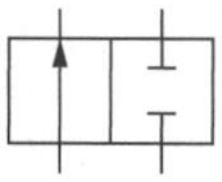	2-position, 2-port, open at the start of cycle	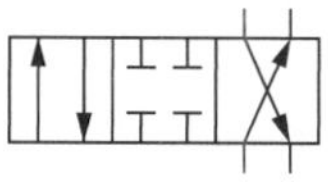	3-position, 4-port, activated left (flow through right)
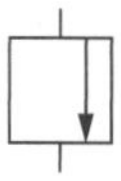	infinite position, normally closed	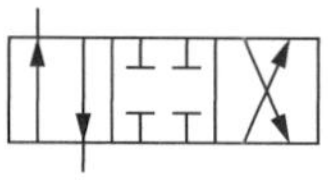	3-position, 4-port, activated right (flow through left)

Table 6 Electrical Circuit Symbols

Sources of Electric Potential

 cell

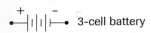 3-cell battery

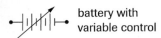 battery with variable control

 DC generator

 AC generator

Electrical Loads

 resistor (fixed)

 resistor (variable)

 lamp

 motor

Electric Meters

 ammeter

 voltmeter

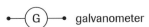 galvanometer

ohmmeter

Wiring and Connectors

wires making a connection

wires crossing (no connection)

switch (open)

switch (closed)

fuse

circuit breaker

ground

2-wire polarized outlet

3-wire polarized outlet

Table 7 Electronic Circuit Symbols

 capacitor (fixed)

 diode

 photodiode

light-emitting diode

rectifier (semiconductor)

amplifier

 transistor (n-p-n)

 transistor (p-n-p)

 integrated circuit

saw-tooth source

square-wave source

triangular source

This section includes numeric and short answers to questions in Section Questions, Chapter and Unit Self Quizzes, and Chapter and Unit Reviews.

Chapter 1

Section 1.1 Questions, p. 16

1. 25 km/h

2. (a) 8.39 m/s [forward];
   8.67 m/s [forward]

4. (a) 55.2 Hz
   (b) −8.00%

6. (b) 1.1×10^2 cm/s
   (c) 44 cm [E]

Section 1.3 Questions, p. 25

2. 19 (km/h)/s [forward]

3. (a) 2.3 m/s^2 [forward]
   (b) 0.24 m/s^2 [W]

8. (a) 2.8×10^3 m/s^2;
   93 m/s^2
   (b) $2.9 \times 10^2 g$; $9.5g$

Section 1.7 Questions, p. 45

1. 2.0 m/s^2 [S]

2. 7.0×10^2 N [W]

3. 6.69×10^3 N [forward]

4. 3.0 kg

5. (a) 6.3×10^2 N [down]
   (b) 4.4 N [down]
   (c) 2.8×10^4 N [down]

6. (a) 0.10 kg
   (b) 4.3 kg
   (c) 1.8×10^3 kg

Section 1.8 Questions, p. 51

4. (b) 6.3×10^4 N [up]
   (c) 1.9 m/s^2 [up]

Section 1.9 Questions, p. 55

2. (a) 0.058
   (b) 0.92

3. (a) 1.5×10^2 N
   (b) 0.31

4. 62 N

5. (b) 0.12 m/s^2

Chapter 1 Self Quiz, p. 65

1. T

2. F

3. F

4. T

5. T

6. T

7. F

8. F

9. (a)

10. (d)

11. (c)

12. (a)

13. (d)

14. (b)

Chapter 1 Review, pp. 66–67

3. (a) 52.43 m/s; 188.7 km/h
   (b) 0.0 m/s

5. 5.1 m/s^2 [S]

9. (a) 2.8 N [forward]
   (b) 2.3 m/s^2 [forward]

10. 6.0×10^{-4} N

11. 1.5×10^4 N

12. 0.064

13. (a) 34 N
    (b) 12 N

14. 0.34

17. (b) 46 N [W]
    (c) 3.1 m/s^2 [W]
    (d) 0.082

22. (b) 66 cm/s [E]

Chapter 2

Section 2.3 Questions, p. 87

2. (a) 8.1×10^2 N·m
   (b) 78 N·m

3. 96 N

4. (a) 0.75 m
   (b) 1.6×10^2 kg

5. 3.2×10^2 N

6. (a) 49 N
   (b) 1.7×10^2 N

7. (b) 1.2 m
   (c) 1.5 m

8. 2.0×10^2 N

9. (a) 6.0 cm
   (b) 2.0 cm

Section 2.4 Questions, p. 96

3. (a) 4.0; 3.0; 75%
   (b) 0.24; 0.20; 82%
   (c) 7.0 N; 0.12 m; 82%
   (d) 0.35 m; 2.6×10^3 N;
   93%

4. (a) 3.7×10^2 N; 0.13

5. (a) 5.2
   (b) 4.6
   (c) 88%
   (d) 65 kg

6. (a) 3.9
   (b) 2.1
   (c) 54%

Chapter 2 Self Quiz, p. 110

1. F

2. T

3. T

4. F

5. T

6. F

7. T

8. T

9. (c)

10. (b)

11. (a)

12. (a)

13. (d)

14. (c)

15. (d)

Chapter 2 Review, pp. 111–113

6. (b) 5.1×10^2 N

7. (a) 9.3×10^2 N
   (b) 0.47 m

8. (b) 19 N
   (c) 1.9 kg

10. (b) 5.5×10^2 N
    (c) 0.10

11. (a) 8.0
    (b) 9.0
    (c) 89%

12. (a) 3.0; 0.77; 0.33
    (c) 4.7 cm; 18 cm; 42 cm

Unit 1 Self Quiz, pp. 116–117

1. (a) F
   (b) E
   (c) A
   (d) B
   (e) D

3. (a) D
   (b) E
   (c) G
   (d) C
   (e) F
   (f) I

4. T

5. F

6. T

7. F

8. F

9. F

10. T

11. F

12. (c)

13. (d)

14. (c)

15. (b)

16. (b)

17. (a)

18. (c)

19. (b)

20. (c)

21. (b)

Unit 1 Review, pp. 118–121

1. 0.77 Hz

2. 46 km/h

4. 2.4 (km/h)/s

5. (a) 1.5 m/s^2 [N]
   (b) 1.8×10^4 N [N]

11. (b) 3.9 m/s^2 [right]

12. 0.35

15. (a) 4.1 m/s^2 [E]
    (b) 2.5×10^2 N [E]
    (c) 61 kg

18. (a) 0.57 m
    (b) 35 kg

D

19. (a) 1.0; 0.99; 99%
    (b) 1.0; 0.68; 68%
22. (a) 2.0 m/s [E];
       1.0 m/s [E]
    (b) 1.0×10^1 m/s^2 [W]
    (c) 5.0 m/s^2 [E]

Chapter 3

Section 3.2 Questions, pp. 138–139

1. 2.5×10^2 J
3. 2.9×10^7 J; 29 MJ
4. (a) 2.2×10^2 N
    (b) 22 kg
5. -3.3×10^5 J
7. 4.8 m
9. (a) 2.4 J
10. (b) 15 N/m
     (c) 0.30 J
     (d) 1.2 J
     (e) 0.9 J

Section 3.4 Questions, p. 146

2. 1.4×10^3 J; 1.4 kJ
3. 0.17 kg
4. 3.0 m
6. 18 m/s
8. (a) 4 times; 9 times

Section 3.6 Questions, p. 154

7. (a) 1.72×10^4 J
    (b) 5.2×10^3 J
    (c) 13 m/s

Chapter 3 Self Quiz, p. 168

1. T
2. T
3. T
4. F
5. T
6. F
7. T
8. F
9. F
10. F
11. (a)
12. (a)

13. (d)
14. (c)
15. (d)
16. (c)
17. (c)

Chapter 3 Review, pp. 169–171

6. (a) 1.4×10^3 N
    (b) 2.5×10^4 J
7. (a) 2.6×10^3 N
    (b) 5.5×10^4 m/s^2
8. (a) 6.2×10^2 N
    (b) -5.8×10^3 J
9. (a) 2.8×10^4 J
    (b) 2.8×10^4 J
10. 2.1 kg
11. 3.0 m
12. (a) 27 m/s
     (b) 1.0×10^3 J
13. increases by a factor of 16
14. (a) 7.8×10^2 J
     (b) 3.1×10^3 J
     (c) 2.3×10^3 J
15. 1.6 kg
23. (a) 7.1×10^2 J
     (b) 4.3×10^2 J
24. (a) 9.5 m/s

Chapter 4

Section 4.1 Questions, p. 177

1. (a) 42 W
    (b) 5.8×10^4 J
    (c) 1.5×10^3 s
    (d) 1.1×10^3 W
    (e) 3.6×10^7 J
2. (a) 5.5×10^3 J
    (b) 1.7×10^3 W
3. (a) 1.7×10^2 J
    (b) 9.3×10^3 J
    (c) 2.1×10^2 W
4. (a) 3.3×10^3 s; 56 min
5. 6.1×10^3 MJ

Section 4.2 Questions, pp. 187–188

1. (a) 0.66; 34%
    (b) 0.050; 95%

(c) 4.5×10^3 J; 75%
(d) 84%; 52 kJ
2. 0.68, or 68%
3. 0.63, or 63%
4. (a) 3.3×10^3 J
    (b) 1.7×10^2 J
5. (a) 7.5×10^2 J
    (b) 7.3×10^2 J
    (c) 0.97, or 97%

Section 4.4 Questions, p. 199

3. (a) 0.068, or 6.8%
4. (a) 99.0%
    (b) 91.7%
    (c) 848 kW
    (d) 938 kW

Chapter 4 Self Quiz, p. 202

1. F
2. F
3. T
4. F
5. T
6. T
7. F
8. F
9. (b)
10. (b)
11. (a)
12. (d)
13. (c)
14. (c)
15. (d)
16. (d)

Chapter 4 Review, pp. 203–205

2. 15 W
3. 4.3×10^7 J; 43 MJ
4. (a) 4.3×10^7 J
    (b) 7.2×10^5 W; 0.72 MW
6. (b) 0.60; 0.60; 0.95
7. 0.97, or 97%
8. (b) 5.2×10^4 s; 14 h
9. (a) 0.072 J
    (b) 0.056 J
    (c) 0.77, or 77%
13. 19%

Unit 2 Self Quiz, pp. 208–209

3. (a) G
    (b) F
    (c) D
    (d) H
    (e) B
4. T
5. T
6. T
7. T
8. F
9. T
10. F
11. F
12. T
13. (b)
14. (b)
15. (c)
16. (d)
17. (c)
18. (b)
19. (d)
20. (d)
21. (a)
22. (b)

Unit 2 Review, pp. 210-213

1. (b) 9.6 J
    (c) 96%
4. (c) 3.8×10^{14} J
5. (a) 7.3×10^2 J
    (b) 2.2×10^5 J
    (c) 6.5×10^3 J
6. 12.4 m/s
7. 4.9×10^{-3} kg
8. (a) 1.8×10^2 m/s
    (b) 1.1×10^8 J;
       1.1×10^2 MJ
10. (a) 0.056 J
     (d) 1.5 m/s
11. (a) 7.0 m/s
12. (b) 3.6×10^2 J;
       -2.5×10^2 J;
       -1.1×10^2 J
     (c) 0.0 J
16. 2.5×10^2 W

17. 3.6×10^5 J

20. 2.2×10^4 W

22. 15%

27. (a) 2.9×10^5 J
    (b) 3.1×10^2 s; 5.1 min

Chapter 5

Section 5.1 Questions, p. 225

4. (a) 5.9×10^2 g/L
   (b) 0.57 g/L
   (c) 1.0×10^3 kg/m^3
   (d) 3.24 kg
   (e) 6.25×10^{-4} m^3

5. (a) 0.500 m; 0.200 m; 0.150 m
   (b) 1.50×10^{-2} m^3
   (c) 8.53×10^3 kg/m^3

7. (b) 68 g
   (c) 1.3 g/mL

Section 5.2 Questions, p. 236

3. (a) 86 Pa
   (b) 2.5 kPa
   (c) 5.7×10^4 N
   (d) 0.63 m^2

4. (a) 3.6×10^{-2} m^2
   (b) 1.4×10^4 N
   (c) 9.5×10^4 Pa; 95 kPa

5. (a) 0.15 m^2
   (b) 31 N
   (c) 3.2 kg

6. 3.1×10^{-2} m^2

8. 1.7×10^7 N

9. (a) 7.4×10^{-6} m^3/s
   (b) 1.6×10^{-4} m^3

10. (a) 4.8 L/min
    (b) 25 L/min

11. 0.17 m^3/s

Section 5.4 Questions, p. 246

4. (a) 213 kPa
   (b) 102 kPa
   (c) 103 kPa

5. (b) 7.93 m

6. 16.7 kPa

7. (a) 3.11×10^6 Pa; 3.11×10^3 kPa

(b) 3.21×10^3 kPa

9. (a) 4.1 m

Section 5.5 Questions, p. 252

2. 1.20×10^2 N; 1.05×10^3 N

3. 7.5×10^2 N

Section 5.8 Questions, p. 266

1. (a) 6.3 L
   (b) 4.9 L
   (c) 7.5 s
   (d) 5.8×10^4 N
   (e) 3.8×10^4 J
   (f) 9.0×10^3 W

2. (a) 9.2×10^{-2} J
   (b) 7.7×10^{-2} W

Chapter 5 Self Quiz, p. 272

1. F
2. F
3. F
4. T
5. T
6. T
7. T
8. T
9. T
10. (a)
11. (c)
12. (c)
13. (d)
14. (b)
15. (c)

Chapter 5 Review, pp. 273–275

1. (a) 1.3 kg/m^3
   (b) 1.1×10^3 kg/m^3

10. (a) 3.9×10^4 Pa
    (b) 3.7×10^3 Pa

11. (a) 9.6×10^{-5} m^3
    (b) 1.0×10^1 m/s

12. (a) 3.6×10^5 N
    (b) 2.4×10^5 N

13. 1.17×10^8 Pa; 117 MPa

16. (a) 5.4×10^2 Pa
    (b) 4.8×10^2 N

19. (a) 16 L
    (b) 15 L
    (c) 43 s
    (d) 2.5×10^5 N
    (e) 1.4×10^5 J
    (f) 1.0×10^4 W

Chapter 6

Chapter 6 Self Quiz, p. 298

1. T
2. F
3. T
4. T
5. F
6. F
7. T
8. F
9. (a)
10. (c)
11. (a)
12. (b)
13. (d)
14. (b)
15. (b)

Unit 3 Self Quiz, pp. 304–305

3. (a) G
   (b) J
   (c) H
   (d) C
   (e) B

4. T
5. F
6. T
7. T
8. T
9. F
10. F
11. T
12. T
13. F
14. F
15. F
16. F
17. T

18. (b)
19. (d)
20. (a)
21. (d)
22. (b)
23. (d)
24. (b)
25. (b)
26. (a)
27. (b)

Unit 3 Review, pp. 306–309

3. (a) $D_A = 1.3$ g/L; $D_B = 1.3$ kg/m^3

8. (a) 0.22 m^2

9. (a) 3.6 m^3
   (b) 59 m/s

13. 103.2 kPa; 107.6 kPa

15. (a) 2.5×10^4 N

18. (a) 3.0×10^3 Pa
    (b) 105 kPa

21. 1.1×10^2 kPa

Chapter 7

Section 7.2 Questions, p. 323

2. (a) 2.51×10^{-3} A
   (b) 995 mA
   (c) 0.855 A

Section 7.3 Questions, p. 327

2. (a) 0.125 V
   (b) 126 V; 0.126 kV

Section 7.4 Questions, p. 332

2. (a) 6.0 Ω
   (b) 11 Ω
   (c) 9.01 V
   (d) 0.40 A

3. 12 Ω

4. 1.3 A

5. 1.2×10^2 V

8. (a) 15 Ω ± 10%
   (b) yellow, violet, brown, gold

9. (a) 0.20 mA; 2.4×10^2 mA

Section 7.6 Questions, p. 343

1. (a) 18 V
   (b) 6.0 V
2. 9.4 A
4. 42 V
6. (a) 900 Ω
   (b) 60 Ω
8. $I_1 = 0.60$ A;
   $V_1 = V_2 = V_3 = 36$ V;
   $V_4 = 84$ V; $I_2 = 1.8$ A;
   $I_3 = 0.60$ A; $I_4 = 3.0$ A

Section 7.8 Questions, p. 350

4. (b) 10 Ω; 5.0 Ω; 3.3 Ω;
       2.5 Ω; 2.0 Ω
   (c) 4 lamps

Section 7.9 Questions, pp. 356–357

1. (a) 1.8 W; 45 J
   (b) 3.0 A; 2.7×10^2 J
   (c) 1.2×10^2 V;
       1.6×10^2 s
2. (a) 1.8 W
   (b) 5.0×10^3 W
   (c) 1.1×10^2 W
3. 18 V
4. 0.020 A
5. (a) 16 A
   (b) 17 MJ
   (c) 43¢
6. (a) 3.0×10^2 kW·h
   (b) $29.40
7. (b) 48 420 kW·h; $55.68
8. (a) 1.8×10^3 W
   (b) 30
10. (a) $P = I^2R$
    (b) 2.0×10^4 W
    (c) 2.0 W

Chapter 7 Self Quiz, p. 363

1. F
2. T
3. T
4. T
5. F
6. T
7. F
8. F
9. T
10. F
11. (b)
12. (d)
13. (a)
14. (d)
15. (a)
16. (a)
17. (a)

Chapter 7 Review, pp. 364–365

5. 1.1 A
6. 5.0 V
7. (a) 1.2×10^3 Ω
   (b) 6.8×10^{-2} W
   (c) 61 J
10. (a) 36 V
    (b) 0.48 A
    (c) 5.0×10^1 Ω
11. (a) 1.5 A
    (b) $R_1 = 2.0$ Ω;
        $R_2 = 6.0$ Ω
    (c) 1.5 Ω
12. For R_1: $I = 0.60$ A;
             $\Delta V = 3.6$ V;
    For R_2: $I = 1.8$ A;
             $\Delta V = 3.6$ V;
    For R_3: $I = 0.60$ A;
             $\Delta V = 3.6$ V;
    For R_4: $I = 3.0$ A;
             $\Delta V = 8.4$ V
13. (a) 27.0 V
    (b) 9.0 V
16. 0.030 A
17. 14 V
18. (a) 3.2 MJ
    (b) 11 ¢
19. (a) 2.88×10^3 kW·h
    (b) $253.44
20. (a) $R_A = 3.0 \times 10^1$ Ω;
        $R_B = 1.0 \times 10^1$ Ω

Chapter 8

Section 8.9 Questions, p. 402

7. (a) 1 GB
   (b) 2 GB
   (c) 6.5×10^7 Hz

Chapter 8 Self Quiz, p. 406

1. T
2. F
3. F
4. T
5. F
6. T
7. F
8. T
9. F
10. T
11. (a)
12. (b)
13. (c)
14. (d)
15. (a)
16. (d)
17. (d)

Chapter 8 Review, pp. 407–408

12. (b) 8×10^{-6} mm;
        8×10^{-9} m
    (c) +97 mV
    (d) 2×10^{-5} μF;
        2×10^{-11} F

Unit 4 Self Quiz, pp. 412–413

4. F
5. F
6. T
7. F
8. F
9. F
10. T
11. T
12. T
13. F
14. T
15. T
16. F
17. F
18. T
19. T
20. (c)
21. (c)
22. (d)
23. (a)
24. (c)
25. (d)
26. (c)
27. (d)
28. (c)
29. (a)
30. (a)
31. (c)

Unit 4 Review, pp. 414–417

6. (a) 12 Ω ± 5%;
       18 Ω ± 5%
   (b) 30 Ω; 7.2 W
7. (a) 0.30 A; 4.0×10^2 Ω
   (b) 9.6 Ω; 6.0×10^3 W
   (c) 1.2×10^2 V; 8.0 Ω
   (d) 4.2 A; 29 Ω
   (e) 3.0 V; 6.0×10^{-3} W
8. (a) 1.8 V
   (b) 0.10 A
   (c) 9.0×10^1 Ω
   (d) 12 Ω; 24 Ω; 36 Ω
9. (a) 0.50 A; 1.0 A; 1.8 A
   (b) 48 Ω
   (c) 6.9 Ω
10. (a) $R_t = 2.0$ W; $I_t = 3.0$ A;
        $I_1 = 1.5$ A;
        $I_2 = I_3 = 1.5$ A;
        $\Delta V_1 = 6.0$ V,
        $\Delta V_2 = 4.5$ V,
        $\Delta V_3 = 1.5$ V
    (b) $R_t = 3.0$ Ω; $I_t = 1.5$ A;
        $I_1 = 1.5$ A; $I_2 = 1.2$ A;
        $I_3 = 0.30$ A; $I_4 = 1.5$ A;
        $\Delta V_1 = 1.5$ V,
        $\Delta V_2 = \Delta V_3 = 2.4$ V,
        $\Delta V_4 = 0.60$ V
11. 9 bulbs
12. 1.4 W
13. 28 mA
14. 12 V
15. (a) 3.6 MJ
    (b) 9.0 ¢/(kW·h)

Chapter 9

Section 9.1 Questions, p. 427

2. 4.0 cm

3. (a) 0.60 Hz; 1.7 s
   (b) 1.8 Hz; 0.56 s
   (c) 0.25 Hz; 4.0 s

4. (b) 0.46 s
   (c) 2.2 Hz

5. 5.0×10^{-11} s

6. 2.5×10^2 Hz

Section 9.4 Questions, p. 438

2. (a) 8.2 cm
   (b) 4.5 cm
   (c) 2.2 m

3. (a) 4.2×10^7 m/s
   (b) 1.7×10^6 m/s

4. 14 cm

5. (a) 2.0×10^3 Hz; 2.0 kHz
   (b) 5.0×10^{-4} s

8. (a) 5.0 min; 8.9 min

Section 9.6 Questions, p. 449

2. (b) 42 cm
   (c) 15 Hz

3. (a) 0.70 Hz
   (b) 1.4 Hz
   (c) 2.1 Hz

Section 9.8 Questions, p. 463

4. (a) 2 Hz; 3 Hz; 5 Hz

Section 9.9 Questions, p. 469

2. (a) 240.0 Hz
   (b) 720.0 Hz

4. 1.35 m

Chapter 9 Self Quiz, p. 479

1. T
2. F
3. T
4. F
5. F
6. T
7. F
8. T
9. T
10. (c)
11. (d)
12. (c)
13. (b)
14. (a)
15. (d)
16. (a)
17. (d)

Chapter 9 Review, pp. 480–483

1. (b) 3.5 cm

2. (a) 0.72 s; 1.4 Hz
   (b) 0.33 s; 3.0 Hz

6. 1.5×10^3 m/s

7. 6.0 m/s

8. (c) 0.25 m

10. (a) 18 m/s
    (b) 4.5 m
    (c) 1; 2

17. 2 Hz; 5 Hz; 7 Hz

18. (a) 437 Hz

19. (a) 440.0 Hz
    (b) 880.0 Hz

20. 1.4×10^3 m/s

32. (a) 2: 697 Hz and 1336 Hz; 6: 770 Hz and 1477 Hz; 7: 852 Hz and 1209 Hz; #: 941 Hz and 1477 Hz
    (b) 1.17×10^{-3} s; 6.77×10^{-4} s

Chapter 10

Section 10.1 Questions, p. 493

6. 1.50×10^{11} m

7. (a) 5.00×10^{22} Hz; 2.00×10^{-23} s
   (b) 3.33×10^{-8} m; 1.11×10^{-16} s
   (c) 2.00×10^3 Hz; 1.50×10^5 m

Section 10.3 Questions, pp. 505–506

1. (a) 23.0°

(b) 45.0°
(c) 0.0°

2. $\theta_i = \theta_r = 52°$

3. (a) 20 cm/s
   (b) 40 cm/s

10. 3

Section 10.4 Questions, p. 515

2. (a) 1.34
   (b) 1.73

3. (a) 1.97×10^8 m/s
   (b) 1.85×10^8 m/s

Section 10.6 Questions, p. 529

2. (a) 6.33×10^{-7} m
   (b) 4.74×10^{14} Hz

Chapter 10 Self Quiz, p. 533

1. T
2. F
3. F
4. T
5. F
6. T
7. T
8. F
9. F
10. T
11. (d)
12. (c)
13. (d)
14. (b)
15. (d)
16. (a)
17. (a)

Chapter 10 Review, pp. 534–535

3. (a) 3.00×10^8 m/s
   (b) 3.00×10^2 m

4. (a) 6.56×10^{-7} m

11. (a) 0°
    (b) 0°

13. 1.38

15. (a) 1.94×10^8 m/s
    (b) 2.75×10^8 m/s

16. (a) 50.0°
    (b) 1.53
    (c) 1.96×10^8 m/s

Unit 5 Self Quiz, pp. 538–539

3. (a) C
   (b) A
   (c) F
   (d) H
   (e) J

(f) M
(g) L
(h) O

4. T
5. F
6. F
7. T
8. F
9. T
10. F
11. T
12. F
13. T
14. F
15. F
16. (d)
17. (b)
18. (a)
19. (b)
20. (c)
21. (c)
22. (c)

Unit 5 Review, pp. 540–543

2. (a) 4.4×10^2 Hz; 2.3×10^{-3} s

6. 512 Hz

7. (a) 3.26 m

8. 244 Hz; 20.6 m

10. (a) 8.0 m

11. 67 m/s

18. 89.3 m

19. 300.0 Hz; 600.0 Hz; 1.200×10^3 Hz

25. 6.06×10^{-7} m; 4.95×10^{14} Hz

27. (a) 1.46
    (b) 2.05×10^8 m/s

35. 000 100 010 000; 000 000 010 001; 111 000 010 001

Glossary

A

absolute pressure the true pressure, or the sum of the atmospheric and gauge pressures; symbol p_{abs}

accelerated motion motion with changing velocity; the change can be in direction, speed, or both

acceleration the rate of change of velocity; it is a vector quantity, symbol $\vec{a}$

acceptor atom in a semiconductor, a doping atom with three valence electrons; see also *doping*

active solar heating an energy-transformation technology that absorbs solar energy and converts it into thermal energy, which is then distributed where needed in the structure; compare with *passive solar heating*

actual mechanical advantage (AMA) the ratio of the load force to the effort force for a machine; $\dfrac{F_L}{F_E}$

actuator a device that transforms fluid forces into mechanical forces

air resistance friction experienced by an object as it moves through the air; symbol $\vec{F}_{air}$

alternating current (AC) an electric current that reverses direction periodically; compare with *direct current*

altimeter an instrument used to determine the altitude of an airplane above sea level using atmospheric pressure

ammeter the instrument used to measure electric current

ampere (amp) the SI unit of measurement of electric current; symbol A

amplitude modulation (AM) the method of controlling radio signals by varying the amplitude of the electromagnetic carrier wave, which has a constant frequency

amplitude the maximum displacement of a vibration from its normal rest position; symbol A

analog signal a signal that varies continuously with time; compare with *digital signal*

analog-to-digital conversion changing analog signals to digital signals

aneroid barometer an instrument used to measure atmospheric pressure without the use of a liquid

angle of incidence the angle between an incoming, or incident, ray and the normal; symbol θ_i

angle of reflection the angle between a reflected ray and the normal; symbol θ_r

antinode in a standing wave, the position of maximum displacement from the rest axis

applied force a general name for any contact force; symbol $\vec{F}_A$

atmospheric pressure

atmospheric pressure the pressure exerted by air molecules and other particles above Earth's surface; symbol p_{atm}

audible range the range of frequencies that an ear can detect

average acceleration the change of velocity divided by the time interval for the change $\left(\dfrac{\Delta\vec{v}}{\Delta t}\right)$; it is a vector quantity, symbol $\vec{a}_{av}$

average speed the ratio of the total distance travelled to the total time of travel $\left(\dfrac{\Delta d}{\Delta t}\right)$; it is a scalar quantity, symbol v_{av}

average velocity the ratio of the displacement to the time interval for the displacement $\left(\dfrac{\Delta\vec{d}}{\Delta t}\right)$; it is a vector quantity, symbol $\vec{v}_{av}$

B

barometer an instrument used to measure atmospheric pressure

base in a transistor, the centre section; made of one type of semiconductor; compare with *collector* and *emitter*

beat frequency the number of beats heard per second; measured in hertz

beats the series of loud and soft sounds produced by the interference of sounds of two nearly identical frequencies

Bernoulli's principle states that where the speed of a fluid is low, the pressure is high; where the speed of a fluid is high, the pressure is low

biomass energy the chemical potential energy stored in plant material and in animal waste

biomechanical system a system of a living body; the arm is an example of a biomechanical system

block and tackle a system of two sets of pulleys and one cable, with the upper set fixed and the load attached to the lower movable set

C

capacitance the ratio of the charge on a capacitor to the voltage across its plates; unit farad, F

capacitor a device used to store electric charge and thus electrical energy

central processing unit (CPU) the electronic device that controls the operations of a computer

chain hoist a system of two fixed pulleys and one movable pulley linked by an endless chain; pulling on the chain raises or lowers a load attached to the movable pulley

charge-coupled device (CCD) a semiconductor chip that converts light into electrical energy

closed circuit an electrical circuit forming a complete path for the current

coefficient of friction the ratio of the magnitude of the force of friction between two surfaces to the magnitude of the normal force between the surfaces $\left(\dfrac{F_f}{F_N}\right)$; symbol μ (no unit)

coefficient of kinetic friction the ratio of the magnitude of the force of kinetic friction to the magnitude of the normal force $\left(\dfrac{F_K}{F_N}\right)$; symbol μ_K (no unit)

coefficient of static friction the ratio of the magnitude of the maximum force of static friction to the magnitude of the normal force $\left(\dfrac{F_S}{F_N}\right)$; symbol μ_S (no unit)

cogeneration the process of using waste thermal energy for the production of heat or electrical energy

collector in a transistor, the thicker outside section; made of the same type of semiconductor as the emitter; compare with *base*

compound machine a machine made of two or more simple machines

compressibility the ability of the particles in a substance to be pressed closer together

compression the part of a longitudinal wave where the particles are close together; compare with *rarefaction*

conduction the process by which the collision of atoms and electrons transfers heat through a material or between two materials in contact

constant acceleration straight-line motion in which the speed changes by the same amount each second

constructive interference the increase in displacement that results when pulses or waves interfere

control a switch for starting and stopping the current in an electrical circuit

convection the process of transferring heat by a circulating path of fluid particles

conventional current the flow of positive charges in a circuit; symbol I; compare with *electron flow*

converging mirror a curved mirror that causes the reflected light to come together; compare with *diverging mirror*

crest in a transverse wave, the part above the rest axis

critical angle the minimum angle of incidence of light inside a transparent medium that results in total internal reflection; symbol θ_c

current electricity the flow of electric charges

cycle one complete vibration

D

density the mass per unit volume of a substance; it is a scalar quantity with SI units of kg/m^3; symbol D

destructive interference the reduction in displacement that results when pulses or waves meet

diffraction the bending of waves as they travel around objects or through openings

diffuse reflection the reflection of light off a rough surface; does not produce an image

digital signal a signal represented as on and off pulses, bits, or zeroes and ones; compare with *analog signal*

digital-to-analog conversion changing digital signals to analog signals

diode an electronic device formed by joining two differently doped semiconductors; has two leads

direct current (DC) an electric current in a single direction; compare with *alternating current*

displacement the change in position of an object, including the object's direction; it is a vector quantity; for example, the displacement of an object that moves 15 cm north is represented as $\vec{\Delta d} = 15$ cm [N]

diverging mirror a curved mirror that causes the reflected light to spread outward; compare with *converging mirror*

donor atom in a semiconductor, a doping atom with five valence electrons

doping the process of adding impurities to semiconductors to improve conduction properties

drag the forces that act against an object's motion through a fluid

dynamics the study of the forces that produce motion; compare with *kinematics*

E

echolocation the use of reflected sounds to determine distances

efficiency (eff) a decimal number or percentage that rates how well a device transforms energy; calculated as the ratio of the useful energy output of a device to the energy input required to operate the device; eff $= \dfrac{E_{out}}{E_{in}}$ and % eff $= \dfrac{E_{out}}{E_{in}} \times 100\%$; can also be expressed in terms of power: eff $= \dfrac{P_{out}}{P_{in}}$ and % eff $= \dfrac{P_{out}}{P_{in}} \times 100\%$

effort arm in a lever, the perpendicular distance from the fulcrum to the effort force, symbol d_E

effort force a force applied to one part of a lever to move a load at another part; symbol F_E

electric current a measure of the number of electric charges that pass by a particular point in a circuit each second; current $= \dfrac{\text{charge}}{\text{time}}$; symbol I

electric potential difference either a potential rise or a potential drop; symbol ΔV

electric potential drop a measure of the amount of energy per charge given to a load; symbol ΔV

electric potential rise a measure of the amount of energy per charge given by the source; symbol ΔV

electric resistance a measure of how much an electrical component opposes the flow of electric charges; symbol R

electrical circuit an arrangement of components used to transform electrical energy into some other form of energy in an electrical device

electrical conductor a substance through which electrons can easily move; for example, copper is an excellent conductor, whereas wood is not

electrical insulator a material that prevents the transfer of electric charges

electromagnetic (EM) spectrum the range of radiant energies

electron flow the flow of negative charges in a circuit; symbol e^-; compare with *conventional current*

electronics the study of the controlled conduction of electrons, or current, through gases, solids, and vacuums

emitter in a transistor, the thinner of the outside sections; made of the same type of semiconductor as the collector; compare with *base*

energy resource a raw material, obtained from nature, that can be used to do work; also called an energy source; for example, fossil fuels and solar cells are energy sources

energy the capacity to do work or to accomplish a task

energy transformation the change of energy from one form to another; for example, in a solar cell, radiant energy is transformed into electrical energy

energy-transformation technology a device used to transform energy for a specific purpose

equivalent resistance a single resistance that can replace all the resistances in an electrical circuit while maintaining the same current when connected to the same source; symbol R_{total} or R_t

F

farad the SI unit of capacitance; symbol F

fibre optics the study of the transmission of light in solid, transparent fibres

first-class lever a lever with the fulcrum between the load and the effort force

fluid a substance that flows and takes the shape of its container; liquids and gases are both fluids

fluid dynamics the study of the factors that affect fluids in motion

fluid statics the study of fluids at rest

fluid system an arrangement of components used to transmit and control forces in a fluid

focal length the distance from the focal point to the reflecting surface of a curved mirror; symbol f

focal point in a diverging mirror, the point from which parallel incident rays appear to reflect; in a converging mirror, the point where parallel incident rays reflect and meet

force a push or a pull; it is a vector quantity, symbol $\vec{F}$

force of gravity the force of attraction between all objects; symbol $\vec{F_g}$; only noticeable if the mass of at least one object is huge; for example, gravity keeps more than 60 moons orbiting Jupiter

forced vibration the response of an object to a vibration with a different frequency from the object's resonant frequency

forward bias the connection of the source to a diode such that positive is connected to p and negative to n to allow maximum current

free-body diagram (FBD) a drawing of just the object being analyzed, not the entire situation, that shows all the forces acting on the object; compare with *system diagram*

free electron an electron that has broken away from an atom, for example, in a semiconductor; it has a negative charge

free fall the acceleration of an object near Earth's surface if air resistance is ignored; the average value is $\vec{g} = 9.8$ m/s^2 [down]

frequency modulation (FM) the method of controlling radio signals by varying the frequency of the EM carrier wave, which maintains a constant amplitude

frequency the number of cycles per second; unit hertz; symbol f

friction the force between objects in contact; it is parallel to the contact surfaces and acts in a direction opposite to any motion or attempted motion; symbol $\vec{F_f}$

fuel cell a device that changes chemical potential energy directly into electrical energy

fulcrum a support around which a lever can rotate or pivot

fundamental frequency the lowest resonant frequency of a system that produces a standing wave; also called the first harmonic

G

gauge pressure the difference between the absolute pressure and the atmospheric pressure; symbol p_g

gears toothed wheels of different diameters linked together to change the speed or distance moved

geothermal energy thermal energy or heat taken from beneath Earth's surface

gravitational potential energy the type of energy possessed by an object because of its position above a reference level; symbol E_g; it is a scalar quantity, measured in joules (J)

H

harmonics the frequencies that make up a sound

heat a measure of the energy transferred from a warm body to a cooler body due to a difference in temperature

heat engine an engine that transforms heat from burning fuels into the kinetic energy of the moving parts of a machine

heat pump a device that uses evaporation and condensation to heat a home in winter and cool it in summer

hertz the SI unit of frequency; symbol Hz

hole in a semiconductor, the vacancy left by a free electron; it has a positive charge

holography the process of making a three-dimensional image on a two-dimensional surface using the interference of laser light

hydraulic energy energy generated by harnessing the gravitational potential energy of water (as in a dam)

hydraulic press a device in which a small force on a small piston is transmitted through an enclosed liquid and applies a large force on a large piston

hydraulic system a mechanical system that operates using a liquid under pressure

hydraulics the science and technology of the mechanical properties of liquids

hydroforming the process of using water under high pressure to shape metal components

I

ideal mechanical advantage (IMA) the ratio of the effort arm (or effort distance) to the load arm (or load distance) for a machine; $\dfrac{d_E}{d_L}$

impurities in semiconductors, substances added to a semiconductor material to increase that material's conduction properties

in phase vibrating objects that have the same frequency and are at the same positions in their cycles at the same instant; compare with *out of phase*

inclined plane a ramp; used to increase the load that can be raised by an effort force

index of refraction the ratio of the speed of light in a vacuum to the speed of light in a given medium; symbol n; $n = \dfrac{c}{v}$

inertia a property of matter that causes an object to resist changes in its state of motion; it is directly proportional to the mass of the object

infrared light-emitting diode (ILED) a diode that transforms electrical energy into infrared energy

instantaneous speed the speed at a particular instant; it is a scalar quantity, symbol v

instantaneous velocity the velocity at a particular instant; it is the rate of change of position; it is a vector quantity, symbol $\vec{v}$

integrated circuit (IC) a combined electronic circuit with up to millions of diodes, transistors, resistors, and capacitors; it replaces numerous individual circuits

intensity level represents the loudness of a sound; unit decibel (dB)

J

joule the SI unit of work and of energy; symbol J

K

kinematics the study of motion; compare with *dynamics*

kinetic energy the type of energy due to an object's motion; symbol E_K; it is a scalar quantity, measured in joules (J); calculated using the equation $E_K = \dfrac{mv^2}{2}$

kinetic friction the force that acts against an object's motion; symbol $\vec{F}_K$

Kirchhoff's current rule (KCR) states that at any junction point in an electrical circuit, the total current into the junction equals the total current out of the junction

Kirchhoff's voltage rule (KVR) states that in any complete path in an electrical circuit, the sum of the potential rises equals the sum of the potential drops

L

laminar flow fluid flow in which adjacent regions of fluid flow smoothly over one another

laser a technology that emits light in phase and with one wavelength; acronym for light amplification by the stimulated emission of radiation

law of conservation of energy states that when energy changes from one form to another, no energy is created or destroyed

law of the lever states that when a lever is in static equilibrium, the magnitude of the effort torque equals the magnitude of the load torque

laws of reflection state that the angle of incidence equals the angle of reflection ($\theta_i = \theta_r$), and the incident ray, normal, and reflected ray all lie in the same plane

length (of a pendulum) the distance from its suspension point to the centre of the mass; symbol l

lever a rigid bar that can rotate freely around a fulcrum

light-emitting diode (LED) a diode that transforms electrical energy into visible radiant energy

load a device that transforms electrical energy into another form of energy

load arm the perpendicular distance from the fulcrum to the load force, symbol d_L

load force the force exerted by the load on a lever; symbol F_L

local consumption the production of electrical energy close to the consumer

longitudinal vibration periodic motion that is parallel to the rest axis

M

machine a device that performs tasks by achieving at least one of the five main functions of machines

manometer an instrument used to measure gauge pressure

mass the quantity of matter in an object; it is a scalar quantity measured in kilograms (kg) in SI; symbol m

mechanical energy the sum of gravitational potential energy and kinetic energy; it is a scalar quantity, measured in joules (J)

mechanical resonance the maximum response due to the transfer of energy from one object to another in a mechanical system at the same natural frequency

Musical Instrument Digital Interface (MIDI) a system of electronic devices that can create music

N

net force the vector sum of all the forces acting on an object; also called the resultant force; symbol $\vec{F}_{net}$

newton the SI unit of force; it is derived from the base units of metres, kilograms, and seconds; symbol N; the magnitude of the net force needed to give a 1-kg object an acceleration of magnitude 1 m/s^2

Newton's first law of motion states that an object will maintain its state of rest or constant velocity if the net force acting on it is zero; also called the law of inertia

Newton's second law of motion states that if a net external force exists on an object, the object accelerates in the direction of the net force; the magnitude of the acceleration increases as the net force increases, and it decreases for objects of larger mass; $\vec{F}_{net} = m\vec{a}$

Newton's third law of motion states that for every force, there is a reaction force equal in magnitude but opposite in direction; also called the action–reaction law

node a position of zero amplitude resulting from destructive interference of pulses or waves

nonrenewable resource an energy resource that does not renew itself in a normal human lifetime

normal force the force at right angles to objects in contact; symbol $\vec{F}_N$

n-type semiconductor a crystalline material with free electrons from donor impurities

nuclear fission a nuclear reaction in which the nucleus of an atom is split, transforming nuclear energy into other forms of energy

nuclear fusion the process in which the nuclei of the atoms of light elements, such as hydrogen, join together at extremely high temperatures and densities to become larger nuclei, releasing energy in the process

O

ohm the SI unit of electric resistance; symbol Ω

Ohm's law states that for many devices, the ratio of the electric potential difference across a resistor to the current through it is constant if the temperature remains constant; the constant value is the resistance

opaque material a substance that absorbs or reflects all the light that strikes it, so no image is seen through it

open circuit an electrical circuit that is not complete, so there is no current

out of phase vibrating objects that are at different positions in their cycles at the same instant; compare with *in phase*

overall efficiency the total efficiency of all the energy transformations required to operate an energy-transformation technology; obtained by calculating the product of the efficiencies of all stages; $eff_{overall} = (eff_1)(eff_2)(eff_3) ...$

overloaded circuit an electrical circuit in which the current exceeds the circuit's safe limit

P

parabolic mirror a curved mirror in the shape of a paraboloid

parallel connection an electrical connection in which the current follows two or more paths; compare with *series connection*

pascal the SI unit of pressure; symbol Pa; $1.0 \text{ Pa} = 1.0 \text{ N/m}^2$

Pascal's principle states that pressure applied to an enclosed liquid is transmitted equally to every part of the liquid and to the walls of the container

passive solar heating the design and building of a structure to best take advantage of solar energy at all times of the year; compare with *active solar heating*

percent efficiency the ratio of the AMA to the IMA of a machine, expressed as a percentage; see *efficiency*

period the number of seconds per cycle; symbol T

periodic wave a travelling wave produced by a source vibrating at a constant frequency

photodiode a semiconductor device that converts radiant energy into electrical energy; also called a photocell, a photoelectric cell, or a solar cell

pneumatic system a mechanical system that operates using a gas under pressure

pneumatics the science and technology of the mechanical properties of gases

position the distance of an object from a reference position, including the object's direction; it is a vector quantity; for example, a position 30 m east is represented as $\vec{d} = 30 \text{ m [E]}$

power the rate of doing work or transforming energy, or the rate at which energy is supplied; symbol P; it is a scalar quantity, measured in watts

pressure the magnitude of the force per unit area; the SI unit is the pascal

principle of superposition states that the resulting displacement of two interfering pulses or waves is the algebraic sum of the displacements of the individual pulses or waves

***p*-type semiconductor** a crystalline material with holes from acceptor impurities

pulley a wheel with a grooved rim in which a rope or cable can run

pure semiconductor a semiconductor made of crystalline material with no impurities

Q

quality in music, the pleasantness of a sound related to the waveform of the sound

R

radiation the process in which energy is transferred by means of electromagnetic waves

rarefaction the part of a longitudinal wave where the particles are spread apart; compare with *compression*

rectifier circuit an electronic component that converts AC to DC

reference level the level to which a raised object may fall

refraction the bending of light as it travels at an angle from one medium to another of different density

regular reflection the reflection of light off a smooth, shiny surface; allows us to see an image

renewable resource an energy resource that renews itself in a normal human lifetime

resistor a device that has a known resistance

resonant frequency the natural frequency of a vibrating object

resonator the case, box, or sounding board of a stringed instrument

reverse bias the connection of the source to a diode such that negative is connected to p and positive to n to prevent current

robot an automated device, often computer-controlled, that performs a task

S

sampling rate the frequency at which samples of the analog signal are taken during an analog-to-digital conversion

sampling resolution the maximum number of possible states available for a digital signal; also called resolution

scalar quantity a quantity that has magnitude (or size) only (including any units); time is a scalar quantity

screw an inclined plane wrapped around a central shaft that increases the applied or effort force

second-class lever a lever with the load between the fulcrum and the effort force

semiconductor a solid, crystalline substance that conducts electric current better than an insulator but not as well as a conductor

semiconductor device an electronic component made of a semiconductor

series connection an electrical connection in which the current moves in one path; compare with *parallel connection*

short circuit an error in which a conductor is connected across a source or other component, bypassing the load; sometimes just called a "short"

Snell's law of refraction states that the ratio of the sine of the angle of incidence to the sine of the angle of refraction is constant, and is equal to the index of refraction, n; $n = \dfrac{\sin \theta_i}{\sin \theta_R}$

solar cells electronic devices that transform light energy into electrical energy directly; also called photovoltaic cells, photoelectric cells, or simply photocells

solar energy radiant energy from the Sun

sonar sound navigation and ranging

source an energy-transformation device that transforms one form of energy into electrical energy; also called an energy source

standing wave a pattern of loops and nodes created by the interference of periodic waves of equal wavelength and amplitude

static friction the force that prevents a stationary object from starting to move; symbol $\vec{F}_S$

static pressure head the height of the fluid in a column above a position with a specific pressure

streamlining the process of reducing the turbulence experienced by an object moving rapidly relative to a fluid

supercrest the result when a crest interferes with a crest

supertrough the result when a trough interferes with a trough

sympathetic vibration the response of an object to another vibration with the same resonant frequency

system diagram a drawing of all the objects in the situation under analysis; compare with *free-body diagram*

T

temperature a measure of the average kinetic energy of the atoms or molecules of a substance

tension the force exerted by strings, ropes, fibres, and cables; symbol $\vec{F}_T$

thermal energy the total kinetic energy and potential energy possessed by the atoms or molecules of a substance

third-class lever a lever with the effort force exerted between the fulcrum and the load

tidal energy energy generated by the rising and falling of ocean tides

torque the turning effect caused by a force on a rigid object around an axis or a fulcrum, symbol T; it is measured in newton-metres (N•m); it can also be called a "moment of force"

torsional vibration periodic motion in which the object or system twists around the rest axis

total internal reflection the reflection of light that strikes the interior surface of a transparent medium at angles equal to or greater than the critical angle

transducer a device used to transform energy in an electrical or electronic system

transistor an electronic control device that amplifies a small input signal into a larger output signal

translucent material a substance in which some light is transmitted and some is scattered, so no clear image is seen through it

transparent material a substance in which light can be easily transmitted, so a clear image is seen through it

transverse vibration periodic motion that is perpendicular to the rest axis

trough the part of a transverse wave below the rest axis

turbulent flow fluid flow in which a disturbance resists the fluid's motion; it results when fluids cannot move smoothly around or through objects

U

ultrasonic sound sound with a frequency higher than the normal audible range of a human

universal wave equation the equation for the speed of all waves in terms of the frequency of the wave's source and wavelength of the wave; $v = f\lambda$; also $c = f\lambda$

V

vector quantity a quantity that has magnitude (including any units) and direction; for example, a car moving at velocity 80 km/h westward would be represented as $\vec{v} = 80$ km/h [W]

vibration the periodic motion of a particle or mechanical system; also called an oscillation

vibrator in music, the string of a stringed instrument

viscosity the property of a fluid that determines its resistance to flow; a high viscosity means a high resistance to flow

visible spectrum the set of colours visible to the human eye

volt the SI unit of electric potential difference; symbol V

voltmeter the instrument used to measure electric potential difference

volume flow rate the volume per second of a fluid flow; symbol q_v

W

water jet cutting the process of using water under high pressure and at a high speed to cut manufactured components

watt the SI unit of power; symbol W; 1 W is equivalent to 1 J/s

wave a disturbance, caused by a vibration, that transfers energy over a distance

waveform the instantaneous displacement of interfering waves

wavelength the distance between adjacent points on a wave that are in phase; symbol λ

wedge a double inclined plane that increases the applied or effort force

weight the force of gravity on an object; it is a vector quantity measured in newtons, symbol $\vec{F}_g$

wheel and axle a large-diameter, rigid disk connected to a small-diameter rod

wind energy energy generated by harnessing the kinetic energy of wind

work in physics, the amount of energy transferred to an object by a force applied over a distance; symbol W

Glossary

▶ Index

Index (side tab)

▶ **Credits**

Photo Credits

Unit 1 Opener: © Photodisc; Inset: © Ariel Skelley/Corbis/Magma.

Chapter 1: p. 7: © CP Picture Archive/Moe Doiron; p. 8: © Bruce Coleman Collection/First Light; p. 10: Science Kit & Boreal Laboratories. www.sciencekit.com; p. 12: © Duomo/Corbis/Magma; p. 23: Paramount Canada's Wonderland; p. 26: © CP Picture Archive/Jacques Boissinot; p. 27: © James Gritz/Photodisc; p. 28: Left: Dave Starrett, Right: Dave Starrett; p. 34: Science Kit & Boreal Laboratories. www.sciencekit.com; p. 35: © Leonard de Selva/Corbis/Magma; p. 40: NASA; p. 42: Corel; p. 48: Left: Dave Starrett, Centre: Science Kit & Boreal Laboratories. www.sciencekit.com, Right: Science Kit & Boreal Laboratories. www.sciencekit.com; p. 51: NASA; p. 52: © Jens Meyer/Associated Press; p. 58: Top: © Photodisc, Centre: © C. Lee/PhotoLink/PhotoDisc, Bottom: © Dean Conger/Corbis/Magma; p. 60: © Tom Kinsbergen/Science Photo Library; p. 61: © Jan Hinsch/SPL/Photo Researchers; p. 67: © Corel.

Chapter 2: p. 68: © Ed Eckstein/Corbis/Magma; p. 69: Dave Starrett; p. 76: Left: © ThinkStock LLC/Index Stock Imagery, Right: © Photodisc; p. 96: Dave Starrett; p. 100: © NASA/Science Photo Library; p. 101: Dave Starrett; p. 105: Top Left: © Photodisc, Top Right: © Corbis/Magma, Bottom Left: © Andrew Farquhar/Valan Photos, Bottom Right: © Photodisc; p. 110: © Steve Broulis/Superstock; p. 111: Left: Science Kit & Boreal Laboratories. www.sciencekit.com, Centre: Science Kit & Boreal Laboratories. www.sciencekit.com, Right: Science Kit & Boreal Laboratories. www.sciencekit.com; p. 112: Science Kit & Boreal Laboratories. www.sciencekit.com; p. 113: Science Kit & Boreal Laboratories. www.sciencekit.com; p. 114: Top Left: © Conor Caffrey/Science Photo Library, Top Right: © Frank Conaway/IndexStock Imagery, Bottom Left: © CP Picture Archive/Don Gaudette, Bottom Right: © Philip Kaake/Corbis/Magma; p. 117: Dave Starrett; p. 118: CNE, Toronto; p. 119: Top: Science Kit & Boreal Laboratories. www.sciencekit.com, Bottom: Science Kit & Boreal Laboratories. www.sciencekit.com; p. 120: © Corbis/Magma; p. 121: Left: © Brett Froomer/Image Bank, Right: © AFP/Corbis/Magma.

Unit 2 Opener: © Ron Stroud/Masterfile, Inset: © Stone.

Chapter 3: p. 126: © Mark Richards/Photo Edit; p. 127: Dave Starrett; p. 128: Left: © CP Picture Archive/Frank Gunn, Right: © Photodisc; p. 129: Top: Dave Starrett, Bottom: © CP Picture Archive/Tom Hanson; p. 131: Top Left: Science Kit & Boreal Laboratories. www.sciencekit.com, Top Right: Science Kit & Boreal Laboratories. www.sciencekit.com, Centre Left: Science Kit & Boreal Laboratories. www.sciencekit.com, Centre Right: Science Kit & Boreal Laboratories. www.sciencekit.com, Bottom: © Michael S. Yamashita/Corbis/Magma; p. 132: © Lester Lefkowitz/Stock Market/First Light; p. 137: © CP Picture Archive/Adrian Wyld; p. 141: Take Stock Inc.; p.142: © Photo Researchers; p. 143: © Jeremy Woodhouse/Photodisc; p. 145: Top: © A.J. Copley/Visuals Unlimited, Bottom: © Roger Ressmeyer/Corbis/Magma; p. 146: © CP Picture Archive/COA CPC; p. 147: Dave Starrett; p. 150: © Simon Lewis/Science Photo Library; p. 151: Top Right: © Richard Cummins/Corbis/Magma, Bottom Left: Dave Starrett, Bottom Right: Science Kit and Boreal Laboratories. www.sciencekit.com; p. 159: Left: © CP Picture Archive/Greg Baker, Centre: © Ulrike Welsch/Photo Edit, Right: © Roger Ressmeyer/Corbis/Magma; p. 161: Top Left: Vision Quest Windelectric Inc., Top Right: Courtesy of Nova Scotia Power Inc., Bottom: Al Hirsch; p. 162: Left: Ballard

Power Systems, Inc., Right: Ballard Power Systems, Inc.; p. 169: © Associated Press/AP.

Chapter 4: p. 173: Top: © Photodisc, Bottom: Dave Starrett; p. 174: Top: © CP Picture Archive/Blaise Edwards, Bottom: © Bettmann/Corbis/Magma; p. 178: © Zefa Visual Media/Germany/Index Stock Imagery; p. 185: © David Leah/Science Photo Library; p. 186: Top: Science Kit & Boreal Laboratories. www.sciencekit.com, Bottom: Science Kit & Boreal Laboratories. www.sciencekit.com; p. 194: Top: © Eyewire, Bottom: © CP Picture Archive/Kevin Frayer; p. 198: © CP Picture Archive/Kevin Frayer; p. 203: Town of Fountain Hills.

Unit 2: p. 206: Left: © Pidgeon Thomas/Corbis Sygma/Magma, Right: © Daniel Mirer/Corbis/Magma; p. 207: Al Hirsch; p. 210: Left: © Tony Freeman/Photo Edit, Right: © A. Marsh/First Light; p. 211: Photo provided courtesy of Bombardier Inc.; p. 212: © Sally Vanderlaan/Visuals Unlimited; p. 213: © Bettman/Corbis/Magma.

Unit 3 Opener: © Eyewire, Inset Top: © CC Studio/Science Photo Library, Bottom Inset: © Peter Skinner/Photo Researchers; p. 217: © Taxi.

Chapter 5: p. 218: ©Dan Steinberg/Associated Press; p. 221: Top: © Garry Watson/Science Photo Library, Bottom: © Jonathan Blair/Corbis/Magma; p. 226: Top: Zefa Visual Media/Index Stock Imagery, Bottom: Science Kit & Boreal Laboratories. www.sciencekit.com; p. 229: © CP Picture Archive/Phill Snel; p. 230: Photo courtesy of Southwest Research Institute ®; p. 234: Dave Starrett; p. 236: Top: © CP Picture Archive/Jack Smith, Bottom: Dave Starrett; p. 237: © Ralph White/Corbis/Magma; p. 239: Maximilian Stock Ltd/Science Photo Library; p. 240: Science Kit & Boreal Laboratories. www.sciencekit.com; p. 243: Science Kit & Boreal Laboratories. www.sciencekit.com; p. 245: Corel; p. 247: © Phil Cantor/Index Stock Imagery; p. 249: Science Kit & Boreal Laboratories. www.sciencekit.com; p. 250: Top: © CP Picture Archive/Terry Easterby, Bottom: Arni Katz/Index Stock Imagery; p. 252: Corel; p. 253: Left: Science Kit & Boreal Laboratories. www.sciencekit.com, Right: Dave Starrett; p. 254: Dave Starrett; p. 258: Left: © Bruce Ando/Index Stock Imagery, Right: Novastock/Index Stock Imagery; p. 275: Science Kit & Boreal Laboratories. www.sciencekit.com.

Chapter 6: p. 276: © Corbis/Magma; p. 277: © CP Picture Archive/Hans Deryk; p. 278: © Kelly-Mooney Photography/Corbis/Magma; p. 279: NASA; p. 280: Top: National Research Council of Canada, Bottom: © CP Picture Archive/TRSun; p. 282: © CP Picture Archive/Adrian Wyld; p. 283: © Roger Ressmeyer/Corbis/Magma; p. 285: National Research Council of Canada; p. 286: © Nicholas Pinturas/Stone; p. 287: (c) Corbis/Magma; p. 295: © Mark L. Stephenson/Corbis/Magma; p. 296: © Peter Harholdt/Corbis/Magma; p. 299: National Research Council of Canada.

Unit 3: p. 306: Science Kit & Boreal Laboratories. www.sciencekit.com; p. 308: © CP Picture Archive/Adrian Wyld.

Unit 4 Opener: © Stone, Inset: Take Stock Inc.; p. 313: Rosenfeld Images Ltd/Science Photo Library.

Chapter 7: p. 314: © Photodisc; p. 315: Top: Science Kit & Boreal Laboratories. www.sciencekit.com, Bottom: Science Kit & Boreal Laboratories. www.sciencekit.com; p. 316: © AFP/Corbis/Magma; p. 318: Left: Science Kit & Boreal Laboratories. www.sciencekit.com, Right: Dave Starrett; p. 326: Left: © Michael Newman/Photo Edit,

Right: Onne van der Wal/Corbis/Magma; p. 327: Michael Newman/Photo Edit; p. 329: Left: Science Kit & Boreal Laboratories. www.sciencekit.com, Centre: Science Kit & Boreal Laboratories. www.sciencekit.com, Right: Science Kit & Boreal Laboratories. www.sciencekit.com; p. 332: © Royalty-Free/Corbis/Magma; p. 338: © Photodisc; p. 346: Left: Rubberball Productions, Right: © Stone; p. 348: Science Kit & Boreal Laboratories. www.sciencekit.com; p. 349: Top: CSA International, Bottom: Underwriters' Laboratories; p. 350: Left: Dick Hemingway, Right: Brian Heimbecker; p. 351: Left: Courtesy of Natural Resources Canada, Centre: © Photodisc, Right: © Photodisc; p. 352: Chinapac International; p. 357: Science Kit & Boreal Laboratories. www.sciencekit.com; p. 358: Science Kit & Boreal Laboratories. www.sciencekit.com; p. 364: Top Left: © Image Network, Top Right: Nelson photo, Bottom Left: Val Wilkinson/VALAN Photos, Bottom Right: © Amy Etra/Photo Edit.

Chapter 8: p. 367: © Stone/Lonnie Duka; p. 368: © Mark Wagner/Stone; p. 373: © Volker Steger/Science Photo Library; p. 381: Ton Kinsbergen/Science Photo Library; p. 387: Left: Science Kit & Boreal Laboratories. www.sciencekit.com, Right: Photo courtesy lonestardigital.com; p. 390: © G. K. & Vikki Hart/Image Bank; p. 398: Top Left: Take Stock Inc., Top Right: Take Stock Inc., Bottom Left: © CP Picture Archive/Marcio Jose Sanchez, Bottom Right: Take Stock Inc.; p. 400: Courtesy of Casio.

Unit 4: p. 408: Top Left: © Michael Keller/Index Stock Imagery, Top Right: © David Toase/Photodisc, Bottom Left: © Siede Preis/Photodisc, Bottom Right: Courtesy of QWIPTECH; p. 413: © Photodisc.

Unit 5 opener: Courtesy of 21st Century Airships Inc., Inset: © John Madere/Corbis/Magma. p. 419: Top: © Digital Vision, Bottom: © Zefa Visual Media, Germany/Index Stock Imagery.

Chapter 9: p. 420: © Ira Montgomery/Image Bank; p. 422: Duo trapeze act from the 'O' show by Cirque du Soleil (R), Photo: Veronique Vial, Jean-Francois Gratton, Costumes: Dominique Lemieux, (c) Cirque du Soleil Inc.; p. 424: © Photodisc; p. 426: ©Photodisc; p. 427: ©Richard Hutchings/Photo Edit; p. 436: © Sergio Dorantes/Corbis/Magma; p. 439: © Tony Freeman/Photo Edit; p. 444: © Photodisc; p. 445: All: Associated Press; p. 454: © Allan Pappe/PhotoDisc; p. 459: © Comstock/Russ Kinne; p. 460: Dept. of Clinical Radiology/Salisbury District Hospital/Science Photo Library; p. 462: © Sami Sarkis/Photodisc; p. 466: Top: © C Squared Studios/PhotoDisc, Bottom: © Bob Krist/CORBIS/Magma; p. 467: Left: © Kelly & Mooney Photography/CORBIS/Magma, Right: © CMCD/PhotoDisc; p. 470: © C Squared Studios/PhotoDisc; p. 471: Corel; p. 482: Left: Ontario Science Centre, Right: © Larry Stepanowicz/Visuals Unlimited.

Chapter 10: p. 485: Courtesy of autoindex.com; p. 486: Top: Ken Kay/Fundamental Photographs, Bottom: M. Cagnet, M. Francon, J.C. Thierr, Atlas of Optical Phenomena: Supplement, Springer-Verlag. (c) 1971; p. 487: Science Kit & Boreal Laboratories. www.sciencekit.com; p. 489: Take Stock Inc.; p. 491: Take Stock Inc.; p. 494: Dave Starrett; p. 498: © PhotoEdit/Myrleen Ferguson Cate; p. 499: Omni Photo Communications Inc/Index Stock Imagery; p. 500: Left: © Vince Steano/ Corbis/Magma, Right: National Research Council Canada; p. 502: Mark Gibson/Index Stock Imagery; p. 503: Photo courtesy of 21st Century Airships; p. 504: Photo courtesy of Telesat Canada; p. 507: © Charles O'Rear/CORBIS/Magma; p. 514: Left: Take Stock Inc., Right Top: © Spencer Grant/Photo Researchers; Right Bottom: Science Kit & Boreal Laboratories. www.sciencekit.com; p. 519: Left: © Ed Young/Corbis/Magma, Right: © Kim Steele/PhotoDisc; p. 521: Left: Science Kit & Boreal Laboratories. www.sciencekit.com, Right: Paul Barefoot/Holophile; p. 522: © Bettmann/Corbis/Magma; p. 523: © CP Picture Archive/Rod Currie; p. 525: © Corbis/Magma; p. 526: Top: © Reuters NewMedia Inc./Corbis/Magma, Bottom: © Ed Lallo/Index Stock Imagery; p. 534: Dave Starrett; p. 535: © Andrew Farquar/Valan Photos.

Unit 5: p. 536: © CP Picture Archive/Greg Agnew; p. 537: © AFP/Corbis/Magma; p. 543: © CP Picture Archive/DigitAgent.

Appendix: p. 544: © NASA; p. 550: Science Kit & Boreal Laboratories. www.sciencekit.com.